HANDBOOK OF POWER ELECTRONICS IN AUTONOMOUS AND ELECTRIC VEHICLES

HANDBOOK OF POWER ELECTRONICS IN AUTONOMOUS AND ELECTRIC VEHICLES

Edited by

MUHAMMAD H. RASHID

Life Fellow IEEE, Fellow IET (UK), Professor of Electrical Engineering, Florida Polytechnic University, USA

ELSEVIER

ACADEMIC PRESS

An imprint of Elsevier

Academic Press is an imprint of Elsevier
125 London Wall, London EC2Y 5AS, United Kingdom
525 B Street, Suite 1650, San Diego, CA 92101, United States
50 Hampshire Street, 5th Floor, Cambridge, MA 02139, United States

ISBN 978-0-323-99545-0

For information on all Academic Press publications
visit our website at https://www.elsevier.com/books-and-journals

Publisher: Megan Ball
Acquisitions Editor: Rachel Pomery
Editorial Project Manager: Joshua Mearns
Production Project Manager: Kamesh R
Cover Designer: Vicky Pearson Esser

Typeset by STRAIVE, India

Contents

Contributors

Mehdi Ahmadi Jirdehi Department of Electrical Engineering, Kermanshah University of Technology, Kermanshah, Iran

Mohammad S. Alam Texas A&M University—Kingsville, Kingsville, TX, United States

Ayse Kocalmis Bilhan Electrical and Electronics Engineering Department, Faculty of Engineering and Architecture, Nevsehir Haci Bektas Veli University, Nevsehir, Turkey

Weiqiang Dong Southern Arkansas University, Magnolia, AR, United States

Ahmed El Hajjaji University of Picardie Jules Verne, Amiens, France

Abdelkader El Kamel Centrale Lille, France

Hamed Faghihian Department of Mechanical Engineering, University of South Florida, Tampa, FL, United States

M. Nandhini Gayathri School of Electrical and Electronics Engineering, SASTRA Deemed to be University, Thanjavur, Tamil Nadu, India

Syed Rafay Hasan Tennessee Technological University, Cookeville, TN, United States

Shahin Hedayati Kia University of Picardie Jules Verne, Amiens, France

James Holland Department of Computer Science and Engineering, University of South Florida, Tampa, FL, United States

Mahmoud Ibrahim Tallinn University of Technology, Tallinn, Estonia

Ersan Kabalci Electrical and Electronics Engineering Department, Faculty of Engineering and Architecture, Nevsehir Haci Bektas Veli University, Nevsehir, Turkey

Ants Kallaste Tallinn University of Technology, Tallinn, Estonia

A. Kannabhiran School of Electrical and Electronics Engineering, SASTRA Deemed to be University, Thanjavur, Tamil Nadu, India

Houshang Karimi Department of Electrical Engineering and Computer Science, York University, Toronto, ON, Canada

Mohammad Reza Khalghani Electrical and Computer Engineering Department, Florida Polytechnic University, Lakeland, FL, United States

Faiq Khalid Vienna University of Technology, Vienna, Austria

Pathan Meeravali Khan Mahindra University, Hyderabad, India

S. Kodeeswaran Department of Electric Power and Energy Conversion, Saveetha School of Engineering, SIMATS Deemed to be University, Chennai; School of Electrical and Electronics Engineering, SASTRA Deemed to be University, Thanjavur, Tamil Nadu, India

B. Mohan Krishna Mahindra University, Hyderabad, India

N.B. Shanthi Kumar Mahindra University, Hyderabad, India

Sreedhar Madichetty Mahindra University, Hyderabad, India

Mohammad Reza Maghami Institute for the Future of Knowledge (IFK), The University of Johannesburg, Johannesburg, South Africa

Hasan Mehrjerdi Electrical and Computer Engineering Department, The George Washington University, Washington, DC, United States

Arthur Guseni Oliver Mutambara Institute for the Future of Knowledge (IFK), The University of Johannesburg, Johannesburg, South Africa

Anushree Ramanath Schneider Electric, San Jose, CA, United States

Muhammad H. Rashid Florida Polytechnic University, Lakeland, FL, United States

Anton Rassõlkin Tallinn University of Technology, Tallinn, Estonia

Morteza Rezaei Larijani University of Picardie Jules Verne, Amiens, France; Sharif University of Technology, Tehran, Iran

P. Sanjeevikumar Department of Electrical Engineering, IT and Cybernetic, University of South-Eastern Norway, Porsgrunn, Norway

Arman Sargolzaei Department of Mechanical Engineering, University of South Florida, Tampa, FL, United States

Mohammad Shaterabadi Department of Electrical Engineering and Computer Science, York University, Toronto, ON, Canada

Vahid Sohrabi Tabar Faculty of Electrical and Computer Engineering, University of Tabriz, Tabriz, Iran

Tambiara Tabassum Electrical and Computer Engineering Department, Florida Polytechnic University, Lakeland, FL, United States

Onur Toker Florida Polytechnic University, Lakeland, FL, United States

Mehmet Uzunoglu Tyzen Trade LLC, Fulshear, TX, United States

Toomas Vaimann Tallinn University of Technology, Tallinn, Estonia

Anurag Verma Department of Electrical Engineering, Institute of Engineering and Technology, Lucknow, Uttar Pradesh, India

Bulent Vural Medikont Electronics, Istanbul, Turkey

MohammadReza Zolghadri Sharif University of Technology, Tehran, Iran

CHAPTER

1

Introduction to autonomous vehicles

Hamed Faghihian[a], *James Holland*[b], *and Arman Sargolzaei*[a]

[a]Department of Mechanical Engineering, University of South Florida, Tampa, FL, United States

[b]Department of Computer Science and Engineering, University of South Florida, Tampa, FL, United States

1.1 Introduction

The distinction between conventional automobiles and autonomous vehicles (AVs) is that the latter are capable of operating without human involvement. Within vehicle autonomy, there exist various levels of autonomy implemented on individual vehicles that affect the level of involvement required by the human occupants. In 2016, the Society of Automotive Engineers (SAE) formally defined six levels of autonomy [1] (shown in Fig. 1).

Level 0 (No Automation): At this level of autonomy, the driver is fully responsible for controlling the vehicle and would not receive any assistance from driver aid systems, such as automatic emergency brakes, cruise control, etc. This first level is often referred to as completely manual driving, whereas vehicles of this category are referred to as conventional vehicles.

Level 1 (Driver Assist): At this level of autonomy, the vehicle is equipped with at least one driver aid system enabling features, such as cruise control. However, at this level, the driver is still in charge of all primary driving tasks and is fully responsible for piloting the vehicle. Therefore, the driver is required to maintain proper situational awareness of their surroundings and of other drivers.

Level 2 (Partial Driving Automation): Vehicles of this class are equipped with at least two driver assistance features, which are used in tandem to enable partial autonomy. Generally, one of these two features controls the vehicle's forward movement, while the other is responsible for controlling lateral movements. The most well-known example of this level of autonomy is the combination of adaptive cruise control (ACC) and Lane-Keep Assist systems to drastically reduce the driver's task-load during highway travel. At this level of autonomy, which is constrained to only an individual lane, the driver should continue to monitor their surroundings and be ready to take control of their vehicle at any time. Systems that enable this level of autonomy are known as Advanced Driver Assistance Systems (ADAS).

Level 3 (Conditional Driving Automation): At this level of Autonomy, the vehicle should be able to detect environmental phenomena and make decisions based on the circumstances it encounters. However, the vehicle's perception of its surroundings is not adequate enough to allow the driver for the driver to become distracted from the act of driving. The transition from Level 2 to Level 3 is a pivotal technological leap from relying on limited, traditional technologies to leveraging a variety of cutting-edge detection and decision-making techniques to enable ADAS. This era in AV development could become the turning point for this technology. From the viewpoint of drivers, however, this tidal shift is hardly noticeable as the driver is still required to pay attention and remain ready to take control of the car at any moment. At this level of automation, the vehicle must use sensors and other hardware to perceive and react to the external environment.

Level 4 (Advanced Automation): The fourth level of automation is known as the high automation level and is where the vehicle is capable of executing almost all of the driving tasks required in certain scenarios. Due to this, the vehicle can perform all driving tasks automatically unless the driver overrides the autonomous driving system or if legislation prohibits the car from operating in autonomous mode in certain areas. The most important progress from level 3 to level 4 of autonomy is that the car does not require continuous monitoring from the driver as it is capable of safely coming to a stop if unable to operate in certain situations or if being limited (or prohibited) in certain areas. This progress is very significant from the point of view of ordinary people since this level of autonomy lets the driver become fully distracted during vehicle autonomous driving while still being able to override the autonomous mode at any moment.

Handbook of Power Electronics in Autonomous and Electric Vehicles
https://doi.org/10.1016/B978-0-323-99545-0.00018-X

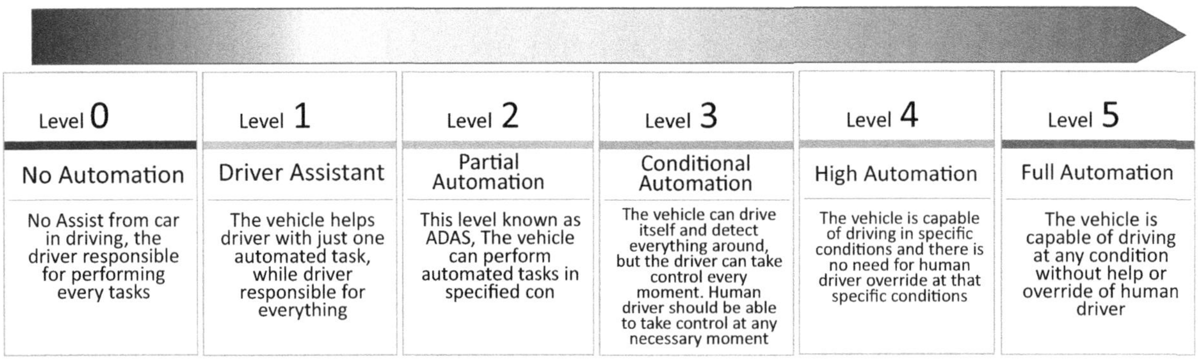

FIG. 1 Levels of autonomy in vehicles.

Level 5 (Full Self-Driving): At this level of autonomy, the vehicle itself is responsible for any driving task, anywhere! As such, this vehicle does not need pedals, a steering wheel, and not even a driver. It can safely go anywhere that a vehicle operated by a qualified driver can go. At this level, drivers do not need to nor can they disengage from autonomous mode. This mode is widely referred to as full self-driving (FSD).

1.1.1 Motivation

AVs had long held the interest of the scientific community, with the first demonstration of vehicle autonomy occurring in 1925 when Francis Houdina piloted his radio-controlled car through the streets of Manhattan [2]. While originally aimed at ushering in an era of convenient transportation, AVs have become a beacon of hope for improving roadway safety for motorists and pedestrians. In 2016, the NHTSA found that human error accounts for over 94% of crash fatalities [3]. While AVs are poised to greatly decrease traffic fatalities, connected autonomous vehicles (CAVs) are the most viable solution for eliminating the vast majority of crashes and fatalities [4]. This chapter addresses how the theory and application of AVs have evolved, what and where gaps in this technology exist, and what future issues may emerge in this field. Studying the evolution and history of innovation targeted at enabling AVs demonstrates the rapid progression of technology that society has experienced in a relatively short period. Predicting and planning for failures and pitfalls today will allow for a better traversal of them in the future.

1.1.2 History of autonomous vehicles

The field of automated vehicles launched by Francis Houdina's radio-controlled marvel has exploded over the last 70 years. During this time, researchers have worked to develop theories and fundamental methods of control that are critical for supporting the idea of self-driving cars. In the late 1950s, automated steering and speed control of automobiles was unveiled on the test track for future "electronic highways" by General Motors and RCA [5,6]. This led to the start of university research focused on this area at Ohio State University in 1962 [7]. Following this trend, many research centers around the world began gaining interest in AV research as one depicted in Fig. 2. Two such other examples are MITI Mechanical Engineering Laboratory's announcements that they were successful in controlling an unmanned vehicle using an inductive cable [8] as well as success in automatic control of steering using machine vision [9]. In 1986, European automakers started the PROMETHEUS program (PRograM for European Traffic with Highest Efficiency and Unprecedented Safety). This program aimed to promote the development of new technologies for the European automotive industry to improve its readiness for emerging technologies. Major Western European countries funded this program. To meet the program goals, many hardware and sensor development projects have been conducted to enable automated vehicles and artificial intelligence (AI) technologies. This program lasted over 8 years, finally culminating with the first presentation of their automated vehicles. The "Cooperative Driving" exhibit possessed four main functions, including ACC, Intelligent Maneuver Control (IMC), Emergency Warning (EW), and Medium-Range Preformation (MRP) for informing the driver about future hazards [10]. In addition to these, another concept, called "Copilot," was presented. With Copilot, the driver benefited from suggestions provided by the vehicle to ensure safe driving [11].

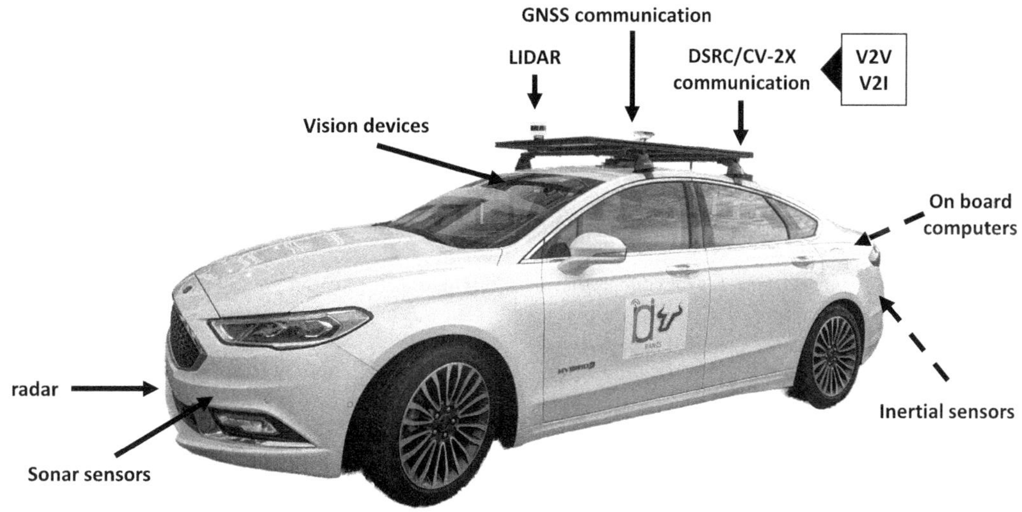

FIG. 2 Illustration of an ongoing project toward Connected and Autonomous vehicles, RANCS Lab., 2023.

At the same time as these automated vehicle developments in Europe, the California Department of Transportation (Caltrans) and the Institute of Transportation studies at the University of California in Berkeley established programs of their own. The program, named PATH (Program on Advanced Technology for the Highways), was aimed at addressing three aspects of transportation issues; these included propulsion technology, highway automation, and vehicular driving assistance. The PATH program was the first program to develop fully automated vehicles from 1986 to 1993 [12,13]. The program's researchers' findings encouraged other states to shape and support a national movement for developing new automotive technology, leading to the Mobility 2000 program [14,15].

The Mobility 2000 project was active from 1988 to 1991, with the main responsibility of proposing detailed programs for the development of intelligent transportation systems (ITS). The main concept of this system was based on the vehicle sensing its location by using permanent magnets installed along the road. This was one of the first works in which vehicle longitudinal dynamics were taken into account for controlling the position of the vehicle. Along with using vehicle dynamics, Peng and Tomizuka [16] used road curvature as a feed-forward with an optimal controller to increase ride quality. This work was followed up by [17] where a PID controller was used to show that the control of both ride quality and curvature tracking was feasible even during high levels of lateral acceleration. The PATH program has continued to verify this method with different control methods, such as fuzzy logic controllers [18], and also with different maneuver scenarios [19]. The findings of the PATH program were continued by Narendren and Hedrick [20] in 1994 with the idea of "auto-tracking." With the auto-tracking method, the vehicle movement is measured and compared relative to other road vehicles instead of using implanted sensors for localizing the test vehicle. Following this study, several works have been done to verify the robustness of the auto-tracking method but none of them were capable of being used for the leader vehicle in a sequence of vehicles [21,22]. In addition to all of these efforts, the National Highway Traffic Safety Administration (NHTSA) has supported projects seeking to improve the crash performance of vehicles, leading to work in the area of crash avoidance system development [23]. This was pursued by the Federal Highway Administration (FHWA) to develop Automated Highway systems, and some published findings were reported in [24,25]. These efforts have also been funded by other government organizations in both Europe and Japan [26,27].

Nowadays, automated vehicles are no longer merely a dream as this technology is rapidly integrated into modern vehicles in the form of ADAS like ACC and forward collision prevention. Furthermore, due to the sheer number of vehicles operating on today's roadways, self-driving technology is quickly becoming an obligation. The current state of motoring results in unnecessary danger to road users, excessive contamination of the environment through emissions and other automotive-related pollution, as well as an increased reliance upon imported fuels. AVs, particularly electric vehicles (EVs), are proving to be a viable solution to these and more issues facing our nation and humanity as a whole. Electrification of the automotive industry is a paradigm shift on its own, one in which automakers will be required to manufacture more sustainable, efficient, and safer vehicles by leveraging cutting-edge technologies. Therefore, studying the current state of the automotive industry is beneficial for predicting the future state of this and other industries.

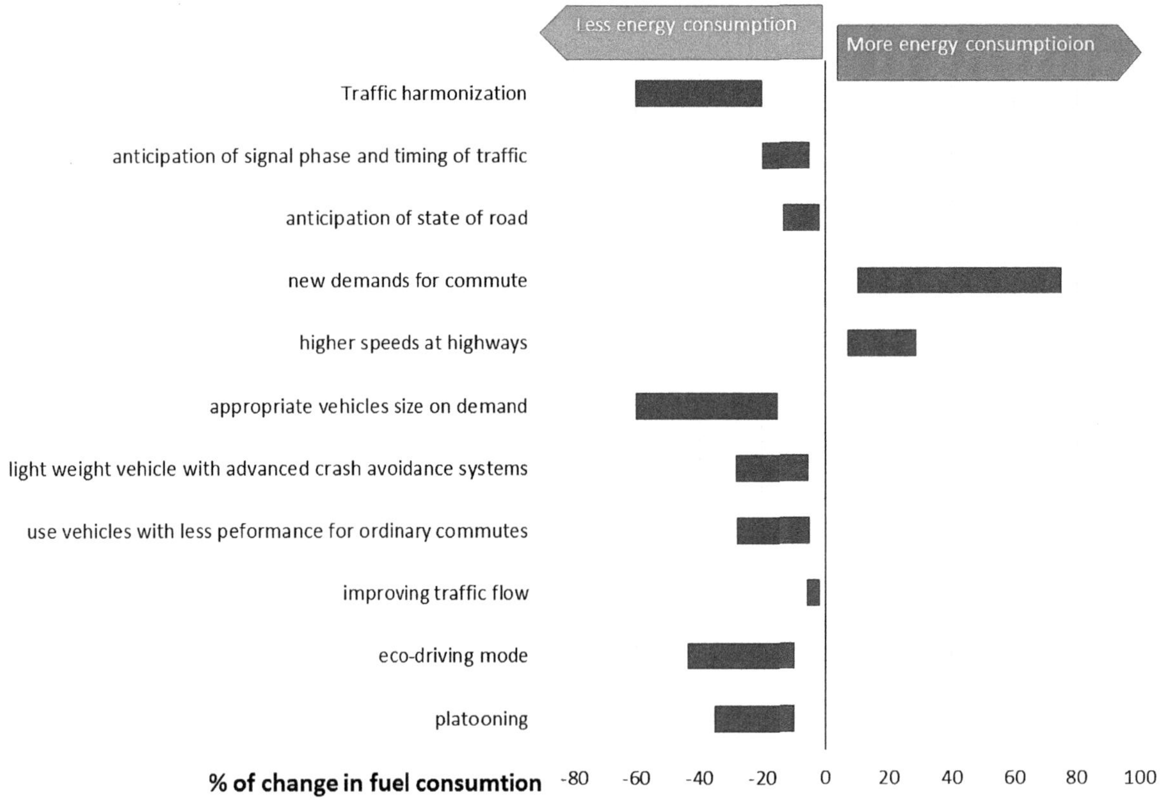

FIG. 3 A Waymo's all-electric Jaguar with 360 vision system with its cameras, radar panels and Lidar sits atop. The figure extracted from Ref. [43].

1.1.3 State of the industry

Over the course of nearly a century, machines and devices from agricultural harvesters to vacuum cleaners have been imbued with autonomous capabilities; yet the dream of the self-driving car is still to be realized [28]. In April 2020, there were over 250 companies working in the AV space [29]. Of those companies, a small subset of 30 firms has spent over $16 billion on developing AVs with disproportionately little to show for their efforts [27]. AV development has been notoriously unprofitable due to incredibly small profit margins that are simply a result of the cost of pursuing this endeavor [30]. Financial challenges facing the industry further mounted during the global pandemic of 2020, causing widespread layoffs [31]. Select firms were able to pivot their AV development approaches and techniques to adapt to the new workflow of lockdowns and remote work [32]. One such adaptation of firms was the choice to pivot the focus of AV adoption from ride-sharing and transportation to food and goods delivery [24]. Waymo, for example, raised over $2 billion in the early days of the pandemic due to this change in messaging [25]. This allowed for the larger players in this space to weather the storm and, in many cases, grow larger during a period of greatly decreased progress [33]. This has spurned what can be regarded as a contraction of the AV space, exhibited by several mergers, acquisitions, and fleet sell-offs [34–39].

Beginning in 2019, Google spin-off company, Waymo, has been providing fully autonomous services to passengers across the United States, demonstrating the company's clear lead in the space [40]. However, early this year, several top executives stepped away from the firm [41,42]. Fig. 3 depicts a Waymo driving on a road in San Francisco, California in 2021.

1.2 Impacts of autonomous vehicles

There are seemingly innumerable benefits that could be brought about by AVs. The first that comes to mind is exponentially increasing roadway safety. It is well regarded that AVs, and particularly CAVs, will have a huge impact on decreasing crash fatalities with some research indicating as much as a 96% crash reduction [4].

Another benefit often discussed, particularly in recent years, is the positive impact that AVs will have on the environment. This impact is brought about through improved efficiency through precise control of the vehicle and the ability to optimize the actions of individual AVs as well as platoons. At the current rate of electrification, it is also likely that the majority of AVs will be powered by electric drivetrains. EVs inherently are less damaging to the environment than traditional automobiles, from assembly to decommissioning.

In addition to improving roadway safety and leading the charge in saving the planet, AVs will also have major economic benefits. The first of these is aiding in moving the US away from imported fuel reliance and toward domestic, renewable energy production. The influence of AVs on the demand for renewable energies will further increase the jobs created in this sector in addition to those in the areas of battery-related research, development, and production. Due to the high-tech nature of AVs, the US tech sector will likely also grow in order to meet the growing demand for features and capabilities, creating even more professional employment opportunities.

1.3 Autonomous vehicles: Review of control methods

AVs require a control algorithm capable of controlling the vehicle within a complex and dynamic time-varying system. The control system should be designed to react to unpredictable situations. This means that the controller should be able to react as quickly as the average person. The control method of an AV has a close relationship with the level of automation and, by the end of this section, how each controller is developed at each level of automation. Originally, the control algorithms for AVs were mostly rule-based algorithms. As such, these controllers were mostly dependent on data gathered by the vehicle sensors [44]. In such a controller, the vehicle decision-making algorithm is based on an online controller [45], which was difficult for computing systems of the era to run effectively [46]. To minimize the required computations of the controller, another approach was presented where parameters were tuned at each time step instead of during the whole update [47]. All of these approaches were rule-based approaches where the controller designer was required to define generalized rules for the controller to obey in each scenario [48].

Controlling an automated vehicle and its navigation, however, is another challenge. The traditional thought process for controlling AVs is to leverage the path-tracking algorithms heavily utilized in robotics [49]. Some of these techniques use predictive control [50], fuzzy controllers [51], or pure-pursuit algorithm. Pure-pursuit algorithms are geometric methods, which mostly based on calculating a look-ahead point and controlling the vehicle to reach it [52]. Using pure-pursuit algorithms, Ollero et al. [53] improved upon this approach by adding an upstream fuzzy-logic controller to set look-ahead points and tune pure-pursuit parameters. Several other researchers proposed different methods of path tracking based on the idea of a pure-pursuit algorithm. For example, Zhao [54] proposed an adaptive-PID controller to increase the tracking capabilities and flexibility of a vehicle control system operating in an unknown environment.

Another approach to controlling the movements of an AV is to compartmentalize the primary functions of an AV into longitudinal and lateral movements. In terms of longitudinal movements, the first step is to enable the ability of a car to follow another in an automated procedure [55]. To this end, several approaches have been developed. The majority of work was limited to controlling the throttle and brake of the vehicle to avoid collision [56,57]. Other works have been done in this manner by using feedback from sensors, like GPS, camera, and Lidar [58–60]. Cai et al. [61] used a neuro-fuzzy logic controller to control the vehicle's longitudinal distance from the vehicle in front. This chapter was one of the first ones that address the problem by using AI. In terms of lateral movement control of AVs, Lane-keeping and lane-changing behaviors are the two main functions expected from an AV. Lane-keeping behavior is to keep the vehicle's direction despite different situations and distributions. Zhaoxue and Pengfei proposed, in Ref. [62], a method for detecting the lane of the AV through computer vision. The vision system is responsible for detecting the lane and its characteristics in addition to detecting the trajectory, any curvature, and the vehicle's exact position relative to the road [63]. Originally, the primary function of a lane-keeping system was to detect the lane to warn the driver of lane departure [64,65]. In 2009, Marino et al. presented a PID-based controller for controlling the steering of a vehicle to follow a path based on the lane-keeping system. The controller used a visual system to determine the distance between the center line of the lane and the vehicle, attempting to compensate for this distance with respect to the feedback from the yaw rate sensor and the visual system.

Another lateral movement control issue for an AV is that of lane-changing maneuvers. This maneuver is necessary for AVs to safely overtake or avoid slow or disabled vehicles. In lane-changing maneuvers, it is important to consider the possible circumstances and obstacles, such as traffic signals and other vehicle dynamics [66]. Therefore, lane-changing maneuvers demand further consideration, such as vehicle perception and position estimation and transition between lane-keeping and lane-changing during the start and end of the maneuver. O'Brien et al. presented a nonlinear controller along with a dynamic model for controlling a vehicle to maintain a path or track curvature. In this work, it was shown that lane-changing maneuver stability is achievable with an H-infinity controller [67]. Additional related works have been done to test and examine other control algorithms for this purpose, such as adaptive-PID-based controllers [68]. Foote et al. presented a controller that consists of a feed-forward signal from the path curvature along with a PID controller for manipulating the steering system [69].

Most of the previously presented works are based on classical, rules-based control methods. However, the complexity of driving demands AI-based controllers that can adapt to different circumstances [70]. An AI-based controller is capable of learning and updating its parameters based on data received. The AV controller is responsible for driving or assisting the driver in complicated situations. Therefore, it is difficult—if not impossible—to design a traditional controller capable of considering a myriad of situations. The AI-based controller can be used in a vast range of applications as it mimics a human-like learning behavior [71]. Since the lane-changing maneuver is much more complicated than a simple lane-keeping maneuver, many studies suggest using AI-based controllers for this application. Pomerleau et al. presented a controller based on neural networks. This controller was known as the "Rapidly Adapting Lateral Positioning Handler (RALPH)" [72]. In the work reported in Ref. [47], driving mode and driving behavior in a lane-change maneuver are controlled by a learning algorithm. The controller shapes trajectories using similar maneuver data. Each trajectory is designed in the form of a polynomial, and its coefficients are tuned by means of a neural network learning algorithm. In another instance of using AI-based controllers, Yao et al. presented a trajectory prediction approach based on the K-nearest neighbor algorithm instead of designing the trajectory by a traditional, rules-based approach [73]. In an AV, it is not only important to do a lane-change maneuver, but also it is important to detect obstacles or a slow-moving vehicle in front of where the AV would maneuver and perform a lane-change to avoid a collision or deceleration. In Ref. [74], the author discussed a vision-based method for detecting obstacles in front of the vehicle to avoid crashes. In Ref. [75], an obstacle avoidance system was presented. In this work, obstacles are detected by means of a visual camera and by utilizing an "end-to-end" learning method. This study showed a feasible method for detecting obstacles in front of any vehicle. Two previous works showed how an AV could detect any obstacle, and these methods were developed to be used in emergency situations. Still, it was impossible to initiate the lane-change procedure in a normal driving situation for a desired overtaking or fast driving by means of those approaches. However, later in 2017, Vallon et al. [76] presented an approach for an autonomous lane-changing algorithm using an SVM-based learning method with a predictive controller that was responsible for completing the maneuver safely and comfortably.

1.3.1 Using AI-based technologies in AVs

Up to the third level of autonomy, the automated vehicle's systems only enable features that support the driver in performing largely all the required tasks, referred to as advanced driver aid systems (ADAS). These features mostly rely upon more conventional control strategies to achieve safe and efficient operation. However, at the fourth (high automation) and fifth (full automation) levels of autonomy, traditional, rule-based methods prove inadequate for safe and reliable operation. At full automation, the vehicle should be able to operate itself in every circumstance, requiring a control algorithm with adequate perception and decision-making capabilities. Providing all these capabilities and qualifications requires employing AI in many aspects of an AV system [77].

The AV controller can be divided into two sections. The first one, perception, is responsible for mapping, localization, and detection, while the other one, Decision-making, handles the propulsion of the vehicle toward the goal, which has been derived from path planning and decision-making algorithms. The fully automated vehicle should be able to behave like a human driver in different situations. It should be able to recognize the environment, detect obstacles, and obtain a realistic perception of the environment and of upcoming events. Another aspect of AV controllers is the vehicle dynamics controller for propulsion which includes controlling the steering wheel and acceleration/deceleration required during different maneuvers.

Using AI-based control algorithms enables better perception and localization and decreases the number of expensive sensors, such as LIDAR (Light Detection And Ranging) and high-accuracy GPS sensors, that are required by the system. Traditional rule-based controllers are dependent on collecting accurate data from the environment. On the other hand, AI-based control methods can leverage learning algorithms to detect the environment and accomplish localization and mapping [45]. Different review articles analyzed the efficacy of using AI in AVs for perception tasks [78–80], localization and mapping applications [78,79,81], pedestrian detection [80,82], as well as planning and decision-making [83]. The AI-based learning methods, such as convolutional neural network [84] and deep neural network [85], are capable of reducing the demand for expensive and sophisticated sensors. Assuming that the vehicle possesses an adequate understanding of the surrounding environment, it should navigate toward the goal point. Eraqi et al. [86] proposed an "end-to-end" learning algorithm responsible for steering a car in traffic. Other similar works have been done using other AI-based controllers, which are mainly dependent on learning algorithms to investigate the driver behavior for controlling the vehicle [87].

1.4 Energy efficiency in autonomous vehicles

Energy efficiency is one of the most important challenges being solved through emerging technologies. With CAVs, energy consumption can be optimized at every level of driving, from how these vehicles operate and behave to the drivetrains and propulsion systems that power them. These energy savings are not solely limited to CAVs either. Connected vehicles are capable of transmitting data that can notify surrounding drivers such that they change their driving behavior to optimize efficiency, resulting in direct energy cost savings. With these new tools, it is possible to control how a vehicle moves, behaves, and communicates with other nearby motorists to increase safety and comfort and use the least energy possible. State of the art in energy efficiency of CAVs can be divided into three main parts, including optimization of energy consumption in automated vehicles, energy consumption optimization using vehicle-to-vehicle (V2V) communications, and energy consumption optimization using vehicle-to-infrastructures (V2I) communication.

In 2017, the National Highway Traffic Safety Administration (NHTSA) released a Federal Motor Vehicle Safety Standard (FMVSS) focused on increasing the proliferation rate of V2V communication technology on light-duty vehicles and to standardize the message transmission protocols [88]. While this obligation's main goals aim to increase the safety of transportation, the connection between vehicles can also be leveraged for tackling energy optimization challenges among a group of vehicles. Furthermore, in 2016, the FHWA published a provision and guideline to address deployment regarding V2I communications. This sought to address requirements and to help encourage the effective and efficient use of this new emerging technology [89]. In recent years, it has become abundantly clear that there is a great race among auto manufacturers to establish Automated Vehicles (AVs). Therefore, it is paramount to increase the efficiency of vehicles using these new technologies.

Existing work in optimizing the energy consumption in CAVs has been performed from various aspects, with the goal of each being to discuss the different solutions for increasing CAV efficiency. In this section, existing CAV energy optimization efforts are gathered and classified to present a comprehensive picture of what has been achieved in this area.

One of the important ways to increase the efficiency of CAVs is to acquire information about the road ahead, e.g., knowing about slopes, intersections, road limitations, etc. This will allow the vehicle to have a better reaction in order to optimize energy efficiency. It could be effective and important since the vehicle can adjust its path optimization and velocities with this knowledge. To this end, different studies have been done. In 2010, Asadi et al. [90] proposed a predictive planning algorithm to decrease the fuel consumption of vehicles by means of harmonizing vehicle speed with traffic flow to solve an optimization problem based on traffic constraints and data at each step of the path. This study showed that traffic speed and data could be used to optimize the energy consumption of CAVs. In other words, it is shown that in the absence of online traffic data, historical data is also beneficial for predicting traffic and road situations for optimization of energy consumption [91,92]. These works also demonstrated that knowing the state of the road, even with grades of uncertainty, is highly beneficial for increasing the efficiency of vehicles. It is important to know this since the mechanism for increasing efficiency in these works was to tune the gains of propulsion systems or to assist the controller of the power transmission system in increasing the efficiency of the vehicle. This idea brings the fact that by knowing the slopes of the road and selecting appropriate gains in power transmission systems, the efficiency of the vehicle can be increased. Recognizing the road slopes can be obtained by means of the 3D road map. This idea was studied in Ref. [93,94]. This method of "Look ahead," as it is referred to in existing literature [95], has already been commercialized in Daimler trucks and marketed as "RunSmart predictive cruise" [96]. All of these efforts showed that even without online data from the road situation, it is possible to increase the efficiency of vehicles. However, it is totally reasonable to assume that this method would be more beneficial for CAVS with online data from other vehicles. Another method for increasing the efficiency of CAVs that has gained interest from researchers in recent years is dynamic control which is based on the traffic flows caused by traffic signals and random delays. Using V2I communication is also possible to optimize the traffic flows themselves to further increase efficiency. This has been studied in Refs. [97] and [98]. One of the developments that use infrastructure-based information for increasing efficiency is the Signal Phase and Timing (SPaT) technology which has been presented for the first time in Ref. [98]. The SPaT technology is also used for other purposes, such as pedestrian safety or collision avoidance

[99]. Using this technology for increasing efficiency by means of sharing traffic data and intersection situations and optimization of vehicle speed and acceleration is recently studied. Zifei et al. claim that using SPaT and other optimization can increase efficiency by 8.5%–15.6% [100]. The mechanism for increasing efficiency using SPaT and other V2I communication to transmit data (timing and signals) from intersections to the optimization unit in the CAV. The CAV adjusts speed and acceleration in the manner that it approaches the intersection in the most optimal manner and without wasting energy on unnecessary speed corrections.

Another way to increase CAV efficiency presented in the literature is by anticipating the behavior of approaching vehicles. For example, it is a common practice for drivers to provide a safe gap between their vehicle and the next vehicle. If the approaching vehicle follows close and does not leave a safe gap, the lead vehicle's driver changes its behavior or lets it pass by. If the host car is a CAV, it could anticipate the behavior of the approaching car to take action in advance to ensure safety while conserving energy. Otherwise, the lead vehicle could lose energy when it is entirely possible to increase efficiency by leveraging data from the following car and with optimization of the lead vehicle with the probabilistic determination of the following car's behavior [101]. Proficiency of this anticipation on increasing a CAV efficiency reported between 3% [102] up to 8% for the host vehicle [103]. However, if CAV technologies and optimization algorithms were leveraged, this could be as large as 44% for the follower vehicle.

By surveying the literature on CAV efficiency, one will find that the majority of effort has been dedicated to increasing the efficiency of CAVs in follower car technologies. Follower vehicle technologies are among the most applied technologies in commercially available vehicles, and they can be named the first step for automated vehicles. Therefore, the most effort for increasing the efficiency of CAVs has been made using this technology. Many existing papers discussed the opportunity of increasing efficiency in groupings of CAVs, known as "Platoons" [104]. Also, experimental studies have been done to determine and increase the efficiency of this traffic management method [105]. With recent developments in V2V connectivity, this opportunity for increasing the efficiency of vehicles has attracted increased attention through CACC. CACC technology has been developed with the intention not only to increase safety and efficiency but also to play an important role in furthering vehicle autonomy. CACC technology's effects and influence on the increasing efficiency of CAVs have been the topic of many different papers and articles. Ploeg et al. [106] discussed the positive effects of CACC on harmonizing the traffic flow, which could be beneficial for increasing the overall efficiency of a string of vehicles. Among different papers discussing how CACC technology can be beneficial for CAVs, [107] emphasized how increasing the efficiency of the group of vehicles through "Eco-platooning" and optimization algorithms can increase efficiency by 10.5% overall. One of the most promising control algorithms that have garnered the most attention is those that are capable of solving "lane-change and merge" situations. Lane-changing is one of the most sophisticated operations for an AV to perform. This operation receives lots of attention since lane-changing should be accomplished with a minimum fault and with minimal safety risk. However, in Ref. [108], leveraging lane-changing maneuvers to increase efficiency was discussed. In other studies, the focus was not on the issue of efficiency.

One of the most recent and promising ideas for increasing the fuel efficiency of CAVs is to harmonize CAVs by manipulating the traffic flow. This harmonization could be done either with conventional methods, such as CACC technology [106] or traffic signals and V2I communications [109]. Furthermore, harmonization itself can be achieved as an independent goal with CAVs. One of the most promising studies that experiments have explored was introduced in Ref. [110] as the "Theory of Jam-Absorption Driving (JAD)." According to JAD, an automated car with relatively few control rules can propagate compression and expansion waves along the traffic flow. The result of this is a dispersal of traffic jams that can cause huge amounts of wasted energy consumption. The JAD efficiency in preventing traffic jams using V2V connection and optimal control has been demonstrated in Ref. [111]. It is abundantly clear that JAD technology not only affects motorist safety but can also greatly increase fuel efficiency, which is an indirect benefit of CAVs.

The effect of CAVs on the total energy consumption of society has also been investigated, with a wide range of results being reported. Based on these reports, it can be concluded that CAVs should be expected to decrease total energy consumption by at least 60% but possibly by as much as 200%. This variance is due to the duality of CAV operation being both microscopic as a CAV focuses on its own operation as well as the macroscopic potential for implementing large-scale, efficiency-focused traffic management. This paradox has also been discussed in the U.S. Energy Information Administration's (EIA's) annual energy outlook in 2018 [112]. Fig. 4 discusses how a CAV can lead to increased or decreased energy consumption. Based on this diagram and other works, it is clear that a CAV can increase each vehicle's efficiency through several means while also increasing energy demands through large-scale deployment and utilization. Other studies suggest that increasing the penetration of CAVs further increases the efficiency of CAVs, both in gasoline-powered CAVs and those powered by electric motors. The results of these studies are reported in [113] and performed by Iowa State University. In Fig. 5, a comparison of decreased fuel consumption under varying penetration rates for three different driving scenarios of CAVs is depicted.

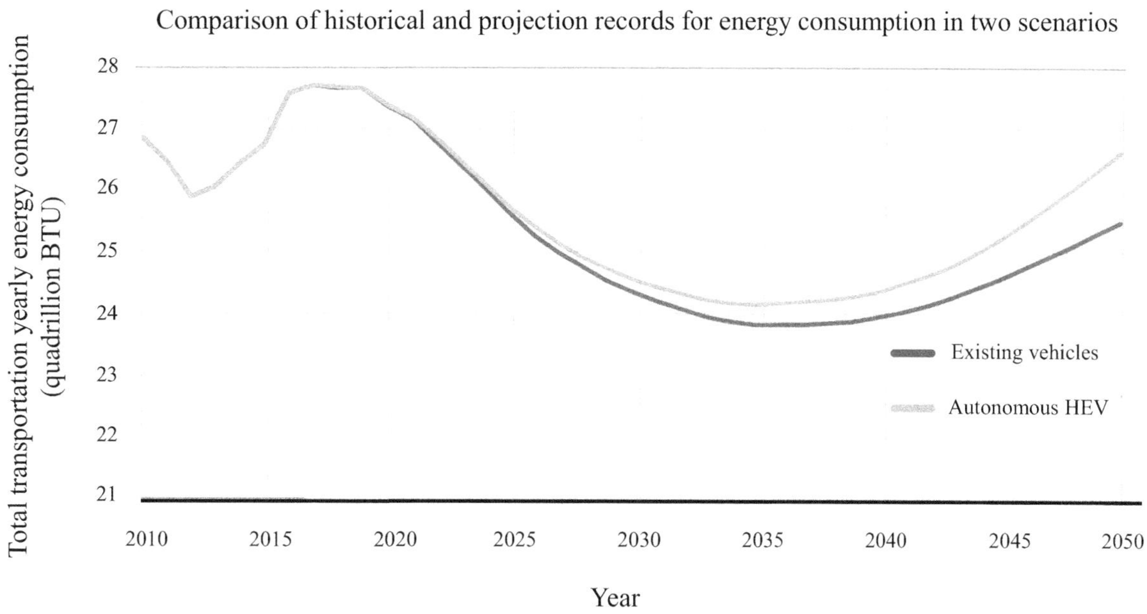

FIG. 4 Effect of different mechanisms in CAV on energy consumption. The figure extracted from Ref. [114].

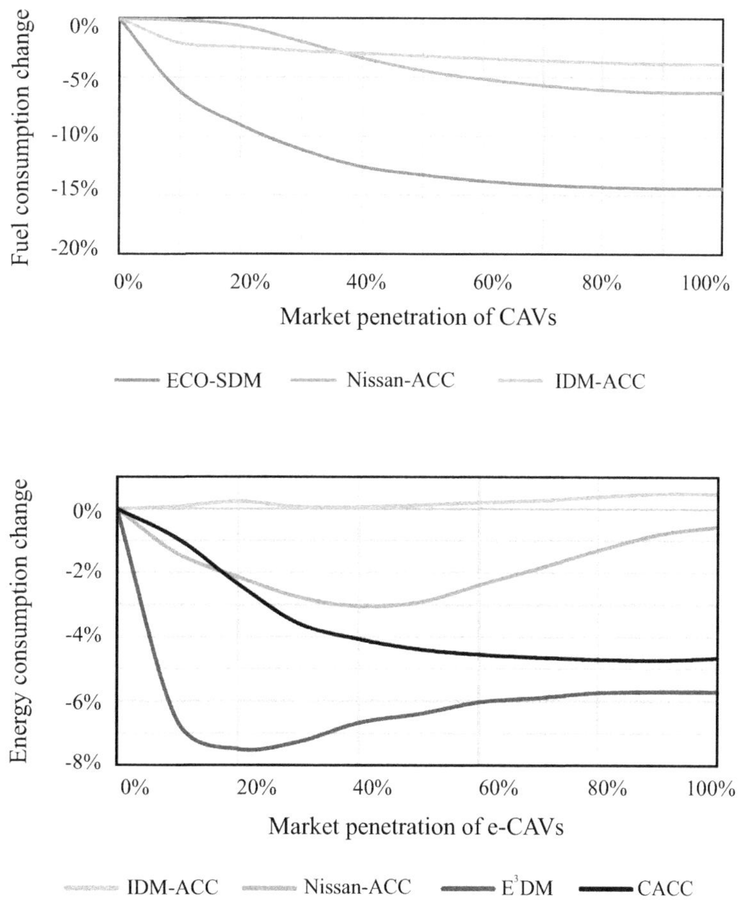

FIG. 5 Comparison of decreasing energy consumption by increasing penetration of CAVs in a:Gasoline CAVs and b:e-CAVs under three different driving scenarios. The figure extracted from Ref. [113].

1.5 Challenges

While corporate optimism around CAVs is bursting at the seams, a growing majority of Americans express concern about the driverless revolution [115–117]. The Partners for Automated Vehicle Education's (PAVE's) recent polling of Americans revealed that nearly 3 in 4 feel that the technology is far from ready and nearly half stating that they would never use self-driving services [118]. These results are consistent with the trend of mistrust that followed several high-profile CAV incidents [119,120].

Public uncertainty is but one of the many challenges facing the industry, closely followed by security and privacy concerns. Security and privacy concerns are deeply intertwined due to the nature of AVs and even modern vehicles. Most vehicles available today are equipped with various hardware and sensors that enable features that not only aid drivers but also record vast amounts of personal data. Related to security is the reliability of these vehicles to operate safely. The testing and verification of AV systems prove to be a lengthy and tedious task that experts estimate to take hundreds of years to complete. In this section, these issues and more will be discussed further in more detail.

1.5.1 Safety and security challenges

Traditionally, automobiles have largely been dependent on mechanical systems and human operators to properly function and safely operate in its environment. Where electronics were implemented, these systems were primarily used to optimize operational tasks, such as fuel injector timing, ignition control, and automatic headlights. In recent decades, however, technology from the humble GPS to cutting-edge ADASs have pervaded the automotive industry. The integration of this variety of technology has also had the unintended side effect of creating vast, largely insecure attack surfaces that adversaries could easily exploit [121]. This concern is validated as automotive cyber security incidents doubled in 2019 and are up over 600% since 2016 [122]. This growth is likely to continue as the number of potential targets increases and the overall vulnerability of each target grows [123].

Automobiles are already considered to be deeply important to secure due to privacy and data related concerns. In terms of AVs, however, the security of these systems directly impacts the safety of motorists and the general public. If an AV were to become compromised, it could lead to dire circumstances for the occupants and even all those around them [124–127]. Ensuring that AVs are capable of piloting themselves is useless without also guaranteeing that these vehicles be able to securely communicate and control themselves and each other.

1.5.2 Testing and verification challenges

For testing automated and AV technology, there are several approaches [128–133]. Simulated testing is comprised of offline and online methods that occur in real time. Offline testing seeks to optimize testing in terms of test execution speed while real-time simulations focus on testing accuracy within a bounded response time [128]. Online methods also ensure access to testing data, which accelerates the validation phase of the testing process and increases certainty during development [129]. Real-time simulations consist of Software-in-the-Loop (SiL), Hardware-in-the-Loop (HiL), and Vehicle-in-the-Loop (ViL) testing.

SiL is often used for situations where traditional testing methods are infeasible due to factors like cost or safety [130]. However, SiL is unable to match the accuracy and granularity of other methods that incorporate physical hardware and components. On the other hand, HiL subjects physical components to virtual stimuli to explore how a system reacts in real time [131]. As the physical system is included, HiL is able to achieve higher quality results when compared to SiL testing. In terms of automotive testing, however, HiL lacks the inclusion of the real-time motion and dynamics of the test vehicle. ViL solves this shortcoming by integrating the entire physical vehicle into the simulated testing environment. The downside of ViL testing, however, is that this method can only consider the events and data generated within the simulation, leading to lower accuracy and resolution compared to real-world testing. Real-world testing is regarded as the best in terms of accuracy and resolution due to the inclusion of vehicle dynamics and motion. Unfortunately, this method is costly in terms of time and capital [134–136]. Real-world testing is also incapable of investigating the many unsafe or infeasible scenarios, particularly edge cases that involve pedestrians.

There is also limited work in the area of combining these aforementioned methods into comprehensive testing platforms. The result of work in this area could revolutionize the way in which AVs and even conventional automobiles are tested. Work conducted by Dr. Sargozlaei et al. seeks to create a generalized approach for testing AVs. Their method would subject a full-size vehicle to testing scenarios that occur within a dynamic, virtual environment. The "virtual stimuli" generated during testing are then relayed to sensors and other perception hardware on the vehicle to prompt the subject to carry out actions and reactions within the real world. As such, the research platform integrates SiL, HiL, ViL, and real-world testing in a manner that allows each method to counterbalance the shortcomings of one another.

1.5.3 Other challenges

Aside from the previously discussed, there are other more abstract challenges facing AVs. Of these, cost and production, as well as technological limitations and public perception will be discussed.

1.5.3.1 *Production and cost*

In 2020, AV research costs exceeded $16 billion and key firms in this area have continued to invest capital to bring vehicle autonomy to market [137]. Research and development is largely to blame as the investment into hardware and personnel does not yield a return, in the short term. For example, vehicles specialized for AV and CAV research often cost from $50,000 to $250,000 or more. Producing even a minimally viable CAV would require sensors for perception, GPS and IMU for localization, V2X communication systems, and hardware for processing this data to act upon it—in addition to the cost of the passenger vehicle it is integrated into. Due to this, the costs of even a basic CAV would likely start at $50,000. CAVs in their current state are not scalable due to the vast on-board computation resources required and associated cost. At this price point, adoption of CAVs would be slow which will hinder their performance, impact, and reputation. While lower-cost CAVs are likely on the horizon, paving the way for their deployment in the present is vital for ensuring future success. Research into low-resource, collaborative training, and decision-making approaches is vital to ushering in this new age of transit.

1.5.3.2 *Technological limitations*

The current state of AVs, while at the bleeding edge of science, is still insufficient for enabling certain aspects of vehicle autonomy today and for the foreseeable future. For example, AVs were originally conceived as FSD vehicles that navigated the world, independently from one another. However, in more recent years, mainstream thought has shifted toward collaborative autonomy, where CAVs maintain constant communication with other vehicles and intelligent infrastructure systems (ITS) to enable safe, widespread autonomy. Much of this shift comes from the simple fact that CAVs can better perceive their surroundings and have better decision-making due to constant communication. This does, however, introduce security and privacy challenges as previously discussed.

1.5.3.3 *Electrification*

As electrification continues to rapidly pervade the automotive industry, it is assumed that AVs will be EVs. This brings about a host of benefits and challenges, the first of these challenges being charging. The energy production necessary to charge a majority EV population is exponentially greater than what the US currently possesses. Furthermore, the time required to charge an EV can still take several hours. While there are high-speed chargers and EVs available, these are in the minority of presently existing EVs. If one were to assume that charge times are an irrelevant issue, the availability of EV charging stations is still nowhere close to the same availability as traditional fueling stations. These are, however, issues that will be solved with time and society has already made strides in pursuing this effort.

1.6 Summary and conclusion

This chapter demonstrates a complete chronological development of CAVs, from simple primitive ideas to different available technologies and solutions. We discussed that the CAVs require different infrastructures and developed control techniques. It is impossible to imagine CAVs without the development of infrastructures and also without determining protocols for shared infrastructures. This means that for having CAVs developed in public, it is not only the automakers' role to determine and develop technologies but also other society members and partners should take part in determining procedures. Up to now, much work has been done, but to make CAVs much more applicable, it is important to work on V2V communication and also V2I communication.

The most important part of an AV is its ability to make an appropriate decision in every circumstance in mixed traffic. The driver capabilities already do this task. Therefore, to achieve a fully autonomous vehicle, it is very important to have a decision algorithm and dynamic controller that is as capable as a human can. To this end, the revolutionary idea was to use AI-based control in CAVs. However, by using ordinary model-based control methods, which are so-called "traditional controllers," some goals are achieved, but it has been proved that these control methods are limited and cannot work for a fully automated vehicle. Model-based controllers could operate and deal with recognized or predicted situations, but in vastly different situations, especially those in a crowded city, these model-based controllers cannot perform well. The AV should be able to not only make an appropriate decision in a specific situation, but also it needs to develop its decision-making capabilities during a driving cycle. This is what happens right now when a driver is learning to drive a car. Using AI-based algorithms somehow mimics drivers' behaviors in learning and dealing with an unpredicted situation. Therefore, recent works on CAV path controls have been moved into using AI-based controllers instead of developing complicated dynamics models.

Another progress in controlling CAVs was in increasing CAVs energy efficiency. Since AVs will definitely increase demands for transportation, it is important to design and develop CAVs with more efficiency to compensate for this increase in transportation demand and increase total transportation efficiency. A CAV can benefit from several aspects for increasing efficiency. These may include knowledge about coming traffic and road and path conditions. It is clear that being aware of the future situation is essential, but without taking action, it could not be useful. Therefore, it's necessary to develop controllers to use collected data to increase efficiency. This topic is among the most topics in CAVs, but it seems it can be developed much more by using AI-based controllers, especially with using connectivity capabilities of CAVs to manage the flow of vehicles as well as manage every vehicle's efficiency.

The CAVs have other challenges that should be considered. Among all of these challenges, it is important to consider cyber-attacks and interruptions in CAV connectivity. A CAV as a connectable instrument could have interfered unintentionally or on purpose. This interference could lead to a disaster since it can affect the driving path of a vehicle with false data. Therefore, it is necessary to study the security of CAVs in the future and design secure control methods.

Acknowledgment

Partial support of this research was provided by the National Science Foundation under Grant No. CNS-1919855 and ECCS-EPCN-2241718. Any opinions, findings, conclusions, or recommendations expressed in this material are those of the author(s) and do not necessarily reflect the views of the sponsoring agency.

References

[1] SAE International, Taxonomy and Definitions for Terms Related to Driving Automation Systems for On-Road Motor Vehicles, SAE, 2016.
[2] R. Felton, The man who tested the first driverless car in 1925 had a bizarre feud with Harry Houdini. February 14 (2017), https://jalopnik.com/the-man-who-tested-the-first-driverless-car-in-1925-had-1792312207.
[3] NHTSA, 2015 Motor vehicle crashes: overview, US National Highway Traffic Safety Administration, 2016, pp. 1–9.
[4] A. Papadoulis, M. Quddus, M. Imprialou, Evaluating the safety impact of connected and autonomous vehicles on motorways, Accid. Anal. Prev. 124 (2019) 12–22.
[5] S. Gorman, Electricity grid in us penetrated by spies, Wall Street J. 8 (2009).
[6] A. Abbaspour, P. Aboutalebi, K.K. Yen, A. Sargolzaei, Neural adaptive observer-based sensor and actuator fault detection in nonlinear systems: application in UAV, ISA Trans. 67 (2017) 317–329.
[7] R.E. Fenton, R.J. Mayhan, Automated highway studies at the Ohio state university-an overview, IEEE Trans. Veh. Technol. 40 (1) (1991) 100–113.
[8] R. Oshima, Control system for automobile driving, in: Tokyo IFAC Symposium, 1965, pp. 347–357.
[9] T. Fujioka, A case study on an automated driving highway system in Japan, in: Transportation Research Board 72nd Annual Meeting, 1993.
[10] F. Broqua, Cooperative driving: basic concepts and a first assessment of "intelligent cruise control" strategies, in: DRIVE Conference 1991, Brussels, Belgium. Advanced Telematics in Road Transport, Vol. II, 1991.
[11] M. De Saint Blancard, B. Dubuisson, M. Rombaut, Pro-lab2: a vehicle with driving assistance, in: Towards an Intelligent Transport System. Proceedings of the First World Congress on Applications of Transport Telematics and Intelligent Vehicle-Highway Systems, November 30–3 December 1994, Paris, Vol. 4, 1994.
[12] S.E. Shladover, The California path program of IVHS research and its approach to vehicle-highway automation, in: Proceedings of the Intelligent Vehicles92 Symposium, IEEE, 1992, pp. 347–352.
[13] K.S. Chang, J. Karl Hedrick, W.-B. Zhang, P. Varaiya, M. Tomizuka, S.E. Shladover, Automated highway system experiments in the path program, J. Intell. Transp. Syst. 1 (1) (1993) 63–87.
[14] S.E. Shladover, California and the roots of IVHS, IVHS Rev. (1993) 27–34.
[15] L. Saxton, et al., Mobility 2000 and the Roots of IVHS, Tech. Rep., United States, Federal Highway Administration, 1999.
[16] H. Peng, M. Tomizuka, Preview control for vehicle lateral guidance in highway automation, J. Dyn. Sys. Meas. Control. 115 (4) (1993) 679–686.
[17] H. Peng, W. Zhang, M. Tomizuka, S. Shladover, A reusability study of vehicle lateral control system, Veh. Syst. Dyn. 23 (1) (1994) 259–278.
[18] T. Hessburg, M. Tomizuka, Fuzzy logic control for lateral vehicle guidance, IEEE Control. Syst. Mag. 14 (4) (1994) 55–63.
[19] W. Chee, M. Tomizuka, Lane change maneuver for AHS applications, JSAE Rev. 2 (16) (1995) 219.
[20] V. Narendran, J.K. Hedrick, Autonomous lateral control of vehicles in an automated highway system, Veh. Syst. Dyn. 23 (1) (1994) 307–324.
[21] T. Fujioka, K. Suzuki, Lateral autonomous driving by sliding control, JSAE Rev. 2 (16) (1995) 219.

[22] J. Ackermann, W. Sienel, Robust control for automatic steering, in: 1990 American Control Conference, IEEE, 1990, pp. 795–800.
[23] W.A. Leasure, NHTSA's IVHS Collision Avoidance Research Program: Strategic Plan and Status Update, 2000.
[24] Hall, Pandemic shifts focus for autonomous vehicle companies, 2020, Available at https://www.governing.com/next/Pandemic-Shifts-Focus-for-Autonomous-Vehicle-Companies.html.
[25] K. Korosec, Waymo brings in $2.25 billion from outside investors, alphabet, 2020, Available at https://social.techcrunch.com/2020/03/02/waymo-brings-in-2-25-billion-from-outside-investors-alphabet/.
[26] S. Tsugawa, N. Watanabe, H. Fujii, Super smart vehicle system-its concept and preliminary works, in: Vehicle Navigation and Information Systems Conference, 1991, Vol. 2, IEEE, 1991, pp. 269–277.
[27] A. Efrati, Money pit: self-driving cars' $16 billion cash burn, 2020, Available at https://www.theinformation.com/articles/money-pit-self-driving-cars-16-billion-cash-burn.
[28] M. Weber, Where to? A history of autonomous vehicles, 2014, Available at https://computerhistory.org/blog/where-to-a-history-of-autonomous-vehicles/.
[29] B. Grey, Top 30 autonomous vehicle technology and car companies, 2020, Available at https://www.greyb.com/autonomous-vehicle-companies/.
[30] T. LeBlanc, State of the industry 2019: your impending pro AV profit problems, 2018, Available at https://www.commercialintegrator.com/businessresources/research/stat-of-the-industry-av-2019.
[31] C. Srinivasa, Leveraging e/av patent portfolios during a pandemic, 2021, Available at https://www.natlawreview.com/article/leveraging-eav-patent-portfolios-during-pandemic.
[32] C. Metz, E. Griffith, This was supposed to be the year driverless cars went mainstream, The New York Times (2020).
[33] Hyperdrive, The state of the self-driving car race 2020, 2020, Available at https://www.bloomberg.com/features/2020-self-driving-car-race/.
[34] A.J. Hawkins, Uber's fraught and deadly pursuit of self-driving cars is over, 2020, Available at https://www.theverge.com/2020/12/7/22158745/uber-selling-autonomous-vehicle-business-aurora-innovation. abstractNote=The deal marks the end of Uber's fraught and deadly pursuit of robot taxis.
[35] A.J. Hawkins, The autonomous vehicle world is shrinking — it's overdue, 2021, Available at https://www.theverge.com/22423489/autonomous-vehicle-consolidation-acquisition-lyft-uber.
[36] A.J. Hawkins, Cruise acquires voyage in another autonomous vehicle merger, 2021, Available at https://www.theverge.com/2021/3/15/22331794/cruise-acquires-voyage-self-driving-car-merger.
[37] A.J. Hawkins, Toyota is buying lyft's autonomous car division for $550 million, 2021, Available at https://www.theverge.com/2021/4/26/22404406/toyota-lyft-autonomous-vehicle-acquisition-amount-deal. abstractNote=The latest acquisition in the AV world.
[38] K. Korosec, Nuro acquires autonomous trucking startup ike, 2020, Available at https://social.techcrunch.com/2020/12/23/nuro-acquires-autonomous-trucking-startup-ike/. abstractNote=Autonomous delivery company Nuro has acquired Ike, a startup founded by veterans of Apple, Google and Uber Advanced Technologies Group that aimed to commercialize self-driving trucks. The deal, the latest in a busy season of acquisitions and consolidation in the autonomous vehicle industry, brings ....
[39] L. Chapman, D. Hull, Uber sells cash-burning self-driving unit to aurora to focus on turning a profit, Available at https://financialpost.com/transportation/autos/uber-sells-cash-burning-self-driving-unit-to-aurora-to-focus-on-turning-a-profit.
[40] J. Krafcik, Waymo is opening its fully driverless service to the general public in phoenix, 2020, Available at https://blog.waymo.com/2020/10/waymo-is-opening-its-fully-driverless.html.
[41] C. Gartenberg, Waymo ceo john krafcik steps down, replaced by two co-ceos, 2021, Available at https://www.theverge.com/2021/4/2/22364317/waymo-ceo-john-krafcik-stepping-down-self-driving-cars-google-alphabet. abstractNote=Waymo's first CEO is hitting the road.
[42] K. Korosec, Waymo to lose its cfo and head of automotive partnerships, 2021, Available at https://social.techcrunch.com/2021/05/12/waymos-cfo-and-head-of-automotive-partnerships-are-leaving-the-autonomous-vehicle-company/.
[43] P. Dave, H. Jin, Google Self-Driving Spinoff Waymo Begins Testing with Public in San Francisco, Reuters, 2021.
[44] T. Le-Anh, R.M.B.M. de Koster, A review of design and control of automated guided vehicle systems, Eur. J. Oper. Res. 171 (2006) 1–23.
[45] B. Paden, M. Cáp, S.Z. Yong, D.S. Yershov, E. Frazzoli, A survey of motion planning and control techniques for self-driving urban vehicles, IEEE Trans. Intell. Veh. 1 (2016) 33–55.
[46] M. Pasquier, H.C. Quek, M. Toh, Fuzzylot: a novel self-organising fuzzy-neural rule-based pilot system for automated vehicles, Neural Netw. 14 (8) (2001) 1099–1112.
[47] M. Kuderer, S. Gulati, W. Burgard, Learning driving styles for autonomous vehicles from demonstration, in: 2015 IEEE International Conference on Robotics and Automation (ICRA), IEEE, 2015, pp. 2641–2646.
[48] D. Silver, J.A. Bagnell, A. Stentz, Learning autonomous driving styles and maneuvers from expert demonstration, in: Experimental Robotics, Springer, 2013, pp. 371–386.
[49] M. Samuel, M. Hussein, M.B. Mohamad, A review of some pure-pursuit based path tracking techniques for control of autonomous vehicle, Int. J. Comput. Appl. 135 (1) (2016) 35–38.
[50] A. Ollero, O. Amidi, Predictive path tracking of mobile robots. application to the cmu navlab, in: Proceedings of 5th International Conference on Advanced Robotics, Robots in Unstructured Environments, 91, ICAR, 1991, pp. 1081–1086.
[51] A. García-Cerezo, A. Ollero, J.L. Martínez, Design of a robust high-performance fuzzy path tracker for autonomous vehicles, Int. J. Syst. Sci. 27 (8) (1996) 799–806.
[52] A. Ollero, G. Heredia, Stability analysis of mobile robot path tracking, in: Proceedings 1995 IEEE/RSJ International Conference on Intelligent Robots and Systems. Human Robot Interaction and Cooperative Robots, Vol. 3, IEEE, 1995, pp. 461–466.
[53] A. Ollero, A. García-Cerezo, J. Martinez, Fuzzy supervisory path tracking of mobile reports, Control. Eng. Pract. 2 (2) (1994) 313–319.
[54] P. Zhao, J. Chen, Y. Song, X. Tao, T. Xu, T. Mei, Design of a control system for an autonomous vehicle based on adaptive-pid, Int. J. Adv. Robot. Syst. 9 (2) (2012) 44.
[55] M. Brackstone, M. McDonald, Car-following: a historical review, Transport. Res. F: Traffic Psychol. Behav. 2 (4) (1999) 181–196.
[56] J. Ruan, M. Fu, Y. Li, J. Huang, Study on throttle control of intelligent vehicle longitudinal motion, in: IEEE International Conference on Vehicular Electronics and Safety, IEEE, 2005, pp. 176–181.

[57] J.E. Naranjo, C. González, J. Reviejo, R. García, T. De Pedro, Adaptive fuzzy control for inter-vehicle gap keeping, IEEE Trans. Intell. Transp. Syst. 4 (3) (2003) 132–142.
[58] J.I. Hernandez, C.-Y. Kuo, Steering control of automated vehicles using absolute positioning GPS and magnetic markers, IEEE Trans. Veh. Technol. 52 (1) (2003) 150–161.
[59] A.M. Muad, A. Hussain, S.A. Samad, M.M. Mustaffa, B.Y. Majlis, Implementation of inverse perspective mapping algorithm for the development of an automatic lane tracking system, in: 2004 IEEE Region 10 Conference TENCON 2004, IEEE, 2004, pp. 207–210.
[60] S. Kammel, B. Pitzer, Lidar-based lane marker detection and mapping, in: 2008 IEEE Intelligent Vehicles Symposium, IEEE, 2008, pp. 1137–1142.
[61] L. Cai, A.B. Rad, W.-L. Chan, An intelligent longitudinal controller for application in semiautonomous vehicles, IEEE Trans. Ind. Electron. 57 (4) (2009) 1487–1497.
[62] C. Zhaoxue, S. Pengfei, Efficient method for camera calibration in traffic scenes, Electron. Lett. 40 (6) (2004) 368–369.
[63] J.C. McCall, M.M. Trivedi, An integrated, robust approach to lane marking detection and lane tracking, in: IEEE Intelligent Vehicles Symposium, 2004, IEEE, 2004, pp. 533–537.
[64] J. Wang, S. Schroedl, K. Mezger, R. Ortloff, A. Joos, T. Passegger, Lane keeping based on location technology, IEEE Trans. Intell. Transp. Syst. 6 (3) (2005) 351–356.
[65] R. Risack, N. Mohler, W. Enkelmann, A video-based lane keeping assistant, in: Proceedings of the IEEE Intelligent Vehicles Symposium 2000 (Cat. No. 00TH8511), IEEE, 2000, pp. 356–361.
[66] P. Hidas, Modelling lane changing and merging in microscopic traffic simulation, Transp. Res. Part C Emerging Technol. 10 (5–6) (2002) 351–371.
[67] R.T. O'Brien, P.A. Iglesias, T.J. Urban, Vehicle lateral control for automated highway systems, IEEE Trans. Control Syst. Technol. 4 (3) (1996) 266–273.
[68] S. Chaib, M.S. Netto, S. Mammar, H∞, adaptive, pid and fuzzy control: a comparison of controllers for vehicle lane keeping, in: IEEE Intelligent Vehicles Symposium, 2004, IEEE, 2004, pp. 139–144.
[69] L.B. Cremean, T.B. Foote, J.H. Gillula, G.H. Hines, D. Kogan, K.L. Kriechbaum, J.C. Lamb, J. Leibs, L. Lindzey, C.E. Rasmussen, et al., Alice: an information-rich autonomous vehicle for high-speed desert navigation, J. Field Rob. 23 (9) (2006) 777–810.
[70] A. Likmeta, A.M. Metelli, A. Tirinzoni, R. Giol, M. Restelli, D. Romano, Combining reinforcement learning with rule-based controllers for transparent and general decision-making in autonomous driving, Robot. Auton. Syst. 131 (2020) 103568.
[71] J.E. Naranjo, C. González, R. García, T. de Pedro, J. Revuelto, J. Reviejo, Fuzzy logic based lateral control for GPS map tracking, in: IEEE Intelligent Vehicles Symposium, 2004, IEEE, 2004, pp. 397–400.
[72] D. Pomerleau, Ralph: rapidly adapting lateral position handler, in: Proceedings of the Intelligent Vehicles' 95. Symposium, IEEE, 1995, pp. 506–511.
[73] W. Yao, H. Zhao, P. Bonnifait, H. Zha, Lane change trajectory prediction by using recorded human driving data, in: 2013 IEEE Intelligent Vehicles Symposium (IV), IEEE, 2013, pp. 430–436.
[74] G. Freitas, B. Hamner, M. Bergerman, S. Singh, A practical obstacle detection system for autonomous orchard vehicles, in: 2012 IEEE/RSJ International Conference on Intelligent Robots and Systems, IEEE, 2012, pp. 3391–3398.
[75] U. Muller, J. Ben, E. Cosatto, B. Flepp, Y.L. Cun, Off-road obstacle avoidance through end-to-end learning, in: Advances in Neural Information Processing Systems, Citeseer, 2006, pp. 739–746.
[76] C. Vallon, Z. Ercan, A. Carvalho, F. Borrelli, A machine learning approach for personalized autonomous lane change initiation and control, in: 2017 IEEE Intelligent Vehicles Symposium (IV), IEEE, 2017, pp. 1590–1595.
[77] Y. Ma, Z. Wang, H. Yang, L. Yang, Artificial intelligence applications in the development of autonomous vehicles: a survey, IEEE/CAA J. Autom. Sin. 7 (2) (2020) 315–329.
[78] J. Van Brummelen, M. O'Brien, D. Gruyer, H. Najjaran, Autonomous vehicle perception: the technology of today and tomorrow, Transp. Res. Part C Emerging Technol. 89 (2018) 384–406.
[79] J. Janai, F. Güney, A. Behl, A. Geiger, et al., Computer vision for autonomous vehicles: problems, datasets and state of the art, Found. Trends Comput. Graph. Vis. 12 (1–3) (2020) 1–308.
[80] R. Benenson, M. Omran, J. Hosang, B. Schiele, Ten years of pedestrian detection, what have we learned? in: European Conference on Computer Vision, Springer, 2014, pp. 613–627.
[81] S. Kuutti, S. Fallah, K. Katsaros, M. Dianati, F. Mccullough, A. Mouzakitis, A survey of the state-of-the-art localization techniques and their potentials for autonomous vehicle applications, IEEE Internet Things J. 5 (2) (2018) 829–846.
[82] S. Zhang, R. Benenson, M. Omran, J. Hosang, B. Schiele, How far are we from solving pedestrian detection? in: Proceedings of the IEEE Conference on Computer Vision and Pattern Recognition, 2016, pp. 1259–1267.
[83] W. Schwarting, J. Alonso-Mora, D. Rus, Planning and decision-making for autonomous vehicles, Annu. Rev. Control Rob. Auton. Syst. 1 (2018) 187–210.
[84] D.J. Fagnant, K. Kockelman, Preparing a nation for autonomous vehicles: opportunities, barriers and policy recommendations, Transp. Res. A Policy Pract. 77 (2015) 167–181.
[85] D. CireAan, U. Meier, J. Masci, J. Schmidhuber, Multi-column deep neural network for traffic sign classification, Neural Netw. 32 (2012) 333–338.
[86] H.M. Eraqi, M.N. Moustafa, J. Honer, End-to-end deep learning for steering autonomous vehicles considering temporal dependencies, arXiv preprint arXiv:1710.03804 (2017).
[87] X. Di, R. Shi, A survey on autonomous vehicle control in the era of mixed-autonomy: from physics-based to ai-guided driving policy learning, Transp. Res. Part C Emerging Technol. 125 (2021) 103008.
[88] NHTS Administration, et al., Federal motor vehicle safety standards; v2v communications, Fed. Regist. 82 (8) (2017) 3854–4019.
[89] M. Shulman, S. Geisler, Crash Avoidance Metrics Partnership, United States Federal Highway Administration, Development of Vehicle-to-Infrastructure Applications Program Second Annual Report, Report Number: FHWA-JPO-16-480, 2016. https://rosap.ntl.bts.gov/view/dot/3620.

[90] B. Asadi, C. Zhang, A. Vahidi, The role of traffic flow preview for planning fuel optimal vehicle velocity, in: Dynamic Systems and Control Conference, 44182, 2010, pp. 813–819.
[91] S. Schepmann, A. Vahidi, Heavy vehicle fuel economy improvement using ultracapacitor power assist and preview-based mpc energy management, in: Proceedings of the 2011 American Control Conference, IEEE, 2011, pp. 2707–2712.
[92] N. Wan, C. Zhang, A. Vahidi, Probabilistic anticipation and control in autonomous car following, IEEE Trans. Control Syst. Technol. 27 (1) (2017) 30–38.
[93] C. Zhang, A. Vahidi, P. Pisu, X. Li, K. Tennant, Role of terrain preview in energy management of hybrid electric vehicles, IEEE Trans. Veh. Technol. 59 (3) (2009) 1139–1147.
[94] S. Terwen, M. Back, V. Krebs, Predictive powertrain control for heavy duty trucks, IFAC Proc. Vol. 37 (22) (2004) 105–110.
[95] E. Hellström, J. Åslund, L. Nielsen, Design of an efficient algorithm for fuel-optimal look-ahead control, Control. Eng. Pract. 18 (11) (2010) 1318–1327.
[96] MultiMedia LLC, Diamler Use Predictive Cruise, 2009. https://northamerica.daimlertruck.com/company/press-releases/pressdetail/freightliner-trucks-launches-runsmart-predictive-cruise-2009-03-19.
[97] B. Asadi, A. Vahidi, Predictive use of traffic signal state for fuel saving, IFAC Proc. Vol. 42 (15) (2009) 484–489.
[98] S. Mandava, K. Boriboonsomsin, M. Barth, Arterial velocity planning based on traffic signal information under light traffic conditions, in: 2009 12th International IEEE Conference on Intelligent Transportation Systems, IEEE, 2009, pp. 1–6.
[99] M. Maile, L. Delgrossi, Cooperative intersection collision avoidance system for violations (CICAS-V) for avoidance of violation-based intersection crashes, in: Enhanced Safety of Vehicles, 2009–0118, 2009.
[100] Z. Nie, H. Farzaneh, Real-time dynamic predictive cruise control for enhancing eco-driving of electric vehicles, considering traffic constraints and signal phase and timing (spat) information, using artificial-neural-network-based energy consumption model, Energy 241 (2022) 122888.
[101] M. Bichi, G. Ripaccioli, S. Di Cairano, D. Bernardini, A. Bemporad, I.V. Kolmanovsky, Stochastic model predictive control with driver behavior learning for improved powertrain control, in: 49th IEEE Conference on Decision and Control (CDC), IEEE, 2010, pp. 6077–6082.
[102] K. McDonough, I. Kolmanovsky, D. Filev, S. Szwabowski, D. Yanakiev, J. Michelini, Stochastic fuel efficient optimal control of vehicle speed, in: Optimization and Optimal Control in Automotive Systems, Springer, 2014, pp. 147–162.
[103] J. Han, A. Sciarretta, L.L. Ojeda, G. De Nunzio, L. Thibault, Safe-and eco-driving control for connected and automated electric vehicles using analytical state-constrained optimal solution, IEEE Trans. Intell. Veh. 3 (2) (2018) 163–172.
[104] R. Kunze, R. Ramakers, K. Henning, S. Jeschke, Organization and operation of electronically coupled truck platoons on German motorways, in: Automation, Communication and Cybernetics in Science and Engineering 2009/2010, Springer, 2011, pp. 427–439.
[105] A. Al Alam, A. Gattami, K.H. Johansson, An experimental study on the fuel reduction potential of heavy duty vehicle platooning, in: 13th International IEEE Conference on Intelligent Transportation Systems, IEEE, 2010, pp. 306–311.
[106] J. Ploeg, B.T. Scheepers, E. Van Nunen, N. Van de Wouw, H. Nijmeijer, Design and experimental evaluation of cooperative adaptive cruise control, in: 2011 14th International IEEE Conference on Intelligent Transportation Systems (ITSC), IEEE, 2011, pp. 260–265.
[107] A. Lelouvier, J. Guanetti, F. Borrelli, Eco-platooning of autonomous electrical vehicles using distributed model predictive control, Parameters 2 (2017) 4.
[108] T. Awal, M. Murshed, M. Ali, An efficient cooperative lane-changing algorithm for sensor-and communication-enabled automated vehicles, in: 2015 IEEE Intelligent Vehicles Symposium (IV), IEEE, 2015, pp. 1328–1333.
[109] N. Wan, A. Vahidi, A. Luckow, Optimal speed advisory for connected vehicles in arterial roads and the impact on mixed traffic, Transp. Res. Part C Emerging Technol. 69 (2016) 548–563.
[110] R. Nishi, A. Tomoeda, K. Shimura, K. Nishinari, Theory of jam-absorption driving, Transp. Res. B Methodol. 50 (2013) 116–129.
[111] S. Wang, Z. Li, Z. Cao, A. Jolfaei, Q. Cao, Jam-absorption driving strategy for improving safety near oscillations in a connected vehicle environment considering consequential jams, IEEE Intell. Transp. Syst. Mag. 14 (2) (2021) 41–52.
[112] N. Chase, J. Maples, M. Schipper, Autonomous Vehicles: Uncertainties and Energy Implications, Annual Energy Outlook 2018, EIA, https://www.eia.gov/outlooks/aeo/av.php.
[113] J. Dong, C. Lu, L. Hu, et al., Estimating Energy Efficiency of Connected and Autonomous Vehicles in a Mixed Fleet, Tech. Rep.,, University of Iowa Center for Transportation Research and Education, 2018.
[114] Z. Wadud, D. MacKenzie, P. Leiby, Help or hindrance? The travel, energy and carbon impacts of highly automated vehicles, Transp. Res. A Policy Pract. 86 (2016) 1–18.
[115] A. Walker, Are self-driving cars safe for our cities? 2016, Available at https://archive.curbed.com/2016/9/21/12991696/driverless-cars-safety-pros-cons.
[116] A.J. Hawkins, Americans still deeply skeptical about driverless cars: poll, Available at https://www.theverge.com/2018/1/12/16883510/self-driving-car-poll-congress-bill-safety.
[117] F. Richter, Consumer concerns about self-driving cars, 2018, Available at https://www.statista.com/chart/5950/concerns-about-self-driving-cars/.
[118] Partners for Automated Vehicle Education (PAVE), Americans wary of AVS but say education and experience with technology can build trust, 2020, Available at https://pavecampaign.org/pave-poll-americans-wary-of-avs-but-say-education-and-experience-with-technology-can-build-trust/.
[119] AAA, Three-quarters of Americans “afraid” to ride in a self-driving vehicle, 2016, Available at https://newsroom.aaa.com/2016/03/three-quarters-of-americans-afraid-to-ride-in-a-self-driving-vehicle/.
[120] Caravanpublic opinion poll: Driverless cars, 2018. https://saferoads.org/wp-content/uploads/2018/01/AV-Poll-Report-January-2018-FINAL.pdf.
[121] Argus Cyber Security, The attack surface of a connected vehicle, 2021. https://argus-sec.com/attack-surface/.
[122] Upstream Security, Upstream Security’s 2020 Global Automotive Cybersecurity Report, 2020. https://upstream.auto/upstream-security-global-automotive-cybersecurity-report-2020/.
[123] Intelligence, P&S, Automotive Cybersecurity Market Research Report: By Application (ADAS & Safety System, Infotainment, Body Electronics, Powertrain, Telematics), Vehicle Type (Passenger Cars, Commercial Vehicles), Form (In-Vehicle Services, External Cloud Services)- Global

Industry Analysis and Growth Forecast to 2030, 2020. https://www.psmarketresearch.com/market-analysis/automotive-cybersecurity-market.
[124] M. Victorio, A. Sargolzaei, M.R. Khalghani, A secure control design for networked control systems with linear dynamics under a time-delay switch attack, Electronics 10 (3) (2021) 322.
[125] N. Khoshavi, G. Tristani, A. Sargolzaei, Blockchain applications to improve operation and security of transportation systems: a survey, Electronics 10 (5) (2021) 629.
[126] A. Sargolzaei, A. Abbaspour, M.A. Al Faruque, A. Salah Eddin, K. Yen, Security challenges of networked control systems, in: Sustainable Interdependent Networks, Springer, 2018, pp. 77–95.
[127] A. Sargolzaei, C.D. Crane, A. Abbaspour, S. Noei, A machine learning approach for fault detection in vehicular cyber-physical systems, in: Machine Learning and Applications (ICMLA), 2016 15th IEEE International Conference on, IEEE, 2016, pp. 636–640.
[128] J. Bélanger, P. Venne, J.-N. Paquin, et al., The what, where and why of real-time simulation, Planet Rt 1 (1) (2010) 25–29.
[129] O. Gietelink, J. Ploeg, B.D. Schutter, M. Verhaegen, Development of advanced driver assistance systems with vehicle hardware-in-the-loop simulations, Veh. Syst. Dyn. 44 (2006) 569–590.
[130] Advantages and disadvantages of simulation. https://www.concentricmarket.com/blog/advantages-and-disadvantages-of-simulation.
[131] D. Bullock, B. Johnson, R. Wells, M. Kyte, Z. Li, Hardware-in-the-loop simulation, Transp. Res. Part C Emerging Technol. 12 (2004) 73–89.
[132] S.S. Banerjee, S. Jha, J. Cyriac, Z.T. Kalbarczyk, R.K. Iyer, Hands off the wheel in autonomous vehicles?: A systems perspective on over a million miles of field data, in: 2018 48th Annual IEEE/IFIP International Conference on Dependable Systems and Networks (DSN), 2018, pp. 586–597.
[133] A. Ala'J, A. Sargolzaei, M.I. Akbaş, Autonomous vehicles scenario testing framework and model of computation: on generation and coverage, IEEE Access 9 (2021) 60617–60628.
[134] A.J. Hawkins, Toyota will test self-driving car 'edge cases' at new proving ground in Michigan, Available at https://www.theverge.com/2018/5/3/17314778/toyota-self-driving-car-test-proving-ground-michigan.
[135] A.J. Hawkins, Waymo gave me a ride in a car with no driver, Available at https://www.theverge.com/2017/10/31/16579180/waymo-self-driving-test-facility-castle-google.
[136] D. Muoio, Uber built a fake city in Pittsburgh with roaming mannequins to test its self-driving cars, Available at https://www.businessinsider.com/ubers-fake-city-pittsburgh-self-driving-cars-2017-10.
[137] R. Baldwin, Self-driving-car research has cost $16 billion. what do we have to show for it? 2020. https://www.caranddriver.com/news/a30857661/autonomous-car-self-driving-research-expensive/.

CHAPTER

2

Introduction to autonomous systems

Anurag Verma

Department of Electrical Engineering, Institute of Engineering and Technology, Lucknow, Uttar Pradesh, India

2.1 Historical evolution of autonomous systems

The concept of autonomous systems has its roots in various domains, including robotics, aviation, and automotive industries [1–4]. Early robotics research aimed to create machines capable of performing tasks independently, leading to the development of industrial robots that could automate repetitive processes in manufacturing [2,5–7]. The aviation industry pioneered autopilot systems that could stabilize and control aircraft, reducing the pilot's workload during flight. In recent years, the automotive sector has made significant strides in developing self-driving cars, aiming to revolutionize transportation [8].

The history of autonomous systems is a fascinating journey that spans multiple disciplines, from engineering and robotics to artificial intelligence (AI) [9,10]. The concept of autonomous systems involves the development of machines and software capable of performing tasks and making decisions without continuous human intervention [11,12]. Here's a brief overview of the history of autonomous systems:

Early concepts (antiquity to 20th century): The idea of machines operating independently dates back to ancient times, with myths and stories featuring automations. However, serious considerations of autonomous machines began in the 20th century [13].

Cybernetics and feedback control (1940s–1950s): The field of cybernetics, pioneered by Norbert Wiener and others, explored the concept of feedback systems and self-regulation in machines. This laid the theoretical foundation for autonomous systems [14].

Early robotics (1960s–1970s): The first industrial robots were introduced, designed for specific tasks in controlled environments. These early robotic systems were preprogrammed and lacked the ability to adapt to changing conditions [15,16].

AI and expert systems (1970s–1980s): Advances in AI led to the development of expert systems, capable of making decisions based on predefined rules. However, these systems were limited to specific domains and lacked adaptability [17].

Mobile robotics and unmanned vehicles (1980s–1990s): Researchers began exploring mobile robotics, focusing on creating autonomous vehicles capable of navigating dynamic environments. This era saw the development of unmanned aerial vehicles (UAVs) and unmanned ground vehicles (UGVs) [18,19].

Machine learning and neural networks (1990s–2000s): The integration of machine learning techniques, including neural networks, allowed autonomous systems to learn from data and improve their performance over time. This era marked a shift from rule-based systems to more adaptive and intelligent approaches [20].

Autonomous vehicles and drones (2010s–2020s): The past decade witnessed significant advancements in autonomous vehicles, including self-driving cars and drones [21]. Companies like Tesla, Waymo, and others have made strides in developing vehicles capable of navigating complex urban environments [22].

Robotics in various industries (present): Autonomous systems are increasingly integrated into various industries, including agriculture, healthcare, manufacturing, and logistics. Robots and AI-powered systems are performing tasks ranging from surgery to warehouse operations [23].

Challenges and ethical considerations: The proliferation of autonomous systems has brought forth challenges related to safety, security, and ethical considerations. The development of robust regulatory frameworks and ethical guidelines is crucial for the responsible deployment of autonomous technologies [24].

Handbook of Power Electronics in Autonomous and Electric Vehicles
https://doi.org/10.1016/B978-0-323-99545-0.00017-8

The history of autonomous systems is a dynamic and ongoing narrative, with continuous advancements shaping the future of robotics and AI. As technology continues to evolve, the capabilities of autonomous systems are expected to expand, influencing various aspects of daily life and industries.

2.1.1 Autonomous systems

The world is on the brink of a technological revolution that promises to reshape industries, enhance productivity, and transform our daily lives [25]. At the heart of this transformation lies autonomous systems, a field that marries cutting-edge technology with AI to create machines capable of making decisions and taking actions without direct human intervention [26]. An autonomous system can be defined as a self-governing entity that performs tasks, solves problems, or interacts with its environment using a combination of sensors, actuators, and intelligent algorithms. In the realm of robotics and autonomous systems [27], we find a fascinating array of platforms ranging from automobiles and airplanes to robots and unmanned aerial vehicles (UAVs). These platforms are designed to move and operate in physical environments, all driven by a shared goal: achieving desired actions. To accomplish these tasks effectively, a diverse range of tools and algorithms is at our disposal. This book chapter explores the rich landscape of robotics and autonomous systems, delving into the art of designing, modeling, simulating, estimating, navigating, and controlling these platforms. We'll also discuss methods to monitor and interact with the physical environment, ultimately enabling us to bridge the gap between the digital and physical worlds.

2.1.1.1 Self-governing nature

At the heart of autonomous systems lies their ability to operate independently. Unlike traditional machines that rely on constant human control, autonomous systems have the capacity to make decisions and execute actions based on their internal algorithms and sensor inputs [28]. This self-governing nature empowers them to adapt to changing circumstances and fulfill their objectives with a degree of autonomy. Autonomous systems, whether they are robots, unmanned vehicles, or any other intelligent machinery, are characterized by their remarkable ability to operate independently, free from constant human control [29]. This self-governing nature is a fundamental concept in the field of robotics and automation and is key to their effectiveness and adaptability. In this explanation, we will delve into the core aspects of this self-governing nature and explore how autonomous systems make decisions and execute actions based on their internal algorithms and sensor inputs.

2.1.1.2 Internal algorithms

At the heart of autonomous systems are sophisticated internal algorithms [20]. These algorithms are a set of instructions and decision-making processes that are embedded within the system's software or hardware. They are meticulously designed to enable the system to analyze data, process information, and generate responses. These algorithms encompass various tasks, such as perception, decision-making, path planning, and motion control. For example, a self-driving car employs algorithms to interpret sensor data, recognize road signs, and determine safe driving actions [30].

2.1.1.3 Sensor inputs

Autonomous systems are equipped with an array of sensors that provide real-time data about their surroundings [31]. These sensors can include cameras, Light Detection and Ranging (LIDAR), Radio Detecting and Ranging (RADAR), ultrasonic sensors, Global Positioning System (GPS), and a range of other sensory devices [32]. The data collected from these sensors form the basis for the system's perception of the environment. The system constantly receives and processes this sensor data to understand its surroundings, detect obstacles, identify targets, and monitor its own state. The data from sensors are crucial for making informed decisions [33].

2.1.1.4 Decision-making

The self-governing nature of autonomous systems lies in their ability to make decisions autonomously [34]. Based on the data collected from sensors and their internal algorithms, these systems can assess situations, interpret information, and make real-time decisions [35]. These decisions can range from simple tasks like avoiding obstacles to complex actions like planning routes or interacting with the environment [33]. For instance, a drone may use its sensors to adjust its flight path to avoid collisions, while an autonomous vacuum cleaner decides how to navigate a room and clean efficiently [36].

2.1.1.5 *Adaptation to changing circumstances*

One of the defining characteristics of autonomous systems is their adaptability [37]. They can respond effectively to changing circumstances and unforeseen events. When an obstacle appears in their path, they can modify their course. When weather conditions change, they can adjust their operations. This adaptability is a result of their ability to constantly monitor and analyze sensor data, update their internal algorithms, and make decisions accordingly [38].

2.1.1.6 *Degree of autonomy*

The level of autonomy in autonomous systems can vary [39]. Some systems may operate with minimal human intervention, while others may require periodic oversight or intervention in complex or ambiguous situations [40]. The degree of autonomy is determined by the specific application and the trust in the system's capabilities [39]. For instance, autonomous drones used for surveying may require less human intervention than a surgical robot where the consequences of errors can be critical [41].

In conclusion, the self-governing nature of autonomous systems is central to their functionality. These systems rely on internal algorithms and sensor inputs to make decisions and execute actions independently. This autonomy allows them to adapt to changing circumstances, carry out tasks with precision, and operate efficiently in a wide range of applications. The development of autonomous systems continues to advance, offering promising solutions in fields such as transportation, healthcare, agriculture, and many more.

2.1.2 Decision-making and action-taking

Autonomous systems are equipped with advanced algorithms that empower them to make informed decisions [35]. These decisions, in turn, drive the actions taken by the system's actuators. Whether it's a self-driving car navigating through traffic or a robotic arm assembling intricate components, the decision-making and action-taking processes are intrinsic to the autonomy of these systems [30].

Sensing the environment: Autonomous systems are equipped with various sensors that act as their "sensory organs." These sensors continuously gather data from the system's environment. For example, in a self-driving car, sensors might include cameras, LIDAR, RADAR, and other technologies that provide information about the road, traffic, and obstacles. In a robotic arm, sensors may detect the position and orientation of objects [42,43].

Decision-making algorithms: The collected data is then processed by sophisticated algorithms, often powered by AI and machine learning (ML). These algorithms analyze the information, identify patterns, and make decisions based on predefined rules or learned behaviors [35].

In the context of a self-driving car, decision-making algorithms might determine actions such as slowing down, changing lanes, or stopping at a traffic signal based on the perceived environment [30,43]. In a robotic arm, the algorithms could decide the precise movements needed to assemble components [44] (Table 1).

TABLE 1 Algorithms at the helm: Navigating the autonomous frontier with intelligent decision-making [45–47].

Algorithm type	Description
Rule-based systems	Use predefined rules based on expert knowledge for decision-making
Finite state machines (FSM)	Model the system as a set of states, transitions, and actions, suitable for discrete decisions
Markov decision processes (MDP)	Model decision-making as a stochastic process, considering state transitions and expected rewards
Reinforcement learning	Train the system through trial and error, learning from rewards or penalties
Deep neural networks	Utilize deep learning techniques, such as CNNs and RNNs, for decision-making tasks
Genetic algorithms	Evolve a population of candidate solutions over generations to find optimal solutions
Fuzzy logic	Model decision criteria with linguistic variables and rules, dealing with uncertainty
Potential fields	Guide the system through attractive and repulsive forces in the environment
A* Algorithm	Path-finding algorithm for planning routes in environments with known terrain
Monte Carlo methods	Use statistical sampling for decision-making in complex and uncertain environments

TABLE 2 Empowering autonomy: Informed decision-making, where real-time data shapes precision, safety, and efficiency in every algorithmic move [48–52].

Aspect	Description
Process	Informed decision-making
Input	Real-time data from sensors
Objective	Optimize the system's performance, safety, and efficiency
Factors considered	- Real-time environmental conditions - System performance metrics - Safety considerations
Algorithmic operation	- Continuous analysis of incoming sensor data - Identification of patterns and relevant features - Decision-making based on predefined rules or learned behaviors
Outcome	Decisions that are responsive to the current state of the environment, tailored for optimal system behavior
Benefits	- Enhanced system performance - Improved safety measures - Increased operational efficiency
Role in autonomy	Central to the autonomy of the system, allowing it to adapt and respond to dynamic environmental changes

Informed decision-making: The decisions made by these algorithms are informed by the real-time data from the sensors. The goal is to make decisions that optimize the system's performance, safety, and efficiency in response to the current environmental conditions [33].

This tabular format provides a structured overview of the key aspects related to informed decision-making in autonomous systems as tabulated in Table 2.

Action-taking by actuators: Once a decision is reached, the system needs to act on it. This is where the actuators come into play. Actuators are the physical components that carry out the actions determined by the decision-making algorithms [53].

In the case of a self-driving car, actuators could include the motors that control acceleration, braking, and steering. For a robotic arm, the actuators would be the motors or mechanisms responsible for moving the arm's joints and grippers [5,8] (Table 3).

Continuous feedback loop: The concept of the continuous feedback loop in autonomous systems encapsulates the dynamic and adaptive nature of their operation. In this intricate dance between the system and its environment, the process unfolds as a dynamic sequence of actions and reactions [55]. As the autonomous system takes actions based on its decisions, it influences and, in turn, is influenced by the environment. This recognition of the dynamic nature of the surroundings is crucial, as factors such as moving objects, alterations in lighting, or the emergence of new obstacles constantly reshape the system's operational landscape [56]. To navigate this dynamic terrain effectively, the system relies on its sensors as vigilant sentinels, continuously gathering real-time data about the evolving state of the environment [57]. This updated sensory information then feeds into the continuous feedback loop, creating an iterative exchange where the system's decisions are informed by the most recent environmental data. This real-time adaptation is a hallmark of the continuous feedback loop, enabling the system to make agile and informed decisions that align with the current conditions. It's a symphony of responsiveness, ensuring that the autonomous system can fluidly adjust its behavior, navigate complex scenarios, and interact seamlessly with its surroundings [58].

The process is dynamic and continuous. As the system takes actions, the environment may change, requiring the sensors to provide updated data (Table 4). This continuous feedback loop allows the system to adapt its decisions and actions in real-time [57].

Autonomy and adaptability: The ability of autonomous systems to make informed decisions and take actions based on these decisions is what grants them autonomy. These systems can navigate complex environments, respond to dynamic situations, and perform tasks without direct human intervention.

In summary, the synergy between decision-making algorithms and actuators is the essence of autonomy in autonomous systems. It allows them to perceive their environment, interpret that information, make decisions, and execute actions accordingly. This dynamic and adaptive process enables autonomous systems to operate effectively in diverse and changing conditions.

TABLE 3 Harmony in autonomy: Navigating the dynamic landscape with informed decision-making algorithms and precise action-taking by actuators [33,47,52,54].

Aspect	Informed decision-making	Action-taking by actuators
Process	Informed decision-making	Action-taking by actuators
Input	Real-time data from sensors	Decisions made by algorithms
Objective	Optimize system's performance, safety, and efficiency	Execute actions based on decisions
Factors considered	Real-time environmental conditions	Direct link to decisions
	System performance metrics	Translates decisions into physical responses
	Safety considerations	
Algorithmic operation	Continuous analysis of incoming sensor data	Directly linked to decisions made by algorithms
	Identification of patterns and relevant features	Collaborates closely with sensors; acts based on real-time data received
	Decision-making based on predefined rules or learned behaviors	
Outcome	Decisions responsive to the current state of the environment	Critical for translating decisions into physical responses; enables the system to interact with the environment
Benefits	Enhanced system performance	Direct execution of chosen course of action
	Improved safety measures	Real-time responsiveness to changes
	Increased operational efficiency	
Role in autonomy	Central to the autonomy of the system; allows it to adapt and respond to dynamic environmental changes	Critical for enabling physical interaction and dynamic responses in autonomous systems
Recent literature	*Title*: "Empowering Autonomy: Advances in Informed Decision-Making Algorithms"	*Title*: "Advancements in Actuator Technologies for Autonomous Systems"
	Authors: Johnson, A., et al.	*Authors*: Smith, J., et al.
	Published: Journal of Autonomous Systems, 2023	*Published*: Journal of Robotics and Automation, 2022
	Summary: This paper explores recent developments in decision-making algorithms, emphasizing their role in enhancing autonomy	*Summary*: The paper discusses recent innovations in actuator technologies, highlighting their role in enhancing the responsiveness and efficiency of autonomous systems

TABLE 4 Dynamic autonomy unleashed: The symphony of continuous feedback loop, where sensors and actions harmonize, enabling real-time adaptation to the ever-changing world [33,35,43,44,58–61].

Aspect	Explanation
Process dynamics	The process within autonomous systems is dynamic, meaning it is ongoing and subject to constant change. As the system operates, it takes actions based on its decisions, influencing the environment and potentially altering the conditions
Environmental changes	The dynamic nature of the process acknowledges that the environment in which the autonomous system operates is not static. Factors such as the movement of objects, changes in lighting, or the presence of new obstacles can alter the surroundings
Sensor data update	In response to these changes, the system relies on its sensors to continuously provide updated data. Sensors act as the system's "eyes and ears," collecting real-time information about the evolving state of the environment
Continuous feedback loop	The continuous feedback loop is the iterative exchange of information between the system's actions and the updated sensory data. As the system takes actions and the environment changes, the sensors feed new information back into the decision-making process
Adaptive decision-making	This loop enables the system to adapt its decisions in real time. The updated data from sensors informs the decision-making algorithms, allowing the system to make more informed and relevant choices based on the current state of the environment
Real-time adaptation	Real-time adaptation is a key feature facilitated by the continuous feedback loop. It ensures that the autonomous system can dynamically adjust its behavior, enhancing its ability to navigate, interact, and respond effectively to the ever-changing surroundings

2.1.3 Sensor-actuator synergy

Central to the functionality of autonomous systems is their seamless integration of sensors and actuators. Sensors act as the "sensory organs," providing real-time data about the environment [35]. This data is then processed and interpreted by the system's algorithms to inform decision-making. Subsequently, the actuators, which can range from motors to robotic arms, execute the chosen actions based on these decisions. This synergy between sensors and actuators is fundamental to the system's ability to interact with and navigate its surroundings [32].

Sensors (sensory organs): Sensors are like the system's eyes, ears, and touch receptors. They gather information from the environment, such as detecting objects, measuring distances, or identifying temperatures.

These sensors continuously feed real-time data to the system, allowing it to build an understanding of its surroundings. It's like the system's way of "seeing" and "sensing" what's happening in the world [62].

Algorithms (brain): The system's algorithms can be thought of as its "brain." They process and analyze the data received from sensors. These algorithms make sense of the information, identify patterns, and make decisions based on predefined rules or learned behaviors [63].

Actuators (muscles): Actuators are the system's muscles or tools for interaction. Once the algorithms have made decisions, the actuators carry out the necessary actions. These actions could include moving, grabbing, or manipulating objects in the environment.

Just as our muscles follow the instructions from our brain to perform specific tasks, actuators execute the commands generated by the system's algorithms.

Synergy: The synergy between sensors and actuators is crucial. The information gathered by sensors guides the decision-making process of the system's algorithms. The decisions, in turn, trigger actions through the actuators.

For example, if a sensor detects an obstacle, the algorithm might decide to change the system's path. The actuators then receive this command and adjust the system's movement accordingly.

2.1.4 Harnessing artificial intelligence

Autonomous systems harness the power of AI to achieve their autonomy. Through ML algorithms, neural networks, and other AI techniques, these systems can learn from their experiences, adapt to new situations, and continuously improve their performance. This integration of AI enables them to handle complex tasks that were once the domain of human expertise.

2.1.5 Examples of autonomous systems

To better understand the concept of autonomous systems, consider some real-world examples [1]:

- *Self-driving cars:* These vehicles utilize an array of sensors and AI algorithms to navigate roads, make split-second decisions, and ensure passenger safety.
- *Industrial robots:* Autonomous robotic arms perform intricate tasks on assembly lines, enhancing production efficiency and precision.
- *Unmanned aerial vehicles (drones):* Drones equipped with autonomous capabilities can perform aerial surveillance, deliveries, and even search and rescue missions.
- *Autonomous vacuum cleaners:* These devices use sensors to navigate around obstacles and efficiently clean indoor spaces.

2.1.6 The purpose of autonomy

The ultimate purpose of developing autonomous systems is to augment human capabilities and enhance various aspects of our lives. By relieving humans from mundane or dangerous tasks, these systems contribute to increased efficiency, improved safety, and expanded possibilities across industries ranging from transportation and manufacturing to healthcare and space exploration.

Table 5 represents the key aspects of defining autonomous systems in a structured and concise manner, making it easier to understand the concept and its various elements.

TABLE 5 Key aspects of defining autonomous systems [4,30,43].

Aspect	Description
Definition	An autonomous system is a self-governing entity that performs tasks, solves problems, or interacts with its environment
Key characteristics	• Operates without direct human intervention • Makes decisions and takes actions based on its internal algorithms and sensors • Utilizes a combination of sensors and actuators • Employs AI to perform tasks
Core components	• Sensors: Provide information about the environment • Actuators: Execute actions based on decisions • Algorithms: Drive decision-making and problem-solving processes
Examples	Self-driving cars, industrial robots, unmanned aerial vehicles (drones), autonomous vacuum cleaners
Purpose	To enable machines to function independently, enhancing efficiency, safety, and convenience

2.2 Key components of autonomous systems

Autonomous systems comprise several essential components that work in harmony to achieve autonomy:

2.2.1 Sensors

Sensors act as the "eyes" and "ears" of autonomous systems, providing them with information about their surroundings. These can include cameras, LIDAR, RADAR, ultrasonic sensors, and more. By processing data from these sensors, autonomous systems can perceive and understand their environment [62].

2.2.2 Perception algorithms

Perception algorithms analyze data from sensors to extract meaningful information. Computer vision algorithms, for example, can identify objects, pedestrians, traffic signs, and lane markings from camera images. These algorithms play a crucial role in enabling autonomous systems to make sense of the world around them [63].

2.2.3 Decision-making algorithms

Decision-making algorithms take sensor data and generate appropriate actions. These algorithms often involve complex computations, considering various factors such as safety, efficiency, and mission objectives. Reinforcement learning, deep learning, and rule-based systems are common approaches used in decision-making.

2.2.4 Actuators

Actuators are the "muscles" of autonomous systems, responsible for carrying out physical actions based on decisions made by the system. These can include motors, servos, pneumatic systems, and more. Actuators execute tasks such as steering a car, adjusting the throttle, or manipulating robotic arms.

2.3 Levels of autonomy

Autonomous systems are often categorized into levels of autonomy, as defined by various industry standards. The levels range from no automation (Level 0) to full autonomy (Level 5) [39]:

- *Level 0 (no automation):* The human operator performs all tasks with no assistance from the system.
- *Level 1 (driver assistance):* The system can assist the human operator with specific tasks, such as steering or braking.
- *Level 2 (partial automation):* The system can control both steering and acceleration/deceleration simultaneously under certain conditions. The human must remain engaged and monitor the environment.

- *Level 3 (conditional automation):* The system can handle certain tasks and monitor the environment in specific conditions. The human can disengage but must be ready to take over when requested by the system.
- *Level 4 (high automation):* The system can handle all tasks within defined operational domains or conditions. Human intervention might be required outside these conditions.
- *Level 5 (full automation):* The system can perform all tasks and operate in all conditions that a human driver could handle.

2.4 Applications of autonomous systems

The applications of autonomous systems are vast and continually expanding. Autonomous systems have a wide range of applications across various industries, bringing about transformative changes in how tasks are performed. Here's an explanation of some key applications of autonomous systems:

1. *Autonomous vehicles:*
   - *Description:* Self-driving cars, trucks, and drones are designed to navigate and operate without human intervention. They use sensors, cameras, and advanced algorithms to interpret the environment, making transportation safer and more efficient.
2. *Agriculture and farming:*
   - *Description:* Autonomous tractors, drones, and robotic systems are employed for precision farming. They can optimize planting, watering, and harvesting processes, leading to increased efficiency and reduced resource usage.
3. *Healthcare:*
   - *Description:* Surgical robots assist surgeons in performing minimally invasive procedures with precision. Autonomous medical devices can monitor patient conditions and administer treatments, enhancing the overall quality of healthcare.
4. *Manufacturing and warehousing:*
   - *Description:* Autonomous robots are used in manufacturing processes, assembly lines, and warehouses for tasks like material handling, inventory management, and quality control. These systems contribute to increased productivity and reduced operational costs.
5. *Surveillance and security:*
   - *Description:* Autonomous drones and ground-based robots are deployed for surveillance and security purposes. They can monitor large areas, identify potential threats, and enhance the effectiveness of security operations.
6. *Space exploration:*
   - *Description:* Autonomous rovers and spacecraft play a crucial role in space exploration. These systems can navigate extra-terrestrial environments, collect data, and conduct experiments without direct human control.
7. *Environmental monitoring:*
   - *Description:* Autonomous sensors and drones are used for tracking environmental changes, monitoring pollution levels, and studying ecosystems. These systems provide valuable data for environmental research and conservation efforts.
8. *Search and rescue:*
   - *Description:* Autonomous drones and robots are employed in search and rescue missions to locate and assist individuals in emergency situations. Their agility and ability to cover challenging terrains make them valuable in disaster response efforts.
9. *Retail and customer service:*
   - *Description:* Autonomous robots are used in retail settings for tasks like inventory management, restocking shelves, and assisting customers. In the realm of e-commerce, autonomous delivery vehicles contribute to efficient and timely order fulfillment.
10. *Infrastructure inspection:*
    - *Description:* Drones and ground-based robots are utilized for inspecting critical infrastructure such as bridges, pipelines, and power lines. Autonomous systems can perform inspections more safely and cost effectively than traditional methods.

11. *Education and research:*
    - *Description:* Autonomous systems are employed as educational tools for teaching programming, robotics, and AI concepts. In research, these systems can be used to conduct experiments, gather data, and explore new scientific frontiers.

These applications showcase the versatility of autonomous systems, providing solutions to diverse challenges and revolutionizing industries through increased efficiency, safety, and innovation.

2.5 Challenges and future directions

Ensuring safety, addressing ethical concerns, achieving interoperability, securing data, and adapting to complex environments are key challenges for autonomous systems. The future involves advancing artificial general intelligence (AGI), emphasizing human-centric design, embracing explainable AI (XAI), exploring swarm robotics, and continuous learning. Energy efficiency and global collaboration are also crucial focus areas. These advancements aim to shape a future where autonomous systems are more versatile, efficient, and integrated into various aspects of daily life.

2.5.1 Challenges of autonomous systems

1. *Safety and reliability:* Ensuring the safety and reliability of autonomous systems, especially in critical applications like autonomous vehicles, remains a significant challenge. Issues such as system failures, sensor inaccuracies, and unpredictable environmental factors must be addressed.
2. *Ethical and legal concerns:* The deployment of autonomous systems raises ethical questions, particularly in situations where decisions involve human safety. Establishing legal frameworks and ethical guidelines for autonomous technologies, including liability and accountability, is an ongoing challenge.
3. *Interoperability:* Many autonomous systems are developed by different manufacturers and may not seamlessly integrate or communicate with one another. Achieving interoperability is crucial for creating comprehensive and efficient autonomous ecosystems, especially in smart cities and interconnected industries.
4. *Data security and privacy:* Autonomous systems rely heavily on data, often collected in real-time. Ensuring the security and privacy of this data is a challenge, as unauthorized access or data breaches can have serious consequences, both in terms of safety and individual privacy.
5. *Adaptability to complex environments:* Autonomous systems face challenges when operating in dynamic and complex environments. Adapting to unforeseen scenarios, diverse weather conditions, and handling novel situations remains a key challenge, especially for autonomous vehicles and drones.
6. *Human-autonomy collaboration:* Integrating autonomous systems into environments where humans are also present requires effective collaboration. Ensuring a smooth interaction between autonomous systems and humans is essential for user acceptance and safety.
7. *Regulatory hurdles:* Developing comprehensive and standardized regulations for autonomous technologies is a complex task. Regulatory bodies worldwide are working to define frameworks that ensure safety without stifling innovation.

2.5.2 Future directions of autonomous systems

1. *Artificial general intelligence (AGI):* Research is ongoing to develop autonomous systems with more generalized intelligence, capable of understanding and adapting to a broader range of tasks and environments.
2. *Edge computing and AI:* Leveraging edge computing to process data closer to the source and incorporating advanced AI algorithms allows for faster decision-making, reducing latency and enhancing the efficiency of autonomous systems.
3. *Human-centric design:* Future autonomous systems will focus on human- centric design, emphasizing user experience, understanding human behavior, and creating interfaces that facilitate seamless collaboration between humans and machines.
4. *Explainable AI (XAI):* Ensuring that autonomous systems are transparent and can provide explanations for their decisions is gaining importance. Explainable AI helps build trust and enables users to understand the reasoning behind autonomous actions.

5. *Swarm robotics:* Research in swarm robotics explores the coordination of large numbers of autonomous robots to work collaboratively. This approach draws inspiration from natural swarm behavior, such as seen in ants or bees, to achieve complex tasks collectively.
6. *Continuous learning:* Implementing systems that can continuously learn and adapt over time will be a key focus. This involves the development of algorithms that can improve performance through experience and exposure to new data.
7. *Energy efficiency:* Enhancing the energy efficiency of autonomous systems is crucial, especially for applications like drones and electric vehicles. Research is ongoing to develop more energy-efficient hardware and algorithms to extend the operational capabilities of autonomous systems.
8. *Global collaboration:* Encouraging collaboration among researchers, industries, and governments globally is essential for establishing standards, sharing knowledge, and addressing challenges that are not confined by geographical boundaries.

The future of autonomous systems holds great promise, with ongoing advancements aimed at addressing current challenges and pushing the boundaries of what autonomous technologies can achieve. The evolution of these systems will likely play a pivotal role in shaping the future of various industries and everyday life.

2.6 Conclusion

In conclusion, this book chapter provides a thorough and insightful exploration of the multifaceted realm of autonomous systems, spanning the historical evolution, fundamental principles, challenges, and real-world applications. By tracing the trajectory of autonomous systems development, the chapter elucidates key milestones and technological advancements that have propelled the field into its current state. The integration of AI, robotics, and control theory is examined in-depth, shedding light on the core principles that underpin the creation of intelligent entities capable of independent decision-making and action execution.

A notable strength of the chapter lies in its comprehensive consideration of the challenges and opportunities inherent in the development and deployment of autonomous systems. The ethical, legal, and societal implications are thoughtfully addressed, providing readers with a holistic understanding of the broader context in which these systems operate. The critical examination of state-of-the-art technologies and methodologies adds a practical dimension, offering insights into cutting-edge research shaping the field.

The inclusion of real-world applications across diverse domains, such as transportation, healthcare, and manufacturing, further underscores the transformative impact of autonomous systems on society. By emphasizing a balanced blend of historical context, theoretical foundations, and practical insights, the chapter caters to a wide audience, including researchers, practitioners, and students entering the dynamic domain of autonomous systems.

The highlighted keywords—Communication system control, Space vehicles, AI, Automatic control, Decision making, Hardware, Remotely operated vehicles, Unmanned Aerial Vehicles—capture the breadth of topics covered in the chapter, providing a roadmap for readers to delve deeper into specific areas of interest. Overall, this chapter serves as an essential primer, laying the groundwork for readers to navigate and contribute to the ongoing evolution of autonomous systems in our increasingly autonomous world.

References

[1] D.P. Watson, D.H. Scheidt, Autonomous systems, Johns Hopkins APL Tech. Dig. 26 (4) (2005) 368–376.
[2] F. Jahan, W. Sun, Q. Niyaz, M. Alam, Security modeling of autonomous systems: a survey, ACM Comput. Surv. 52 (5) (2019) 1–34.
[3] M. Fisher, V. Mascardi, K.Y. Rozier, B.H. Schlingloff, M. Winikoff, N. Yorke-Smith, Towards a framework for certification of reliable autonomous systems, Auton. Agent. Multi-Agent Syst. 35 (2021) 1–65.
[4] J. Chen, J. Sun, G. Wang, From unmanned systems to autonomous intelligent systems, Engineering 12 (2022) 16–19.
[5] R. Goel, P. Gupta, Robotics and industry 4.0, in: A Roadmap to Industry 4.0: Smart Production, Sharp Business and Sustainable Development, Springer, 2020, pp. 157–169.
[6] M. Hägele, K. Nilsson, J.N. Pires, R. Bischoff, Industrial robotics, in: Springer Handbook of Robotics, Springer, 2016, pp. 1385–1422.
[7] T. Bock, T. Linner, Robotic Industrialization, Cambridge University Press, 2015.
[8] D.M. Marshall, Dull, dirty, and dangerous: the FAA's regulatory authority over unmanned aircraft operations, Issues Aviat. Law Pol. (2004) 10085.
[9] R.R. Murphy, Introduction to AI Robotics, MIT Press, 2019.
[10] C. Frost, A. Butt, D. Silva, Challenges and opportunities for autonomous systems in space, in: Frontiers of Engineering: Reports on Leading-Edge Engineering from the 2010 Symposium, 2010.

[11] P.J. Antsaklis, K.M. Passino, S.J. Wang, An introduction to autonomous control systems, IEEE Control. Syst. Mag. 11 (4) (1991) 5–13.
[12] H. He, J. Gray, A. Cangelosi, Q. Meng, T.M. McGinnity, J. Mehnen, The challenges and opportunities of human-centered AI for trustworthy robots and autonomous systems, IEEE Trans. Cognit. Dev. Syst. 14 (4) (2021) 1398–1412.
[13] A. Broeckmann, Machine Art in the Twentieth Century, MIT Press, 2016.
[14] J.P. Ivens, Making the Modern Person: The Tyler Rationale, Curriculum Studies, and Cybernetic Systems, The University of Wisconsin-Madison, 2018.
[15] K. Miliutina, Analysis of Russian Policy on the Development of AI in the Military, Center for a New American Security, 2022.
[16] S. Bennett, Modern Records, Maverick Methods: Technology and Process in Popular Music Record Production 1978–2000, Bloomsbury Publishing USA, 2018.
[17] W. Truszkowski, H. Hallock, C. Rouff, J. Karlin, J. Rash, M. Hinchey, R. Sterritt, Autonomous and Autonomic Systems: With Applications to NASA Intelligent Spacecraft Operations and Exploration Systems, Springer Science & Business Media, 2009.
[18] M.A. Peters, Towards a theory of knowledge socialism: cognitive capitalism and the fourth knowledge revolution, in: Knowledge Socialism: The Rise of Peer Production: Collegiality, Collaboration, and Collective Intelligence, Springer, 2020, pp. 15–31.
[19] M.M. Rohde, S.M. Rohde, Applying a virtual product development process to mobile robotics including real-time, interactive simulation, in: 2015 NDIA Ground Vehicle Systems Engineering and Technology Symposium Modeling & Simulation, Testing and Validation (MSTV) Technical Session, Quantam Signal, August 4–6, 2015, Novi, Michigan, 2015.
[20] A.G. Farizawani, M. Puteh, Y. Marina, A. Rivaie, A review of artificial neural network learning rule based on multiple variant of conjugate gradient approaches, J. Phys. Conf. Ser. 1529 (2) (2020) 022040. IOP Publishing.
[21] F. Miatton, Blockchain at the edge: The nexus of capturing new value in 5G, in: 2020 International Conference on Technology and Entrepreneurship-Virtual (ICTE-V), IEEE, 2020, pp. 1–6.
[22] J. Fadaie, The state of modeling, simulation, and data utilization within industry: an autonomous vehicles perspective, arXiv preprint arXiv:1910.06075 (2019).
[23] T. Dhanabalan, A. Sathish, Transforming Indian industries through artificial intelligence and robotics in industry 4.0, Int. J. Mech. Eng. Technol. 9 (10) (2018) 835–845.
[24] R. Leenes, E. Palmerini, B.J. Koops, A. Bertolini, P. Salvini, F. Lucivero, Regulatory challenges of robotics: some guidelines for addressing legal and ethical issues, Law Innov. Technol. 9 (1) (2017) 1–44.
[25] E. Brynjolfsson, A. McAfee, Race Against the Machine: How the Digital Revolution Is Accelerating Innovation, Driving Productivity, and Irreversibly Transforming Employment and the Economy, Brynjolfsson and McAfee, 2011.
[26] A.L. Schuller, At the crossroads of control: the intersection of artificial intelligence in autonomous weapon systems with international humanitarian law, Harv. Natl. Sec. J. 8 (2017) 379.
[27] B. Martin, D.C. Tarraf, T.C. Whitmore, J. DeWeese, C. Kenney, J. Schmid, P. DeLuca, Advancing Autonomous Systems, Rand Corporation, 2019, pp. 9–11.
[28] R.D. Stacey, D. Griffin, P. Shaw, Complexity and Management: Fad or Radical Challenge to Systems Thinking?, Psychology Press, 2000.
[29] P.D. Scharre, The opportunity and challenge of autonomous systems, in: Autonomous Systems: Issues for Defence Policymakers, NATO Communications and Information Agency, 2015, pp. 3–26.
[30] C. Badue, R. Guidolini, R.V. Carneiro, P. Azevedo, V.B. Cardoso, A. Forechi, A.F. De Souza, Self-driving cars: a survey, Expert Syst. Appl. 165 (2021) 113816.
[31] L. Liu, S. Lu, R. Zhong, B. Wu, Y. Yao, Q. Zhang, W. Shi, Computing systems for autonomous driving: state of the art and challenges, IEEE Internet Things J. 8 (8) (2020) 6469–6486.
[32] E. Balestrieri, P. Daponte, L. De Vito, F. Lamonaca, Sensors and measurements for unmanned systems: an overview, Sensors 21 (4) (2021) 1518.
[33] J.R. Busby, J.V. Ernst, T.E. Varnado, Data, data everywhere! Data collection, synthesis, and analysis are important for drawing conclusions and making informed decisions in the realms of society, economics, technology, and politics, Technol. Teach. 69 (3) (2009) 20–25.
[34] T. Smithers, Autonomy in robots and other agents, Brain Cogn. 34 (1) (1997) 88–106.
[35] A.A. Phatak, F.G. Wieland, K. Vempala, F. Volkmar, D. Memmert, Artificial intelligence based body sensor network framework—narrative review: proposing an end-to-end framework using wearable sensors, real-time location systems and artificial intelligence/machine learning algorithms for data collection, data mining and knowledge discovery in sports and healthcare, Sports Med. - Open 7 (2021) 1–15.
[36] A. Hemmati, A.M. Rahmani, The internet of autonomous things applications: a taxonomy, technologies, and future directions, Internet Things 20 (2022) 100635.
[37] J. Sifakis, Autonomous systems—an architectural characterization, arXiv preprint arXiv:1811.10277 (2018).
[38] J. Borenstein, Y. Koren, Real-time obstacle avoidance for fast mobile robots, IEEE Trans. Syst. Man Cybern. 19 (5) (1989) 1179–1187.
[39] P. Antsaklis, Autonomy and metrics of autonomy, Annu. Rev. Control. 49 (2020) 15–26.
[40] D.A. Wiegmann, S.A. Shappell, A Human Error Analysis of Commercial Aviation Accidents Using the Human Factors Analysis and Classification System (HFACS), Defence Technical Information Center, 2001.
[41] M.L. Cummings, D. Britton, Regulating safety-critical autonomous systems: Past, present, and future perspectives, in: Living with Robots, Academic Press, 2020, pp. 119–140.
[42] Y. Li, Ros-Based Sensor Fusion and Motion Planning for Autonomous Vehicles: Application to Automated Parking System, Doctoral dissertation,, Wayne State University, 2019.
[43] H. Tunga, R. Saha, S. Kar, A method of fully autonomous driving in self-driving cars based on machine learning and deep learning, in: Intelligent Multi-modal Data Processing, Wiley, 2021, pp. 131–156.
[44] C.G. Lee, Robot arm kinematics, dynamics, and control, Computer 15 (12) (1982) 62–80.
[45] H.C. Burmeister, M. Constapel, Autonomous collision avoidance at sea: a survey, Front. Rob. AI 8 (2021) 739013.
[46] Y. Wang, Artificial intelligence in educational leadership: a symbiotic role of human-artificial intelligence decision-making, J. Educ. Adm. 59 (3) (2021) 256–270.
[47] Y. Wang, When artificial intelligence meets educational leaders' data- informed decision-making: a cautionary tale, Stud. Educ. Eval. 69 (2021) 100872.

[48] S. Grafanaki, Autonomy challenges in the age of big data, Fordham Intell. Prop. Media Ent. LJ 27 (2016) 803.
[49] K. Yeung, Algorithmic regulation: a critical interrogation, Regul. Gov. 12 (4) (2018) 505–523.
[50] G. Singh, Leveraging ChatGPT for real-time decision-making in autonomous systems, Eduzone 12 (2) (2023) 101–106.
[51] H.S.M. Lim, A. Taeihagh, Algorithmic decision-making in AVs: understanding ethical and technical concerns for smart cities, Sustainability 11 (20) (2019) 5791.
[52] R. Wang, R. Bush-Evans, E. Arden-Close, E. Bolat, J. McAlaney, S. Hodge, K. Phalp, Transparency in persuasive technology, immersive technology, and online marketing: facilitating users' informed decision making and practical implications, Comput. Hum. Behav. 139 (2023) 107545.
[53] A. Haldorai, A review on artificial intelligence in internet of things and cyber physical systems, J. Comput. Nat. Sci. 3 (1) (2023) 012–023.
[54] R.P. França, A.C.B. Monteiro, R. Arthu, Y. Iano, The evolution of robotic systems: overview and its application in modern times, in: Safety, Security, and Reliability of Robotic Systems, Taylor & Francis, CRC Press, 2020, pp. 1–20.
[55] D. Vernon, G. Metta, G. Sandini, A survey of artificial cognitive systems: implications for the autonomous development of mental capabilities in computational agents, IEEE Trans. Evol. Comput. 11 (2) (2007) 151–180.
[56] A. Farina, Ecology, Cognition and Landscape: Linking Natural and Social Systems, Vol. 11, Springer Science & Business Media, 2009.
[57] H.J. de Knegt, J.A. Eikelboom, F. van Langevelde, W.F. Spruyt, H.H. Prins, Timely poacher detection and localization using sentinel animal movement, Sci. Rep. 11 (1) (2021) 4596.
[58] J. Xie, S. Liu, X. Wang, Framework for a closed-loop cooperative human cyber-physical system for the mining industry driven by VR and AR: MHCPS, Comput. Ind. Eng. 168 (2022) 108050.
[59] H. Allioui, Y. Mourdi, Exploring the full potentials of IoT for better financial growth and stability: a comprehensive survey, Sensors 23 (19) (2023) 8015.
[60] H.C. Yen, The Practice of Cinematic Experience in Everyday Life: Moving-Image, Time and the Journey to the Unknown, Doctoral dissertation, Teachers College, Columbia University, 2009.
[61] P. Moisala, Kaija Saariaho, University of Illinois Press, 2010.
[62] A. Verma, S. Prakash, V. Srivastava, A. Kumar, S.C. Mukhopadhyay, Sensing, controlling, and IoT infrastructure in smart building: a review, IEEE Sensors J. 19 (20) (2019) 9036–9046.
[63] F. Lotte, L. Bougrain, A. Cichocki, M. Clerc, M. Congedo, A. Rakotomamonjy, F. Yger, A review of classification algorithms for EEG-based brain–computer interfaces: a 10 year update, J. Neural Eng. 15 (3) (2018) 031005.

CHAPTER

3

Sensors for autonomous vehicles

Weiqiang Dong

Southern Arkansas University, Magnolia, AR, United States

3.1 Introduction

Sensory systems enable organisms to understand the physical structure and temporal dynamics of their environment. A sensory receptor cell is capable of responding to a certain kind of stimulus energy and converting it to electrical brain impulses through a sequence of physiological reactions. These brain impulses permit an internal representation of the surrounding world after sensory processing and integration [1]. Humans have created machines and theories to do similar tasks for the service of mankind. Sensors monitor signals or changes in the environment. The decisions made by the sensors are combined using the detection theory to provide more consistent, accurate, and useful data.

In 2019, the worldwide market for smart sensors was valued at $37.12 billion and was predicted to reach $91.37 billion by 2027, growing at a CAGR of 14.30% from 2020 to 2027 [2]. Sensors are important in today's process technologies [3]. They are used in a variety of applications and cover a wide range of research topics, including medical diagnostics, industrial production, energy systems, and national security. Pressure, temperature, flow, motion, displacement and dimension, force and torque, quantitative analysis, and synthetic sensations are all continually monitored by sensors. The data collected by sensors is supplied into detection algorithms, which analyze the surroundings, form insights, and make decisions under different criteria. Scientists and engineers have worked to give sensors the ability to learn autonomously and adapt to their surroundings so that they can perform complicated jobs. Today, artificial intelligence has become a kind of standard in many areas. Numerous machine learning techniques are proposed and used in sensing. Many fields benefit from the advancement of sensors.

Sensors can be classified into active sensors and passive sensors depending on whether they require external sources of power to operate. Sensors that detect the incoming ambient environmental energy are called passive sensors [4]. Passive sensors, such as temperature detectors and microphones can only detect energy when there is sufficient naturally existing energy. On the contrary, the active sensors release energy into the environment, then monitor the response of the environment. Active sensors include ultrasonic sensors, global positioning systems (GPS), and light detection and ranging (lidar). An advantage of active sensors is the ability to obtain measurements anytime. Camera sensors fall into both the passive and active sensor categories. The camera may be considered a passive sensor if it just receives energy to produce observations. The camera becomes an active sensor when there is insufficient light for the camera to capture the targets, and the camera utilizes a flash to illuminate the objects and record the radiation reflected from them. Sensors can also be classified as proprioceptive sensors and exteroceptive sensors [5]. Proprioceptive sensors measure the internal values of the system. Examples of proprioceptive sensors include a compass and GPS. Exteroceptive sensors, such as radar, thermal camera sensors, and lidar, collect data from the system's surroundings. Additionally, sensors may be divided into contactless sensors and contact sensors. As their names suggest, contact sensors make measurements by making physical contact with the target, while contactless sensors do not [6]. Contact sensors usually are sensitive and affordable. However, the installation and detection of contact sensors can be invasive. One application of contactless sensors is to provide noncontact temperature measurements to maintain social distance during a pandemic [7].

Sensors have been implemented in automotive vehicles for a long time [8]. Traditional sensors, such as the fuel level sensor and engine temperature sensor, have been applied to monitor and control the vehicle's conditions. Piezoresistive pressure sensors have been employed to monitor the tire pressure [9]; Camera sensors can work as reverse parking

Handbook of Power Electronics in Autonomous and Electric Vehicles
https://doi.org/10.1016/B978-0-323-99545-0.00021-X

assistants. The advanced driver assistance systems (ADAS) do not operate autonomously but rather utilize various sensors to provide extra information about the traffic condition to support and aid a driver in making essential decisions [10], e.g., Blind spot detection sends an auditory signal to alert the driver when objects are detected in the blind areas. In autonomous vehicles, sensors such as Lidar sensors, camera sensors, radar-sensors, and ultrasound sensors can be employed for exteroceptive sensing. GPS, inertial measurement units (IMU), and encoders work as proprioceptive sensors [11]. This chapter will mainly discuss the sensors for environment perception for autonomous vehicles.

3.2 Autonomous vehicles sensors and technologies

The properties of a sensor determine its appropriateness for capturing particular sorts of data and limit its range of operational settings [12]. The perception competences can be solved using various types of sensors and a wide range of algorithms. In [13], the authors discuss sensors in current research and commercial autonomous vehicles, autonomous features now available on the market, and localization and mapping techniques utilized in current autonomous vehicles research.

3.2.1 Camera

Camera sensors have been utilized for a long time in various fields such as computer vision and object detection. They have been employed for reversing aid and parking aid on nonautonomous vehicles. Since, the camera sensors are able to identify texture and have relatively high resolution, they are effective in classifying, recognizing, and detecting objects if they are visible. Autonomous vehicles also employ camera sensors to accomplish activities that require color perception, such as the identification of road signs and traffic signals. In addition, the artificial intelligence research using data collected by camera sensors is quite advanced [14,15]. Nonetheless, the performance of the camera sensors is contingent on visibility and illumination conditions. The primary drawbacks of cameras are their sensitivity to low light levels and susceptibility to bad weather conditions. They are significantly affected by shadows, reflections, and intense sunlight and, notably, become filthy quickly. Moreover, the computational requirement is heavy in order to process the images collected by the camera sensor.

3.2.2 The pinhole model

Camera sensors map the 3D points of the real environment into a 2D image space. Note that in this Chapter, we assume the 2D images are digital. A digital image is a representation of a 2D image as a discrete collection of digital values known as picture components or pixels [16]. It is an approximation of a genuine scene.

The pinhole model describes the mathematical relationship between points in 3D space and their projection onto the image plane [17]. A graphical illustration is shown in Fig. 1. In Fig. 2, point C is called *the camera center*, which is the

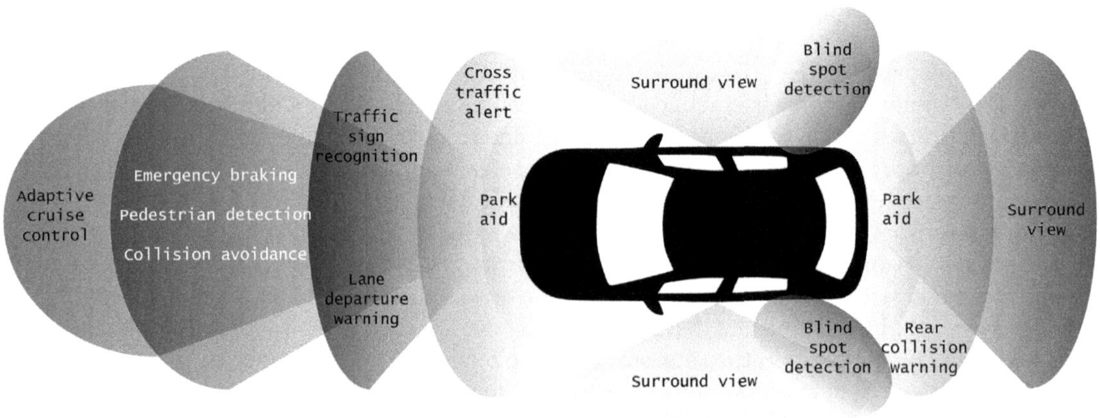

FIG. 1 Camera sensors in a modern vehicle. *Adapted from A. Dubey, Stereo vision-Facing the challenges and seeing the opportunities for ADAS applications,* Texas Instruments Technical Note, *2016.*

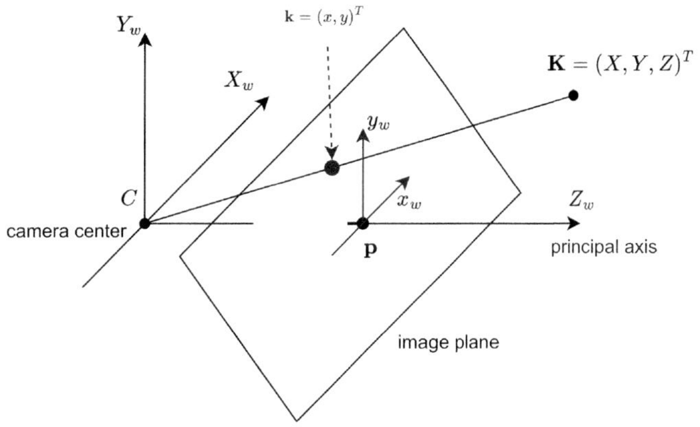

FIG. 2 Pinhole camera geometry. *Adapted from R. Hartley and A. Zisserman,* Multiple View Geometry in Computer Vision. *Cambridge University Press, 2003.*

location of the aperture of the camera. C is used as the origin of a 3D orthogonal coordinate system (X_w, Y_w, Z_w). The *principal axis* is the line from the camera center perpendicular to the *image plane*. Note that the actual image plane should appear behind the camera center. We put it in front of the camera to keep the projected image unflipped [18]. The intersection of the image plane and *principal axis* is called *principal point* $P = (x_0, y_0)$.

A point in the 3D space is $\boldsymbol{K} = (x, y, z)$. Eq. (1) is used to map X to a point $k = (x, y)$ on the 2D image plane. $\boldsymbol{C}$ is the intrinsic matrix, which contains parameters that are internal and fixed to a particular camera, such as the skew coefficient between the x_w axis and the y_w axis, s; the focal length of the x_w axis and the y_w axis, f_x and f_y; and the coordinate of the principal point (x_0, y_0). $[\boldsymbol{R}|\boldsymbol{t}]$ is the extrinsic matrix, which contains the camera parameters that are external to the camera and are subject to change relative to the world frame.

$$\begin{bmatrix} x \\ y \\ 1 \end{bmatrix} = \overbrace{C}^{\text{intrinsic matrix}} \times \overbrace{[\boldsymbol{R}|\boldsymbol{t}]}^{\text{extrinsic matrix}} \times \begin{bmatrix} X \\ Y \\ Z \\ 1 \end{bmatrix}, \boldsymbol{C} = \underbrace{\begin{bmatrix} 1 & 0 & x_0 \\ 0 & 1 & y_0 \\ 0 & 0 & 1 \end{bmatrix}}_{\text{2D translation}} \times \underbrace{\begin{bmatrix} f_x & 0 & 0 \\ 0 & f_y & 0 \\ 0 & 0 & 1 \end{bmatrix}}_{\text{2D scaling}} \times \underbrace{\begin{bmatrix} 1 & s/f_x & 0 \\ 0 & 1 & 0 \\ 0 & 0 & 1 \end{bmatrix}}_{\text{2D sheer}},$$
$$[\boldsymbol{R}|\boldsymbol{t}] = \underbrace{[I|t]}_{\text{3D translation}} \times \underbrace{\begin{bmatrix} R & 0 \\ 0 & 1 \end{bmatrix}}_{\text{3D rotation}}. \tag{1}$$

3.2.3 2D cameras and 3D cameras

The monocular (single-eyed) system captures images that must be processed and evaluated using a single camera sensor. The images recorded by a traditional monocular camera system are 2D width and height. Monocular cameras can estimate the distance based on the basic idea that items closer to the camera seem larger and consequently occupy a larger pixel area inside the image. The monocular camera system can perform a variety of tasks adequately. The technology and underlying algorithms are capable of accomplishing some object detection tasks, such as recognizing lanes, pedestrians, traffic signs, and other cars in lanes [19].

Some 3D camera sensors are capable of perceiving depth information [12,20].

3.2.3.1 *Stereo camera*

A system having two (or more) cameras that are separated is known as a stereo vision system. Stereo vision and human eyes operate according to the same principles [20]. Two cameras are placed next to each other at a certain distance (known as the baseline) and activated simultaneously. If the two cameras are calibrated, the question of determining the depth of an object can be restated to determine the disparity between the images captured by the two cameras for that point. The disparity is defined as the number of pixels a certain point has shifted in the image captured by one camera against the other camera [21,22]. It is possible to calculate and triangulate point-to-point correspondences between the two pictures to determine their respective locations in 3D space.

Stereo vision systems have relatively low costs. However, calculating quick and reliable point-to-point correspondences is difficult, and depth estimate error rises quadratically with camera distance. In addition, the object must be textured in order to recover point-to-point correspondences, which is not necessarily a realistic assumption.

3.2.3.2 *Time-of-flight*

The time-of-flight technique uses the same round-trip time idea as LIDAR sensors. The cameras employ an artificial light signal (supplied by a laser or LED) that passes between the camera and the item. In order to determine distance, this technology monitors the time it takes for light to travel to and bounce off of objects [23]. However, the time-of-flight technology requires the target must possess a reflecting surface. In addition, the time-of-flight technology is difficult to track multiple targets at the same time, and it is sensitive to signal interference [24,25].

3.2.3.3 *Structured light*

Structured light cameras project a known pattern onto the object inside their range of vision, and 3D data is derived by evaluating the projection deformation. The structured light system is comprised of a projector and a camera. The target is projected with a coded stripe pattern, and the camera takes an image to record the distortion caused by the depth variation of the surface. For each visible point in the world, a stripe number (stripe value) and image location correspond (pixel coordinates). Given the structured light system settings, each stripe value creates a plane in the world, while each pixel defines a ray. Therefore, the 3D surface point may be rebuilt using triangulation if the system parameters have been received through system calibration [26,27]. Structured light cameras have a low computational cost and do not rely on textured surfaces. However, the structured light camera requires precise calibration, and its performance deteriorates for far distances.

3.2.3.4 *Laser triangulation*

A laser triangulation device can estimate absolute and relative distances using geometric relations [28]. The laser source project light onto the measured surface of the item through the focusing lens. When the measured object surface moves, the object points also move, causing the object point to project via the receiving lens onto an image sensor and the image point to likewise move. Consequently, the calibration continues to generate calibration curves. Therefore, we can establish the relationship between the surface's displacement and the image point [29]. Due to the need to scan the item, the detection procedure could be slow.

3.2.4 Visible camera sensors and infrared camera sensors

In certain circumstances, the concept of the camera can be expanded to include infrared (IR) cameras. Every object with a temperature greater than absolute zero emits electromagnetic radiation [30,31]. Fig. 3 shows the electromagnetic spectrum. The visible spectrum ranges from 0.38 μm (blue, nearly ultraviolet to the human eye) to 0.75 μm (deep red). Wavelengths longer than the visible spectrum can only be detected by specialized sensors. The IR wavelength ranges from 0.75 to 1000. IR is not visible to the human eye, yet it interacts with things similar to visible light. Thus, IR light is

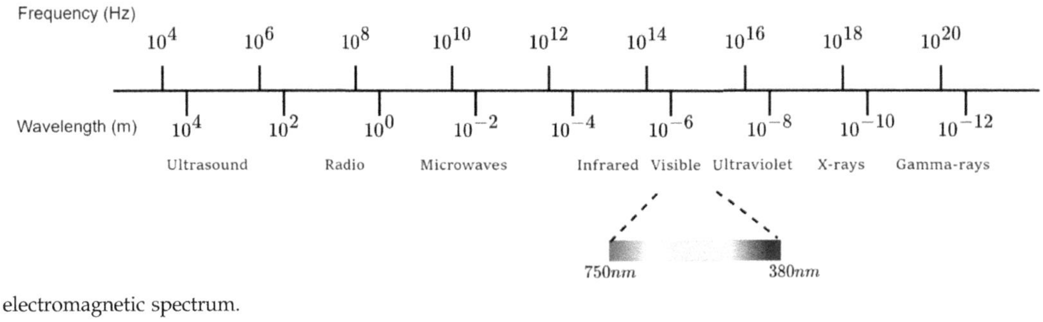

FIG. 3 The electromagnetic spectrum.

reflecting light; it is similar to visible light in that it reflects off of things. As a result of its reflecting nature, images from an IR camera contain shadows and contrast and are comparable to visible images in resolution and detail [32,33]. Hence, IR cameras usually work as night light cameras and thermal cameras in autonomous vehicles [34–38].

This IR spectrum range can be further divided into three smaller regions: The near-infrared (NIR) wavelength ranges from 0.75 to 3 μm; The mid-infrared (MIR) wavelength ranges from 3 to 6 μm; The far-infrared (FIR) wavelength ranges from 6 to 1000 μm. In the night vision system on a vehicle, NIR cameras are usually active sensors. They utilize a NIR light source to illuminate an object and then detect the reflected light. In the meantime, objects release more radiation in the FIR spectrum and appear more prominent in the FIR image than they do in the NIR image. Hence, FIR cameras are usually passive sensors [37,39,40].

3.2.5 CCD camera sensors and CMOS camera sensors

Charged coupled device (CCD) and Complementary Metal-Oxide-Semiconductor (CMOS) are the two most used types of image sensors in digital cameras (complementary metal-oxide-semiconductor). Both image sensors turn light signals and electrons into electronic data that is then transformed into a picture, but their methods for translating light signals to readable data are distinct [41,42]. Upon completion of the exposure, a CCD progressively passes each pixel's charge packet to a common output structure, which transforms the charge to a voltage, buffers it, and sends it off-chip. In a CMOS imager, each pixel performs the charge-to-voltage conversion [43,44].

After the exposure is complete, a CCD progressively passes each pixel's charge packet to a common output structure, which transforms it to a voltage, buffers it, and sends it off-chip. CCDs provide higher picture quality and versatility at the cost of increased system size. CCDs feature a complex manufacturing procedure, greater quantization efficiency, less noise, and superior picture quality in low light settings. They are considered to be the most appropriate technology for high-end imaging applications.

Each pixel in a CMOS imager performs the charge-to-voltage conversion. CMOS imagers are developed to provide a greater integration, a lower power consumption, and a lower price at the sacrifice of image quality and flexibility. They are the preferred technology for lower picture quality requirements.

3.2.6 Lidar

Light detection and ranging is also known as lidar, LIDAR, LiDAR, and LADAR [45]. Lidar is well known as a result of Theodore Maiman's Ruby Laser invention in 1960 [46,47]. It has been applied in many areas, such as topographic mapping, bathymetry, building quality control, forestry mapping, and precision agriculture. Recently, lidar systems have also been deployed on mobile phones. In 2026, the worldwide LiDAR market is expected to reach $6.71 billion. In the future years, the market will expand significantly as a result of rising investments in the creation of newer products [48]. The lidar system has three main components. First, it requires a light source that can produce light at a target. The light source is commonly a NIR laser. Second, a detector is used in order to gather the light that is reflected back from the target. The light that is collected is then processed in order to determine the distance, velocity, and form an object. Third, the lidar system employs a beam deflection mechanism to direct the light in all desired directions.

The lidar sensor is capable of providing information in a 3D way. Additionally, it is able to work during the day and the night. The lidar sensor has a wide field of view (FoV). It is a sensor that has the capability of having a very high resolution and is known for its high level of accuracy. Due to these advantages, most autonomous vehicle companies, e.g., Waymo, Volvo, and Baidu, are now using lidar. However, the lidar sensor has several downsides. It is quite probable that the cost of a lidar system will be much higher than the cost of other sensors. The data obtained by the lidar system do not give any information on the appearance and color of the object being scanned. Hence, it cannot be used for traffic sign/light detection. Because of the reflection of the water, they may lose their effectiveness under conditions of rain and fog. Moreover, the power consumption of the lidar sensor is higher than other sensors. Some autonomous vehicle companies, such as Tesla, do not use lidar. Elon Musk, the CEO of Tesla, stated that lidar is a "fool's errand" and that anyone who relies on it is "doomed" at Tesla's Autonomy Day in 2019.

3.2.7 Measuring principle

Different lidar systems may use different measuring principles.

3.2.7.1 Time-of-flight lidar

The primary measuring principle utilized by lidar for imaging is time-of-flight (TOF). The Lidar sensor emits millions of light pulses per second and measures the time it takes for the pulses to be reflected back [49]. Then, these data may be utilized to construct 3D representations of the surrounding environment. Fig. 4 shows a graphical illustration of the simplified diagram of the pulsed TOF lidar. The light is emitted by the laser source and is reflected by the surface of a target.

When using the TOF lidar, the distance to the target can be calculated as:

$$d = \frac{1}{2}ct. \tag{2}$$

In Eq. (2), d is the distance to the target, c is the speed of the light in the medium between the lidar and the target, where $c = 3 \times \frac{10^8 \text{m}}{s}$ in air, and t is the time required for the light to travel from the transmitter to the target and back to the receiver. The lidar-based range measurement is equal to measuring the round-trip delay of light reaching the target, based on this equation and the fact that the speed of light is often known with high precision. TOF lidar is the most prevalent form of lidar system now in use due to its elegance in simplicity and implementation. The necessity to sample numerous points every second imposes restrictions on the detecting time for a specific target and, consequently, on the integration time, number of pulses, and minimum pulse power. Due to the inherent ambiguity of whether the received signal is shifted by a period (or many periods) relative to the transmitted signal, periodicity between pulses creates ambiguities at long range [50].

3.2.7.2 AMCW lidar

Amplitude-modulated continuous-wave lidar (AMCW), also known as indirect TOF lidar, functions similarly to TOF lidar in that the receiver measures the target delay. To determine the range, it employs the phase-shift caused in an intensity-modulated periodic signal during its round-trip to the target. The emitted light consists of either sinusoidal or square waves. After reflection from the target, the receiver catches the light signal that has been returned. The distance is inferred from the phase shift that occurs between the reflected and emitted signals. When using the AMCW lidar, the distance to the target can be calculated as:

$$d = c\frac{1}{2}\frac{\Delta\phi}{2\pi}\frac{1}{f_m}. \tag{3}$$

In Eq. (3), d is the distance to the target, c is the speed of the light, $\Delta\phi$ is the phase shift that occurs between the received signal and the transmitted signal, and f_m is the frequency of the emitted light.

AMCW lidar transmitters utilize continuous-wave or quasi-continuous-wave laser diodes or LEDs. In the electrical world, the intensity may be controlled by altering the bias current of the diode. The ease of use of these lidars makes them an attractive option for short-range applications. However, the modulation waveform lacks sharp pulses and has a far lower frequency content [51]. Therefore, AMCW lidars cannot provide fine range resolution and the capacity to identify multiple targets. In addition, similar to the TOF lidar, there is a compromise between detection range and detection resolution.

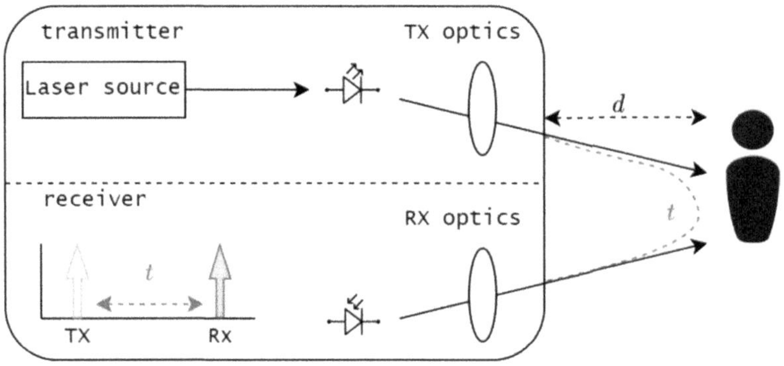

FIG. 4 Pulsed time-of-flight (TOF) measuring principle.

3.2.7.3 FMCW lidar

FMCW lidars are essentially distinct from AMCW and pulsed systems. Both pulsed TOF and AMCW lidars depend on light intensity modulation. In the receivers of these lidars, photons are often regarded as particles, with their arrival timings encoding range information. In contrast, FMCW lidars use the light's wave characteristics. FMCW lidar utilizes coherent optics to amplify the signal's information. The frequency of the light emitted by the transmitter is linearly modulated with respect to time. Typically, a diode laser serves as the source to allow coherent detection. The source light is divided into two portions. One portion of the source light serves as the transmitted signal, which is sent to the target. The other portion of the source light is employed as the local oscillator. For a stationary target with little Doppler impact on the transmitted signal, the delay between the returned light and the source results in a constant frequency difference between the transmitted signal and the returned signal. The frequency difference is directly proportional to the signal traveling time and, therefore, the distance to the target. The local oscillator is mixed with the collected light in a waveguide to measure the frequency difference. When using the FMCW lidar, the distance to the target can be calculated as:

$$d = c\frac{1}{2}\frac{1}{\gamma}f_d. \tag{4}$$

In Eq. (4), d is the distance to the target, c is the speed of the light, f_d is the frequency difference between the signal at the local oscillator and the returned signal, γ is the slope of frequency modulation against time, expressed in Hertz per second.

For nonstationary targets having a velocity component parallel to the beam, a Doppler shift is produced that raises/lowers the frequency of the returned signal for targets moving toward/away from the source. The resulting beat frequency will not only be connected to the distance but also to the relative velocity of the target. The Doppler frequency takes into consideration the velocity component, which affects the up or down sweep of the beat frequency. As a result, FMCW lidar can detect distance and relative velocity simultaneously.

In addition to a wider range resolution, FMCW lidars have better ambient light immunity. The reason is that the phase and frequency coherence of the received signal and the local oscillator are required to generate the interference pattern. FMCW lidars also employ coherent optics to improve the information of the signal, resulting in an improved signal-to-noise ratio. Moreover, the coherent detection strategy of FMCW lidars provides superior sensitivity and resilience against environmental disturbances compared to pulsed and AMCW lidars.

3.2.8 Lidar operating at 905 and 1550 nm

The lidar system employs active sensors that emit their own source of energy for lighting. The LiDAR sensors available today are capable of transmitting 160,000 pulses per second. The detection range can be more than 200 m. The major wavelengths that are employed in lidars are 905 and 1550 nm, which are near-infrared lights. The availability of 905 nm pulsed lidar systems led to their selection as the technology of choice for the first phases of autonomous vehicle research [52–54]. The availability of 905 nm pulsed lidar systems led to their selection as the technology of choice for the first phases of autonomous vehicle research. Nevertheless, all laser products that are sold commercially in the United States are required to obtain eye-safety certification by complying with the eye-safety performance standard established by the Food and Drug Administration (FDA), which is in line with the standard established by the International Electrotechnical Commission (IEC) 60825-1. Eye safety concerns require that lidar systems operating at 905 nm operate at shorter ranges than those operating at 1550 nm. This reduces light emission at that 905, which has the consequence of reducing the range of many automotive lidars to around 100 m. This range is sufficient for detecting dangers and coming to a halt in autonomous vehicles traveling between 20 and 30 miles per hour in urban areas; however, it is not sufficient for self-driving automobiles traveling at highway speeds. The lidar systems operating at 1550 nm can employ a more powerful laser without causing as much harm to the retina since the human cornea is able to absorb light at this wavelength, and no light at that wavelength can reach the retina. However, at a wavelength of 1400 nm, the water in the atmosphere starts to absorb energy. Hence, the power consumption for the 1550 nm lidar is much higher than the 905 lidar, especially under rain, snow, and fog conditions. Nowadays, the automotive manufacturing industry is quickly transitioning its product lines to include electric automobiles. Because of this, the usage of sensors that have large power needs, such as the lidar systems that employ wavelengths of 1550 nm, will be less appealing to use in electric vehicles. The cost of the lidar systems operating at 1550 nm is much higher than the lidar systems operating at 905 nm.

3.2.9 Beam deflection mechanism

1-D, 2-D, and 3-D lidar are the three basic types of lidar sensors that are applicable to a wide variety of applications. A 1-D lidar sensor does not have a beam deflection mechanism. It estimates the distance to the nearest object along a line in front of it. 1-D lidar sensors are similar to ultrasonics but more reliable and less vulnerable to interference. It is also known as a laser range finder (LRF). 2-D lidar sensors and 3-D lidar sensors have their beam deflection mechanism. 2-D lidar sensors offer extra measurements about the angle of the targeted objects. Hence they can provide information on both X and Y coordinates. Similarly, 3-D lidar sensors emit laser beams along the vertical axis to detect the height of objects. The lidar system employs different beam deflection mechanisms, also known as scanning methods, in order to collect 2-D or 3-D information from all directions in the desired FoV.

3.2.9.1 *Mechanical lidar*

The mechanical lidar sensor employs a motorized scanner to obtain the information from all directions. The idea is intuitive. One laser can only be used to make measurements in one direction. Then, a set of lasers are stacked together to make measurements in one vertical line. A motor is then used to rotate the "vertical line" to make the measurement from all directions.

The mechanical lidar sensor is the most prevalent form of lidar sensor. Velodyne introduced the HDL-64E lidar sensor in 2007, which was based on a motorized scanner and a stack of several lasers and photodetectors. This technology has been the standard in the market for autonomous vehicles for 10 years. The precision of the point cloud that is obtained in each turn is directly proportional to the number of lasers that are deployed in the lidar sensor. There are already 3D lidar sensors on the market that combine anywhere from 4 to 128 lasers or channels, with a horizontal FOV that spans 360 degrees and a vertical FOV that varies anywhere from 20 to 45 degrees, all while maintaining an accuracy of a few centimeters.

However, these types of LiDAR are not effective with their use of power and are susceptible to mechanical stress as well as wear. Furthermore, their vertical resolution is unchangeable and is reliant on the number of transmitter and receiver channels. Hence, a high vertical resolution always comes at an increased cost. In addition, the existing lidar technologies based on motorized scanners have poor performance or are expensive and cumbersome.

3.2.9.2 *Solid state lidar*

On the contrary, the solid state lidar (SSL) sensors do not make use of spinning lenses, which eliminates the risk of a mechanical failure. Since the SSL sensor does not have a spinning component, its size is usually smaller than the mechanical lidar [42,49,55,56]. There are some different types of SSL sensors.

Optical-phased arrays, more commonly referred to as OPAs, are a type of solid-state beam steering technology that enables nonmechanical steering of optical beams. OPA consists of a collection of light-emitting components whose phase may be altered. The power of the laser is distributed among a number of transmitters, each of which has its own independent phase-control system. It is possible to create and direct a laser beam by the use of dynamic adjustments to the relative phase shifts among the transmitters. OPAs are advantageous in that they have a straightforward construction, a compact size, a more reasonable cost, high stability, random-access pointing, and a strong capacity for optical power management. On the other hand, there is still a disadvantage caused by the insertion loss of the laser power in the OPA. In addition, the OPA lidar sensor has a relatively small horizontal FoV, which means the autonomous vehicle requires multiple OPA lidar sensors to achieve a horizontal FoV of 360 degrees.

The flash lidar operates on a similar concept as the time-of-flight camera, which relies on a flood illumination of the whole scene. It first sends out a large piece of laser radiation, which covers the detecting region in a brief amount of time, and then it absorbs that radiation using a sensor that possesses a high degree of sensitivity in order to finish constructing the image of the environment. It operates more similarly to a camera. Due to the fact that the laser beam is able to spread straight in all directions, a single flash may illuminate the entire scene. The system then uses an array of photodetectors to collect the reflected laser beams in different directions. Flood illumination indicates that each pixel of the photodetector array will only receive a small fraction of the returning laser power. This will result in a low signal-to-noise ratio (SNR), which will either significantly limit the distance measurement range or require a very high laser power. In addition, the size and density of the detector array in a detector array-based nonscanning LiDAR system both have a role in determining the resolution of the measurement.

Micro-Electro-Mechanical lidar system, often known as MEMS, makes use of a very small mirror measuring just a few millimeters wide in order to point a stationary laser beam in a variety of various directions. A mirror of this size would have a very low moment of inertia, which would allow it to move extremely fast, which allows it to produce a two-dimensional scanning pattern in a fraction of a second. MEMS lidar systems have been attracting a growing amount of interest for their usage in automotive applications as a result of the prospective benefits that they provide (in particular, the fact that they are lightweight, small, and have a low power consumption). MEMS-based lidar imaging devices have concurrently demonstrated the practicability of the technology in a variety of contexts.

3.2.10 Data representation

There are three typical forms of data representations [57,58]: (1) point cloud, (2) feature-based, and (3) grid. Point cloud-based techniques employ raw sensor data to describe the environment. This method creates an accurate representation, but it demands a substantial amount of memory and computer resources. Typically, a voxel-based filtering process is used in the raw point cloud to limit the number of points in order to relieve this issue. Feature-based techniques employ locally distinct characteristics to describe sensor data. Commonly utilized characteristics include lines and surfaces. This method is the most memory-efficient, but it is frequently overly abstract, and its accuracy is dependent on the nature of the point cloud, as not all environment characteristics can be adequately modeled. Grid-based approaches divide the area into small grid pieces known as cells, each of which holds information about the sensory space it covers. Grid-based solutions are memory-efficient, simple to construct, and independent of specified characteristics, which makes them an effective approach for representing sensor data in autonomous cars. Nevertheless, evaluating the size of the discretization is not simple.

3.2.11 Radar

Radio Detection and Ranging is what "radar" stands for. The radar sensor is an active sensor. It operates on a very similar premise to that of lidar; the main distinction lies in the fact that it makes use of radio waves rather than lasers. The antenna of a radar system fulfills the role of both a radar receiver and a radar transmitter. Nevertheless, in contrast to light waves, radio waves are less likely to be absorbed when coming into touch with the thing being analyzed. Hence, the effective working distance of radar is significantly longer. The military's use of radar technology is perhaps the one that is most known to the general public. It is planned to equip military aircraft and warships with radar so that they can assess altitude and also identify other surrounding transport equipment and objects.

FMCW radars and pulse radars are the two primary categories that are used to describe vehicle-mounted radars. The pulse radar needs to carefully isolate the transmitted signal when it is receiving the reflected signal, while a high-power signal will be transmitted in a transitory continuous cycle, which leads to high requirements for the hardware and a sophisticated structure. As a result, most current car radars are built on FMCW technology and make use of digital beam forming in order to regulate the path that the emitted wave takes. FMCW involves producing a signal with a known and stable frequency that is modulated by a second continuous signal whose frequency fluctuates up and down (typically using a triangular shape). The difference in frequency that exists between the signals that are emitted and those that are reflected can be used to calculate distance.

The radar sensor can operate at frequencies of 24 GHz, 60 GHz, 77 GHz, and 79 GHz, respectively. Due to the fact that the frequency corresponds to wavelengths of millimeters, these radars are often referred to as Millimeter Wave (MMW) radars. The 77 GHz band radar is the most common one. It provides a greater resolution of range, speed, and angle than other vehicle radars. This frequency band has received widespread approval and is poised to become the standard frequency band for vehicle radars in the near future. The radar sensor can be categorized by the detecting range, such as short-range, medium-range, and long-range radar. A long-range radar may be utilized to detect distant targets or things in front of the car. On the other hand, short-range radar and medium-range radar are utilized for parking assistance or side view detection.

Radars are able to detect obstacles and quantify information about them, such as their relative position based on the amount of time it takes for a wave to travel and their relative velocity based on the Doppler effect that electromagnetic waves have. Since the frequency of radars waves is not impacted by the illumination or the weather, these devices are able to detect whether it is day or night, cloudy, foggy, or snowing outside. Most Blind Spot Warning systems use radar sensors, although others employ cameras instead. The radar sensor has also been commonly used in lane change assistants and cross traffic alerts. However, the radar sensor is sensitive to noise. Therefore, it has a restricted capacity for categorization. The MMW radar has the limitation of being unable to discriminate relatively stationary and static targets, which is one of its drawbacks. It is also difficult for the radar sensor to identify any metal items in the perceived surroundings, such as traffic signs. In addition, because a growing number of vehicles are being installed with FMCW radar, the issue of shared frequency interference is going to develop into a concern.

3.2.12 Ultrasonic sensor

Ultrasonic sensors have been utilized in automobiles for a significant amount of time, and many vehicles have them installed on both the front and the back sides of their bodies. The frequency of the wave runs anything from 20 kHz all the way up to 70 kHz. When the frequency is increased, the sensitivity also increases, and the detection angle becomes narrower. The ultrasonic radar, when it is functioning properly, is able to measure obstacles within a range of 0.2–5 m with an accuracy of 1–3 cm by emitting and receiving ultrasonic waves.

The Ultrasonic Park Assist (UPA) Sensor and the Active Parking Assist (APA) Sensor are both examples of ultrasonic sensors that have found usage in the automobile industry. The UPA radar, also known as the reversing radar, is attached to the front and back bumpers of the vehicle, where it is utilized to measure the distance between the vehicle and any obstructions in its path. The UPA radar has a detection range that is typically between 15 and 250 cm. In order to get a better reading of the side obstacles, the APA sensor is mounted on the side of the vehicle. The APA radar has a detection range between 30 and 500 cm. Since the APA radar has a greater detection range than the UPA radar does, it is more expensive and requires more power.

The ultrasonic sensor is very low-cost, while it offers accurate readings when measuring distances with any substance, regardless of the color of the material, even in dusty surroundings or when the weather is unfavorable. On the other hand, the ultrasonic sensor has a number of drawbacks, one of which is that they are subject to significant influence from disruptions in the sound waves. Because sound can only travel through a medium, changes in the conditions of the environment, such as temperature, can affect how sound travels. In addition to that, it does not function at high speeds while having a poor resolution and a limited detecting range.

3.2.13 GPS and IMU

The Global Positioning System (GPS) and the Inertial Measurement Unit (IMU) sensor are both useful tools for determining the location and attitude of a vehicle. GPS is a regularly utilized sensor in autonomous cars for navigation and localization since it may give important location capabilities. An IMU is a set of sensors that may include an accelerometer, gyroscope, and magnetometer. It gathers information on the movement of the vehicles. The primary problem of these devices is that they cannot offer information on the vehicle's real position. Instead, they only provide the motion information. GPS and IMU are usually combined to provide navigation.

3.3 Sensor fusion

Due to sensing limitations in range, signal characteristics, and operating circumstances, it is difficult for a single kind of sensor to fulfill the objectives of obstacle detection—there is no one sensor that can rule them all. The radar graphs in Fig. 5 show the comparison of the characteristics of several sensors utilized for the perception of autonomous vehicles. For example, because camera sensors have a high resolution and the ability to distinguish between colors, this makes them particularly effective at detecting objects. However, while operating in less-than-ideal weather conditions, their performance suffers. In the meantime, the single camera sensor cannot measure the true distance without a reference distance. Both range and precision are much improved with the mechanical lidar sensors. On the other side, they are rather pricey, and they lack the ability to distinguish between different colors. The computational requirement for using camera sensors and lidar sensors is relatively high. Radar sensors' performance is unaffected by the surrounding atmosphere in any way. Despite this, their capacity to perceive the surrounding world falls short of expectations.

Therefore, sensor fusion is now required in all AD systems in order to overcome the limitations of individual sensor types, hence enhancing the entire system's performance and robustness [59,60]. The study in discusses the security issue of multisensory fusion-based autonomous driving perception [61]. A survey in Ref. [62] covers the primary factors for the onboard multi-sensor design of intelligent ground vehicles in off-road conditions and provides users with guidance for choosing sensors according to their performance needs and application scenarios. Many deep learning methods have been utilized in sensor fusion. The study in Ref. [63] analyzes the strengths and limits of existing deep learning approaches through comparative analysis and explores the research difficulties in terms of computing, architecture selection, goal formulation, generalization, verification and validation, and safety.

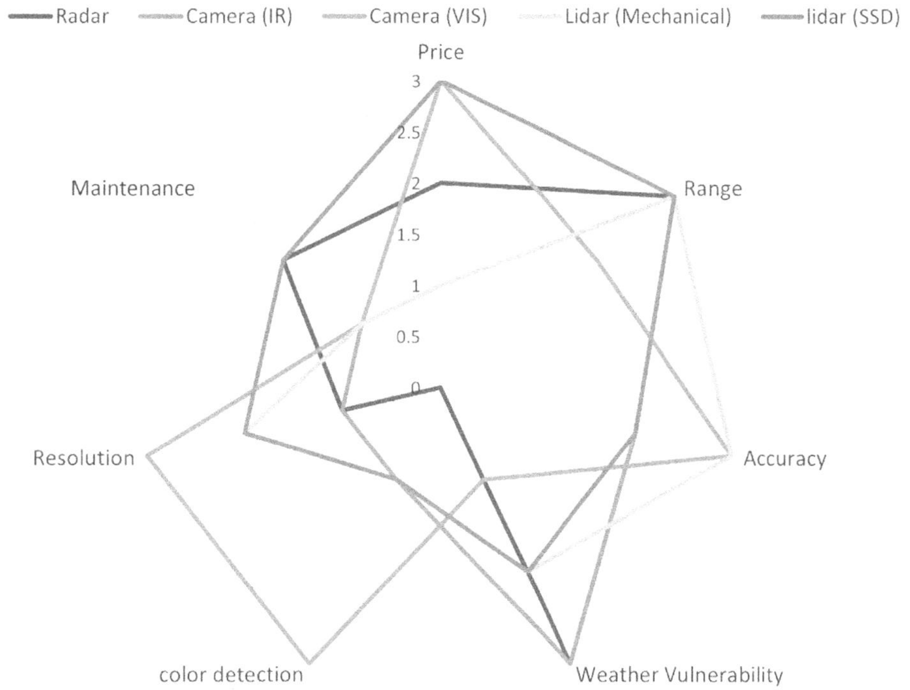

FIG. 5 Characteristics of sensors utilized for the perception of autonomous vehicles.

3.3.1 Data fusion

By expanding the number of sensors and fusing various sensors, autonomous driving capabilities may be enhanced. Multiple sensors of the same type or of various sorts collect information in distinct categories and from distinct vantage points on the environment. This information may be redundant and contradictory, or it may complement one another. In the end, the control center can only give single right instruction, requiring the fusion of information from various sensors in order to make an optimal determination.

Sensor fusion can be divided into three categories: high-level fusion (HLF), mid-level fusion (MLF), and low-level fusion (LLF). HLF is the most common fusion technology among car OEMs. Each sensor generates its own local decision independently. The data fusion center collects all the local decisions and integrates them into a global decision [64,65]. HLF techniques are frequently utilized due to their greater simplicity and faster speed compared to LLF and MLF procedures. However, the information collected by different sensors is dependent, which implies that HLF method can be inaccurate. In contrast, low-level fusion (LLF) integrates data from many sensor types at the level of raw data, retaining all information and potentially enhancing the accuracy of object recognition. Complexity is inherent to LLF. For the proper perception of the surroundings, extremely precise extrinsic calibration of the sensors is necessary. In addition, time synchronization is necessary for the sensor. Deep neural networks may utilize the multimodal input data for object detection. Fusion and detection algorithms must be able to operate in real-time under highway driving conditions. The potential benefits of LLF have gained interest in recent years. MLF, also known as feature-level fusion, is an abstraction level positioned between LLF and HLF. It combines many target characteristics retrieved from the associated sensor data.

3.3.2 Calibration

Sensor calibration provides the autonomous system with information on the position and orientation of the sensors in real-world coordinates by comparing the relative positions of known objects observed by the sensors. The accuracy and precision of the calibration have a direct impact on the detection accuracy of the sensor as well as the subsequent processing phases, which include sensor fusion, motion planning, and path tracking.

Sensor calibration can be divided into intrinsic calibration and extrinsic calibration. Intrinsic calibration addresses the intrinsic properties of a specific sensor, such as the principal point coordinate of the camera sensor. The lidar sensors are usually calibrated by the manufacturer. Based on the size of the calibration objects, we may generally divide the camera sensor calibration methods into four groups, i.e., 3-D reference object-based calibration, 2-D plane-based calibration, 1-D line-based calibration, and self-calibration [66]. Calibration based on a 3-D reference item gives the maximum accuracy. Whereas the 2D plane-based calibration approach is the most preferred due to its usability and precision. If we wish to calibrate multiple cameras concurrently, the 1-D line-based calibration approach is preferable. Self-calibration requires the estimation of a large number of parameters, resulting in a significantly more difficult mathematical problem. Self-calibration will be utilized when other techniques of calibration are impossible. Zhang's algorithm, which is based on the image of the absolute conic, is one of the most popular camera calibration methods [67].

The extrinsic calibration determines the position and orientation of a sensor with respect to the world or to another sensor [68]. The extrinsic parameters of the camera sensors have been discussed in Eq. (1). The most common extrinsic calibration task of the sensors on autonomous vehicles is camera-lidar calibration. In multisensor systems, however, the extrinsic calibration of many sensors employing different physical measurement techniques might be a difficulty. The data collected by the camera sensors are usually RGB pixel array. The lidar data is the 3D point cloud. The radar data is usually 2D. GPS and IMU provide the position and attitude information of the vehicle.

Extrinsic calibration can be divided into targetless calibration and target-based calibration. The traditional extrinsic calibration methods are target-based. The target-based calibration methods usually require a fixed target, such as a checkerboard, an ordinary box, or a trihedron [69,70]. The targetless calibration method does not require an artificial target or a structured environment [71]. One of the famous calibration methods is hand-eye calibration, which first appeared and got its name from the robotics community [72].

Sensors are installed at different locations of the vehicle. For example, the lidar sensor is usually mounted on the top of the vehicle, while the camera can be installed at the front of the car. Occlusion issues appear because each sensor has a different view [73]. One option is to consider each point to be a rectangle, and if the points that are inside that rectangle have varying depths, consider those points to be occluded.

3.3.3 Synchronization

In the process of sensor fusion, synchronization is an essential component for re-creating an accurate and full picture of the surrounding environment. The purpose of synchronizing sensors is to guarantee those sensor readings that have the same timestamp refer to the same event. Unsynchronized sensors might create a misleading and unclear representation of the surroundings, which could lead to results that are potentially fatal given the high speed at which the autonomous vehicle can drive. In addition, different sensors can sometimes have different frequencies of detection, which is due to the different structures and the way that the observations are collected (e.g., a LiDAR could scan the environment around the vehicle 20 times per second, while a camera could do so at 30 Hz or higher). Most of the sensors are equipped with their very own internal clocks, which may be utilized to synchronize the data collected with the world clock (e.g., GPS clock). One such possibility is to utilize a single accurate clock as the trigger for all of the sensors.

3.4 Still challenging

According to records that were made public on Wednesday by the U.S. safety regulators, automakers reported roughly 400 accidents involving vehicles equipped with partially automated driver-assist systems [74]. There are still many obstacles to overcome in the process of designing the sensors for autonomous cars.

3.4.1 Weather/road conditions

Under ideal working conditions, AV sensor technology should function as intended in terms of sensing the surrounding environment and carrying out the required activities. Nonetheless, unfavorable weather conditions (such as rain, snow, and fog) can pose significant obstacles to AV sensors and the algorithms that extract data from them [75]. For instance, the performance of camera sensors deteriorates when the vehicle is operating on a snowing evening. The state of the roads may be exceedingly unexpected and would differ from one location to the next. There are roadways that are both paved and clearly designated. In other instances, the road conditions are quite hazardous. For example, there is no lane marks or clearly defined lanes.

3.4.2 Security

The outcomes of laboratory experiments indicate that blindness, jamming, replay, relay, and spoofing attacks are effective. [76]. By introducing a small number of unnoticeable perturbations to an image, the resulting adversarial instances can lead a classifier to make certain errors [77,78]. A physically realizable, adversarial 3D-printed object can deceive an autonomous driving system into failing to identify it, so causing a collision [79]. It is possible to control the spoofing attack strategically in order to trick the machine learning model used in autonomous vehicles [80].

3.4.3 Computation

The challenge of the computation comes from detection speed and computation speed. Human reaction time is about 300 ms. The sensors need to realize object detection and other tasks in a fraction of the human's reaction time. In the meantime, autonomous vehicle highly depends on deep learning. Deep neural networks, such as those used in self-driving vehicles, need a mind-boggling amount of computer power. This is because deep neural networks are extremely complex. In addition to that, they need a substantial amount of data sets. It is necessary to train neural networks using datasets that are representative of everything that may possibly happen on the road, in the weather, and in other situations. In the real world, this translates into petabytes of data for training purposes [81]. The total data generated by the sensors of an autonomous vehicle can reach up to 19 Tb per hour [82].

References

[1] J.L. Kelley, L. Chapuis, W.I. Davies, S.P. Collin, Sensory system responses to human-induced environmental change, Front. Ecol. Evol. 6 (2018) 95.

[2] V.K.S. Sachan n.d. "Smart Sensor Market by Type (Touch Sensor, Image Sensor, Temperature Sensor, Motion Sensor, Position Sensor, and Pressure Sensor) and End Use (Automotive Industry, Consumer Electronics, Industrial, Infrastructure, Healthcare, and Others): Global Opportunity Analysis and Industry Forecast." Available online https://www.alliedmarketresearch.com/smart-sensors-market (Accessed 11 June 2022).

[3] J. Vetelino, A. Reghu, Introduction to Sensors, CRC Press, 2017.

[4] Passive vs. Active Sensing. n.d. Available online https://www.nrcan.gc.ca/maps-tools-publications/satellite-imagery-air-photos/remote-sensing-tutorials/introduction/passive-vs-active-sensing/14639 (Accessed 11 June 2022).

[5] R. Siegwart, I.R. Nourbakhsh, D. Scaramuzza, Introduction to Autonomous Mobile Robots, MIT Press, 2011.

[6] Contact vs. Non-contact Measurement and Linear Displacement Sensors. n.d. Available online https://mtiinstruments.com/contact-vs-non-contact-measurement/ (Accessed 12 June 2022).

[7] Sensors: Contactless Sensors Help to Detect Covid. n.d. Available online https://www.electronicsweekly.com/news/sensors-contactless-sensors-help-detect-covid-2020-09/ (Accessed 12 June 2022).

[8] W.J. Fleming, Overview of automotive sensors, IEEE Sensors J. 1 (4) (2001) 296–308.

[9] C. Wei, W. Zhou, Q. Wang, X. Xia, X. Li, TPMS (tire-pressure monitoring system) sensors: monolithic integration of surface-micromachined piezoresistive pressure sensor and self-testable accelerometer, Microelectron. Eng. 91 (2012) 167–173.

[10] A. Ziebinski, R. Cupek, H. Erdogan, S. Waechter, A survey of ADAS technologies for the future perspective of sensor fusion, in: International Conference on Computational Collective Intelligence, Springer, 2016, pp. 135–146.

[11] S. Campbell, et al., Sensor technology in autonomous vehicles: a review, in: 2018 29th Irish Signals and Systems Conference (ISSC), IEEE, 2018, pp. 1–4.

[12] E. Marti, M.A. De Miguel, F. Garcia, J. Perez, A review of sensor technologies for perception in automated driving, IEEE Intell. Transp. Syst. Mag. 11 (4) (2019) 94–108.

[13] J. Van Brummelen, M. O'Brien, D. Gruyer, H. Najjaran, Autonomous vehicle perception: the technology of today and tomorrow, Transportation Research Part C: Emerging Technologies 89 (2018) 384–406.

[14] L. Ding, Y. Wang, R. Laganière, D. Huang, X. Luo, H. Zhang, A robust and fast multispectral pedestrian detection deep network, Knowl.-Based Syst. 227 (2021) 106990.

[15] X. Yang, Y. Wang, R. Laganière, A scale-aware YOLO model for pedestrian detection, in: International Symposium on Visual Computing, Springer, 2020, pp. 15–26.

[16] K.R. Castleman, Digital Image Processing, Prentice Hall Press, 1996.

[17] R. Hartley, A. Zisserman, Multiple View Geometry in Computer Vision, Cambridge University Press, 2003.

[18] K. Simek. n.d. Dissecting the Camera Matrix, Part 3: The Intrinsic Matrix. Available online https://ksimek.github.io/2013/08/13/intrinsic/ (Accessed 25 June 2022).

[19] A. Dubey, Stereo Vision-Facing the Challenges and Seeing the Opportunities for ADAS Applications, Texas Instruments Technical Note, 2016.

[20] M. Morariu. n.d. A Brief Introduction to 3D Cameras. Available online https://tech.preferred.jp/en/blog/a-brief-introduction-to-3d-cameras/ (Accessed 24 June 2022).

[21] M. Abedin-Nasab, Handbook of Robotic and Image-Guided Surgery, Elsevier, 2019.

[22] R.A. Hamzah, H. Ibrahim, A.H.A. Hassan, Stereo matching algorithm for 3D surface reconstruction based on triangulation principle, in: 2016 1st International Conference on Information Technology, Information Systems and Electrical Engineering (ICITISEE), IEEE, 2016, pp. 119–124.

[23] L. Li, Time-of-Flight Camera–An Introduction, Technical white paper, no. SLOA190B, 2014.

[24] E.-T. Baek, H.-J. Yang, S.-H. Kim, G. Lee, H. Jeong, Distance error correction in time-of-flight cameras using asynchronous integration time, Sensors 20 (4) (2020) 1156.

[25] J.P. Rolland, L.D. Davis, Y. Baillot, A survey of tracking technologies for virtual environments, in: Fundamentals of Wearable Computers and Augmented Reality, CRC Press, 2001, pp. 83–128.
[26] R.J. Valkenburg, A.M. McIvor, Accurate 3D measurement using a structured light system, Image Vis. Comput. 16 (2) (1998) 99–110.
[27] X. Chen, J. Xi, Y. Jin, J. Sun, Accurate calibration for a camera–projector measurement system based on structured light projection, Opt. Lasers Eng. 47 (3–4) (2009) 310–319.
[28] P. Kienle, et al., Optical setup for error compensation in a laser triangulation system, Sensors 20 (17) (2020) 4949. [Online]. Available: https://www.mdpi.com/1424-8220/20/17/4949.
[29] S. Li, X. Jia, M. Chen, Y. Yang, Error analysis and correction for color in laser triangulation measurement, Optik 168 (2018) 165–173.
[30] J.C. Cuevas, F.J. García-Vidal, Radiative heat transfer, ACS Photonics 5 (10) (2018) 3896–3915.
[31] B.B. Lahiri, S. Bagavathiappan, T. Jayakumar, J. Philip, Medical applications of infrared thermography: a review, Infrared Phys. Technol. 55 (4) (2012) 221–235.
[32] What Is the Electromagnetic Spectrum and Shortwave Infrared (SWIR). n.d. Available online https://www.sensorsinc.com/technology/why-swir (Accessed 12 June 2022).
[33] Near, Mid, and Far Infrared. n.d. Available online http://www.icc.dur.ac.uk/~tt/Lectures/Galaxies/Images/Infrared/Regions/irregion s.html (Accessed).
[34] S. Du, P. Zhang, B. Zhang, H. Xu, Weak and occluded vehicle detection in complex infrared environment based on improved YOLOv4, IEEE Access 9 (2021) 25671–25680.
[35] X. Zhang, X. Zhu, Vehicle detection in the aerial infrared images via an improved YOLOv3 network, in: 2019 IEEE 4th International Conference on Signal and Image Processing (ICSIP), IEEE, 2019, pp. 372–376.
[36] X. Liu, T. Yang, J. Li, Real-time ground vehicle detection in aerial infrared imagery based on convolutional neural network, Electronics 7 (6) (2018) 78.
[37] A.S. Ahire, Night vision system in BMW, International Review of Applied Engineering Research 4 (1) (2014) 1–10.
[38] R. Vandersmissen, N. XenICs, Night-vision camera combines thermal and low-light-level images, Photonik Int 2 (2008) 2–4.
[39] R. Srivastava, Recent Development in Optoelectronic Devices, BoD–Books on Demand, 2018.
[40] J. Kallhammer, The road ahead for car night-vision, Nat. Photonics 5 (2006) 12–13.
[41] K. Kainth, B. Singh, Analysis of CCD and CMOS sensor based images from technical and photographic aspects, in: International Conference of Advance Research & Innovation (ICARI), 2020.
[42] F. Rosique, P.J. Navarro, C. Fernández, A. Padilla, A systematic review of perception system and simulators for autonomous vehicles research, Sensors 19 (3) (2019) 648.
[43] D. Litwiller, Ccd vs. cmos, Photonics Spectra 35 (1) (2001) 154–158.
[44] CCD vs CMOS. n.d. Available online https://www.teledynedalsa.com/en/learn/knowledge-center/ccd-vs-cmos/ (Accessed 21 June 2022).
[45] P.F. McManamon, LiDAR Technologies and Systems, SPIE Press, 2019.
[46] T.H. Maiman, Stimulated Optical Radiation in Ruby, vol. 187, Nature, 1960, pp. 493–494.
[47] X. Wang, H. Pan, K. Guo, X. Yang, S. Luo, The evolution of LiDAR and its application in high precision measurement, IOP Conference Series: Earth and Environmental Science 502 (1) (2020) 012008. IOP Publishing.
[48] With 22.7% CAGR, LiDAR Market Size to Surpass USD 6.71 Billion by 2028. n.d. Available online https://www.globenewswire.com/en/news-release/2022/06/07/2457470/0/en/With-22-7-CAGR-LiDAR-Market-Size-to-Surpass-USD-6-71-Billion-by-2028.html (Accessed 25 June 2022).
[49] S. Royo, M. Ballesta-Garcia, An overview of lidar imaging systems for autonomous vehicles, Appl. Sci. 9 (19) (2019) 4093.
[50] J. Doylend, S. Gupta, An overview of silicon photonics for LIDAR, Silicon Photonics XV 11285 (2020) 109–115.
[51] B. Behroozpour, P.A. Sandborn, M.C. Wu, B.E. Boser, Lidar system architectures and circuits, IEEE Commun. Mag. 55 (10) (2017) 135–142.
[52] J. Hecht. n.d. "Automotive Lidar: Safety Questions Raised About 1550 nm Lidar." Available online https://www.laserfocusworld.com/blogs/article/14040682/safety-questions-raised-about-1550-nm-lidar (Accessed).
[53] V. Lidar. n.d. A Guide to Lidar Wavelengths for Autonomous Vehicles and Driver Assistance. Available online https://velodynelidar.com/blog/guide-to-lidar-wavelengths/#:~:text=Current%20state%2Dof%2Dthe%2D,the%20availability%20of%20sensor%20components (Accessed).
[54] S. Rangwala. n.d. The LiDAR Range Wars - Mine Is Longer Than Yours. Available online https://www.forbes.com/sites/sabbirrangwala/2021/05/27/the-lidar-range-wars-mine-is-longer-than-yours/?sh=5cbdfce13141 (Accessed).
[55] D. Wang, C. Watkins, H. Xie, MEMS mirrors for LiDAR: a review, Micromachines 11 (5) (2020) 456.
[56] T. Raj, F.H. Hashim, A.B. Huddin, M.F. Ibrahim, A. Hussain, A survey on LiDAR scanning mechanisms, Electronics 9 (5) (2020) 741.
[57] A. Asvadi, C. Premebida, P. Peixoto, U. Nunes, 3D Lidar-based static and moving obstacle detection in driving environments: an approach based on voxels and multi-region ground planes, Robot. Auton. Syst. 83 (2016) 299–311.
[58] S.D. Pendleton, et al., Perception, planning, control, and coordination for autonomous vehicles, Mach. Des. 5 (1) (2017) 6.
[59] D.J. Yeong, G. Velasco-Hernandez, J. Barry, J. Walsh, Sensor and sensor fusion technology in autonomous vehicles: a review, Sensors 21 (6) (2021) 2140.
[60] J. Fayyad, M.A. Jaradat, D. Gruyer, H. Najjaran, Deep learning sensor fusion for autonomous vehicle perception and localization: a review, Sensors 20 (15) (2020) 4220.
[61] Z. Huang, C. Lv, Y. Xing, J. Wu, Multi-modal sensor fusion-based deep neural network for end-to-end autonomous driving with scene understanding, IEEE Sensors J. 21 (10) (2020) 11781–11790.
[62] J.-W. Hu, et al., A survey on multi-sensor fusion based obstacle detection for intelligent ground vehicles in off-road environments, Front. Inf. Technol. Electron. Eng. 21 (5) (2020) 675–692.
[63] S. Kuutti, R. Bowden, Y. Jin, P. Barber, S. Fallah, A survey of deep learning applications to autonomous vehicle control, IEEE Trans. Intell. Transp. Syst. 22 (2) (2020) 712–733.
[64] W. Dong, M. Kam, Detection performance vs. complexity in parallel decentralized Bayesian decision fusion, in: 2017 51st Annual Conference on Information Sciences and Systems (CISS), IEEE, 2017, pp. 1–6.

[65] W. Dong, M. Kam, Dependent randomization in parallel binary decision fusion, IEEE/CAA Journal of Automatica Sinica 8 (2) (2021) 361–376.
[66] Y.-Y. Chuang, Camera Calibration, Citeseer, 2005.
[67] Z. Zhang, A flexible new technique for camera calibration, IEEE Trans. Pattern Anal. Mach. Intell. 22 (11) (2000) 1330–1334.
[68] F.M. Mirzaei, Extrinsic and Intrinsic Sensor Calibration, University of Minnesota, 2013.
[69] J.-K. Huang, J.W. Grizzle, Improvements to target-based 3D LiDAR to camera calibration, IEEE Access 8 (2020) 134101–134110.
[70] Y. Zhu, C. Li, Y. Zhang, Online camera-lidar calibration with sensor semantic information, in: 2020 IEEE International Conference on Robotics and Automation (ICRA), IEEE, 2020, pp. 4970–4976.
[71] M.Á. Muñoz-Bañón, F.A. Candelas, F. Torres, Targetless camera-LiDAR calibration in unstructured environments, IEEE Access 8 (2020) 143692–143705.
[72] Hand-Eye Calibration. n.d. Available online https://campar.in.tum.de/Chair/HandEyeCalibration (Accessed 12 June 2022).
[73] E. Vezeteu, Stereo-Camera–LiDAR Calibration for Autonomous Driving (MS thesis), 2021.
[74] Nearly 400 Car Crashes in 11 Months Involved Automated Tech, Companies Tell Regulators. n.d. Available on https://www.npr.org/2022/06/15/1105252793/nearly-400-car-crashes-in-11-months-involved-automated-tech-companies-tell-regul (Accessed 13 June 2020).
[75] J. Vargas, S. Alsweiss, O. Toker, R. Razdan, J. Santos, An overview of autonomous vehicles sensors and their vulnerability to weather conditions, Sensors 21 (16) (2021) 5397.
[76] J. Petit, B. Stottelaar, M. Feiri, F. Kargl, Remote attacks on automated vehicles sensors: experiments on camera and lidar, Black Hat Europe 11 (2015) (2015) 995.
[77] R. Duan, X. Ma, Y. Wang, J. Bailey, A.K. Qin, Y. Yang, Adversarial camouflage: Hiding physical-world attacks with natural styles, in: Proceedings of the IEEE/CVF Conference on Computer Vision and Pattern Recognition, 2020, pp. 1000–1008.
[78] S.T. Jan, J. Messou, Y.-C. Lin, J.-B. Huang, G. Wang, Connecting the digital and physical world: improving the robustness of adversarial attacks, in: Proceedings of the AAAI Conference on Artificial Intelligence, 33, no. 01, 2019, pp. 962–969.
[79] Y. Cao, et al., Invisible for both camera and lidar: Security of multi-sensor fusion based perception in autonomous driving under physical-world attacks, in: 2021 IEEE Symposium on Security and Privacy (SP), IEEE, 2021, pp. 176–194.
[80] Y. Cao, et al., Adversarial sensor attack on lidar-based perception in autonomous driving, in: Proceedings of the 2019 ACM SIGSAC Conference on Computer and Communications Security, 2019, pp. 2267–2281.
[81] A. Gupta. n.d. Challenges in Training Algorithms for Autonomous Cars. Available online https://www.iiot-world.com/artificial-intelligence-ml/autonomous-vehicles/challenges-in-training-algorithms-for-autonomous-cars/ (Accessed 7 June 2022).
[82] S. Wright. n.d. Autonomous Cars Generate More Than 300 TB of Data per Year. Available online https://www.tuxera.com/blog/autonomous-cars-300-tb-of-data-per-year/#:~:text=Today%2C%20even%20at%20lower%20levels,possible%20will%20become%20increasingly%20complex (Accessed 1 June 2022).

CHAPTER

4

Radar architectures and cyberattacks from an autonomous vehicles perspective

Onur Toker

Florida Polytechnic University, Lakeland, FL, United States

4.1 Early history of electromagnetism and radio waves

The history of electromagnetic waves, radio, and radar starts with the theoretical work of James Clerk Maxwell. In 1861, Maxwell published a four-part paper titled "On Physical Lines of Force," Maxwell [1], and established the foundations of the classical electrodynamics. For a detailed history of Maxwell's great discovery, see Sarkar et al. [2]; Mitolo and Araneo [3]; Nahin [4]; Rautio [5] and references therein. Maxwell's equations are basically a coupled set of partial differential equations and admit a wave-like solution. This theoretical observation led to the experimental demonstration of radio waves. In 1888, Heinrich Rudolf Hertz used a spark gap transmitter and experimentally demonstrated radio waves. His step-up was a 50 MHz frequency spark-gap transmitter and a loop antenna with a spark gap, see Susskind [6]; Kraus [7]; Simpson [8]; Blanchard [9]; Mathis and Emeritus [10]; Elliott [11] and references therein for further details. Historically, Hertz considered his work simply as an experimental demonstration of Maxwell's theory and did not consider his invention as something of great practical importance.

Independent of these developments, French scientist Édouard Branly published a paper in 1890 and described the details of a device called coherer. Coherer is a tube with two electrodes with loose metal powder in between. Coherer was used as a simple radio signal detector in the early days of radio and wireless telegraphy and remained widely used until better detector technologies are developed, see Dilhac [12]; Gandhi et al. [13]; Bose [14], and references therein.

In the late 1890s, Marconi started to experiment with different spark gap transmitters and began working on long-distance radio transmission and wireless telegraphy. In 1895, he achieved radio transmission over a distance of 2 miles, and in 1902, over the Atlantic Ocean achieving wireless transmission over a distance of 2100 miles, see Marconi [15]; Falciasecca and Valotti [16]; Beynon [17]; Corazza [18]; Brittain [19]. Basically, spark gap transmitters and coherers started the wireless telegraphy and the wireless communication revolution.

To be able to build better radio transceivers, we had to wait for the invention of the vacuum tube. The story of vacuum tube-based radios starts with the vacuum tube diode (a.k.a. Fleming valve), invented John Ambrose Fleming in 1904, see Sarkar et al. [20]; Hong [21]; Brittain [22]. Two years later, in 1906, Lee De Forest invented the first electronic amplifying device called the triode, which is basically Fleming's diode with a third electrode added, see Delogne [23]; Brittain [24]; Pestrikov [25]. Triode was functionally similar to a transistor, and its invention led to various oscillator and amplifier circuits, and the development of much better vacuum tube-based radio transmitters and receivers. Lee De Forest is widely regarded as the inventor of electronics, Delogne [23].

In 1904, Christian Hülsmeyer demonstrated the use of radio waves to detect the presence of ships over a long distance, see Griffiths et al. [26]. This was a spark gap type radio system and is considered as the first practical demonstration of a working radar system. Before and during World War 2 years, engineers realized the need for compact and high-power electromagnetic wave sources. This was especially important for air-borne radar applications. Before the invention of transistors in 1947, there was another remarkable discovery, the magnetron. In 1940, John Randall and Harry Boot working at the University of Birmingham, discovered the magnetron that can produce hundreds of watts output at GHz frequencies, Brown [27]. During the World War 2, there were major developments in radar technology. During the same period, engineers realized the importance of jamming radar systems, and prevention techniques against such attacks. These research efforts can be viewed as the beginning of cyberattacks on cyber-physical systems.

Handbook of Power Electronics in Autonomous and Electric Vehicles
https://doi.org/10.1016/B978-0-323-99545-0.00005-1

TABLE 1 IEEE Std 521-1984 based radar frequency bands.

Radar designation	Frequency range
HF	3–30 MHz
VHF	30–300 MHz
UHF	300–1000 MHz
L	1–2 GHz
S	2–4 GHz
C	4–8 GHz
X	8–12 GHz
K_u	12–18 GHz
K	18–27 GHz
K_a	27–40 GHz
V	40–75 GHz
W	75–110 GHz
mm	110–300 GHz

4.2 Electromagnetic spectrum

The IEEE Std 521-1984 documentation, IEEE [28], defines various frequency bands for radar applications. This is summarized in Table 1.

The frequency, f in Hertz (Hz), and wavelength, λ, in meters (m) of electromagnetic waves are related to each other by the formula

$$\lambda = c/f$$

where c is the speed of light. Earth's atmosphere has a complex electromagnetic absorption pattern. Certain frequencies can reach the ionosphere and get reflected there, whereas certain others are strongly absorbed by oxygen or water vapor, see Altshuler and Marr [29]; Paulson [30] and references therein. Around 20 GHz levels, electromagnetic absorption due to water vapor is high. But we have a low absorption window in the spectrum, called the K_a band, between 27 and 40 GHz. Around 60 GHz, electromagnetic absorption due to oxygen is high, but again we have a low absorption window in the spectrum, called the W band, between 75 and 110 GHz.

Lower frequency radars require larger antennas and have lower distance and/or angular accuracy. But, high-power radios are easier to build and operate at lower frequencies. During World War 2 years, radars were operating in the HF portion of the spectrum, see Blanchard and van Genderen [31]; Brown [32], and had hundreds of kW of peak output power. The British radar system, Chain Home, used an analog oscilloscope and an amplifier. Early automotive radars were operating in the 24 GHz range, see Figs. 1 and 2, but most of the modern automotive radars operate at 77 GHz, see Fig. 3. These shorter wavelengths provide various advantages for radar sensors and their use in autonomous vehicles (AVs) and advanced driver assistance systems (ADASs).

4.3 Classification of radars

Radar technology started relatively recently, about 100 years ago during World War 2 years. During this very short time period, we have seen unprecedented developments like synthetic aperture radars (SAR), see-through wall radars, ground penetrating radars (GPR), and software-defined radars, just to name a few.

Radar systems can be classified in various different ways, but primarily a radar can be used either for generating an image or map or for generating a numerical value. The first type is called the Imaging Radar, and they usually generate a two-dimensional map-like output. Some of the multitransmit and multireceive antenna radar systems can even generate a coarse three-dimensional map plus velocity estimates for all of the reflector objects. On the other extreme, a

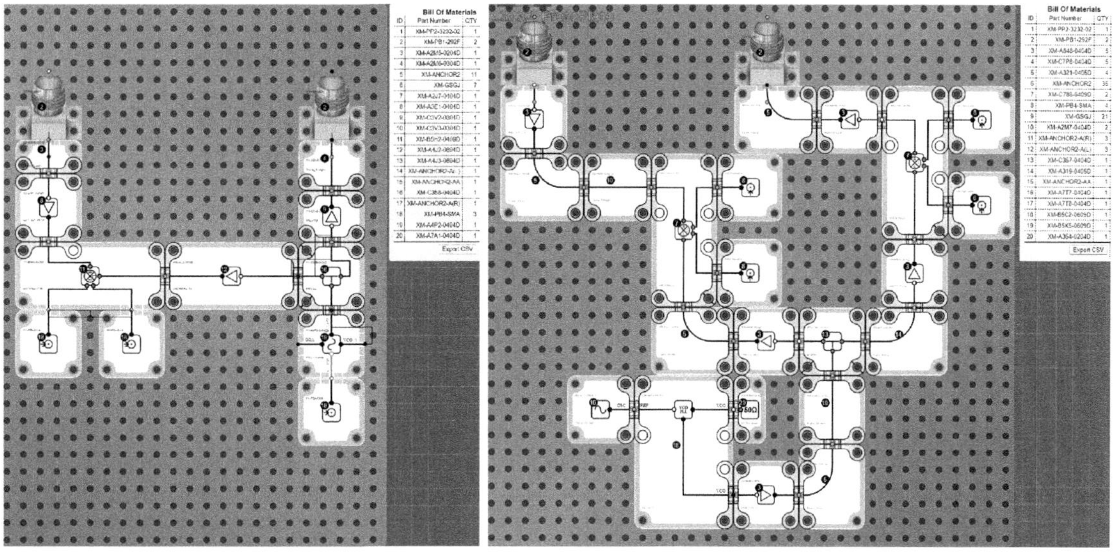

FIG. 1 On the left, system block diagram of a 24 GHz FMCW radar; on the right, system block diagram of a 24–28 GHz transceiver with IQ modulator and demodulator, see Toker [33].

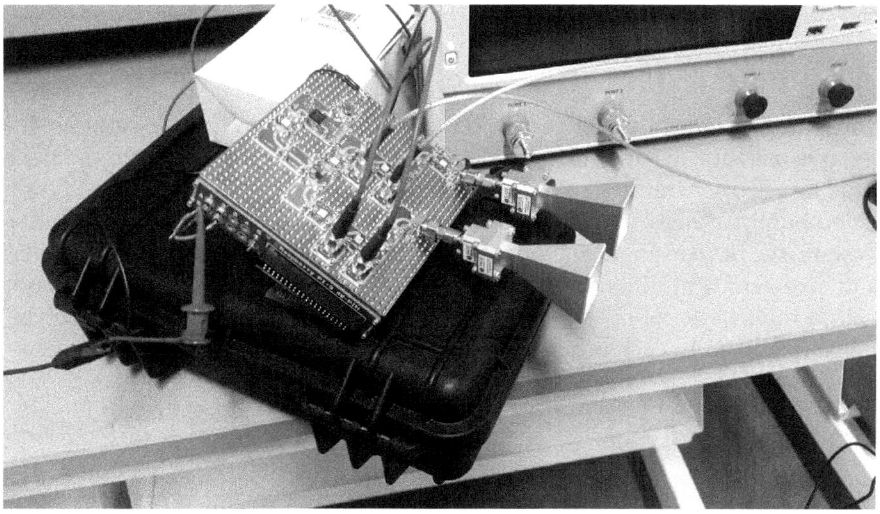

FIG. 2 Photo of 24–28 GHz transceiver built by the author using X-Microwave parts. This system has both IQ modulator and demodulator, and TX and RX horn antennas. *Red(dark gray in the printed version)*-colored SMA's are IQ inputs for the TX chain, and the other two SMA's are for IQ outputs of the RX chain. The baseband signal I/O is through an Ettus USRP B210 that is not visible in the photo.

static transmit and receive antenna system can scan only one direction, and generate only one dimensional range profile. In general, for imaging radars, the output is essentially a map-like data structure. The second type is called non-imaging radars, and typical examples include radar speed sensors, radar-based distance sensors, etc. Compared to lower-cost ultrasonic sensors used in industrial automation, radar sensors are more expensive but provide much better accuracy due to their shorter wavelengths.

Another important classification is based on whether the radar is pulsed or continuous wave radar. The World War 2 British radar system, Chain Home, was a pulsed radar. Each pulse had a frequency around 50 MHz with 6–25 μs duration, and pulses were repeated around 25 times per second. Pulsed radars basically emit high-power radio signals for a short period of time, wait for echos, and repeat this several times in each second. On the contrary, AWR1642, see Fig. 3, is a continuous wave radar system. It operates at 77 GHz and emits a chirp signal continuously, see Toker et al. [34]. To be more specific, each chirp signal is generated in a 160 μs time duration, chirps are repeated 128 times in a 40 ms time window followed by a blank period during which no radio signal is generated. Overall, this whole process is repeated 25 times per second. The AWR1642 has a 10 MHz internal analog-to-digital converter, and achieves a couple of cm level accuracy in range estimation. Pulsed radars do not offer such levels of accuracy, but because of

FIG. 3 Texas Instruments 77 GHz radar development kit AWR1642. It has two TX antennas (lower right) and four RX antennas (upper right), see Toker et al. [34].

their pulsed nature, their range is much higher compared to continuous wave radar systems. In a pulsed radar, the time needed for the echo is used for distance calculation. Whereas for continuous wave radars, a more complicated one- or two-dimensional fast Fourier transform (FFT) like digital signal processing approach is needed for range and velocity estimation. Another important classification is mono-static, bi-static, and multistatic radar types, see Fig. 4. In a mono-static radar system, we have a single antenna working both as a transmitter and receiver. In a bi-static radar system, transmitter and receiver are at different physical locations but they are linked to each other and operate as a single entity. Finally, in a multistatic radar, we have a single transmitter but multiple receiver stations. As in the bi-static case, all stations are linked to each other through a high-speed data link and operate as a single system.

4.4 Radar range equation

In this section, we will present the Radar Range Equation, and define the Radar Cross Section. Consider a transmit antenna operating with a transmitted signal level of P_T and antenna gain G_T. If the receiver antenna has gain G_R, the received signal power at the receive antenna will be

$$P_R = \frac{P_T G_T}{4\pi R^2} \frac{\sigma A_R}{4\pi R^2}$$

where σ is the radar cross section of the target, A_R is the effective area of the receiving antenna, and R is the distance between the radar antenna and the target. Note that the effective area of an antenna, A, and the antenna gain G, are related to each other by $G = (4\pi/\lambda^2)A$. In almost all cases, the actual received signal power at the receive antenna will be less than this theoretical value due to atmospheric absorption. It is important to note that, the received signal power decays with the fourth power of distance.

Radar range equation is a very fundamental relation in understanding the limitations of radar systems. The receiver noise is the other fundamental factor limiting the performance of radar systems. The receiver noise level is defined as $N = kT_s B_n$, where $k = 1.38 \times 10^{-23}$ J K, is the Boltzman constant, T_s is the system noise temperature, and B_n is the noise bandwidth of the receiver. The signal-to-noise ratio of the radar system, SNR, defined as $P_R N$, is a fundamental performance metric for radar systems. As the SNR gets lower and lower, detecting a target becomes more and more difficult and error-prone. For a radar system, another commonly used performance metric is the risk level, which is defined as the sum of the probability of a false alarm, i.e., falsely detecting a target when there is no target object, and the probability of miss, i.e., not detecting a target when there is a target object.

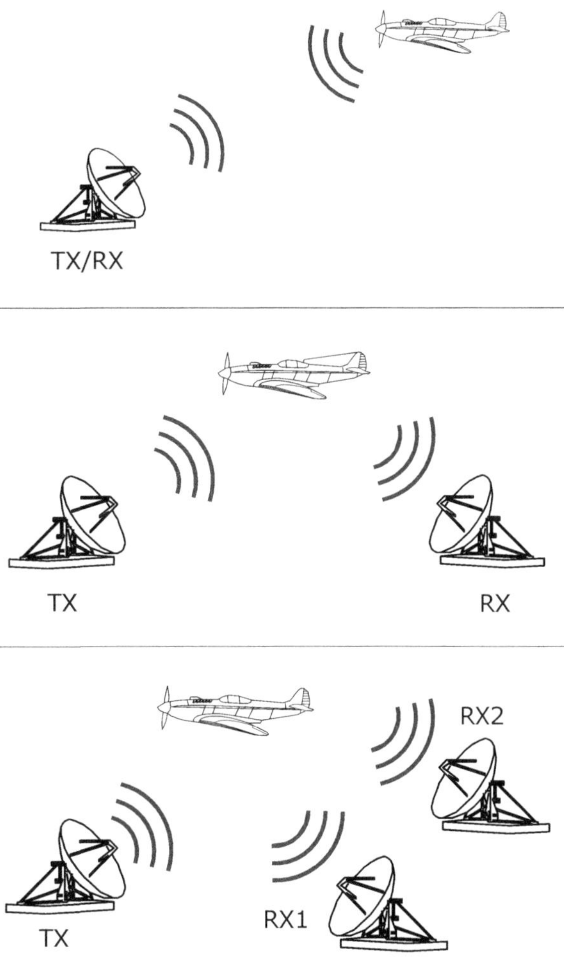

FIG. 4 In a mono-static radar (the figure on the top), the transmit antenna TX and the receive antenna RX are physically the same. In a bi-static radar (the figure in the middle), the transmit antenna TX and the receive antenna RX are physically at different locations, but linked through a high-speed data-link. In a multi-static radar (the figure at the bottom), there is a single transmit antenna TX and multiple receive antennas RX1, RX2, … that are physically at different locations but linked to each other through a high-speed data-link.

4.5 Pulsed radar

Pulsed radar works on a very simple principle, a high-power radio pulse is generated and then we wait for the reception of an echo. Since electromagnetic waves travel at the speed of light, $c \approx 3 \times 10^8\,\mathrm{m/s}$, the echo duration can be computed as

$$\tau_0 = \frac{2R}{c}$$

where R is the distance between the radar antenna and the reflector, see Fig. 5. But if we have multi-path propagation, then we may have multiple echos τ_1, τ_2, with $\tau_k = d_k/c$ where d_k is the total distance of the transmit path plus the receive path for the kth path.

In a pulsed radar, the range will be equal to $R = c\tau/2$. In other words, the range is approximately 150 m multiplied by the echo time in microseconds. By using a single and fixed antenna, it is not possible to estimate both the altitude and azimuth. However, using multiple antennas, both the altitude and azimuth of the reflecting object can be estimated by using trigonometric calculations.

In a pulsed radar system, there are two very critical system parameters called pulse repetition interval, τ_{PRI}, and pulse duration, τ_P. For a pulsed radar, the minimum target distance is

$$R_{min} = \frac{c\tau_P}{2}$$

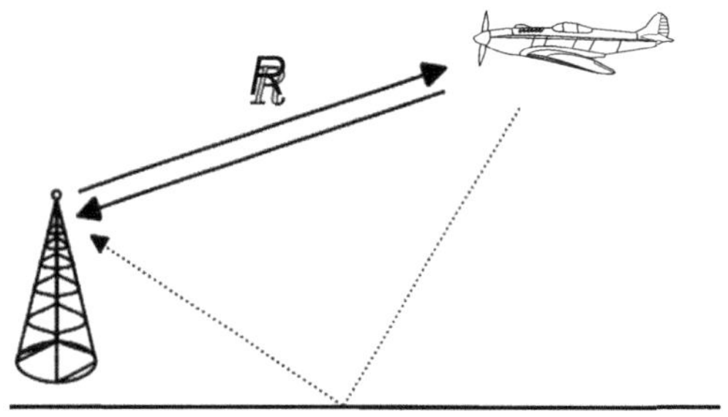

FIG. 5 A pulsed radar with multi-path propagation resulting multiple echos. The first receive path is shown in *solid line*, and the second receive path is shown in *dotted line*.

and the maximum target detection range is

$$R_{\max} = \frac{c(\tau_{\mathrm{PRI}} - \tau_{\mathrm{P}})}{2}$$

In a pulsed radar system, pulse repetition interval, τ_{PRI}, must be larger than the maximum possible echo duration; otherwise, there will be ambiguity in target distance.

4.6 Doppler radar

Doppler radar can be either a pulsed or continuous wave radar. For the pulsed Doppler radar, the transmitted signal will be

$$x_{\mathrm{T}}(t) = \cos(2\pi f_{\mathrm{o}} t)\,(u(t) - u(t - \tau_{\mathrm{P}}))$$

and the received signal will be

$$x_{\mathrm{R}}(t) = \cos(2\pi f_{\mathrm{o}}(t - \tau(t)))\,(u(t - \tau(t)) - u(t - \tau_{\mathrm{P}} - \tau(t)))$$

where $\tau(t) = 2R(t)/c$, and $R(t)$ is the target distance. If the phase of the received waveform is differentiated, we get the frequency of the received wave as

$$f_{\mathrm{R}} = f_{\mathrm{o}}\left(1 - \frac{2}{c}\right)\dot{R}(t)$$

Basically, by analyzing the frequency of the received waveform, it is possible to determine the velocity, v, of the object by using the equation

$$v = \frac{c}{2f_{\mathrm{o}}}(f_{\mathrm{o}} - f_{\mathrm{R}}) = \frac{\lambda}{2}(f_{\mathrm{o}} - f_{\mathrm{R}}),$$

where $f_{\mathrm{o}} - f_{\mathrm{R}}$ is the Doppler shift in frequency, and λ is the wavelength in meters. As in the pulsed radar, target distance can be estimated by using $R = c\tau(t)/2$.

There is a continuous wave version of the Doppler radar that measures only the target speed, see Fig. 6. In this version, a copy of the transmitted waveform is multiplied with the received waveform and then low-pass filtered, i.e., a mixer and a low-pass filter circuit is used. The transmitted and received waveforms are $x_{\mathrm{T}}(t) = \cos(2\pi f_{\mathrm{o}} t)$, $x_{\mathrm{R}}(t) = \cos(2\pi f_{\mathrm{o}}(t - \tau(t)))$, and the mixer output after low-pass filtering is $x_{\mathrm{MIX}}(t) = \cos(2\pi f_{\mathrm{o}} \tau(t))$. Therefore, the frequency of the

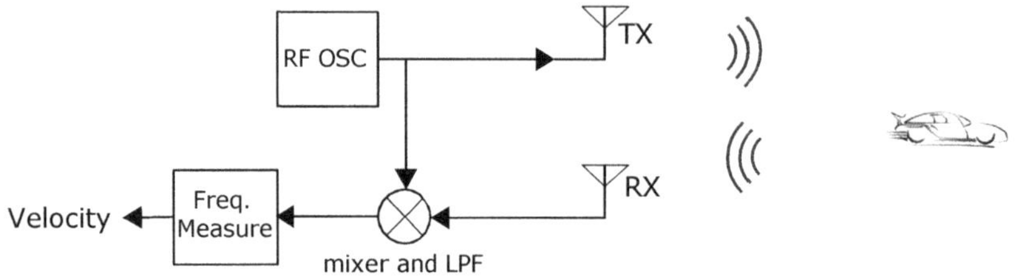

FIG. 6 Doppler radar block diagram.

output signal will be $f_{\text{MIX}}(t) = \frac{2f_0}{c}\dot{R}(t)$, and the velocity estimate will be

$$v = \frac{\lambda}{2} f_{\text{MIX}}$$

In other words, by measuring the mixer output frequency, the velocity of the target can be estimated. Note that the Doppler radar cannot determine the sign of the velocity, whereas FMCW radars can.

4.7 FMCW radar

Frequency Modulated Continuous Wave (FMCW) radar has a system architecture similar to the Doppler radar, but instead of a fixed RF oscillator, we have a voltage-controlled oscillator, VCO, see Fig. 7.

The VCO output is a sinusoidal signal of frequency $f(t) = f_0 + Bq(t)$ where f_0 is the VCO output frequency at the minimum input voltage, $q = 0$, and B is the VCO bandwidth. Such signals are called chirp signals, and $S = B/T_c$ is called the slope of the chirp signal. The VCO input signal, $q(t)$, is

$$q(t) = t/T_c, \quad 0 \leq t < T_c$$

where T_c is the chirp duration. In an FMCW radar system, the transmitted signal (The chirp signal) is, $x_T(t)$,

$$x_T(t) = A_T \cos\left(2\pi \int^t f(\tau)d\tau + \theta_T\right),$$

where A_T is the amplitude, and θ_T is the phase. The received signal (The echo) from a target at a distance (t) will then be:

$$x_R(t) = A_R \cos\left(2\pi \int^{t-t_r} f(\tau)d\tau + \theta_R\right)$$

where $t_r = 2R(t)/c$ is the round-trip time and c is the speed of light. In an FMCW radar, the transmitted and received signals are multiplied and low-pass filtered. The final output is called the beat signal, $s(t)$,

$$s(t) \approx A \cos\left(2\pi f(t)\, t_r + \theta\right)$$

where A is the amplitude, and θ is the phase. For a positive sloped chirp, the beat signal frequency will be

$$f_b = \frac{2SR}{c} + \frac{2f_0 v}{c},$$

where R is the target distance, and v is the velocity of the target.

In an FMCW radar, the beat signal is sampled at frequency $f_s = 1/T_s$, and then converted into digital format. For each chirp duration of T_c seconds, we collect $N = T_c f_s$ samples. We repeat this M times, and generate an $M \times N$ matrix, see Toker [33]; Li et al. [35]. Computing the 2D FFT of this matrix gives us a heat-map-like output for range and velocity estimation, see Fig. 8.

The range resolution of an FMCW radar is

$$\delta R = \frac{c}{2B} \frac{c}{2S}$$

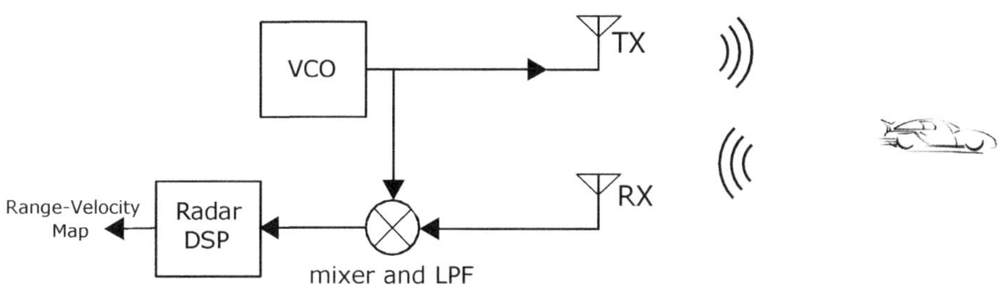

FIG. 7 Frequency Modulated Continuous Wave (FMCW) radar block diagram.

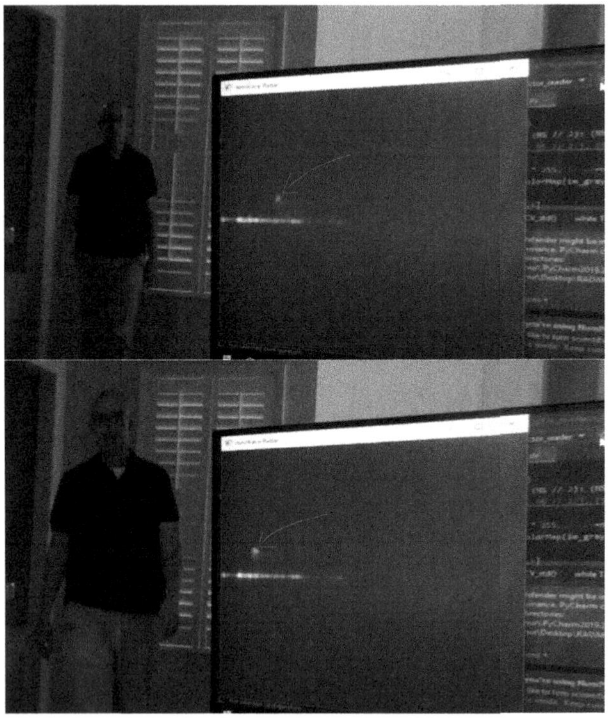

FIG. 8 Radar signal processing using the Texas Instruments 77GHz radar AWR1642, see the YouTube link https://youtu.be/jNuBgl6KMCs. The horizontal axis is the range, the vertical axis is the velocity, and the origin is on the left at the center. The clutter at the center horizontal line is due to reflections from the background. The *dot* marked with the *red(gray in the printed version) arrow* is the detected moving object. Note that, as the object gets closer to the radar, the *dot* moves toward the left.

Basically, for better range resolution we need higher VCO bandwidth. The maximum possible range is

$$R_{\max} = \frac{f_s c}{2S}$$

Therefore, increasing the sampling frequency will increase the maximum unambiguous range. If we have M chirp signals with a time separation of T_c, the resolution in velocity will be

$$\delta v = \frac{\lambda}{2MT_c}$$

and the maximum unambiguous velocity value will be

$$v_{\max} = \frac{\lambda}{4T_c}$$

In FMCW radar systems, a frame is defined as M chirps combined together. The $M \times N$ matrix is called the radar "cube." If we have multiple receive antennas, we form an $M \times N\ r$ tensor where r is the number of receive antennas. A 3D FFT analysis can be used to estimate range, velocity, and direction, see Li et al. [35].

4.8 OFDM radar

The OFDM (Ortogonal Frequency Division Multiplexing) radar is basically using a communication system for radar purposes. Consider a transmitter using the OFDM modulation to send digital information to a nearby receiver. The transmitter can listen for the echos generated by the nearby targets due to its own transmitted waveform and estimate the channel. This is the main idea behind the OFDM radar, see Braun [36]; Ozkaptan et al. [37] and references therein.

Consider the channel from the transmit antenna TX to the receive antenna RX,

$$H(s) = \sum_k a_k e^{-h_k s} L_k(s),$$

where each term represents a reflector on a path from TX to RX, a_k's are complex numbers, and $L_k(s)$ terms represent the frequency-dependent attenuation. In an OFDM radar, we simply estimate the impulse response of the baseband transfer function $H_b(s)$, or its discretized version. Impulse response estimation is done by first estimating the frequency response and then using an inversion formula.

If the carrier frequency is quite high, one can assume that $L_q(s)$ terms are almost constant in $[f_c - B, f_c + B]$ where f_c is the carrier frequency and B is the bandwidth. In this case, the baseband channel transfer function will be

$$H_b(s) = \sum_k e^{-j2\pi f_c h_k} \sqrt{2} a_k e^{-h_k s}$$

see Toker and Ozdemir [38]. By computing the impulse response of the baseband channel, or its discretized version, it is possible to estimate the location of all reflector objects, see Braun [36]; Ozkaptan et al. [37]; Toker and Ozdemir [38]. Basically, each reflector will appear as a peak in the impulse response and the location of the peak will depend on the target distance. By using a more detailed channel analysis it is also possible to estimate both the range and velocity of moving targets, see Sit et al. [39]; Wang et al. [40].In summary, the main advantage of an OFDM radar is the use of an already existing communication system for range and velocity estimation. Since most connected vehicles are expected to have high frequency and high bandwidth transceivers for vehicle-to-vehicle communication, this technique allows implementation of radar sensors simply by using the existing RF hardware.

4.9 Other types of radars

In this section, we will review SARs, GPRs, UWB See-Through-Wall Radars, and Software Defined Radars.

4.9.1 Synthetic aperture radar

SAR is a radar imaging technique that uses a moving radar system to create high-resolution images that are not achievable by conventional radar systems. Unlike traditional radars that use a single static TX/RX antenna configuration, SAR utilizes the motion of the radar antenna and complex signal processing techniques to emulate the effect of a larger antenna commonly known as the synthetic aperture. The moving antenna emits radar signals toward the environment to be imaged and receives the echos. As the antenna moves, the SAR system collects echos from different observation points. The collected radar data are then processed to generate a SAR image. The data processing involves compensating for the platform's motion, noise suppression, and image processing to enhance the final radar image quality. In a SAR system, the unknowns are the locations of radar reflectors and their reflectivity coefficients. By collecting several measurement results, it is possible to solve an inversion problem and estimate the locations and reflectivities of all targets. This extended aperture approach allows SAR systems to achieve a high resolution that is not limited by the physical size of the actual antenna. SARs can also provide three-dimensional information about the environment. By acquiring data from multiple viewing angles, a SAR system can generate even digital elevation models for topographic mapping.

For further details of SARs and processing algorithms, see Inggs [41]; Chen et al. [42]; Moreira et al. [43] and references therein.

4.9.2 Ground penetrating radar

GPR is a technique that uses radar pulses to generate map of underground objects and their properties. In GPR, a transmitter emits short pulses of electromagnetic waves into the ground. A GPR system uses typically the microwave frequency range and scans the underground by using a variety of different frequencies and waveforms. Different wavelengths have different absorption properties and hence offer varying penetration and resolution capabilities. GPR-generated radio waves travel into the subsurface and interact with different materials, such as soil, rocks, concrete, or buried objects. When the radar waves encounter changes in the dielectric properties of the subsurface objects, part of the energy is reflected back to the surface.

By analyzing the time it takes for the radar pulses to travel to the target and return, as well as the strength and shape of the reflected signals, GPR can provide information about the depth, location, and properties of subsurface objects. The data collected by GPR are typically represented as a two-dimensional or three-dimensional image, known as the GPR profile. GPR has various applications in different fields including Archaeology, Construction, Geology, and Military. GPR is a nondestructive imaging method and is highly advantageous compared to more expensive excavation or drilling options.

For further details of GPRs, see Catapano et al. [44]; Kikuta and Tanaka [45]; Wicks et al. [46]; Comite et al. [47]; Sun et al. [48] and references therein.

4.9.3 UWB see-through-wall radars

Ultra-Wideband (UWB) See-Through-Wall Radar is a radar system that utilizes ultra-wideband signals to penetrate walls, allowing detection and imaging capabilities that are not achievable by conventional radar systems. A UWB radar system emits an extremely short-duration pulse of electromagnetic wave with a wide bandwidth. Then the UWB radar system analyzes the time delay, amplitude, and phase of the reflected echos and generates a detailed radar image using complex signal processing algorithms. UWB See-Through-Wall Radar has applications including military, law enforcement, security, and surveillance. The effectiveness of UWB See-Through-Wall Radar can be influenced by various factors, including the type and thickness of the wall material, the presence of clutter or other objects, and environmental conditions. For further details on UWB See-Through-Wall Radars, see Yang and Fathy [49]; Saad et al. [50]; Wang et al. [51]; Yang and Fathy [52]; Liu et al. [53]; Mahfouz et al. [54] and references therein.

4.9.4 Software-defined radar

Software-defined radar is a radar system that uses a software-defined radio to generate radar images. A typical software-defined radar excites the channel at different frequencies and generates a channel frequency response curve over a wide-bandwidth, Toker and Ozdemir [55]. By using this, baseband channel frequency response and impulse response data can be generated. Similar to the OFDM radar, impulse response will have peaks for each reflector and the peak location will depend on the reflector distance, see Prager et al. [56]; Alawsh et al. [57]; Friedt and Feng [58,59]. Compared to other types of radars, scanning of the whole bandwidth takes longer because for each frequency used in the scanning process, the local oscillator must be re-adjusted and a small time delay is needed for the oscillator to stabilize. However, a software-defined radar system can achieve very high bandwidth scans that are typically not possible with voltage-controlled oscillator-based systems. This wide bandwidth allows significantly improved distance resolution, and the ability to differentiate two nearby reflectors.

4.10 Radar jamming and cyberattacks

The history of radar jamming and cyberattacks goes back to World War 2 years, see Blanchard and van Genderen [31]; Brown [32] and references therein. Radar jamming and cyberattacks are a vast topic with very rich literature. A comprehensive discussion of these ideas and tools is beyond the scope of this work. In this section, we will simply focus on physical layer cyberattacks for automotive radars.

A standard frequency-modulated continuous wave radar has no protection against physical-layer cyberattacks. The report Bhat [60] and Alland et al. [61] has a detailed analysis of different possible attacks for automotive radars. The author introduced a physical-layer attack resilient automotive radar in Toker et al. [62] and used simulation tools to demonstrate its resilience. In Toker and Alsweiss [63], an improved attack-resilient design was presented, and its effectiveness was demonstrated using real radar data. In Toker [64], author extended these attack resilient designs further and established a theoretical proof. One important observation made in Toker [64] is the following: Either the attack signal is too small and will not be detected, but it will have a negligible impact on the radar range-velocity heatmap; or with a very high probability, the attack will be detected, see Toker [64] for full details. For an attack-resilient design, it is also important to bind the false alarm rate, and the design introduced in Toker [64] has this desirable feature as well. The author also introduced a different attack-resilient method for OFDM radars, see Toker and Ozdemir [38], and presented both experimental data and mathematical analysis to demonstrate its effectiveness. All experimental data are obtained in Toker and Alsweiss [63] and Toker [64] are based on the real-time radar data recording tools developed in Toker and Kuhn [65]; Toker et al. [34].

A good of number attack resilient systems use an estimator, and generate an alarm if the difference between the actual and estimated values are larger than a selected threshold value. The rationale behind this approach is that a large difference between the actual and estimated values is an indication of an unusual thing. For such systems, if we choose a small threshold, several false alarms will be generated. A large threshold value will result a high probability of miss. Depending on how this estimator error is measured and thresholded, it is possible to develop a variety of cyberattack detectors, see Yang et al. [66]; Kordestani et al. [67]; Ye and Farley [68]; Sargolzaei et al. [69–71]; Abbaspour et al. [72]. In the literature, both analytical detectors using probabilistic models and artificial intelligence (AI)-based detectors are widely studied. It is almost impossible to eliminate false alarms, because natural system noise can sometimes cause unusually large differences between actual and estimated values. These estimator-type detectors are commonly used for communication layer attacks, and as long as a good estimator model is developed, they can be pretty effective in detecting cyberattacks.

Physical layer attacks are different from communication later attacks, see Dutta et al. [73]; Kapoor et al. [74] for physical-layer attacks on automotive radar sensors. In the first paper, Dutta et al. [73], an estimator-based approach is used, and in the second one, Kapoor et al. [74], a completely different method called spatiotemporal challenge response (STCR) approach is introduced. The good thing about the STCR approach is that it does not need an accurate estimator model. In Shoukry et al. [75]; Dutta et al. [73], a related method called physical challenge response authentication (PyCRA) is developed for cyberattack resilience sensor design. Both STCR and PyCRA do not need an accurate estimator for cyberattack resilient operation. In a typical cyberphysical system, both communication layer and physical layer attacks are possible; therefore, both estimator based techniques are needed for resilience against communication layer attacks, and PyCRA-type methods are needed for physical-layer attack resilience. See also Petit and Shladover [76]; Oruganti et al. [77]; Petit and Murray [78] and references therein for a discussion of different attack-resilient designs.

4.11 Conclusion

This chapter presented a review of existing radar architectures, radar jamming, and cybersecurity problems with radars from an AV/ADAS perspective. AV/ADAS systems use cameras, lidars, and radars, and they all have their own advantages. Radars sensors are in general lower cost and computationally less demanding compared to other alternatives, and they are quite effective in moving object. Early automotive radars were mostly 24 GHz, but they are all being phased out by the emergence of millimeter-wave 77 GHz systems that provide much better resolution. The chapter presented real photos and architectural block diagrams of Texas Instruments 77 GHz millimeter-wave radars, and X-Microwave 24–28 GHz RF transceivers used by the author. The millimeter-wave radar system used by the author is an FMCW radar and works with an in-house developed real-time Ethernet data capture system. On the other hand, the 24–28 GHz transceiver is driven by an Ettus USRP B210 for I/Q input and output and is capable of working as an OFDM radar or as a software-defined radar.

Acknowledgment

Partial support of this research was provided by the National Science Foundation under Grant No. CNS-1919855, and by the Florida Polytechnic University. Any opinions, findings, conclusions, or recommendations expressed in this material are those of the author(s) and do not necessarily reflect the views of the sponsoring agency.

References

[1] J. Maxwell, On physical lines of force, Philos. Mag. 90 (2010) 11–23, https://doi.org/10.1080/14786431003659180.
[2] T.K. Sarkar, M. Salazar-Palma, D.L. Sengupta, James Clerk Maxwell: The Founder of Electrical Engineering, in: 2010 Second Region 8 IEEE Conference on the History of Communications, 2010, pp. 1–7, https://doi.org/10.1109/HISTELCON.2010.5735323.
[3] M. Mitolo, R. Araneo, A brief history of Maxwell's equations [History], IEEE Ind. Appl. Mag. 25 (2019) 8–13, https://doi.org/10.1109/MIAS.2019.2898096.
[4] P. Nahin, Maxwell's grand unification, IEEE Spectr. 29 (1992) 45, https://doi.org/10.1109/6.123329.
[5] J.C. Rautio, The long road to Maxwell's equations, IEEE Spectr. 51 (2014) 36–56, https://doi.org/10.1109/MSPEC.2014.6964925.
[6] C. Susskind, Heinrich hertz: a short life, IEEE Trans. Microwave Theory Tech. 36 (1988) 802–805, https://doi.org/10.1109/22.3599.
[7] J. Kraus, Heinrich Hertz-theorist and experimenter, IEEE Trans. Microwave Theory Tech. 36 (1988) 824–829, https://doi.org/10.1109/22.3601.
[8] T. Simpson, On hertz and the speed of light, in: 2017 IEEE International Symposium on Antennas and Propagation & USNC/URSI National Radio Science Meeting, 2017, pp. 1195–1196, https://doi.org/10.1109/APUSNCURSINRSM.2017.8072640.
[9] J. Blanchard, Hertz, the discoverer of electric waves, Proc. Inst. Radio Eng. 26 (1938) 505–515, https://doi.org/10.1109/JRPROC.1938.228325.
[10] W. Mathis, Emeritus, Heinrich Hertz [Pioneers in CAS], IEEE Circuits Syst. Mag. 23 (2023), https://doi.org/10.1109/MCAS.2023.3234728. 79–78.
[11] R. Elliott, The history of electromagnetics as hertz would have known it, IEEE Trans. Microwave Theory Tech. 36 (1988) 806–823, https://doi.org/10.1109/22.3600.
[12] J.M. Dilhac, Edouard Branly, the Coherer, and the Branly effect [History of Communications], IEEE Commun. Mag. 47 (2009) 20–26, https://doi.org/10.1109/MCOM.2009.5277448.
[13] G. Gandhi, V. Aggarwal, L. Chua, Coherer is the elusive memristor, in: 2014 IEEE International Symposium on Circuits and Systems (ISCAS), 2014, pp. 2245–2248, https://doi.org/10.1109/ISCAS.2014.6865617.
[14] J. Bose, On a self-recovering coherer and the study of the cohering action of different metals, Proc. IEEE 86 (1998) 244–247, https://doi.org/10.1109/JPROC.1998.658776.
[15] F. Marconi, Marconi and the invention of wireless communications, in: Proceedings of 1995 SBMO/IEEE MTT-S International Microwave and Optoelectronics Conference, 1995, p. 5, https://doi.org/10.1109/SBMOMO.1995.509589.
[16] G. Falciasecca, B. Valotti, Guglielmo Marconi: the pioneer of wireless communications, in: 2009 European Microwave Conference (EuMC), 2009, pp. 544–546, https://doi.org/10.23919/EUMC.2009.5296358.
[17] W.J.G. Beynon, Marconi, radio waves, and the ionosphere, Radio Sci. 10 (1975) 657–664, https://doi.org/10.1029/RS010i007p00657.
[18] G. Corazza, Marconi's history [radiocommunication], Proc. IEEE 86 (1998) 1307–1311, https://doi.org/10.1109/5.681364.

[19] J. Brittain, Electrical engineering Hall of Fame: Guglielmo Marconi, Proc. IEEE 92 (2004) 1501–1504, https://doi.org/10.1109/JPROC.2004.832946.
[20] T.K. Sarkar, R. Mailloux, A.A. Oliner, M. Salazar-Palma, D.L. Sengupta, Sir John Ambrose Fleming His Involvement in the Development of Wireless, 2006, pp. 311–326, https://doi.org/10.1002/0471783021.ch10.
[21] S. Hong, Transforming an Effect into an Artifact: The Thermionic Valve, MIT Press, 2001, pp. 119–154. https://doi.org/10.7551/mitpress/7255.003.0007.
[22] J.E. Brittain, Electrical engineering Hall of Fame: John A. Fleming, Proc. IEEE 95 (2007) 313–315, https://doi.org/10.1109/JPROC.2006.887329.
[23] P. Delogne, Lee De Forest, the inventor of electronics: a tribute to be paid, Proc. IEEE 86 (1998) 1878–1880, https://doi.org/10.1109/JPROC.1998.705533.
[24] J. Brittain, Electrical engineering Hall of Fame: Lee de Forest, Proc. IEEE 93 (2005) 198–202, https://doi.org/10.1109/JPROC.2004.839598.
[25] V.M. Pestrikov, Cathode switch—the first amplifier tube, in: 2012 22nd International Crimean Conference "Microwave & Telecommunication Technology", 2012, pp. 32–34.
[26] H. Griffiths, P. Knott, W. Koch, Christian Hülsmeyer: invention and demonstration of radar, 1904, IEEE Aerosp. Electron. Syst. Mag. 34 (2019) 56–60, https://doi.org/10.1109/MAES.2019.2934814.
[27] W. Brown, The microwave magnetron and its derivatives, IEEE Trans. Electron Devices 31 (1984) 1595–1605, https://doi.org/10.1109/T-ED.1984.21757.
[28] IEEE, IEEE Standard Letter Designations for Radar-Frequency Bands, IEEE Std 521-1984, 1984, pp. 1–8, https://doi.org/10.1109/IEEESTD.1984.81588.
[29] E. Altshuler, R. Marr, A comparison of experimental and theoretical values of atmospheric absorption at the longer millimeter wavelengths, IEEE Trans. Antennas Propag. 36 (1988) 1471–1480, https://doi.org/10.1109/8.8635.
[30] K.S. Paulson, The effect of variation in atmospheric absorption on millimetric, terrestrial, telecommunications links, Radio Sci. 40 (2005) 1–9, https://doi.org/10.1029/2004RS003218.
[31] Y. Blanchard, P. van Genderen, A German radar chain, facing the British Chain Home during WW II, in: 2014 11th European Radar Conference, 2014, pp. 29–32, https://doi.org/10.1109/EuRAD.2014.6991199.
[32] L. Brown, Airborne radar: its beginnings in World War II, IEEE Aerosp. Electron. Syst. Mag. 19 (2004) 3–8, https://doi.org/10.1109/MAES.2004.1346832.
[33] O. Toker, An FPGA based 24 GHz radar testbed for physical-layer cyberattack research, in: 2021 IEEE Radar Conference (RadarConf21), 2021, pp. 1–4, https://doi.org/10.1109/RadarConf2147009.2021.9455199.
[34] O. Toker, S. Alsweiss, M. Abid, A computer vision based testbed for 77 GHz mm wave radar sensors, in: 2020 SoutheastCon, 2020, pp. 1–7, https://doi.org/10.1109/SoutheastCon44009.2020.9249680.
[35] X. Li, X. Wang, Q. Yang, S. Fu, Signal processing for TDM MIMO FMCW millimeter-wave radar sensors, IEEE Access 9 (2021) 167959–167971, https://doi.org/10.1109/ACCESS.2021.3137387.
[36] M. Braun, OFDM Radar Algorithms in Mobile Communication Networks, Ph.D. thesis, Karlsruher Instituts für Technologie, 2014.
[37] C.D. Ozkaptan, E. Ekici, O. Altintas, C. Wang, OFDM pilot-based radar for joint vehicular communication and radar systems, in: 2018 IEEE Vehicular Networking Conference (VNC), 2018, pp. 1–8.
[38] O. Toker, O. Ozdemir, Physical-layer cyberattack resilient OFDM automotive radars, in: 2020 IEEE Vehicular Networking Conference (VNC), 2020, pp. 1–4, https://doi.org/10.1109/VNC51378.2020.9318366.
[39] Y.L. Sit, T.T. Nguyen, C. Sturm, T. Zwick, 2D radar imaging with velocity estimation using a MIMO OFDM-based radar for automotive applications, in: 2013 European Radar Conference, 2013, pp. 145–148.
[40] X. Wang, Z. Zhang, H.E. Najafabadi, Range and velocity estimation for multi-symbol OFDM-based integrated radar and communications systems, in: 2021 CIE International Conference on Radar (Radar), 2021, pp. 2457–2460, https://doi.org/10.1109/Radar53847.2021.10028123.
[41] M. Inggs, Synthetic aperture radar during the 50 years of the aerospace and electronic systems society, IEEE Aerosp. Electron. Syst. Mag. 38 (2023) 22–31, https://doi.org/10.1109/MAES.2022.3216263.
[42] J. Chen, M. Xing, H. Yu, B. Liang, J. Peng, G.C. Sun, Motion compensation/autofocus in airborne synthetic aperture radar: a review, IEEE Geosci. Remote Sens. Mag. 10 (2022) 185–206, https://doi.org/10.1109/MGRS.2021.3113982.
[43] A. Moreira, P. Prats-Iraola, M. Younis, G. Krieger, I. Hajnsek, K.P. Papathanassiou, A tutorial on synthetic aperture radar, IEEE Geosci. Remote Sens. Mag. 1 (2013) 6–43, https://doi.org/10.1109/MGRS.2013.2248301.
[44] I. Catapano, G. Gennarelli, G. Ludeno, C. Noviello, G. Esposito, F. Soldovieri, Contactless ground penetrating radar imaging: state of the art, challenges, and microwave tomography-based data processing, IEEE Geosci. Remote Sens. Mag. 10 (2022) 251–273, https://doi.org/10.1109/MGRS.2021.3082170.
[45] T. Kikuta, H. Tanaka, Ground probing radar system, IEEE Aerosp. Electron. Syst. Mag. 5 (1990) 23–26, https://doi.org/10.1109/62.54640.
[46] M.C. Wicks, J.D. Norgard, T.N. Cushman, Adaptive tomographic sensors for below ground imaging, IEEE Aerosp. Electron. Syst. Mag. 25 (2010) 24–28, https://doi.org/10.1109/MAES.2010.5486538.
[47] D. Comite, F. Ahmad, M.G. Amin, T. Dogaru, Forward-looking ground-penetrating radar: subsurface target imaging and detection: a review, IEEE Geosci. Remote Sens. Mag. 9 (2021) 173–190, https://doi.org/10.1109/MGRS.2020.3048368.
[48] M. Sun, J. Pan, C. Le Bastard, Y. Wang, J. Li, Advanced signal processing methods for ground-penetrating radar: applications to civil engineering, IEEE Signal Process. Mag. 36 (2019) 74–84, https://doi.org/10.1109/MSP.2019.2900454.
[49] Y. Yang, A. Fathy, Design and implementation of a low-cost real-time ultra-wide band see-through-wall imaging radar system, in: 2007 IEEE/MTT-S International Microwave Symposium, 2007, pp. 1467–1470, https://doi.org/10.1109/MWSYM.2007.380529.
[50] M. Saad, A. Maali, M.S. Azzaz, I. Kakouche, An experimental platform of impulse UWB radar for through-wall imaging based on FPGA, in: 2020 1st International Conference on Communications, Control Systems and Signal Processing (CCSSP), 2020, pp. 198–201, https://doi.org/10.1109/CCSSP49278.2020.9151818.
[51] Y. Wang, Y. Yang, A.E. Fathy, Reconfigurable ultra-wide band see-through-wall imaging radar system, in: 2009 IEEE Antennas and Propagation Society International Symposium, 2009, pp. 1–4, https://doi.org/10.1109/APS.2009.5171851.

[52] Y. Yang, A. Fathy, See-through-wall imaging using ultra wideband short-pulse radar system, in: 2005 IEEE Antennas and Propagation Society International Symposium, vol. 3B, 2005, pp. 334–337, https://doi.org/10.1109/APS.2005.1552508.

[53] M. Liu, J. Xiao, P. Luo, Z. Zhu, Y. Yang, Ultrawideband power-switchable transmitter with 17.7-dBm output power for see-through-wall radar, IEEE Trans. Very Large Scale Integr. VLSI Syst. 28 (2020) 1331–1335, https://doi.org/10.1109/TVLSI.2020.2972687.

[54] M. Mahfouz, A. Fathy, Y. Yang, E. Ali, A. Badawi, See-through-wall imaging using ultra wideband pulse systems, in: 34th Applied Imagery and Pattern Recognition Workshop (AIPR'05), 2005, pp. 6–53, https://doi.org/10.1109/AIPR.2005.40.

[55] O. Toker, O. Ozdemir, A synthetic wide-bandwidth radar system using software defined radios, Eng. Proc. 2 (2020), https://doi.org/10.3390/ecsa-7-08174. https://www.mdpi.com/2673-4591/2/1/9.

[56] S. Prager, T. Thrivikraman, M.S. Haynes, J. Stang, D. Hawkins, M. Moghaddam, Ultrawideband synthesis for high-range-resolution software-defined radar, IEEE Trans. Instrum. Meas. 69 (2020) 3789–3803.

[57] S.A. Alawsh, O.O.A. Basheer, A.O. Sirag, A.H. Muqaibel, Range and angle measurements based on software defined radio platform, in: 2019 IEEE 10th GCC Conference Exhibition (GCC), 2019, pp. 1–5.

[58] J.M. Friedt, W. Feng, Noise RADAR implementation using software defined radio hardware, GNU/Linux Magazine France, 2020. http://jmfriedt.free.fr/sdra_radar.pdf.

[59] J.M. Friedt, W. Feng, Software defined radio based synthetic aperture noise and OFDM (WiFi) RADAR mapping, in: GNU Radio Conference, 2020. http://jmfriedt.free.fr/grcon2020_radar.pdf.

[60] C. Bhat, Cybersecurity Challenges and Pathways in the Context of Connected Vehicle Systems, Technical Report 134, Data-Supported Transportation Operations & Planning Center (D-STOP), Austin, TX, 2018. https://ctr.utexas.edu/wp-content/uploads/134.pdf.

[61] S. Alland, W. Stark, M. Ali, M. Hegde, Interference in automotive radar systems: characteristics, mitigation techniques, and current and future research, IEEE Signal Proc. Mag. 36 (2019) 45–59.

[62] O. Toker, S. Alsweiss, J. Vargas, R. Razdan, Design of an automotive radar sensor firmware resilient to cyberattacks, in: Proceedings of the 2020 IEEE SoutheastCon, Raleigh, NC, 2020.

[63] O. Toker, S. Alsweiss, Design of a cyberattack resilient 77 GHz automotive radar sensor, Electronics 9 (2020) 573, https://doi.org/10.3390/electronics9040573.

[64] O. Toker, Physical-layer cyberattack and interference resilient automotive radars, IEEE Access 8 (2020) 215531–215543, https://doi.org/10.1109/ACCESS.2020.3041621.

[65] O. Toker, B. Kuhn, A python based testbed for real-time testing and visualization using TI's 77 GHz automotive radars, in: Proceedings of the 2019 IEEE Vehicular Networking Conference, Los Angeles, CA, 2019.

[66] C. Yang, L. Feng, H. Zhang, S. He, Z. Shi, A novel data fusion algorithm to combat false data injection attacks in networked radar systems, IEEE Trans. Signal Inf. Process. Netw. 4 (2018) 125–136, https://doi.org/10.1109/TSIPN.2018.2790361.

[67] M. Kordestani, A. Chibakhsh, M. Saif, A control oriented cyber-secure strategy based on multiple sensor fusion, in: Proceedings of the 2019 IEEE International Conference on Systems, Man and Cybernetics (SMC), Bari, Italy, 2019, https://doi.org/10.1109/SMC.2019.8914241.

[68] N. Ye, T. Farley, Resilient tracking control of networked control systems under cyber attacks, IEEE Trans. Cybern. (2019) 1–13, https://doi.org/10.1109/TCYB.2019.2948427.

[69] A. Sargolzaei, K. Yazdani, A. Abbaspour, C.D. Crane III, W.E. Dixon, Detection and mitigation of false data injection attacks in networked control systems, IEEE Trans. Industr. Inform. 16 (2020) 4281–4292.

[70] A. Sargolzaei, K.K. Yen, M.N. Abdelghani, Preventing time-delay switch attack on load frequency control in distributed power systems, IEEE Trans. Smart Grid 7 (2016) 1176–1185.

[71] A. Sargolzaei, K.K. Yen, M.N. Abdelghani, S. Sargolzaei, B. Carbunar, Resilient design of networked control systems under time delay switch attacks, application in smart grid, IEEE Access 5 (2017) 15901–15912.

[72] A. Abbaspour, A. Sargolzaei, P. Forouzannezhad, K.K. Yen, A.I. Sarwat, Resilient control design for load frequency control system under false data injection attacks, IEEE Trans. Ind. Electron. 67 (2020) 7951–7962.

[73] R.G. Dutta, X. Guo, T. Zhang, K. Kwiat, C. Kamhoua, L. Njilla, Y. Jin, Estimation of safe sensor measurements of autonomous system under attack, in: Proceedings of the 54th ACM/EDAC/IEEE Design Automation Conference, Austin, TX, 2017, https://doi.org/10.1145/3061639.3062241.

[74] P. Kapoor, A. Vora, K.D. Kang, Detecting and mitigating spoofing attack against an automotive radar, in: Proceedings of the 2018 IEEE Vehicular Technology Conference, Chicago, IL, 2018, https://doi.org/10.1109/VTCFall.2018.8690734.

[75] Y. Shoukry, P. Martin, Y. Yona, S. Diggavi, M. Srivastava, Attack resilience and recovery using physical challenge response authentication for active sensors under integrity attacks, arXiv:1605.02062v2 (2016). https://arxiv.org/pdf/1605.02062.

[76] J. Petit, S.E. Shladover, Potential cyberattacks on automated vehicles, IEEE Trans. Intell. Transp. Syst. 16 (2015) 546–556, https://doi.org/10.1109/TITS.2014.2342271.

[77] P.S. Oruganti, M. Appel, Q. Ahmed, Hardware-in-loop based automotive embedded systems cybersecurity evaluation testbed, in: Proceedings of the ACM Workshop on Automotive Cybersecurity, 2019, pp. 41–44.

[78] J. Petit, V. Murray, Human, machine, sensor, infrastructure: all together against cyberattacks in AV, in: G. Meyer, S. Beiker (Eds.), Road Vehicle Automation 7, Springer International Publishing, Cham, 2020, pp. 120–131.

[79] S.M. Patole, M. Torlak, D. Wang, M. Ali, Automotive radars: a review of signal processing techniques, IEEE Signal Proc. Mag. 34 (2017) 22–35.

CHAPTER

5

Artificial intelligence for autonomous vehicles: Comprehensive outlook

Mohammad Shaterabadi[a], *Mehdi Ahmadi Jirdehi*[b], *Hasan Mehrjerdi*[c], *and Houshang Karimi*[a]

[a]Department of Electrical Engineering and Computer Science, York University, Toronto, ON, Canada
[b]Department of Electrical Engineering, Kermanshah University of Technology, Kermanshah, Iran
[c]Electrical and Computer Engineering Department, The George Washington University, Washington, DC, United States

5.1 Introduction to the understudy subject

In today's rapidly evolving technological landscape, one of the most fascinating and promising developments is the integration of artificial intelligence (AI) into autonomous vehicles (AVs). The intersection of AI and transportation has the potential to revolutionize the way we commute, impacting everything from safety to convenience. Imagine sitting in a car without worrying about steering, braking, or accelerating—the promise of AVs powered by AI. These vehicles are not just about convenience; they hold the potential to reshape transportation systems, making them safer, more efficient, and environment-friendly. As technology advances, the concept of AVs is gaining momentum. These self-driving cars can sense their environment and navigate without human input. One of the key technologies driving the development of AVs is AI. AI can enhance a vehicle's decision-making abilities, perception, and learning capabilities, helping to make them more efficient than their human-driven counterparts. AI technology, these vehicles can analyze real-time data from sensors, cameras, and sources to make decisions about steering, braking, accelerating, and avoiding obstacles [1,2]. During the past years, this subject has attracted more attention than before due to the progress made in other AI-related areas. Therefore, it seems necessary to consider and survey the recently published papers to increase the understanding of the subject. Therefore, in the following sections, the most recent papers have been assessed.

5.2 AI and transportation

The transportation industry is affected by various factors, such as traffic, accidents, and human errors. However, AI integration has become increasingly popular in this field through neural networks and genetic algorithms. AI has impacted transportation in multiple ways, including developing automated vehicles that provide real-time medical assistance, monitor wildlife, collect electronic waste, and gather crime-related data. Companies can make more informed investment decisions by utilizing AI prediction methods for forecasting volumes in road freight transport systems. AI can also control traffic congestion in real-time, predict the paths of pedestrians and cyclists to reduce accidents, improve productivity, and reduce highway accidents through self-driving cars and trucks. Intelligent registration certificates can even be utilized to track owners digitally if they admit to committing crimes on the road. Furthermore, integrating smart grids can help manage electric vehicles and maximize stored energy while preventing power outages [3].

Handbook of Power Electronics in Autonomous and Electric Vehicles
https://doi.org/10.1016/B978-0-323-99545-0.00002-6

5.3 Research background and recent work of the subject

In recent years, many research projects on various aspects of AVs and AI's role in this matter have been done to make progress in this attractive area. Therefore, this section is dedicated to the most important aspects and recent works, which have been done till now. From the security point of view, Osorio and Pinto in [4] discussed the potential security concerns of the predictability of AV system decision-making processes. It explored the risk of manipulation and opportunistic behavior by individuals seeking to exploit the system's predictability. The authors highlighted that malicious, opportunistic, terrorist, criminal, and noncivic actors might attempt to manipulate these systems, potentially endangering lives and stability. The paper suggested that uncertainty is necessary within AV decision processes to prevent such manipulation. Uncertainty reduces the incentives for misbehavior as the risk of unsuccessful manipulation increases. However, introducing uncertainty can also reduce decision quality, impacting system efficiency and societal welfare. The text addressed possible solutions, including internal and external uncertainty. Internal uncertainty involves adding noise to the decision process, which might decrease decision quality. External uncertainty includes keeping specific decision details private, which could be challenging to maintain over time due to rational individuals learning from past decisions. The authors also proposed restricting human interactions with AVs; however, they acknowledge the difficulty of implementation. The article presented a scenario involving an AV choosing between protecting passengers or pedestrians in a collision. It demonstrates that predictability opens the door to intentional manipulation. The authors argued for preventing perfect knowledge of the system's functioning to ensure manipulation-proof outcomes. The article highlighted the vulnerability of AV systems to human manipulation due to their predictability. It emphasizes the necessity of uncertainty to deter malicious behavior while addressing potential drawbacks to decision quality. The authors also suggested exploring alternative solutions and relaxing the utilitarian approach to mitigate the malicious manipulation problem. The rise of AVs introduces new challenges involving AI decision-making, which holds the power to impact life-and-death situations. Unlike human decision-making, AI decisions pose questions about their decision-making process, mediation, and societal implications. Society, policymakers, and users must comprehend these distinctions. Despite AI decisions being contextualizable, challenges remain regarding technology, conceptualization, and understanding by various parties. These challenges are particularly relevant in assessing the benefits and risks of AI choices. AVs are often positioned as technologies that significantly mitigate risks. However, there is a need to explore the potential new risks they might introduce due to limitations in decision-making abilities. These limitations are evident in the inability to comprehend human values and morals when categorizing the driving environment. It is essential to scrutinize the conceptual framing of AV decision-making and its implications on risks and benefits. This article addressed the shortcomings in the current safety and AI ethics discourse and suggested ways to move forward. AVs capitalize on advanced sensors and AI to enhance driving decisions and reduce risks. They can eliminate problems related to human fatigue, misperception, and impaired judgment. These benefits support a safety argument, suggesting that AVs can make better decisions than humans and should be endorsed by policies. However, an alternative viewpoint highlights potential new risks, such as decision errors and limitations. Therefore, it is necessary to define and communicate the advantages of AV decisional intelligence to avoid underutilizing this technology due to misplaced risk perceptions. The introduction of emerging technologies, like AVs, poses governance challenges. Transitioning from traditional regulation to new governance models requires careful risk assessment. AVs exemplify a socially embedded AI technology that demands a shared risk mitigation process involving multiple actors. Framing the technology's meaning, benefits, and risks can determine how stakeholders interact. The main concern in AV risk mitigation is evaluating their capacity to make driving decisions and understanding the societal implications. Different governance actors, policy-writers, and ethicists approach AV decision-making differently. The mentioned paper acknowledged that AVs might provide safer driving in the long run; however, it delves into the shortcomings of empirically-based safety arguments and ethical considerations. A clear understanding of AI decisional capabilities in the context of AVs is essential. The article highlighted the need to balance the potential benefits of AVs in reducing driving risks with concerns about their inherent absence of moral agency. AVs offer numerous societal benefits, from improved mobility to enhanced safety. However, realizing these benefits requires addressing the associated risks. Governance of AVs requires a multistakeholder approach due to the rapid advancements in technology that outpace traditional regulatory processes. Research that illuminates AV decisional contributes to risk mitigation and provides access to societal benefits. The paper stressed the importance of understanding AVs' decisional spectrum, encompassing risk mitigation and potential limitations. Furthermore, the paper underscored the need to consider conceptual differences between human and machine moral decision-making contexts in AVs. It elaborated on the importance of accurately framing AI and AV technologies to ensure effective governance and regulation. The framing process should be iterative and reflexive to account for the complexity of debates around AVs' societal impact. Precision in framing can support positive impacts while accurately assessing and developing technologies with potentially adverse effects [5]. Ajenaghughrure et al. in [6] explored the impact of risk on users' trust before and after interacting with technologies like AVs. It also examined the physiological aspects of trust using electro-dermal activity responses. A group of 18 participants engaged in a hypothetical AV driving game, facing different risk conditions: very high risk, high risk, low risk, and no risk. These conditions were based on automotive safety levels. Participants assessed the trustworthiness of AVs and relied on either AI or a joystick to control the vehicle. Results reveal that trust and delegation of control to AVs increased as risk decreased, and vice versa. The study demonstrates a significant shift in initial trust levels due to interactions with AVs under varying risk conditions. Moreover, users' psychophysiological responses correlate with their trust levels toward AVs. The research emphasizes the importance of trust in the success of user-AV collaborations in the face of inevitable risks posed by AI technologies like AVs. Although AI-driven technologies like AVs offer substantial benefits such as safer roads and reduced traffic, their emergence also introduces risks, as exemplified by incidents where AVs failed to recognize pedestrians. The imperfections of AI contribute to skepticism and distrust among users, impacting their perception of AV safety. This lack of trust has led to misuse and disregard for AV information, contributing to accidents like the Tesla crash caused by the driver's distraction. Trust is crucial for successful human-AI interactions, especially in complex and uncertain conditions. However, few studies address the interplay between risk and trust in AI technologies. Previous experiments highlighted the decline of trust with increased risk. This study explored this relationship using realistic scenarios involving fully AVs. It employed subjective, objective (psychophysiological signals like electro-dermal activity), and behavioral measures to evaluate trust levels. The study formulated hypotheses regarding initial trust, decreasing trust with higher risk, reduced control delegation with increased risk, and the correlation between psychophysiological responses and trust levels. Future research could focus on enhancing users' trust in AVs by developing real-time trust assessment tools for AI algorithms. Despite limitations in sample size and minor differences in game conditions, this study offered valuable insights into the dynamic relationship between risk and trust in AI technologies. The rise of AI has significantly spurred the progress and integration of AVs in the transportation sector. AI, powered by extensive data from various sensors and advanced computing resources, has become a vital element within AVs, enabling them to perceive their surroundings and make informed decisions. In order to achieve full automation in self-driving cars, understanding the role of AI is crucial. Although previous research has explored aspects of AI implementation in AVs, a comprehensive evaluation of current practices is lacking. Therefore, this article aims to bridge this gap by presenting an extensive overview of critical research in this domain, focusing on the use of AI for primary AV applications: perception, localization and mapping, and decision-making. The paper investigated existing practices to uncover how AI is applied and the challenges associated with its integration. It highlighted potential opportunities by combining AI with emerging technologies, including high-definition mapping, big data, high-performance computing, augmented reality (AR)/virtual reality (VR) simulations, and 5G communication for connected AVs. AVs have garnered global attention, offering solutions to transportation challenges like congestion, safety, and energy efficiency. Recent AI and data processing advancements have increased AV research. Prior automated driving concepts were centered on advanced driver assistance systems (ADAS) that aimed to enhance driver safety. According to industry classifications, these systems fall under lower levels of automation (Levels 1 and 2). However, AI breakthroughs, particularly in deep learning (DL), have enabled more sophisticated applications in AVs, such as obstacle recognition, decision-making, and accurate vehicle control. While some studies have explored AI's role in specific AV components, a comprehensive review of the state-of-the-art practices, especially in recent years, is lacking. Therefore, this paper addressed this gap by surveying contemporary AI practices in AV research, exploring challenges, and identifying future research directions. The paper's framework begins with a literature review and a descriptive analysis of existing studies. It then delved into current AI practices in AV development, outlining challenges and issues associated with AI applications. The investigation reveals opportunities for future research, such as utilizing big data, high-performance computing, high-resolution maps for enhanced data processing, AR/VR platforms for accelerated testing, and 5G connectivity for low-latency communication between AVs. As AI continues to evolve with AVs, it is essential to consider the broader impacts on business, economics, and society [7]. AI-empowered machines imitate human actions and are particularly significant in the automotive industry for advancing vehicular technology. AI collaborates with mechatronics to enhance vehicle functionality. AVs employ onboard sensors like laser, radar, Lidar, GPS, and vehicular communication networks to gather scene information. This data are then utilized for path planning and control techniques, enabling autonomous driving in complex environments. Advanced AI algorithms aid vehicle localization, perception, path planning, and motion control. Therefore, the article reviewed state-of-the-art techniques enhancing AV performance. The rapid technological progress has led to innovative city development through interconnecting networks like Intelligent Home Area Networks (IHAN), Intelligent Industrial Area Networks (IIAN), Intelligent Vehicular Communication Networks (IVCN), and Smart Grids (SG). AVs serve as intelligent nodes within the Internet of Vehicles (IoV), facilitating Vehicle-to-Everything (V2X), Vehicle-to-Vehicle (V2V), and Vehicle-to-Infrastructure (V2I) communication. While autonomous driving efforts began in the 1920s, ongoing advancements aim for completely driverless vehicles capable of independent decision-making. The Society of Automobile Engineers (SAE) categorizes automated vehicles into six levels, ranging from human control (level 0) to full autonomy (level 5). Despite investments from companies like Uber, Google, and Tesla, autonomous systems remain a challenging research area. Effective systems must make intelligent, real-time decisions. Researchers focused on improving algorithms for localization, perception, and detection. Localization addresses the vehicle's position, perception concerns environment understanding, predicting environment outcomes, and deciding involves determining actions. These questions are addressed using sensors and algorithms for reliable and safe autonomous driving. AVs rely on sensors for environmental perception. Data from these sensors guide decision-making, considering safety and efficiency. Advanced solutions, including localization, object detection, path planning, and data fusion, are necessary for task completion. DL algorithms, leveraging powerful tools like GPUs and vast datasets, enhance AV performance by ensuring accuracy and speed. The article surveyed AI technologies used in AVs. It delves into the generic structure of AVs, state-of-the-art localization techniques, path planning methods, and motion control strategies. Autonomous decision-making involves processing sensor data to make optimal choices. This can be achieved through the integrated perceive-plan-act approach or end-to-end learning. Developing intelligent algorithms for safe AV operation is crucial for vehicle design. The article provided insights into the comprehensive layout of an AV, presenting various AI algorithms used for perception, localization, path planning, and motion control [8]. Parekh et al. in [9] discussed the growing popularity of vehicular technology and the focus on autonomous driving for intelligent transportation systems. Accurate positioning technologies are needed to ensure safety and reliability, considering uncertainties like pedestrian behavior, random objects, road types, and settings. The study explored various domains and technologies essential for AVs, including environment detection, pedestrian detection, path planning, motion control, and vehicle cyber security. The aim was to analyze existing technologies, compare approaches, and build public trust by addressing these challenges. The article also examined the public perception of AVs. Historically, vehicles were mechanical; however, IoT and embedded systems advancements have turned them into intelligent devices connected to the internet. These innovations have led to fully functional vehicles supporting features like environment sensing, internet connectivity, traffic rule adherence, self-navigation, decision-making, passenger safety, and parking. These AVs offer enhanced safety and convenience in response to population growth and resource optimization needs. The adoption of electric batteries and the promising energy storage system of LSBs are explored as alternatives to traditional fuel sources. MOFs, compounds held by metal ions, show potential catalytic properties for battery cathodes. Challenges like road safety have prompted governments to use surveillance technologies like CCTV and autonomous and connected cars to address human-related accidents. AVs, driven by IoT advancements, sensors, data analysis, wireless communication, and AI, offer benefits like reduced accidents, fuel consumption, congestion, and improved accessibility for older people and disabled ones. The study was motivated by the alarming statistics of road accidents causing millions of deaths annually. AVs can eliminate human error-related accidents by handling tasks like environment sensing, route planning, speed control, and parking. AVs promise societal and economic benefits by reducing accidents, fuel consumption, and congestion while enhancing accessibility for vulnerable populations. The paper aimed to provide an organized overview of automated driving software practices, filling a literature gap and discussing public psychology and emerging trends. It comprehensively covers perception, planning, control, and detection functions, highlighting advancements and challenges in AV technology. The article presented a comparative survey of recent autonomous driving advancements and problem-solving capabilities. Topics covered include Vehicle Cyber security, Psychology, Pedestrian Detection, Motion Control, Path Planning, and Environment Perception. Existing literature reviews have examined various aspects of AVs; however, this study uniquely compiled a comprehensive overview of emerging technologies, challenges, and functions. It outlined future research directions and challenges, emphasizing the importance of nontechnical factors such as consumer trust, governance, and human behavior. The conclusion asserts that AVs are the future, with Level 3 Automation ready for commercialization. Despite advances, the article suggested caution in speculating about higher-level AV commercialization, advocating for robustness and safety in automation levels. Vehicular technology and autonomous driving are gaining popularity, necessitating accurate positioning technologies to handle uncertainties like pedestrian behavior, road types, and random objects. AVs have gained popularity due to their positive impact on society, improving safety, convenience, fuel efficiency, traffic flow, and cost savings. However, before widespread deployment on main roads, issues in their functionality must be minimized. A significant challenge is accurate pedestrian detection to prevent accidents. Therefore, this survey focused on how DL technologies can address these challenges. Pedestrian detection is crucial for AVs to ensure safety by identifying human movement, preventing accidents, and avoiding obstacles. This is achieved through sensor fusion, combining radar, camera, and LiDAR systems. An Advanced Driving Assistance System (ADS) was introduced to prevent accidents, integrating pedestrian detection among its features. Despite this, pedestrian detection faces issues. Researchers have tackled challenges such as poor obstacle detection, occlusion, low resolution, and pedestrian tracking under varying lighting conditions. This survey discussed DL approaches for pedestrian detection, addressing challenges like occlusion, low-quality images, light detection, and obstacle size. It delves into pedestrian sensor importance and traditional and DL approaches, highlighting DL's effectiveness in tracking pedestrians. Notably, YOLOv3, YOLOv4, faster R-CNN, and Mobile Net-SSD models address key challenges, with Faster R-CNN outperforming Mobile Net-SSD under low-visibility conditions. Multispectral images and YOLO were valuable for detecting different-sized pedestrians in various lighting. The paper concluded with insightful discussions and future research directions for enhancing pedestrian detection in AVs [10]. The rise of AV technology is causing significant disruption across various industries. However, to gain public acceptance, AVs must prove their reliability and safety. Validation becomes crucial to build trust and deploy these vehicles for everyday transportation. Efforts in AV validation include real-life testing, test tracks, and simulation. This paper introduced a simulation-based methodology for validating AV decision-making systems. This approach focused on generating scenarios that test the vehicle's decision-making abilities. This method showcased its effectiveness and efficiency by separating validation concerns and utilizing pseudo-random test generation. Successful AV technologies can revolutionize the automotive, transportation, and energy industries. Achieving this potential involves substantial investment in AI technology, which underpins these platforms. ADAS and other automation solutions are available. With the inevitability of AV introduction, ensuring their safety and viability becomes paramount. A robust method to validate and verify AV competence is necessary to garner public trust and regulatory approval. Validation and verification processes demonstrate that an AV is at least as competent as a human driver. Currently, no regulatory body defines the specifications AVs must meet. Unlike regular vehicles, which adhere to standards like ISO 26262, AVs lack established safety standards. Existing validation methods like shadow driving or image-based testing are expensive, slow, and resource-intensive. Therefore, modeling and simulation are indispensable tools to achieve validation goals efficiently. This paper extended a simulation scenario generation method for validating AV decision-making. It focuses on separating low-fidelity simulation testing for decision-making from other testing methods like high-fidelity simulation for perception systems. The approach employed a semantic language to describe testing scenarios, allowing automatic test generation and control through parameterization. The method enabled the definition of road scenarios and actor behaviors, thus comprehensively covering the AV's complex domain. The approach's efficiency lies in its focus on decision-making in AVs, utilizing low-fidelity simulation. This approach, combined with validation features like assertions, classifies scenarios and detects edge cases. It enabled placing the AV under test in a simulation using hardware or software in the loop techniques. This framework collects data for analysis during and after the simulation run. Ensuring the safe deployment of AVs in real traffic remains a challenge. Industry and research are actively working on AV testing; however, most methods have limitations. Standard methods include shadow driving, test tracks, and hardware-in-the-loop testing. Various entities combine these methods to develop testing frameworks; however, no standard framework exists. Test scenario generation is crucial for validation efforts. Different approaches focus on high-fidelity simulation tied closely to test tracks. This paper proposed a method to break down AV components and testing methods to reduce complexity and gain coverage. Components like perception, object recognition, decision-making, and vehicle dynamics were verified using simulation, hardware-in-the-loop, and real-world testing at different levels in the system. Validation and verification are crucial tools to bridge the gap between current AV technology and widespread market use. Creating a semantic language proposed mechanisms for generating road networks and actors for random and constrained test scenarios on a low-fidelity simulator. The approach aims to integrate with testing methods for other AV functionality layers and enable future scenario exchange with existing tools [11]. Nascimento et al. [12] presented a systematic literature review (SLR) to provide insights into the state of AI-based AV safety literature. After analyzing 4870 papers, 59 relevant studies were chosen and categorized into six groups. An AV system model was introduced for discussion. The study emphasized the importance of a comprehensive safety agenda for AI-based AV research. AI is crucial for AV development, enabling them to interpret their surroundings and make driving decisions. However, AV safety is paramount due to its potential impact on lives and the environment. This study addressed the debate on AI's influence on AV safety, mapping the literature and providing a balanced view. The paper comprised five sections detailing the research methodology, data analysis, AV system model, and conclusions. The systematic literature review method followed established procedures to define research questions, source selection, and data mapping. The research questions encompassed AI's impact on system safety, study topics, AI techniques, addressed issues, reported findings, and suggested future studies. The study enriched AV literature by presenting an overview of AI's effects on safety. From the initial 4870 papers, 59 relevant studies were categorized and analyzed using an AV system model. The model's extension incorporates human interaction for more realistic representations. While AI techniques have shown promising results, their real-world applicability in AVs remains debated. Further research and improvements are needed to validate their suitability and address potential hazards. Moreover, infusing safety engineering into AI-based AV studies would enhance their effectiveness. Kamran et al. [13]

evaluated the integration of AI and advanced materials within the automobile industry. In the fourth Industrial Revolution context, where intelligent technologies optimize production, quality, and costs, AI emerges as a significant player. The paper explored AI's diverse applications in the automotive sector, aiming to enhance vehicle intelligence, safety, and efficiency. The goal was to automate driving, reduce manual labor, and elevate overall performance. Various facets of the automotive industry, including vehicle design, manufacturing, and after-sales services, were examined for potential AI integration. The paper highlighted the limitations of machines compared to human decision-making due to the absence of comprehensive understanding and adaptable thinking. As technology advances, the need arises for AI to mimic human cognition. This need led to the evolution of AI systems capable of decision-making akin to humans. The article corroborated the automobile industry's suitability for AI integration, discussing applications beyond manufacturing, such as designing, after-sales services, and even autonomous driving. Tesla's strides toward fully AVs demonstrate AI's potential. However, robust and high-performance vehicles are essential for these technologies to thrive. Thus, exploring advanced materials like carbon fiber and polymer composites gains significance. These materials contribute to lightweight yet strong vehicles, improving power-to-weight ratios and performance. The article envisioned the future of autonomous cars and the importance of understanding AI applications in the automotive sector. It confirmed that theoretical knowledge is fundamental to technological development. The integration of AI technologies promises reduced road accidents and improved safety. Furthermore, advanced materials like polymer composites and high-strength steel enhance vehicle durability. In summary, the synergy of AI and intelligent materials holds great promise for the automotive industry's research and implementation. The emergence of AVs offers substantial benefits such as enhanced safety and reduced energy consumption, pollution, and traffic congestion. Nevertheless, these vehicles face security and privacy challenges that could undermine their potential advantages if unaddressed. From another point of view, the vulnerabilities present in AVs could be exploited by hackers, posing a significant threat to mobility and data protection. Recent research suggests that integrating Blockchain and AI technologies can provide robust defense mechanisms against malicious attacks on AVs. Despite their distinct operational models, when combined, these technologies can mutually reinforce each other, resolving many security and privacy issues of AVs. AI can optimize blockchain's construction for efficiency and security, while blockchain offers data immutability and a trust mechanism for AI-based solutions, rendering them transparent, reliable, and explicable. Although some research has been conducted in this area, the exploration of Blockchain and AI's application for securing AVs remains somewhat limited. Therefore, this paper assessed the potential implementation of a merger between Blockchain and AI solutions to bolster AV security. It begins by categorizing security and privacy threats that could arise from AV applications. Subsequently, it provided an overview of recent literature regarding Blockchain and AI for securing AVs. The paper concluded by highlighting the limitations and challenges in integrating Blockchain and AI with AVs, offering insights based on a systematic review, and suggesting potential directions for future research. AV technology is gaining attention in academia and industry and has already been adopted by various automotive companies like Uber, Google, Apple, Aptiv, BMW, Tesla, and Toshiba. This technology enables driving automation systems to improve energy efficiency and reduce pollution, congestion, and accidents, addressing that most crashes result from human errors. The market for AVs is projected to grow significantly, with millions of new AVs sold yearly, indicating a substantial shift toward automation in transportation. AVs face security risks despite these benefits, making cyber security a prominent concern. AVs are susceptible to cyber-attacks that could compromise their safety and functionality. Malicious manipulation of Global Positioning System (GPS) data can disrupt AV localization, leading to accidents. Furthermore, hackers can demand ransom to regain access to fleets of connected AVs, posing threats to manufacturers and users. Privacy is also at risk, as hackers can access sensitive personal data collected and stored by AVs, including location, biometric data, and device passwords. In order to counter these threats, the integration of AI and Blockchain is proposed. Blockchain has a history of effectively preventing cyber-attacks, while AI is adept at quickly detecting and responding to threats. However, the synergistic potential of Blockchain and AI remains relatively unexplored, with a dearth of comprehensive studies on using these technologies to address AV security and privacy concerns. This paper aims to bridge that gap by presenting a detailed analysis of the vulnerabilities in AV security, reviewing solutions that use BC, AI, or both, and discussing open challenges and future research directions. The paper categorizes critical cyber-attacks on AV components and communication channels and proposes solutions to enhance AV security. It underscored the promise of integrating AI and Blockchain in securing connected AVs. This combination leveraged AI's data processing capabilities and decision-making process with the chain's secure and trusted platform for data storage and sharing through in-vehicle and V2X networks. While integrating these technologies is still early, the proposed solutions are mainly experimental. Challenges in security, performance, scalability, data storage, transparency, interoperability, regulation, and standards must be addressed [14]. Yang et al. in [15] reviewed intelligent tire research, including sensing systems and estimation methods. It offered constructive strategies to enhance tire sensing system design, cater to harsh tire conditions, and maintain continuous tire-road contact measurement. It also delved into implementation requirements for autonomous driving. These intelligent tires, equipped with advanced sensors and recognition methods, open prospects for better vehicle dynamics control and self-driving technology. Tires solely touch roads, thus impacting driving safety for AVs. The contact patch, where the tire and road interact, heavily influences vehicle dynamics. Recognizing road conditions and tire states can optimize paths and vehicle control. Intelligent tires and embedding sensors directly capture road responses and tire dynamics, although tire conditions challenge performance. Intelligent tire systems include sensors, signal processing, and power supply modules. They estimate parameters from tire-road data. This study added value by emphasizing sensor roles, road classification, refining machine learning (ML) methods, and addressing autonomous driving needs. Contact and noncontact sensors play pivotal roles in intelligent tire deployment. They face challenges like power, transmission, and integration. Road classification via intelligent tires expands their utility. This study also highlighted ML's potential while addressing challenges and offering solutions. Current research employed contact and noncontact sensors, coupled with models or ML, to estimate tire-road parameters. Continuous tire-road contact sensing remains a challenge due to power and signal constraints. Innovative sensors, like tactile bionic sensors, could enhance recognition. Intelligent tire data improves typical vehicle dynamics and boosts AV perception, aiding decision-making, trajectory planning, and motion control, enhancing safety and efficiency. Ren et al. in [16] discussed how advanced communication and AI technologies contribute to developing environmentally friendly smart cities. The focus was on connected and automated vehicles (CAVs), which utilize sensors and AI models for automatic decision-making. These technologies are vital for addressing challenges like traffic congestion and pollution. However, the energy-intensive nature of model training and networking poses challenges. In order to overcome these challenges, the authors propose a green intelligence networking scheme (NGIN) using nonfungible tokens (NFTs). NFTs were employed to describe intelligence through metadata, optimizing energy-efficient networking. The architecture, modules, and mechanisms of NGIN were detailed. The authors employed a quantum-inspired reinforcement learning (QRL) algorithm to solve the core problem, demonstrating its convergence and effectiveness through simulations. The need for intelligent transportation systems in modern smart cities is emphasized. CAVs and the IoV are crucial in enhancing traffic information dissemination, controlling vehicles, and reducing energy consumption. These technologies, driven by AI, help make informed decisions, leading to cleaner and more efficient urban environments. Training accurate ML models requires substantial data from onboard sensors. Although the number of sensors is increasing, safety concerns remain. Furthermore, energy-intensive model training and isolation of CAVs hinder efficient learning. Collective learning (CL) and intelligence networking offer solutions. CL leverages group intelligence, improving decision-making. The authors categorize networked intelligence into pretrained models, shared parameters, and enhancing learning. Security and incentive issues in distributed intelligence networking were tackled through blockchain integration. The authors advocated for NFTs, a blockchain-based solution that tokenizes digital assets efficiently. NFTs facilitate the description and networking of intelligence, incentivizing distributed systems. This led to NGIN's proposal, which employs NFTs to optimize energy-efficient intelligence networking for CAVs in smart cities. The contributions of this work included the introduction of NGIN, specifying its architecture, modules, and networking mechanisms. The authors formulated the intelligence networking problem as a discrete Markov decision process (MDP), aiming to maximize rewards while adhering to energy constraints. The QRL algorithm finds the optimal policy. Simulations confirmed the algorithm's convergence and NGIN's superiority over existing schemes. The automotive sector is evolving rapidly due to advancements in AI and pattern recognition, making vehicles smarter and safer. Autonomous driving, in particular, is gaining attention for its ability to eliminate tiresome driving procedures and enhance safety. AVs frequently execute lane changes, and obstacle avoidance is crucial in self-driving technology. These vehicles gather information about obstacles and vehicles in their path, adapt their routes, and adjust speed and steering for secure driving. Effective lateral and longitudinal control is necessary for obstacle avoidance, considering speed, safety, and efficiency. Human-related traffic incidents causing 1.2 million deaths annually underscore the importance of obstacle avoidance techniques, especially for lane changes. Research in this area holds theoretical and practical significance, particularly for electric cars and autonomous driving. The level of automation in AVs has evolved from zero to six, ranging from no automation to full autonomy. This includes features like adaptive cruise control and lane-keeping assistance (Level 1), partial self-control requiring human intervention (Level 2), conditional automation (Level 3), high automation (Level 4), and fully automated transportation (Level 5). The article focused on Level 5 automation, where vehicles operate independently in a connected and autonomous environment. This technology demands bidirectional connectivity, using sensors to report traffic conditions and interact with the local infrastructure. Crewless aerial vehicles (UAVs) are crucial components in this ecosystem, aiding in tasks like business and security. They utilize sensors, cameras, GPS, and embedded systems. Technologies like the Internet of Things (IoT) and the Flying Ad hoc Network (FANET) enhance the benefits of AVs by sharing data between vehicles and drones, improving their scalability and operational security. Effective communication channels between these vehicles are essential to ensure data dependability. Communication technologies such as Wi-Fi, DSRC, LTE, and Zigbee, including 5G, play a role in this setup. Connected autonomous cars (CAVs) possess the potential to revolutionize transportation, creating new avenues for technology and business models. Analyzing the impact of modeling car-following and lane-changing behavior is vital for traffic flow predictions. Achieving dependable, safe, and enjoyable autonomous cars requires integrating cyber-physical systems, enhancing communication, and accommodating different driving styles influenced by social behavior. However, current research on vehicle automation has limitations, particularly regarding complex scenarios like crossings. While IoV and CAV combinations offer benefits, robust training datasets and systematic data collection methods are lacking, hindering widespread CAV adoption. Despite these challenges, the potential for intelligent learning using algorithms like convolutional neural networks and deep reinforcement learning (DRL) holds promise for achieving highly autonomous or fully autonomous driving. Successful implementation of the cyber-physical system in connected AVs can lead to smoother and more effective control, adapting driving styles based on social behavior [17]. AI is a highly sought-after technology with versatile applications across various domains, such as intelligent transportation systems (ITS), medicine, healthcare, military operations, and more. One crucial application is AVs falling under AI in ITS. Vehicular Ad-hoc Networks (VANET) facilitate communication among AVs within the system. The performance of each vehicle relies on shared information, making detecting malicious vehicles crucial to prevent disruptions. Utilizing ML algorithms aids in identifying data flaws. Recent studies using the stacking ML approach achieved a remarkable accuracy of 98.44%. This paper employed the decision tree-based random forest technique, yielding an accuracy of 98.43% and an F1 score of 98.5% on the VeRiMi dataset. Explainable AI (XAI) enhances understanding of complex black-box ML and DL models. An evaluative model interface aids in measuring and explaining the model's performance, ensuring the responsible application of AI in AVs. AVs aim to enhance passenger comfort, road safety, and efficiency, distinct from traditional vehicles driven by human intelligence and prone to errors. Many traffic issues result from human errors, such as driving under the influence or drowsiness, leading to accidents and congestion. Automation seeks to mitigate these risks, allowing software to predict outcomes and act accordingly. AVs achieve this by processing their surroundings, road conditions, and nearby vehicle data to make informed driving decisions. Therefore, the paper proposed a ML approach to VANET misbehavior detection. Applying XAI aids in comprehending complex model results and decisions. Performance insights were gained through evaluation metrics, offering transparency in trust management. Comparative graphs highlighted the superior performance of random forest on the test set. Future directions include combining ML and DL models for enhanced predictions and further exploration of XAI to improve model interpretability. The application of XAI extends beyond VANET systems to other AI components in AVs. Integration of XAI with blockchain technology is also considered for elevated security and trust management [18].

5.4 Definition of AVs

AVs, also known as self-driving vehicles, utilize state-of-the-art AI technology to function and make decisions independently, without human interference. These vehicles have sensors, cameras, and GPS to interpret their surroundings and control their movement, including acceleration, braking, and steering. There are varying degrees of autonomy, ranging from Level 0 (no autonomy) to Level 5 (full autonomy), with many current vehicles falling somewhere in between. Fig. 1 shows the various levels of the mentioned assortment. The advancement of AV technology has the potential to revolutionize transportation by enhancing safety, efficiency, and accessibility for individuals unable to drive. However, there are valid concerns regarding the reliability and safety of this technology, as well as its potential impact on employment in the transportation industry [19].

5.5 Role of AI in AVs

AI has transformed many industries globally, including AVs. AI technology has enabled vehicles to function without human intervention, offering considerable safety, efficiency, and convenience benefits. AI plays a crucial role in AVs by providing the ability to read the environment and make informed decisions based on real-time data. Various sensors, such as cameras, lidars, radars, and ultrasonic sensors, gather data about the surroundings. This data are then processed by an AI algorithm, enabling the vehicle to extract insightful information and understand its surroundings more effectively. Computer vision is a key aspect of AI in AVs, allowing it to see and interpret its surroundings. AI algorithms can detect and recognize objects such as pedestrians, vehicles, traffic signs, and road markings, enabling vehicles to navigate safely and make timely decisions. AI is also essential in path planning and decision-making,

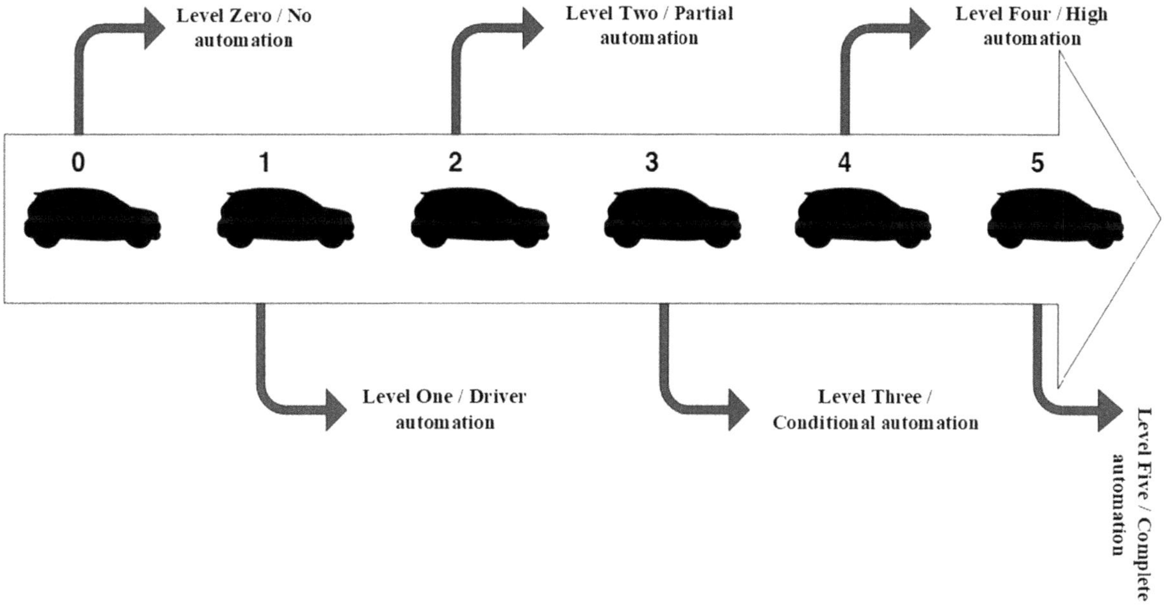

FIG. 1 The various levels of the automation for vehicles.

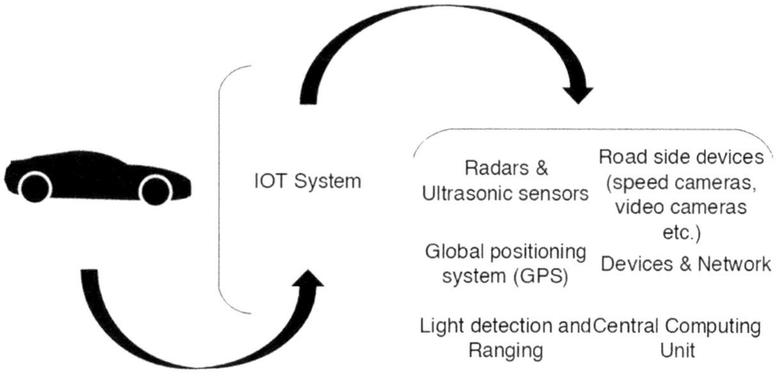

FIG. 2 The interaction of AI and IOT in control of the proposed AV.

enabling the vehicle to use real-time data to determine the best route and make decisions accordingly. ML is also a critical component of AI in AVs, allowing it to learn and improve its efficiency and performance over time. Moreover, a sophisticated control system translates AI decisions into physical actions, enabling the vehicle to operate smoothly and accurately. Overall, AI is fundamental to the safe and efficient operation of AVs, offering solutions to problems such as traffic congestion, accidents, and environmental impact in the future [20,21]. According to the above, AI in corporations with IOT can control the vehicle, and this interaction is depicted in Fig. 2.

5.6 Architectures for self-driving cars based on deep faded learning

There are variables to explain for better understanding and make the text more understandable. Variables can have their values either defined in terms of (t), the discrete-time step, or in terms of (t, $t + k$), the discrete sequence, k denoting the length of the sequence. A state variable $\mathbf{z}$, for instance, can have its value either at discrete time (t), as $\mathbf{z}^{\langle t \rangle}$, or within a sequence interval $\mathbf{z}^{\langle t,\ t+k \rangle}$. Bold symbols indicate vectors and matrices [22,23].

5.6.1 Deep CNNs

In order to better understand and make the paper's text more understandable, we are provided helpful information about some popular approaches for architectures for self-driving cars. Universal nonlinear function approximates are commonly utilized when processing spatial information like images. CNNs, in particular, can extract image features. Before the rise of DL, computer vision systems relied on handcrafted features like HAAR, local binary patterns, or histograms of oriented gradients. However, CNNs can automatically learn an abstraction of the training space, similar to how the mammalian visual cortex functions. In the brain, images are sent from the retina to the visual cortex via the thalamus, where ventral and dorsal information processing occurs. The ventral processing aids in object recognition and visual identification, while the dorsal processing establishes spatial relationships between objects. A CNN mimics the ventral flow by enabling different brain areas to respond to specific features in the visual field. The weights vector of a CNN is expressed in terms of the set of internal connections that determine its functionality, and the set of neuron biases is given by b, where W is the set of weights governing the internal connections. With coefficients learned during training, weights W are organized as image filters. Using local spatial correlations between pixels within an image, a CNN learns translation-invariant convolution filters that capture discriminant image characteristics. In layer k, a multichannel signal representation M_k is the result of channel-wise integration of the representations M_k,c, where c is the form of the signal. Following this process, a signal representation in layer $k+1$ can be generated as follows [24].

$$M_{k+1,l} = \varphi(M_k{}^*W_{k,l} + b_{k,l}) \tag{1}$$

The equation in question involves using a convolutional filter, denoted as $w_{k,l} \in W$, which possesses an equal number of channels as M_k. Meanwhile, the bias term is indicated by $w_{k,l} \in b$, with l indicating the channel index. The * symbol represents the convolution operation itself.

5.6.2 Recurrent neural networks

Recurrent neural networks (RNNs) are a class of artificial neural networks designed for processing data sequences. They are particularly well-suited for tasks involving sequential data, such as natural language processing (NLP), speech recognition, time series forecasting, etc. RNNs can maintain hidden states that capture information about previous elements in a sequence, allowing them to capture temporal dependencies. Moreover, unlike other DL techniques, RNNs are particularly effective at analyzing temporal data sequences. The memory cell of an RNN has a time-dependent feedback loop, which differs from a conventional neural network. During training, basic RNNs are faced with the challenge of vanishing gradients. The gradient signal can be multiplied by many times depending on the steps taken. As a result, a traditional RNN cannot capture long-term dependencies in sequence data. Deep networks, or networks that process long sequences, may have difficulty propagating gradients back to affect the weights at earlier layers. A gradient vanishing network would end up with minimal weight values because gradient vanishing results in the weights not being updated effectively. The following expresses the basic structure and formulation of RNN [7,25].

1. *Basic structure of an RNN:*
   - An RNN consists of a chain of repeating neural network modules, each processing one element of a sequence at a time and maintaining a hidden state that summarizes information about the sequence it has seen so far.
   - At each time step, the RNN takes an input vector (representing the current element in the sequence) and the previous hidden state as inputs, producing an output vector and a new hidden state.
2. *Formulation:*
   - The hidden state at time step t is denoted as $h_{t,}$ and the output at time step t is denoted as y_t
   - The hidden state and output at each time step are computed using the following formulas:

$$h_t = f(W_h.h_{t-1} + W_x.x_t + b_h) \tag{2}$$

Output Computation:

$$y_t = g(W_y.h_t + b_y) \tag{3}$$

Where in the above formulation:
x_t is the input at time step t.
W_h, W_x, and W_y are weight matrices to be learned.
b_h, b_y are bias vectors.
F and g are activation functions, often the hyperbolic tangent (tan h) or the rectified linear unit (ReLU).

3. *Unrolling in time:*
    - In order to process a sequence of length T, the RNN is "unrolled" into T time steps, with the hidden state being updated at each step.
4. *Back propagation through time:*
    - Training an RNN involves using back propagation to update weights and biases based on the error between the predicted and target outputs.
    - Back propagation through time (BPTT) is a variant of back propagation used to train RNNs. It propagates errors backward through time steps.
5. *Vanishing and exploding gradients:*
    - One significant challenge with vanilla RNNs is the vanishing and exploding gradient problem. It occurs when gradients become too small (vanishing) or too large (exploding) as they are propagated back in time. This can make training deep RNNs difficult.
6. *Long short-term memory and gated recurrent unit:*
    - In order to address the vanishing gradient problem, more advanced RNN architectures like LSTMs (long short-term memory) and GRUs (gated recurrent unit) were introduced. These architectures use specialized mechanisms to better capture long-term dependencies in sequences.
7. *Applications of RNNs:*
    - RNNs have been successfully applied in various domains, including:
        - NLP for tasks like text generation, translation, and sentiment analysis.
        - Speech recognition and synthesis.
        - Time series analysis and forecasting.
        - Handwriting recognition.
        - Autonomous robotics.
8. *Limitations:*
    - RNNs have limitations in capturing very long-term dependencies due to vanishing gradients.
    - They can be computationally expensive to train, especially with long sequences.

5.6.3 Deep reinforcement learning

DRL is a ML and AI subfield that combines reinforcement learning (RL) with deep neural networks. It aims to teach agents how to make sequences of decisions in an environment to maximize a cumulative reward signal. DRL has successfully solved complex decision-making tasks like game playing, robotics control, and autonomous driving. Here, an in-depth explanation of DRL, its formulation, and the underlying theory has been provided [26–28]:

1. *Basic components of DRL:*
    - *Agent:* The learner or decision-maker that interacts with the environment.
    - *Environment:* The external system with which the agent interacts.
    - *State (s):* A representation of the environment at a particular time.
    - *Action (a):* A decision or move made by the agent that affects the environment.
    - *Reward (r):* A numerical signal indicating the immediate benefit of an action.
    - *Policy (π):* A strategy or mapping from states to actions, which the agent uses to make decisions.
    - *Value function (V or Q):* A function that estimates the expected cumulative reward under a given policy.
    - *Q-learning:* A popular DRL technique that estimates the expected cumulative reward for a particular action in a given state.
2. *Markov decision process:*
    - DRL is often formulated as a Markov decision process (MDP), a mathematical framework that models decision-making in a stochastic environment.
    - An MDP is defined by a tuple: $<S, A, P, R, \gamma>$
    - S is the set of possible states.
    - A is the set of possible actions.
    - P represents the transition probabilities from one state to another given an action.
    - R is the reward function.
    - γ is the discount factor that determines the trade-off between immediate and future rewards.

3. *Policy gradient methods:*
   - In DRL, deep neural networks are often used to represent policies or value functions.
   - Policy gradient methods optimize the agent's policy directly to maximize expected rewards.
   - Common policy gradient algorithms include REINFORCE, Trust Region Policy Optimization (TRPO), and Proximal Policy Optimization (PPO).
4. *Deep Q-Networks:*
   - Deep Q-Networks (DQN) is a popular DRL algorithm that combines Q-learning with deep neural networks.
   - It uses a neural network to approximate the Q-values for state-action pairs.
   - DQN introduces experience replay and target networks to stabilize training.
5. *Exploration vs. exploitation:*
   - One of the critical challenges in RL is balancing exploration (trying new actions) and exploitation (choosing actions with high expected rewards).
   - Techniques like epsilon-greedy exploration are often used to address this challenge.
6. *Reward shaping and function approximation:*
   - Reward engineering involves designing appropriate reward functions to guide the learning process.
   - Function approximation techniques like deep neural networks estimate value functions or policies.
7. *Challenges and considerations:*
   - DRL can be sample-inefficient and challenging to train.
   - Issues like catastrophic forgetting and overestimation bias in Q-values need to be addressed.
   - Careful exploration strategies are necessary to avoid getting stuck in suboptimal policies.
8. *Applications of DRL:*
   - DRL has been applied to a wide range of domains, including:
   - Playing complex video games, like Atari games.
   - AVs and control.
   - NLP and dialog systems.
   - Healthcare, for optimizing treatment plans.
   - Finance, for portfolio optimization and trading.

5.7 Sensor fusion

Sensor fusion is crucial in various applications, including AVs, robotics, surveillance systems, and AR. By integrating data from diverse sensors, AI systems can overcome the limitations of individual sensors and provide a more holistic understanding of their surroundings, leading to safer and more reliable operations. Integrating multiple sensors through sensor fusion is a fundamental aspect of AI for self-driving cars. Based on Fig. 3, Data from cameras, lidar,

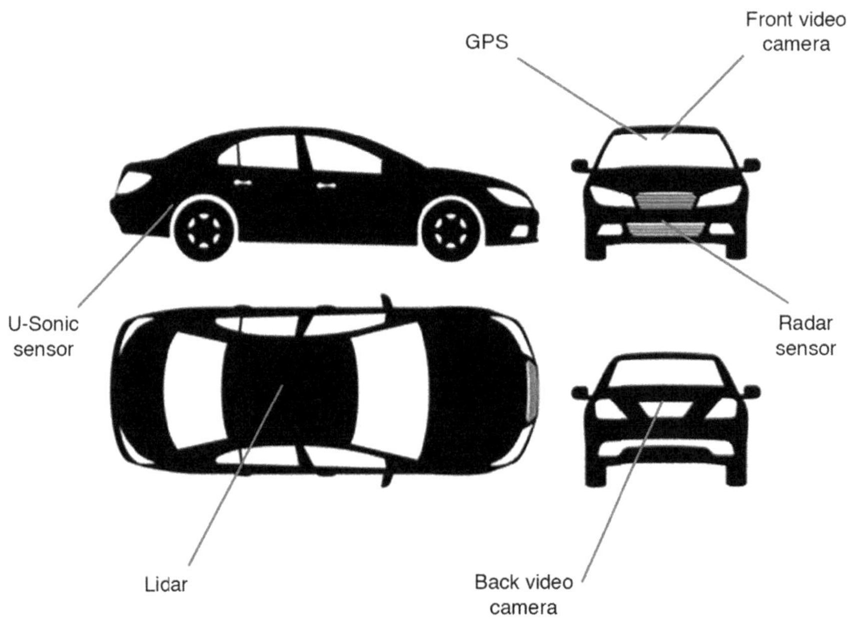

FIG. 3 The diverse sensors and their location on the assumed AV.

radar, and GPS is combined to understand the surrounding environment comprehensively. This improves detection, localization, and mapping accuracy, enabling the system to make better decisions. Sensor fusion compensates for the limitations of individual sensors, making the experience safer and more efficient. The technology has made significant strides in recent years, and it is a crucial component of self-driving cars, allowing them to navigate roads safely and efficiently [29].

5.7.1 What is the sensor fusion?

Sensor fusion is a technique used in AI and robotics to integrate data from multiple sensors or sources to obtain a more accurate and comprehensive understanding of an environment or situation. AI systems can make better-informed decisions and improve their perception and situational awareness.

5.7.2 How does sensor fusion work?

Sensor fusion combines data from different sensors, such as cameras, lidar, radar, GPS, or microphones, and processes this data to create a unified and coherent representation of the environment. The process typically involves the following steps:

1. *Data acquisition:* The AI system collects data from various sensors. Each sensor provides its perspective or information about the environment. For example, cameras capture visual information, while lidar measures distances using laser beams.
2. *Data preprocessing:* The raw data from different sensors may have varying formats, units, and noise levels. Data preprocessing involves cleaning, calibrating, and synchronizing the sensor data to ensure consistency.
3. *Sensor registration:* The data from different sensors must be aligned or registered in a common reference frame. This step ensures that the information from each sensor is correctly positioned relative to the others.
4. *Data fusion*: This is the core step where the AI system combines the sensor data to create a fused representation. Various fusion techniques can be used, such as Bayesian filters (e.g., Kalman filter, Particle filter), neural networks, and statistical methods. These techniques consider the uncertainties associated with each sensor and produce a more accurate and reliable estimate of the environment.
5. *Decision making:* The fused data are then used by the AI system for decision-making, navigation, or other tasks. For example, in AVs, sensor fusion helps identify obstacles, determine vehicle position, and plan a safe path.

5.7.3 Theory behind sensor fusion

Sensor fusion is grounded in several theoretical principles:

1. *Bayesian probability theory:* Bayesian filters are commonly used in sensor fusion to estimate the state of the environment based on sensor measurements and prior beliefs. Bayesian methods model uncertainty and update the belief in the state as new data arrives.
2. *Multisensor data fusion:* This field focuses on combining information from multiple sensors. It involves data association, sensor reliability modeling, and conflict resolution to handle conflicting or uncertain data.
3. *Kalman filtering:* The Kalman filter is widely used in sensor fusion, especially for tracking and estimation. It optimally combines noisy measurements over time to estimate the state of a dynamic system.
4. *Information theory:* Information theory concepts, such as mutual information, quantify the information gained by fusing sensor data. Maximizing information gain is essential for improving decision-making accuracy.

5.7.4 Importance of sensor fusion

AVs rely heavily on sensor fusion technology to operate effectively. In essence, sensor fusion combines data from multiple sensors such as radar, LiDAR, and cameras to provide a comprehensive view of the vehicle's environment. This enables the vehicle's AI to make more accurate decisions. The system can detect and measure distance, speed, and direction by combining sensor data, among other critical variables. A 360-degree view of the surroundings is essential for safe navigation, and sensor fusion technology delivers just that. Collaborative use of sensors is crucial to ensuring optimal vehicle safety and efficiency outcomes. Sensor fusion allows the AI system to understand complex scenarios, anticipate potential hazards, and make decisions that keep drivers and passengers safe [30,31].

5.7.5 Different types of sensors used in AVs

AVs rely heavily on various sensors to accurately perceive their environment and make informed decisions. These sensors detect and measure numerous physical and environmental variables, including distance, speed, acceleration, and temperature. Some of the most commonly used sensors in AVs include cameras, lidars, radars, ultrasonic sensors, and GPS receivers. Cameras are typically used for visual perception and recognition, while lidars and radars utilize laser beams and radio waves to detect objects and their distances precisely. Ultrasonic sensors, however, use sound waves to measure and are commonly employed for parking assistance. Lastly, GPS receivers provide vital location and navigation information to the vehicle. Through these sensors, AVs can detect and respond to their surroundings with high accuracy and efficiency, ultimately making them a safer and more reliable mode of transportation.

5.7.6 Steps involved in sensor fusion

Sensor fusion is an essential component of AI-driven AVs. It involves combining data from various sensors to represent the environment accurately. The primary goal is to achieve real-time data analysis to assist decision-making. The process involves multiple steps: data acquisition, preprocessing, registration, and integration. These steps are critical for achieving precise and efficient sensor fusion. By accurately aligning the data from different sensors, sensor fusion creates a comprehensive perception of the environment, contributing to increased road safety [32,33].

5.8 Perception and localization

The ability to understand and interpret surroundings is a critical component of the success of self-driving cars. Perception involves using cameras, LIDAR, and radar to recognize and comprehend pedestrians, traffic signs, and other vehicles while determining their speed and direction. On the other hand, localization entails identifying the vehicle's exact location in its environment, typically through GPS and other positioning technologies. These two components allow autonomous cars to traverse various terrains safely and efficiently, including highways, city streets, and off-road areas. While significant strides have been made, ongoing innovation in perception and localization technologies is imperative to make self-driving cars a viable transportation option for the future [34,35].

5.8.1 What is the definition of perception and localization?

Perception and localization are essential components of AI systems, particularly in fields like autonomous navigation. Perception is how AI systems acquire information about their surroundings using various sensors, such as cameras, lidar, radar, and microphones. It involves identifying and interpreting objects, obstacles, landmarks, and other relevant environmental information. On the other hand, localization is determining an AI system's position within the environment it perceives. Accurate localization is crucial for navigation and decision-making.

5.8.2 How do these two work?

The following explains the most important steps for perception and localization.

1. *Perception:*
   - *Data acquisition*: Sensors collect data about the environment. For instance, cameras capture images, lidar sends laser beams to measure distances, and microphones record sounds.
   - *Data processing*: The raw sensor data are processed to extract meaningful information. This can involve object detection, feature extraction, image segmentation, and noise reduction.
   - *Object recognition*: AI algorithms analyze the processed data to identify and classify objects, such as pedestrians, cars, traffic signs, or landmarks.
   - *Scene understanding*: Perception systems create a coherent representation of the environment, including objects' spatial arrangement and attributes.
2. *Localization:*
   - *Sensor data integration*: Data from multiple sensors, such as GPS, IMU (Inertial Measurement Unit), and wheel encoders, is used to estimate the AI system's position and orientation.
   - *Map matching*: In cases where a map of the environment is available, localization algorithms compare sensor data to the map to determine the system's location relative to known landmarks or features.

- *Dead reckoning*: IMU and wheel encoders provide information about the system's movement, which can be integrated over time to estimate its position.
- *Particle filters or Kalman filters*: These are common algorithms used for sensor fusion and localization, considering sensor noise and uncertainties.

5.8.3 The theory behind perception and localization

1. *Perception:*
   - *Computer vision:* Perception heavily relies on computer vision, the interdisciplinary image and video analysis field. The computer vision theory includes image processing, feature extraction, object recognition, and DL techniques for visual perception.
   - *Signal processing:* For sensory modalities like lidar and radar, signal processing principles are applied to extract meaningful information from raw sensor data.
2. *Localization:*
   - *Bayesian filtering:* Localization often employs Bayesian filtering methods like the Kalman filter or Particle filter to estimate the system's state (position and orientation) based on sensor data while considering uncertainty.
   - *Simultaneous localization and mapping (SLAM):* In cases where a map is unavailable, SLAM algorithms combine localization and mapping, allowing an AI system to simultaneously estimate its position and build a map of its environment.

5.8.4 Object detection

The detection of objects is a crucial aspect of AV technology. It utilizes advanced computer vision ML algorithms to identify and locate objects in real time, enabling the vehicle's onboard system to make informed decisions and take appropriate action. Object detection is integral to ensuring the safe operation of AVs, as it guarantees the detection and response to obstacles, pedestrians, and other vehicles on the road. The technology leverages cameras, lidar, radar, and other sensors to detect objects and determine their proximity and speed. With the constant evolution of technology, the capabilities of object detection continue to expand, allowing AVs to operate with safety and dependability across a range of driving scenarios.

5.8.5 Classification and recognition

One important aspect of AI for AVs is the ability to classify and recognize objects in the environment. This involves using sensors such as cameras and lidar to gather information about the surroundings and then processing that data to identify objects such as others, pedestrians, and obstacles. DL algorithms and machine vision techniques have dramatically improved accuracy and object recognition. Moreover, classification and recognition play a vital role in decision-making by the AV, enabling it to determine appropriate responses to different situations on the road.

5.9 Mapping

3D maps are a critical aspect of AV technology. These maps allow for accurate navigation by providing detailed information on road layouts, lane markings, traffic signs, and obstacles. To create these maps, lidar, radar, and camera sensors are used to gather data processed by the vehicle's AI system. The resulting map is then used to localize the vehicle and plan routes. As roads and construction sites evolve, the map must be continuously updated to ensure the vehicle can access the most current information.

5.9.1 What does mapping mean?

Mapping in the context of AI and robotics refers to creating a spatial representation of an environment. This representation can include information about the environment's layout, geometry, and features. Mapping is crucial for various applications, including navigation, exploration, and localization.

5.9.2 How does it work?

The principle and general manner of its work is introduced as follows:

1. *Data acquisition:*
   - *Sensors:* Mapping often begins with data acquisition using lidar, cameras, radar, or depth sensors. These sensors collect information about the surroundings, such as distances to objects, visual features, or landmarks.
   - *Robot movement:* In mobile robotics, the robot's movement is tracked using wheel encoders or inertial sensors, providing information about its position and orientation as it moves through the environment.
2. *Data processing:*
   - *Sensor fusion:* If multiple sensors are used, data from these sensors are integrated using techniques like sensor fusion to create a comprehensive and accurate representation of the environment.
   - *Feature extraction:* Algorithms may identify and extract relevant features from sensor data, such as edges, corners, or critical points in images.
   - *Data registration:* Different sensor data streams may need to be registered to a common coordinate system to ensure alignment.
3. *Mapping algorithms:*
   - *Simultaneous localization and mapping (SLAM):* SLAM is a central technique in mapping, where a robot simultaneously estimates its position (localization) and constructs a map of its environment. It combines sensor measurements, robot movement, and uncertainty modeling to achieve this.
   - *Occupancy grid mapping:* This technique divides the environment into a grid and assigns each cell a probability representing the likelihood of being occupied by an obstacle. Laser or lidar scans are used to update the grid over time.
   - *Visual SLAM:* Visual SLAM methods use visual information, such as camera images, to create maps. These approaches often involve feature tracking and triangulation to estimate the position of landmarks.

5.9.3 Theory behind the mapping

1. *Bayesian filtering:*
   - *Kalman filter and particle filter:* These Bayesian filtering techniques are commonly used in SLAM to estimate the vehicle's pose and update the map as new sensor data becomes available.
2. *Occupancy grid mapping:*
   - *Probabilistic models:* The probability theory is applied to model the uncertainty associated with each cell in the occupancy grid. Bayes' theorem is frequently used to update probabilities based on sensor measurements.
3. *Graph theory:*
   - *Graph-based SLAM:* In graph-based SLAM, the environment is represented as a graph where nodes represent robot poses, landmarks, and edges represent measurements or constraints. Graph optimization techniques, like Gauss-Newton or Levenberg-Marquardt, are used to estimate the optimal map and robot trajectory.
4. *Computer vision:*
   - *Feature matching:* In visual SLAM, computer vision theory plays a significant role in feature matching, feature tracking, and visual odometry, which help in estimating the robot's position and building the map.

5.10 Decision-making and control

The AV's AI relies heavily on its decision-making and control system, a crucial component. This system is responsible for receiving input from various sensors, analyzing the data, and making appropriate decisions based on the information received. In order to determine the best course of action, the system uses a combination of algorithms and models such as obstacle avoidance, route planning, and acceleration/deceleration. The system is often trained using ML techniques, including DRL, to improve the quality of its decisions over time. The control aspect of this system translates the decisions made by the AI into physical actions, such as steering, braking, and accelerating. The system's accuracy and reliability are paramount to ensuring the safe operation of AVs [9,36,37].

5.10.1 How does it work?

1. *Perception and sensing:*
   - AI systems begin by perceiving their environment using sensors such as cameras, lidar, radar, and other relevant sensors.
   - Data from these sensors is processed to extract information about the surroundings, including object detection, localization, and recognition.
2. *State estimation:*
   - The AI system estimates its state, including position, orientation, and relevant internal variables. This is often done through sensor fusion and localization techniques.
3. *Decision-making:*
   - Decision-making involves selecting actions or behaviors based on the perceived environment and the system's current state.
   - The process can be rule-based, where predefined rules dictate actions in specific situations, or it can be learned through ML techniques, where the system learns optimal actions from data.
4. *Path planning:*
   - Once decisions are made, the AI system may need to plan a path or trajectory to execute those decisions safely and efficiently.
   - Path planning algorithms consider factors such as obstacles, dynamic changes in the environment, and vehicle dynamics to determine a feasible path.
5. *Control:*
   - Control algorithms translate the desired path or actions into motor commands that drive actuators, such as motors, servos, or thrusters, to control the AI system's movement.
   - Control theory, including proportional-integral-derivative (PID) control and model predictive control (MPC), is crucial in ensuring the precise and stable execution of actions.

5.10.2 The theory behind the decision-making approach

1. *Decision theory:*
   - Decision-making theory, including utility theory and game theory, provides a foundation for understanding how AI systems make choices to optimize objectives while considering uncertainties.
2. *Machine learning:*
   - Reinforcement learning is a subset of ML that involves learning optimal decision-making policies. AI systems learn through interactions with the environment, receiving rewards or penalties based on their actions.
3. *Path planning algorithms:*
   - Graph-based algorithms are used for finding optimal paths in discrete environments.
   - Sampling-based algorithms, such as Rapidly-exploring Random Trees (RRT) and Probabilistic Roadmaps (PRM), are practical for planning in continuous spaces.
4. *Control theory:*
   - PID control theory involves proportional, integral, and derivative terms to regulate the system's output and achieve desired set points.
   - MPC uses predictive models of the system to optimize control inputs over a finite time horizon.
5. *Artificial intelligence:*
   - DRL, a branch of AI, combines deep neural networks with reinforcement learning to enable complex decision-making in high-dimensional state spaces.

5.11 Prediction of driving condition

AVs that use AI technology heavily rely on predicting driving conditions. This is done by analyzing various data sources, including live traffic cameras, weather forecasts, and GPS. Through this analysis, the vehicles can anticipate and respond to various driving conditions, such as traffic levels, the presence of obstacles, and changes in the road due to weather or construction. Accurate prediction of driving conditions can result in a smoother and safer driving experience for passengers. Furthermore, with the help of AI, other factors, such as behavior and route preferences, can be considered. The ability to predict driving conditions is crucial for ensuring AVs can navigate roads safely and efficiently.

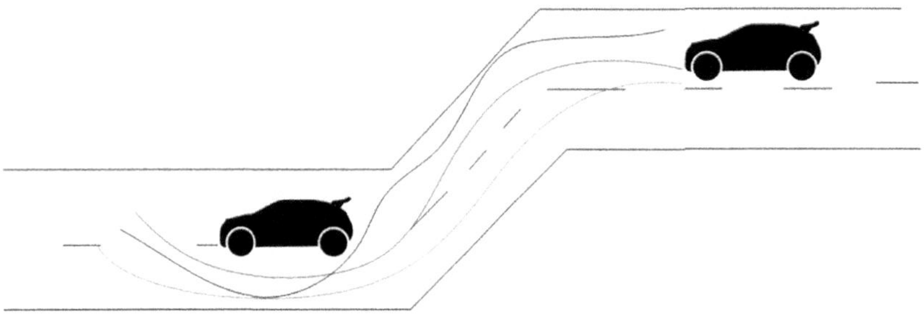

FIG. 4 Path planning of the proposed AV.

5.11.1 Path planning

Path planning is crucial to the success of AV systems. It involves selecting the optimal route for a vehicle to reach its destination while navigating obstacles and adhering to traffic rules and regulations, as shown in Fig. 4. In order to achieve this, path planning algorithms rely on sophisticated sensors, including radar, lidar, cameras, and GPS, to gather data about the environment. This data are then analyzed using advanced techniques to generate a safe and efficient vehicle. Effective path planning is essential for autonomous driving in complex urban and suburban environments, requiring expertise in AI ML, sensor fusion, and real-time processing and communication within the vehicle's control system.

5.11.2 Controlling of motion

The ability to control motion is a crucial aspect of autonomous driving. AI-powered vehicles use real-time data analysis to make acceleration, braking, and steering decisions, considering traffic conditions, weather, and road hazards. These decisions are made by smart sensors and algorithms trained on vast amounts of data, allowing the vehicle to make split-second decisions that prioritize passenger safety and prevent accidents. As AI technologies advance, AVs can become even more adept at navigating complex urban environments with greater accuracy and dependability [13,21].

5.12 Advantages of AVs

- *Safety improvements*
  One of the most significant advantages of AVs is the potential to eliminate human error–a leading cause of accidents. AI systems can react faster and with more precision, minimizing the chances of collisions and promoting road safety.
- *Traffic efficiency*
  AI-driven AVs can potentially reduce traffic congestion by optimizing traffic flow. Vehicles can communicate with each other and adjust speeds, ensuring smoother traffic movement and shorter travel times.
- *Accessibility*
  AVs can provide increased mobility to individuals with disabilities or older people, allowing them to travel without relying on others. This inclusivity is a promising aspect of AI-powered transportation.

5.13 Challenges to overcome

- *Legal and regulatory hurdles*
  The introduction of AVs raises legal and regulatory challenges. Determining liability in the event of accidents, defining safety standards, and addressing privacy concerns are all issues that must be resolved before widespread adoption.

- *Ethical dilemmas*
  AI algorithms must make split-second decisions in life-threatening situations, raising ethical questions. How should an AV prioritize the safety of its occupants versus pedestrians? Resolving these moral dilemmas is crucial.
- *Technical limitations*
  While AI has come a long way, technical challenges still exist. Adverse weather conditions, complex urban environments, and unpredictable human behavior can pose difficulties for AI-driven systems.

5.14 AI and ML in autonomous vehicles

- *Training models*
  ML plays a vital role in training AI models for AVs. These models learn from vast datasets to recognize patterns and accurately predict their surroundings.
- *Real-time adaptability*
  AVs must be able to adapt to dynamic environments in real time. ML enables them to learn from new situations and continuously improve their decision-making processes.

5.15 The future landscape

- *Urban planning changes*
  As AVs become more prevalent, urban planning may evolve. With a reduced need for parking spaces and changes in traffic patterns, cities could be redesigned to be more people-centric.
- *Shared mobility services*
  AVs could lead to the rise of shared mobility services. Instead of owning cars, people might prefer using on-demand AVs, reducing traffic and emissions.
- *Job transformations*
  The widespread adoption of AVs could lead to job transformations in transportation-related sectors. While some driving jobs may decline, new maintenance, AI programming, and monitoring roles may emerge.
- *Environmental benefits*
  By optimizing traffic flow and encouraging shared rides, AVs could contribute to reduced emissions and lower fuel consumption, positively impacting the environment [5,14,38].

5.16 Summary and conclusion

Incorporating AI into AVs is a promising advancement for the transportation industry. AI enables vehicles to operate independently, make their own decisions, and navigate roads more efficiently without human intervention. This chapter delves into various aspects of AI for AVs, including perception, decision-making, and control systems. Research indicates that AI-powered vehicles have the potential to enhance road safety by minimizing human error and response times. These vehicles leverage advanced sensors and ML algorithms to detect and respond to their surroundings more efficiently than human drivers. Moreover, AI allows real-time data analysis, enabling AVs to make informed decisions and adapt to changing road conditions.

Furthermore, incorporating AI in AVs can revolutionize transportation systems by increasing efficiency and reducing traffic congestion. By utilizing intelligent routing algorithms and predictive analytics, travel routes can be optimized, fuel consumption can be minimized, and travel times can be reduced. This benefits both individual users and the environment by decreasing carbon emissions. However, challenges must be addressed before AI can be widely adopted in AVs. Legal and ethical considerations surrounding liability in accidents involving AVs must be addressed, as well as ensuring robust cyber security measures to protect against potential hacking threats.

References

[1] M.A. Jirdehi, M. Shaterabadi, A low-carbon strategy using INVELOX turbines in the presence of real-time energy price uncertainty, Greenh. Gases Sci. Technol. 11 (3) (2021) 461–482.

[2] M. Shaterabadi, M.A. Jirdehi, N. Amiri, Impact of fuel price increasing policies on fuel smuggling and acceptance of green cars: a case study of Iran, J. Eng. 2022 (12) (2022) 1239–1248.

[3] G.D. Sharma, A. Yadav, R. Chopra, Artificial intelligence and effective governance: a review, critique and research agenda, Sustain. Futur. 2 (2020) 100004.

[4] A. Osório, A. Pinto, Information, uncertainty and the manipulability of artificial intelligence autonomous vehicles systems, Int. J. Hum. Comput. Stud. 130 (2019) 40–46.

[5] M. Cunneen, M. Mullins, F. Murphy, Autonomous vehicles and embedded artificial intelligence: the challenges of framing machine driving decisions, Appl. Artif. Intell. 33 (8) (2019) 706–731.

[6] I.B. Ajenaghughrure, S.C. da Costa Sousa, D. Lamas, Risk and trust in artificial intelligence technologies: A case study of autonomous vehicles, in: 2020 13th International Conference on Human System Interaction (HSI), IEEE, 2020, pp. 118–123.

[7] Y. Ma, Z. Wang, H. Yang, L. Yang, Artificial intelligence applications in the development of autonomous vehicles: a survey, IEEE/CAA J. Autom. Sin. 7 (2) (2020) 315–329.

[8] N. Naz, M.K. Ehsan, M.R. Amirzada, M.Y. Ali, M.A. Qureshi, Intelligence of autonomous vehicles: a concise revisit, J. Sens. 2022 (2022).

[9] D. Parekh, et al., A review on autonomous vehicles: progress, methods and challenges, Electronics 11 (14) (2022) 2162.

[10] S. Iftikhar, Z. Zhang, M. Asim, A. Muthanna, A. Koucheryavy, A.A. Abd El-Latif, Deep learning-based pedestrian detection in autonomous vehicles: substantial issues and challenges, Electronics 11 (21) (2022) 3551.

[11] C. Medrano-Berumen, M. İlhan Akbaş, Validation of decision-making in artificial intelligence-based autonomous vehicles, J. Inf. Telecommun. 5 (1) (2021) 83–103.

[12] A.M. Nascimento, et al., A systematic literature review about the impact of artificial intelligence on autonomous vehicle safety, IEEE Trans. Intell. Transp. Syst. 21 (12) (2019) 4928–4946.

[13] S.S. Kamran, A. Haleem, S. Bahl, M. Javaid, C. Prakash, D. Budhhi, Artificial intelligence and advanced materials in automotive industry: potential applications and perspectives, Mater. Today Proc. 62 (2022) 4207–4214.

[14] G. Bendiab, A. Hameurlaine, G. Germanos, N. Kolokotronis, S. Shiaeles, Autonomous vehicles security: challenges and solutions using blockchain and artificial intelligence, IEEE Trans. Intell. Transp. Syst. 24 (2023) 3614–3637.

[15] S. Yang, Y. Chen, R. Shi, R. Wang, Y. Cao, J. Lu, A survey of intelligent tires for tire-road interaction recognition toward autonomous vehicles, IEEE Trans. Intell. Veh. 7 (3) (2022) 520–532.

[16] Y. Ren, R. Xie, F.R. Yu, T. Huang, Y. Liu, Green intelligence networking for connected and autonomous vehicles in smart cities, IEEE Trans. Green Commun. Netw. 6 (3) (2022) 1591–1603.

[17] A. Biswas, et al., State-of-the-art review on recent advancements on lateral control of autonomous vehicles, IEEE Access 10 (2022) 114759–114786.

[18] H. Mankodiya, M.S. Obaidat, R. Gupta, S. Tanwar, XAI-AV: Explainable artificial intelligence for trust management in autonomous vehicles, in: 2021 International Conference on Communications, Computing, Cybersecurity, and Informatics (CCCI), IEEE, 2021, pp. 1–5.

[19] A. Biswas, H.-C. Wang, Autonomous vehicles enabled by the integration of IoT, edge intelligence, 5G, and blockchain, Sensors 23 (4) (2023) 1963.

[20] M. Hengstler, E. Enkel, S. Duelli, Applied artificial intelligence and trust—the case of autonomous vehicles and medical assistance devices, Technol. Forecast. Soc. Chang. 105 (2016) 105–120.

[21] H. Khayyam, B. Javadi, M. Jalili, R.N. Jazar, Artificial intelligence and internet of things for autonomous vehicles, in: R. Jazar, L. Dai (Eds.), Nonlinear Approaches in Engineering Applications: Automotive Applications of Engineering Problems, Springer, Cham, 2020, pp. 39–68.

[22] S.A. Bagloee, M. Tavana, M. Asadi, T. Oliver, Autonomous vehicles: challenges, opportunities, and future implications for transportation policies, J. Mod. Transp. 24 (2016) 284–303.

[23] L. Chen, et al., Milestones in autonomous driving and intelligent vehicles: survey of surveys, IEEE Trans. Intell. Veh. 8 (2) (2022) 1046–1056.

[24] S. Grigorescu, B. Trasnea, T. Cocias, G. Macesanu, A survey of deep learning techniques for autonomous driving, J. Field Robot. 37 (3) (2020) 362–386.

[25] G. Meyer, S. Beiker, Road Vehicle Automation, Springer, 2019.

[26] A. Miglani, N. Kumar, Deep learning models for traffic flow prediction in autonomous vehicles: a review, solutions, and challenges, Veh. Commun. 20 (2019) 100184.

[27] S. Mozaffari, O.Y. Al-Jarrah, M. Dianati, P. Jennings, A. Mouzakitis, Deep learning-based vehicle behavior prediction for autonomous driving applications: a review, IEEE Trans. Intell. Transp. Syst. 23 (1) (2020) 33–47.

[28] W. Schwarting, J. Alonso-Mora, D. Rus, Planning and decision-making for autonomous vehicles, Annu. Rev. Control Robot Auton. Syst. 1 (2018) 187–210.

[29] A. Thakur, S.K. Mishra, Review on vision-based control using artificial intelligence in autonomous ground vehicle, in: Proceedings of the International Conference on Paradigms of Computing, Communication and Data Sciences: PCCDS 2022, Springer, 2023, pp. 617–626.

[30] S. Murugan, A. Sampathkumar, S. Kanaga Suba Raja, S. Ramesh, R. Manikandan, D. Gupta, Autonomous vehicle assisted by heads up display (HUD) with augmented reality based on machine learning techniques, in: Virtual and Augmented Reality for Automobile Industry: Innovation Vision and Applications, Springer, 2022, pp. 45–64.

[31] B. Namatherdhala, N. Mazher, G.K. Sriram, Uses of artificial intelligence in autonomous driving and V2X communication, Int. Res. J. Modern. Eng. Technol. Sci. 4 (7) (2022) 1932–1936.

[32] M. Lakshmanan, Y. Justindhas, A. Moorthy, Survey on autonomous vehicles using artificial intelligence, in: 2022 3rd International Conference on Electronics and Sustainable Communication Systems (ICESC), IEEE, 2022, pp. 1237–1241.

[33] S. Mithas, Z.L. Chen, T.J. Saldanha, A. De Oliveira Silveira, How will artificial intelligence and industry 4.0 emerging technologies transform operations management? Prod. Oper. Manag. 31 (12) (2022) 4475–4487.

[34] D. Fernandez-Llorca, E. Gómez, Trustworthy artificial intelligence requirements in the autonomous driving domain, Computer 56 (2) (2023) 29–39.

[35] A.S. Madhav, A.K. Tyagi, Explainable artificial intelligence (XAI): Connecting artificial decision-making and human trust in autonomous vehicles, in: Proceedings of Third International Conference on Computing, Communications, and Cyber-Security: IC4S 2021, Springer, 2022, pp. 123–136.

[36] N. George, J. Thomas, Authenticating communication of autonomous vehicles with artificial intelligence, in: IOP Conference Series: Materials Science and Engineering, Vol. 396, no. 1, IOP Publishing, 2018, p. 012017.

[37] J.L. Mezrich, Is artificial intelligence (AI) a pipe dream? Why legal issues present significant hurdles to AI autonomy, Am. J. Roentgenol. 219 (1) (2022) 152–156.

[38] G. Bathla, et al., Autonomous vehicles and intelligent automation: applications, challenges, and opportunities, Mob. Inf. Syst. 2022 (2022).

CHAPTER

6

Security for autonomous vehicles

Sreedhar Madichetty[a], *Pathan Meeravali Khan*[a], *and Abdelkader El Kamel*[b]

[a]Mahindra University, Hyderabad, India [b]Centrale Lille, France

6.1 Introduction

Autonomous vehicles (AVs) are currently a focal point in both the automotive industry and academic research. Prominent companies like Tesla, Toyota, BMW, and Uber have incorporated AV technology into their vehicles. This innovation enables automated driving, supplanting the need for human drivers. The advantages include diminished energy consumption, a decrease in road accidents, and a pollution reduction. The advent of advanced technologies and optimized communication systems has given rise to AVs or self-driven vehicles with enhanced functionalities and autonomy levels. These vehicles utilize information from both internal and external systems to provide efficient driving assistance, enhancing overall vehicle and human safety [1]. The Society of Automotive Engineers (SAE) categorizes the level of automation in vehicles, defining 6 levels from 0 (Normal vehicle) to 5 (full automation vehicle). At each level, there is a gradual increase in automation, leading to a reduction in driver involvement and an overall improvement in safety [2]. AVs can be characterized as specific types of Internet of Things (IoT) systems and a subset of cyber-physical systems (CPS). They are complex networks comprising interconnected embedded system components. Unlike traditional vehicles, AVs incorporate a range of automation sensors such as RADAR (Radio detection and ranging technology), cameras, and LiDAR (Light Detection and Ranging), as well as actuators and Electronic Control Units (ECUs) connected to OBC (On-Board Computer) in the vehicle, that assume the role of human driver behavior as shown in Fig. 1. Shows an illustration of AV components.

Within AVs, sensors play a crucial role in monitoring both internal vehicle dynamics (such as speed and location) and the external environment (including traffic conditions). The onboard computer processes information from these internal and external sensors, issuing commands to ECUs that, in turn, control the actuators to perform specific functions. This network of communication between OBC, external sensors, control units (CUs), and actuators forms the internal vehicle communication network. Furthermore, AVs can communicate with other vehicles (V2V), roadside Infrastructure (V2I), the cloud (V2C), and Pedestrians (V2P) using Vehicles-to-Everything (V2X) technology [3] as shown in Fig. 2.

Despite these advancements, it's essential to acknowledge that all components in AVs are susceptible to cyber-attacks and unauthorized access, posing potential risks to vehicle security. So providing cyber security to the AVs is more important. From the AV security point of view, first, the open system interconnection (OSI) model in AV security is discussed.

The OSI model is a conceptual framework that organizes the functions of a telecommunication or computing system into seven distinct abstraction layers (physical layer, data link layer, network layer, transport layer, session layer, presentation layer, and application layer) [4]. As shown in Fig. 3, these layers facilitate the understanding design and communication of complex networks. When applying the OSI model to the cyber security concept of AVs, each layer plays a very important role in the security of the AVs as follows:

1. Physical layer: In the AV context view, the physical layer involves the hardware components such as the sensor, actuator, and communication interface. The security consideration includes protecting these physical components from tampering and ensuring the integrity of hardware.

Handbook of Power Electronics in Autonomous and Electric Vehicles
https://doi.org/10.1016/B978-0-323-99545-0.00014-2

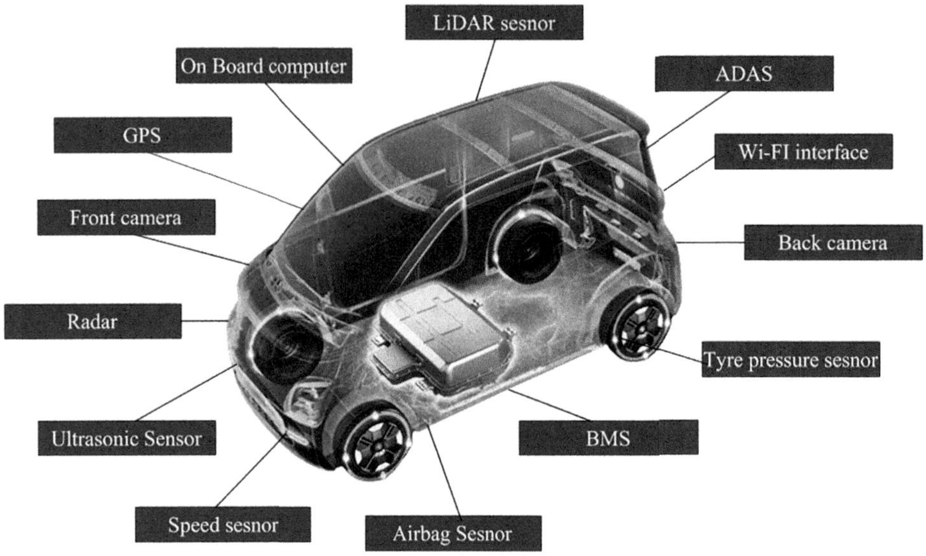

FIG. 1 An illustration of autonomous vehicle components.

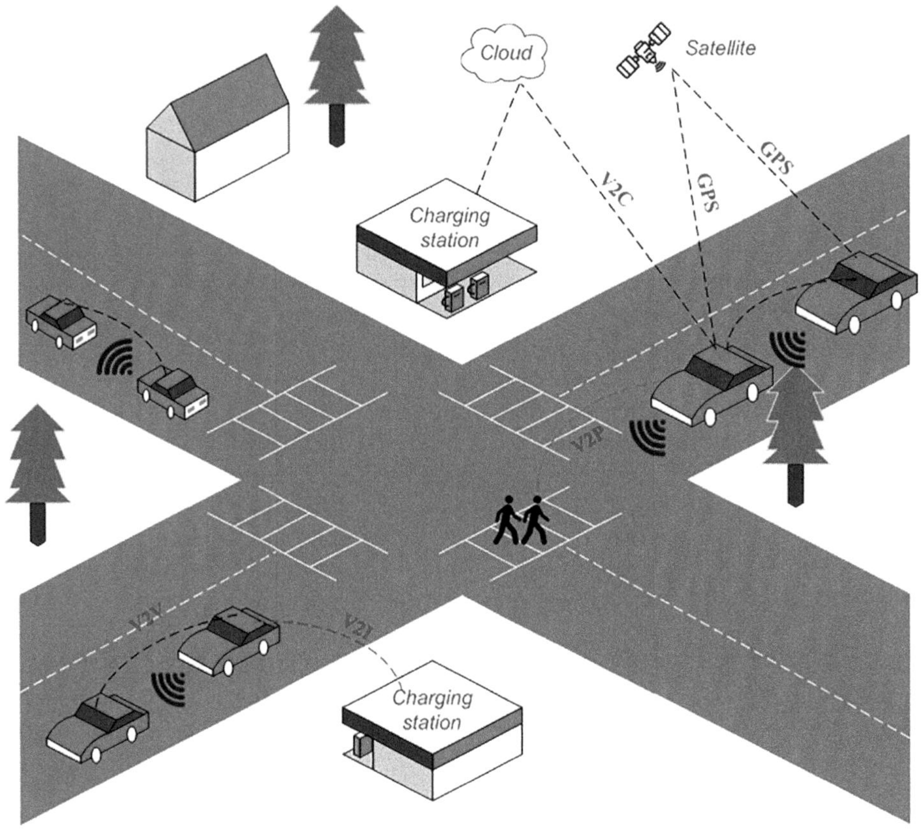

FIG. 2 An illustration of autonomous vehicle communication technology.

2. Data link layer: this layer consideration of node-to-node communication. Like in the AVs, communication channels (control area network (CAN) bus, power line communication (PLC)) between ECUs to ECUs. This layer is involved in ensuring the integrity and reliability of data transmission between components within the vehicle.
3. Network Layer: This layer manages routing and addressing in a network. In the AV context, secure measure involves protecting the network infrastructure and implementing secure routing protocols.
4. Transport layer: This layer ensures end-to-end communication between devices. Security consideration includes integrity, authentication, and encryption to protect data as it moves between different components of the AVs.

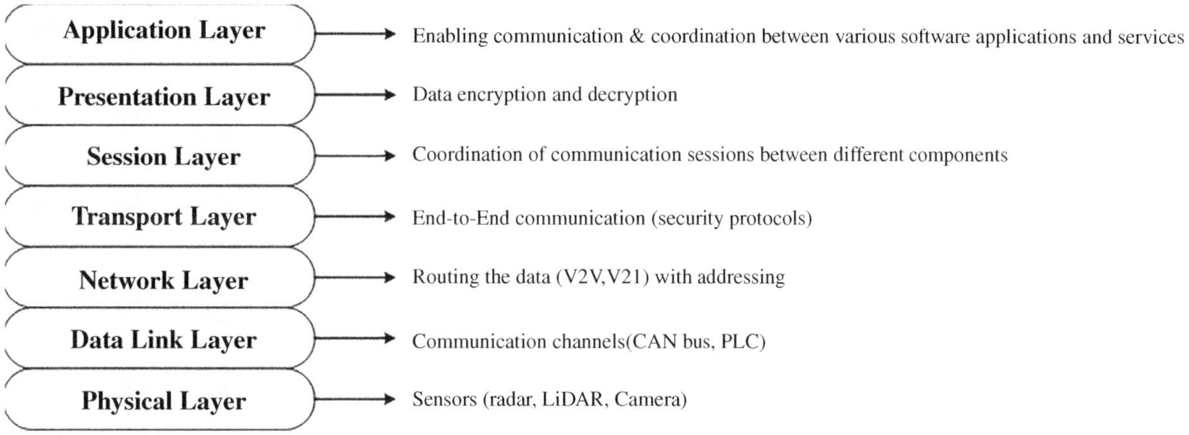

FIG. 3 OSI model in the AVs.

5. Session layer: This layer establishes, maintains, and terminates connections between applications. Security measures at this layer involve managing secure sessions.
6. Presentation layer: Deals with data format translation, compression, and encryption. In the context of AVs, security measures include data encryption and the standardization of the data formats.
7. Application layer: This layer provides network services directly to end users. In the case of AVs, security measures involve securing application-level protocols.

6.2 Different components in autonomous vehicles

In the AVs, four components are explored privacy and securities. These components are (1) sensing components, (2) CUs, (3) in-vehicle communication network, and (4) V2X.

6.2.1 Sensing devices

In AVs, sensors are very crucial components for autonomous driving, they allow to monitoring of vehicle surroundings the environment, and internal vehicle dynamics, collecting the data needed for the automation of the vehicles and their safety [5]. In general, AVs currently generate approximately 25GB of data per hour. With the anticipated rise in automation levels and enhanced safety, these vehicles could produce up to 300TB of data annually. The generated data play a crucial role as it is processed by the OBC system. This processing involves tasks such as mapping a route from one point to another point and transmitting precise instructions to control various aspects of the AV, including steering, speed adjustment, airbag deployment, and brake control. In the AVs, three primary sensors are cameras (mono and stereo cameras), LiDAR, and RADAR. These sensors together provide the vehicle with visuals of its surrounding environment and help it detect speed and distance between the two AVs.

The RADAR detects the object within a certain range by using radio waves. Using this type of sensor to identify the real-time condition of the vehicle. Another sensor is LiDAR used for the short-length laser to achieve higher measurement of the accuracy and better spatial resolution of the RADAR. For detection of the location of the AVs by using global position system (GPS) to provide the position, navigation, and timing (PNT) service to the users. By using GPS, we can identify the exact location of the vehicle in a real-time environment. Additionally, AVs have ultra sensors for measuring the distance between two vehicles with higher frequency sounds. This sound wave compatibility finds the exact distance between the vehicles in solid and liquid waves. Ultra-sonics sensors used short distance.

6.2.2 Control units

The CUs are embedded systems that manage the sensors internal and external of the vehicle and also control automation of the vehicle engine. In AVs, typically small and medium-sized AVs have 50 CUs [6]. Modern luxury vehicles can have more than 150 CUs [7]. Every CU has certain responsibilities for controlling a specific function, such as vehicle body control, brake, speed, media control, door, and window control, etc. CU unit's different modules include a

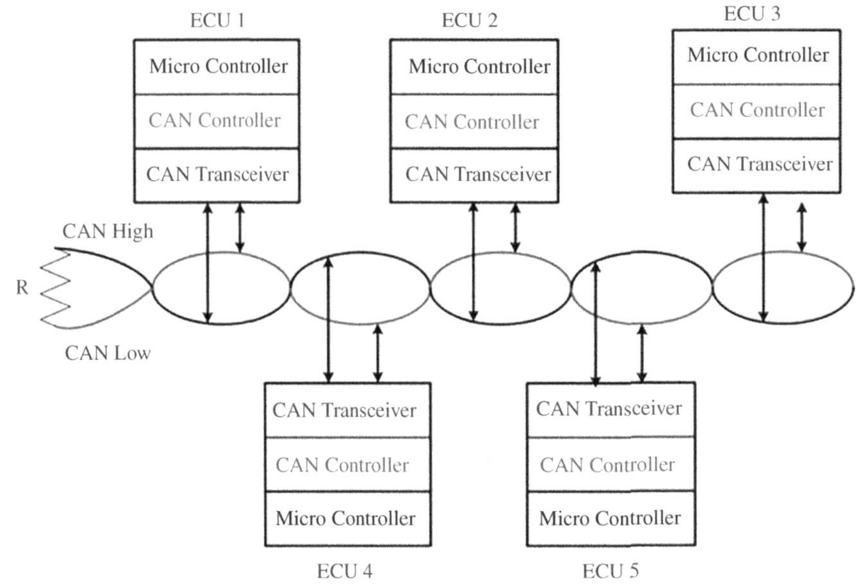

FIG. 4 CAN bus communication between ECUs.

general electronics module, central control module, Engine Control Module (ECM), Brake Control Module (BCM), Transmission Control Module (TCM), Power Train Control Module (PCM), and Body Control Module (BCM). The ECM mainly controls the engine function and speed control. The TCM mainly controls the transmission of the data from sensors. In the PCM, mainly control combines the TCM and ECM.

All the CUs are connected through the nodes with physical wires connection like CAN bus, and PLC for transfer the data from one CU to another CU [8] as shown in Fig. 4. In control area network (CAN bus) consist of two wires (CAN_High, CAN_Low) , these physical wires connection between CUs. By using the differential voltage concept in the CAN bus transfer the data between the CU. In the PLC, the data transfer between CU in the power lines with different frequencies. In the PLC, there is no need for additional wires for data transfer.

6.2.3 Inner-vehicle network communications

In the AVs, an in-vehicle network is used for the transmission of the status of information, and control signals among all sensors, actuators, and all CUs. The in-vehicle network is classified into two types: wired and wireless networks. In wired networks CAN, local area network, Flex ray, Ethernet, Media Oriented Transport System (MOST), PLC, Low Voltage Differential Voltage (LVDS). In wireless technology like bluetooth, zigbee, ultra-wide band (UWB), and wireless fidelity (Wi-Fi). In wired technology, the data transfer by using wire, due to using wire for communication has some limitations like (i) internal wiring harness becomes highly complex, (ii) if the CUs are increasing then corresponding to the wire material increasing, the cost gradually increases, (iii) in some places deployment of the wire technology is very difficult example of the wired structure on the steering wheel and tired parts, and (iv) intervehicle spaces are more heaving. Due to limitations in wired technology, it can be overcome by wireless technology. In this technology, all CUs and sensors are connected without being wired saving energy in the vehicles.

6.2.4 Vehicle to everything (V2X) technology

V2X technology enables AVs to communicate with their surroundings and makes driving safer and more efficient for everyone. In V2V technology, the data transfer takes place between the AVs, whereas in V2C, the communication will happen between AV and to Cloud, and in V2I the communication takes place between the vehicle's Road Side Units (RSU) like charging stations. It utilizes Dedicated Short-Range Communication (DSRC) based on the IEEE 802.11p standard, catering to wireless access in vehicular environments. DSRC facilitates brief information exchanges between DSRC devices, including On-Board Units (OBD) within vehicles, Roadside Units (RSUs), and mobile devices carried by pedestrians. RSUs establish communication with location servers through wired or wireless networks to track information on all vehicles. Additionally, the V2X communication employs the mobile cellular network. In 2017, the mobile industry body 3GPP standardized a suite of technologies known as Cellular Vehicle-to-Everything (C-V2X), specifically designed for communication between AVs and roadside infrastructure.

6.3 Major cyber attacks for autonomous vehicles

In the AVs, several CUs are connected with sensors and transfer information to other CUs by using communication channels; these channels create new opportunities for attackers, and to implement various types of cyber-attacks in the AVs as shown in Fig. 5, which could lead to an access of the vehicle and major safety issues. The main motive of these attackers is to gain control over AVs remotely, steal valuable and confidential information that is used for launching further attacks on the AVs, or interrupt the operation of the AVs. In the AVs, major cyber-attacks are given in the following categories: manipulation-based attacks, identity-based attacks, functionality compromise incidents (service-based attacks), code manipulation intrusions (software-based attacks), and physical access through attacks.

6.3.1 Manipulation-based attacks

The primary objective of such attacks is to gain unauthorized access to the AVs, to compromise the data integrity, and to violate the privacy of the AVs. Different attacks come under the manipulation-based attacks these are Man in the Middle (MITM) attacks, injection attacks, tampering attacks, and modifications attacks.

6.3.1.1 Man in the middle (MITM) attacks

In this attack, the attacker intercepts communication between the AVs and external devices, which are used like Bluetooth, ZigBee, Wi-Fi, and mobile communication protocols [9], and manipulates the data, transmitting between two entities (two AVs in a V2V network, or RSU in V2I network). Fig. 6 shows the MITM attack between AV and RSUs.

6.3.1.2 Injection attacks

This type of attack injects the malicious code into the CUs. CUs are programmed based on Operating System (OS), such as embedded Linux and Android. Due to these attacks, CUs send the information to the third malicious parties' application at the same time [10].

6.3.1.3 Tampering attacks

In this type of attack, the attacker modifies (manipulates, destroys, or edits) data through the unauthorized access of channels in between the trusted AVs [11]. Meanwhile, this type of attack is noticed in the V2V and V2X communications.

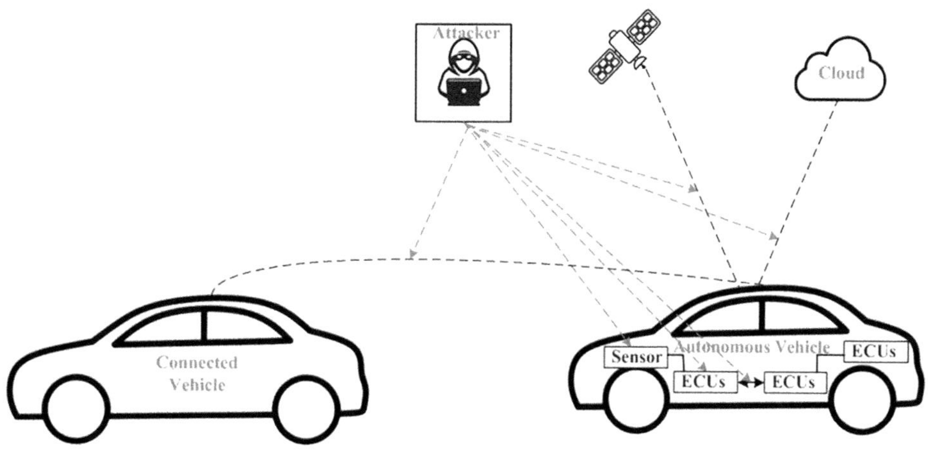

FIG. 5 Different access points in the AVs for cyber attacks.

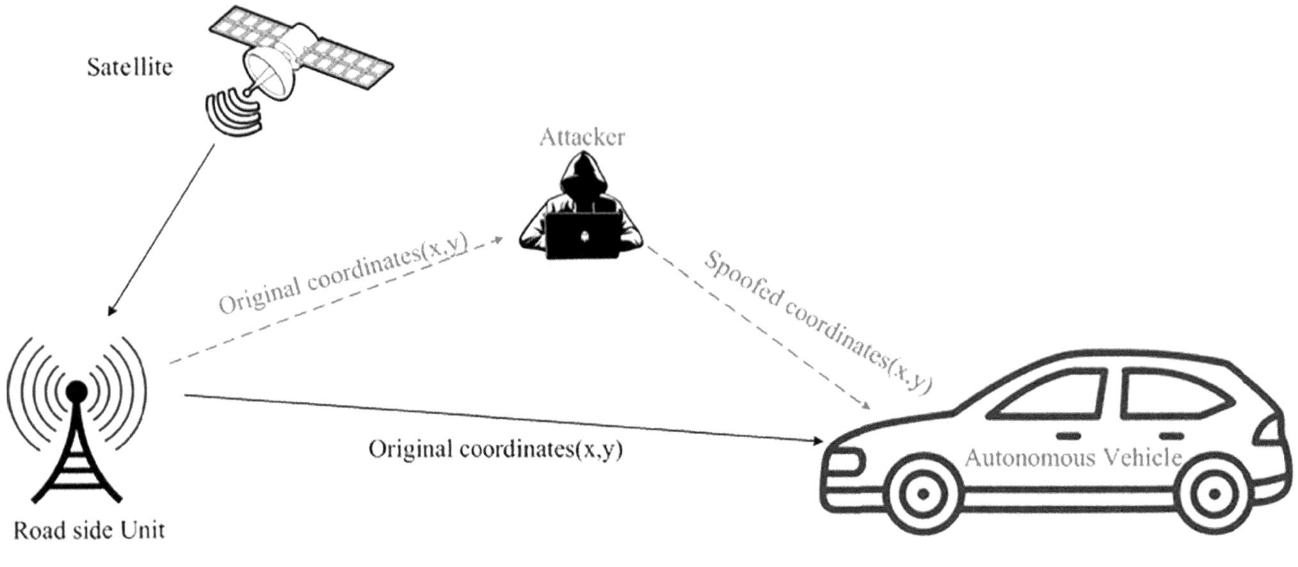

FIG. 6 MITM attack in the between AV and road side units.

6.3.2 Identity-based attacks

Every AV has unique identification for an identified purpose, this helps to identify or recognize the AV and exchange the message. In this attack, the attacker can steal the identity of the AV, to get access to the components of AV and manipulate them [12]. In some cases, sending falsification data to other trusted AVs through the communication channels interrupts the operation of the AVs and traffic flow.

6.3.2.1 *Spoofing attack*

In this attack, attackers gain access to the AVs and send false data. Normally the spoofing attacks are classified as GPS spoofing, LiDAR spoofing, RADAR spoofing, and camera spoofing. In GPS spoofing, the navigation system in the AVs receives false broadcasting data. In LiDAR spoofing, attacks create fake signals that represent objects and inject them into a LiDAR.

6.3.2.2 *Replay attacks*

In this attack, the attackers gain access to the CU, absorb the operation and sniff the information, and modify the message. After accessing the CU, messages are sent continuously other CUs. Replay attacks can exploit communication channels between a vehicle and RSUs or neighboring vehicles to steal confidential encryption keys or passwords, subsequently utilizing this information for unauthorized authentication.

6.3.3 Functionality compromise incidents

The primary objective of such an attack is to interrupt the operation of the AV or to disrupt the AVs in the large traffic flow area. The type of attacks category is as follows:

6.3.3.1 *Denial of service (DoS) attacks*

These attacks are very dangerous in the AVs operation. This attack interrupts the emergency operation of the AVs and stops the sensing information from sensors like LiDAR, camera, and radar. This attack was implemented on the CAN bus, to interrupt the entire communication in the vehicle, as shown in Fig. 7.

6.3.3.2 *Jamming attacks*

The attacker intentionally injects a false message into the communication channels and creates interrupts in the AV, V2V, V2I, and RSU as shown in Fig. 8. It also targets the sensors in the AVs (GPS jamming, LiDAR jamming, and radar jamming) by injecting the falsification message or noise signals in the communication, which causes blindness function in the operations.

FIG. 7 DoS attack between two ECUs.

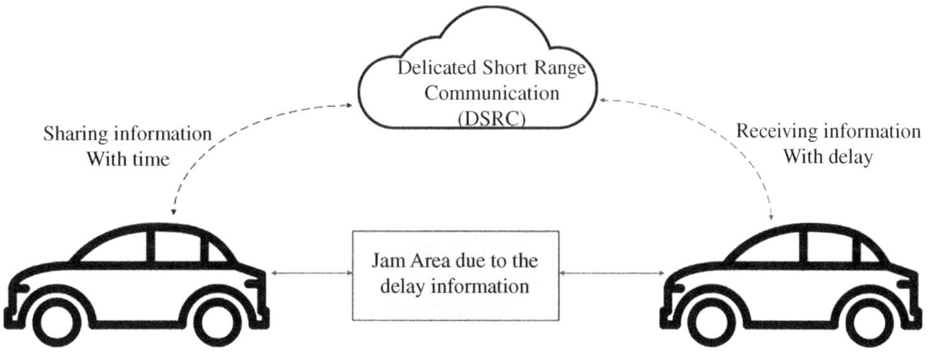

FIG. 8 Jamming attack between AVs and CVs.

6.3.3.3 *Sensor blinding attacks*

This type of attack mainly focuses on the three sensors (LiDAR, radar, and camera) in the AVs. Based on this sensor information and coordination, the AVs move forward. If any sensor blinds, the AV operation malfunctions and creates safety issues.

6.3.3.4 *Spamming attacks*

This type of attack can be implemented in V2V and V2I operations, and it creates spam messages and sends them continuously through the network. Thereby increasing the transmission latency and creating delays in the V2V and V2X communications.

6.3.4 Code manipulation intrusions

Of all the attacks above-mentioned attacks, some attacks come under software-based attacks. In this attack, malicious code or message in the AV and creates a malfunction operation. This includes malware, ransomware, and mobile app attacks.

6.3.4.1 *Malware attacks*

One of the severe threats of the extensive introduction of automation in the vehicle is vehicle hacking by using different forms of malware including viruses, Trojans, rootkits, and backdoors. In this attack, injected malware in the AVs creates malfunction.

6.3.4.2 *Mobile app attacks*

Almost, most the AVs are connected to mobile applications for basic control functions through Bluetooth and WI-FI. The attacker installs a virus or malicious application; it is indirectly connected to a vehicle. By using an application, the attacker may gain unauthorized access to the vehicles.

6.3.5 Physical access through the attacks

Normally, AVs have many direct and indirect physical interfaces that can become potential surfaces of the attacks because they give the attackers direct access to the in-vehicle network. Physical access attacks are classified as direct and indirect access.

6.3.5.1 *Direct access attacks*

In these attacks, attackers directly access the vehicle by using OBD and USB devices. The OBD device is directly connected to the in-vehicle communication (CAN bus). This CAN bus is directly connected to all the CUs in the vehicle. By using OBD-II, the attacker directly injected malicious messages into the in-vehicle communication.

6.3.5.2 *Indirect access attacks*

The in-vehicle network can also be accessed through indirect interfaces without the presence of an attacker. Most AVs are operated by using mobile applications.

6.4 Defense strategies for cyber security in autonomous vehicles

Securing AVs is crucial to ensure their safe and reliable operation during cyber-attacks occurs in various parts of the AVs. This section explains various defense strategies.

6.4.1 Defense for in-vehicle network attacks

To defend against vehicle attacks, based on the type of these attacks, there are the following strategies.

6.4.1.1 *Against remote sensor attacks*

Remote sensor attacks involve manipulating or compromising the sensors of AVs to provide inaccurate or malicious information, potentially leading to unsafe or unintended behavior. Protecting against such attacks is critical for the safe operation of AVs. To defend the remote sensor, there are several efficient solutions including authentication and access control, sensor fusion, and intrusion detection systems for sensors.

In the authentications and access control, it implements a strong authentication mechanism for accessing and configuring sensors. Restrict access control to sensor configuration settings to authorized personnel only. The sensor fusion method integrates data from multiple sensors (such as LiDAR, radar, and cameras). This enhances the overall accuracy of the perception system. In the intrusion detection systems method, adversaries can deploy the algorithms to monitor for abnormal behavior or unauthorized access.

6.4.1.2 *Against GPS spoofing*

GPS spoofing is an attack, which manipulates the GPS signal received by AV to provide false location information. Here are some methods to protect against GPS spoofing. By using signal strength monitoring, monitor the GPS signal strength and their variation. If the GPS signal strength is sudden, exchanges may indicate spoofing attempts. Using Signal Authentication may involve using cryptographic techniques to ensure that the received signals from legitimate.

6.4.1.3 *Against CAN and SAE J1939 buses vulnerabilities*

CAN and SAE J1939 are widely used in the AVs for communication between ECUs. Some methods have to mitigate vulnerabilities in the CAN and SAE J1939 as follows: Network Segmentation, Authentication, and data encryption.

The network authentication isolates critical vehicle systems from less critical ones. This one helps from potential breaches and limits unauthorized access to sensitive parts of the network. In authentication, only authorized devices and ECUs are accessed to participate in communication. In Encryption, the data are encrypted and transmitted to the CAN to prevent eavesdropping and tampering as shown in Fig. 9.

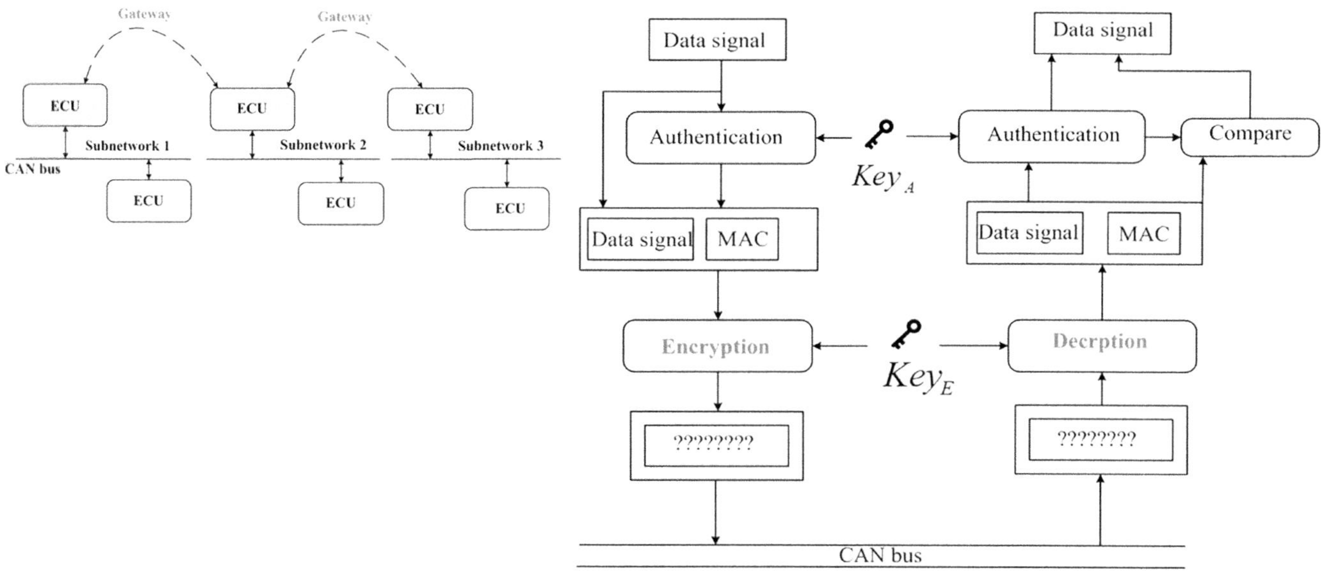

FIG. 9 Network segmentation, authentication, and data encryption in CAN bus security.

6.4.1.4 *Against ECUs software flashing attacks*

ECUs are critical components in vehicles that control various functions, and attacks targeting their software can have server consequences. Encrypted communication is used for software updates to prevent eavesdropping and MITM attacks. The second one is code signing and authentication, this method involves digitally signing firmware updates, and ECUs should verify the signature before applying any changes.

6.4.2 Defense for vehicle-to-everything network attacks

The V2X network is a crucial component of connected and autonomous, enabling communication between V2V, V2I, V2P, and other entities. Securing the V2X network is essential to ensure the safety and reliability of CVs.

6.4.2.1 *Against DoS attacks*

For AVs, a DoS attack could have very safety implications. Here is the measure to guard against DoS attacks in the AVs. Different methods of implementation for defense against DoS attacks are sliding mode, adaptive estimation, bandwidth, and entropy.

6.4.2.2 *Against impersonations attacks*

Impersonation attacks in the context of AVs refer to where an unauthorized entity attempts to mimic or spoof the identity of a legitimate component within the vehicle's communication. To defend the impersonation attacks in the AVs by secure transmission, integrity verifications, authentication, and secure key agreement.

The secure transmission method employs secure and encrypted communication protocols to protect data integrity and confidentiality.

6.4.2.3 *Against replay attacks*

In AVs, replay attacks involve the interception and subsequent retransmission of valid data to deceive a system. For the defense, the replay attacks in the AVs use timestamps and nonce's (random numbers used only once) in communication protocol and implement a token-based authentication mechanism in the AVs, tokens have a limited validity period and should be checked for freshness by the recipient.

6.4.2.4 *Against data falsifications attacks*

Protecting against data falsification attacks is crucial in ensuring the integrity of information within AVs or any system that relies on accurate data. Here are some strategies to mitigate the risk of data falsification attacks: location detection of the attack, dynamic time window, data filtering, and information sharing.

6.4.3 Defense for against other attacks

In addition to specific defenses against V2X network attacks, it is crucial to implement a comprehensive cyber security strategy to protect connected and AVs from a range of potential threats. Here are some general defense measures against various types of attacks.

6.4.3.1 Against infrastructure attacks

Protecting against infrastructure attacks is crucial to ensure the reliability and safety of connected and AVs. Infrastructure attacks can target the communication networks, control systems, and other components that support the functioning of this vehicle. The defense of the infrastructure is as follows: authentications and authorization, cloud-based method.

The cloud-based method is used to detect and mitigate cyber-attacks in the infrastructure of CAVs, providing additional authentication signature mechanisms for vehicle-to-infrastructure communications.

References

[1] J. Cui, L.S. Liew, G. Sabaliauskaite, F. Zhou, A review on safety failures, security attacks, and available countermeasures for autonomous vehicles, Ad Hoc Netw. 90 (2019) 101823.

[2] A. Proch. By 2030, One in 10 Vehicles Will Be Self-Driving Globally. Statista. Accessed September 30, 2021. [Online]. Available: https://www.statista.com/press/p/autonomous_cars_2020/.

[3] How Many Connected Cars Are Sold Worldwide? Verzion. Accessed April 15, 2021. [Online]. Available: https://smartcar.com/blog/connected-cars-worldwide/.

[4] J.D. Day, H. Zimmermann, The OSI reference model, Proc. IEEE 71 (12) (1983) 1334–1340.

[5] A. Chattopadhyay, K.-Y. Lam, Security of autonomous vehicle as a cyber-physical system, in: Proc. 7th Int. Symp. Embedded Comput. Syst. Design (ISED), 2017, pp. 1–6.

[6] K. Kim, J.S. Kim, S. Jeong, J.-H. Park, H.K. Kim, Cybersecurity for autonomous vehicles: review of attacks and defense, Comput. Secur. 103 (2021) 102150.

[7] A. Martínez-Cruz, K.A. Ramírez-Gutiérrez, C. Feregrino-Uribe, A. Morales-Reyes, Security on in-vehicle communication protocols: issues, challenges, and future research directions, Comput. Commun. 180 (2021) 1–20.

[8] S. Guo, et al., DCA: delayed charging attack on the electric shared mobility system, arXiv preprint arXiv:2302.01972 (2023).

[9] X. Sun, F.R. Yu, P. Zhang, A survey on cyber-security of connected and autonomous vehicles (CAVs), IEEE Trans. Intell. Transp. Syst. 23 (7) (2022) 6240–6259.

[10] M. Pham, K. Xiong, A survey on security attacks and defense techniques for connected and autonomous vehicles, Comput. Secur. 109 (2021) 102269.

[11] N. Pitropakis, E. Panaousis, T. Giannetsos, E. Anastasiadis, G. Loukas, A taxonomy and survey of attacks against machine learning, Comput. Sci. Rev. 34 (2019) 100199.

[12] B. Sheehan, F. Murphy, M. Mullins, C. Ryan, Connected and autonomous vehicles: a cyber-risk classification framework, Transp. Res. A Policy Pract. 124 (2019) 523–536.

CHAPTER

7

Security challenges facing autonomous and electric vehicles

James Holland[a] *and Arman Sargolzaei*[b]

[a]Department of Computer Science and Engineering, University of South Florida, Tampa, FL, United States
[b]Department of Mechanical Engineering, University of South Florida, Tampa, FL, United States

7.1 Introduction

Society finds itself at an inflection point where science, technology, and innovation can be leveraged to combat the many, existential issues—namely climate change—that threaten humanity and organic life, itself. For decades, cutting-edge technology has been integrated into nearly every facet of everyday life, from critical infrastructure—like traffic signals and the power grid—to the consumer automobile. In recent years, this has given rise to electrification and advanced driver-aid systems (ADAS) that are quickly pervading the automotive industry. Systems, like lane keep assist and adaptive cruise control, are paving the way toward fully self-driving vehicles in many ways. Firstly, by introducing the public to vehicle autonomy in a way that they can directly benefit from while still being in full control, it is possible to improve the image and reputation of this technology. However, by this same token, it is crucial to *not* embrace immature and unproven technology that risks the safety of the public. Second, current approaches for vehicle autonomy heavily leverage the underlying ADAS hardware and controller for actually piloting the vehicle based on what the sensor suite is perceiving. The continued integration of technology, however, has also had the unintended side effect of creating vast attack surfaces for would-be adversaries to exploit (denoted with), as illustrated in Fig. 1 [1].

As early as 2016, major media outlets started bringing attention to the data privacy issues related to autonomous, electrified vehicles (AEVs) and connected, autonomous, and electrified vehicles (CAEVs) [2,3]. Less than a year later, the Federal Trade Commission and the National Highway Traffic Safety Administration (NHTSA) hosted a workshop to allow for consumer rights groups, experts from industry, the government, and academic representatives to discuss the privacy and security concerns associated with CAEVs [4]. The public's concern and focus on security and privacy is justified by Upstream Security's recent findings that automotive cyber security incidents doubled in 2019 and are up over 600% since 2016 [5]. This growth is likely to continue as the number of potential targets increases and the overall vulnerability of each target grows [6]. The sensitivity of the data collected by a CAV is unique as a would-be attacker could learn everything about your daily routine, your preferred route to home, or when you tend to not be at home [7,8]. Considering this, the House of Representatives passed H.R.3388, more commonly known as the SELF DRIVE act, in 2017 [9]. This bill seeks to set a course for future CAV development [10]. This includes the requirement that CAV manufacturers develop cyber security policies to include ways of detecting vulnerabilities and access control for CAEVs; even so, many question whether this is enough [11].

To this point, we have only discussed security from the perspective of networks and applications, completely ignoring the myriad of circuits, controllers, and actuators responsible for the low-level control that enables AVs and ADAS features. Cyber security is a broad term that applies to many fields, from computer networks to control systems. In each discipline, however, this term means very different things. Networks, for example, take only high-level layers into account, leaving physical-layer components insecure against cyber threats. On the other hand, in control systems, the goal of cyber security is to ensure that the "lights stay on and water keeps flowing" by protecting the control system and processes they monitor from electronic threats [12,13]. Another way to conceptualize this difference is that the goal of network security is information assurance while in control systems the goal is service assurance. This is further

Handbook of Power Electronics in Autonomous and Electric Vehicles
https://doi.org/10.1016/B978-0-323-99545-0.00006-3

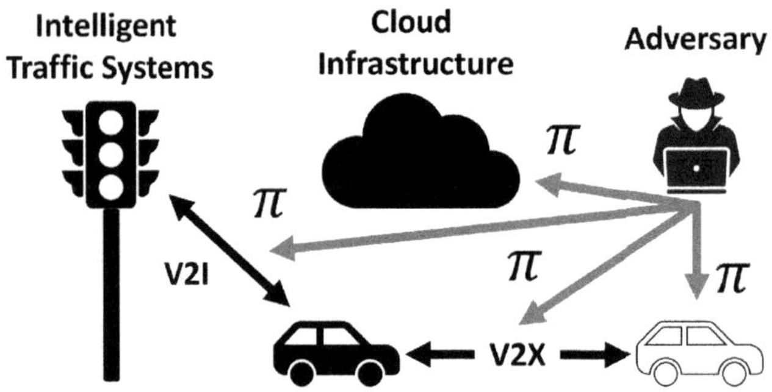

FIG. 1 Potential automotive vulnerabilities in a next-generation transportation network.

illustrated in the security reference models used by each discipline. In networking, the OSI 7-layer model is often employed while control systems generally use this model as well as the Purdue Reference Model.

The remaining sections of this paper are as follows:

- Section 7.2 will provide motivation for improving the security of the autonomous vehicle technology and physical-level components.
- Section 7.3 will outline the current state of security vulnerabilities, particularly in regard to cyber-physical systems, like autonomous vehicles.
- Section 7.4 introduces autonomous vehicle security concerns before discussing individual attack vectors and testing methodologies in detail.
- Section 7.5 discusses the security vulnerabilities of electric vehicle charging stations and introduces specific attacks.
- Section 7.6 provides a generalized approach for modeling CAEVs under attack.
- Section 7.7 discusses two case studies that demonstrate adversarial attacks against traditionally designed connected driving algorithms, like automatic emergency braking and cooperative adaptive cruise control.

7.2 Motivation

Autonomous vehicles (AVs) are poised to revolutionize the way in which society views transportation, travel, and the automobile with a myriad of benefits such as stress-free commutes and safer motoring to rapid food and package delivery services. While AVs have the potential to also drastically reduce the number of traffic fatalities, connected autonomous vehicles (CAEVs) could eliminate as many as 94% of fatalities at near full adoption. Both AVs and CAEVs, however, rely upon technology that is unproven and, until exhaustively tested, unready for wide-scale deployment.

At the current rate of development, it is likely that the technology supporting AVs will exist well before they have been thoroughly tested and proven to be safe and reliable. The burden of proving the readiness of a given AV could require billions of miles of test driving [14]. To accelerate the completion of this task, firms have grown their test fleets and developed cutting-edge virtual environments to test simulated AVs around the clock. Even at this rate, AV testing and verification are still expected to take hundreds of years to complete. This metric, however, makes the critical assumption that an AV is operating under proper circumstances, free of cyber-attacks. This is, of course, an unrealistic assumption for real-world operation. Some of the notable attacks launched against automakers in recent years are shown in Fig. 2. While the possible attack vectors and scenarios of individual attacks may be finite, the potential number of scenarios of combinatorial attacks approaches infinity. Some of the most common are presented in Figs. 3 and 4.

By viewing the problem of AV security from the perspective of a cyber-physical system (CPS), it is possible to understand how security can be ensured while maintaining the safe operation of the vehicle. Through the implementation of hardware-level security and controllers, it is possible to mitigate the effects of not only a cyber-attack but also random environmental noise and disturbance. If only security were implemented at only the network layer and higher, AVs will not be able to operate at full capacity in the best of cases and could result in widespread chaos and disruption.

7.3 The state of insecurity

During the COVID-19 pandemic, cybercrime rates drastically increased with several high-profile data breaches [15–17]. Since 2019, the cost of cyber-crimes has increased every year, topping out at \$1 trillion in 2020 [18]. Cybercriminals can access as much as 93% of company networks by exploiting basic vulnerabilities, like out-of-date

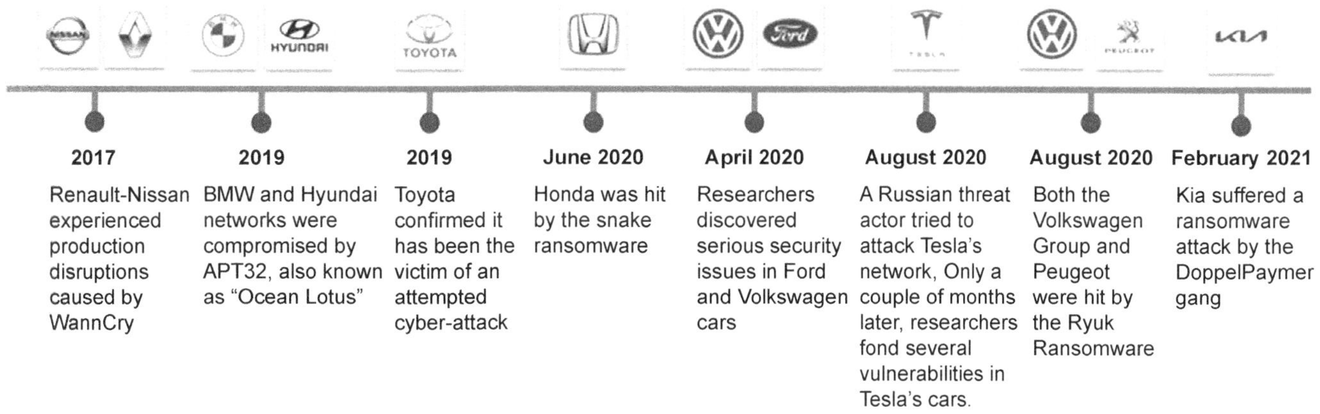

FIG. 2 Timeline of ransomware attacks on automakers since 2017. *Source: Otario.*

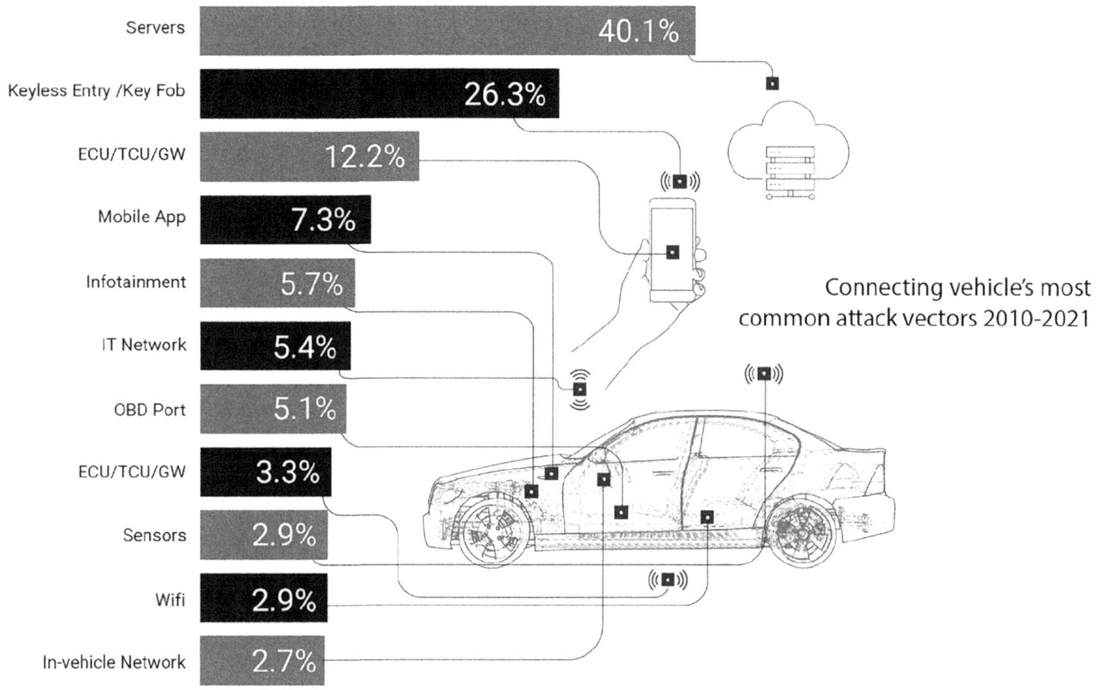

FIG. 3 Autonomous vehicle vulnerabilities sorted by frequency of exploit. *Source: Upstream Security.*

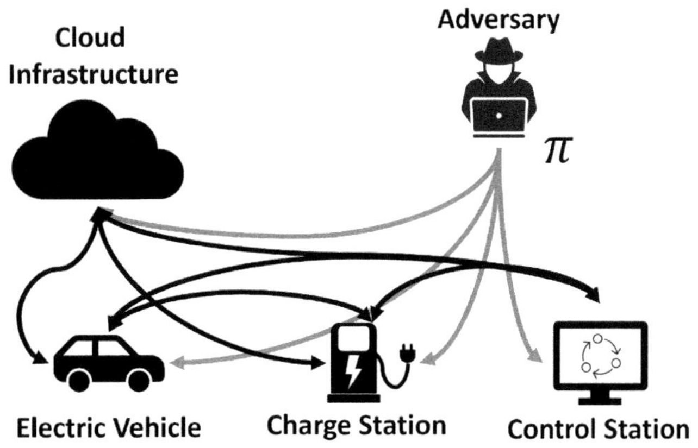

FIG. 4 Electric vehicle infrastructure vulnerabilities.

user credentials [19]. When pressed on the state of their firm's resiliency, executives expressed concerns regarding aspects of their business models, including supply chains and existing security measures [20]. About 44% of surveyed executives view their growing reliance on partners and suppliers exposes them to security threats. Furthermore, over 40% of executives do not think their security efforts have been sufficient with 30% also having expressed that their budgets were inadequate for implementing sufficient security measures.

The issue does not stop with large firms, also spreading to the multitude of small businesses that have moved online to adapt to the COVID-19 pandemic, with less than 50% of these having implemented a cybersecurity policy [21]. Of those, 32% have not changed their policy since the pandemic forced remote and hybrid operations and, as the most common causes of cyber-attacks are malware (22%) and phishing (20%), this introduces unnecessary security vulnerabilities. With cybercrimes costing U.S. businesses more than $6.9 billion in 2021, and only 43% of businesses feel financially prepared to face a cyber-attack in 2022, this situation is unlikely to remedy itself for the foreseeable future.

From 2001 onward, the task of cyber security was primarily implemented at the network layer to guarantee data security in exchange for mission assurance. The result of this is completely insecure level 0 and 1 systems that cannot be upgraded and play an integral role in systems that we all rely upon. Examples of this are lower-level sensor networks such as Highway Addressable Remote Transducer (HART [22]), Profibus [23], Fieldbus [24], etc., that possess no cyber security. Illustrated in Fig. 1 are the most common connected vehicle attack vectors exploited from 2010 to 2021. The most utilized vector is server vulnerabilities closely followed by keyless entry and low-level components.

Imagine a scenario in which an organization possesses state-of-the-art, impenetrable network security but relies upon a completely insecure power supply and no backup generators. A would-be attacker would need only to disconnect power from the facility, perhaps as simple as flipping a switch. Perhaps there isn't even an attacker, as in the Facebook, Instagram, and WhatsApp disruption in 2021 where a software update resulted in locking IT out of the impenetrable server rooms. The focus on building robust systems over resilient and adaptable ones introduces new hurdles for mission success.

7.4 Autonomous vehicle security concerns

Control systems monitor and control physical inputs to maintain or optimize a process and use safety systems to ensure the desired output range. As such, AVs fall under the umbrella of control systems with similar security vulnerabilities. For control systems, networks are implemented in a support capacity to help ensure safety, reliability, and efficiency. Due to this, the components of a control system (i.e., sensors, actuators, etc.) were designed to work in real time, without the need for the IP layer. This distinction introduces the need for enforcing cyber security standards for these low-level components. The remainder of this section is as follows:

- Discussion and definition of adversarial attack vectors
- Discussion of the unique vulnerabilities of specific sensors and vehicle subsystems
- Introduction to the various methods for validating the readiness of an autonomous vehicle platform
- Outlining the challenges facing autonomous vehicle deployment

7.4.1 Adversarial attack vectors

In addition to the faults and delays that occur stochastically, there are also those that are intentionally induced by an adversary seeking to target vulnerabilities in the system. These vulnerabilities include those commonly associated with consumer electronics, such as denial of service (Dos), as well as false data injection (FDI) and other attacks that target the underlying hardware. In order for AVs to reach widespread adoption, they must be able to safely thwart not only the attacks of today but also all those that lie on the horizon. Future-proofing of hardware systems is critical for ensuring that deployed systems are safe and secure for the life cycle of the vehicle. The remainder of this subsection will focus on some specific examples of attacks targeted at CAEVs and how feasible these attacks are for an adversary to implement in the real world.

Injection attacks: Injection attacks are those in which an adversary is able to insert an erroneous signal in place of or on top of the true signal [25]. These attacks can include those that occur within communication channels, sensor inputs, and even actuators and controllers.

Jamming attacks: Jamming attacks are those where an adversary injects an interfering signal on a communication network, such that the signal quality decreases. Attacks of this type include DoS, forms of injection attacks, and any other vector that results in the loss of access to a service, feature, or data.

Sensor manipulation attacks: Sensor manipulation attacks can come in many forms, including both jamming and injection. As such, other attack vectors that are targeted specifically at sensors will be discussed in the following section.

7.4.2 Sensor and system vulnerabilities

CAEVs leverage a host of sensors and systems to record and process data about the world around them to perceive the environment and possible threats or obstacles. As such, the security and resiliency of these components are critical to safe and reliable operation. The remaining subsections will discuss various sensors and vulnerabilities they face.

Global positioning systems (GPS): Global positioning systems (GPS) are capable of high-precision data, tracking the movement of a vehicle within single meter accuracy using satellite networks. However, within the public domain, GPS data are transmitted using an open standard which can be used to generate malicious transmissions that can disrupt communications [26]. This disruption—referred to as spoofing or jamming—can be caused in several ways, with all exploiting the same primary vulnerability: GPS devices are often programmed to give priority to the strongest signal [27,28].

Experimentation performed by the University of Texas demonstrated the viability of GPS spoofing on an *Adaptive Flight Hornet Mini* unmanned air vehicle (UAV) that was equipped with a receiver utilizing the standard *Receiver Autonomous Integrity Monitoring (RAIM)* technology [29]. The GPS spoofer's signal was first calibrated to match the GPS transceiver's true signal before increasing the signal strength (to assume priority) and, then, spoofing a false upward drift. The UAV responded by diving toward the ground until a safety pilot intervened.

Inertial measurement unit (IMU): Inertial measurement units (IMUs) are the combination of a gyroscope and accelerometers to provide velocity, acceleration, and orientation data [30]. These data are often used as feedback by low-level controllers to enable the proper operation of systems that dynamically adapt to the environment, such as cruise control up a gradient. Similar to GPS, by injecting erroneous data into the communications of a low-level controller (i.e., throttle), it is possible to create disruption or, worse, complete instability of the system or vehicle.

Light detection and ranging (LiDAR): Light detection and ranging (LiDAR) sensors are capable of obtaining a map of a vehicle's surroundings for use in localization, obstacle, avoidance, and navigation. LiDAR sensors are also susceptible to spoofing and jamming attacks as there is no guarantee of the validity of the 3D model [30]. Work at the University of Twente demonstrated the viability of using a controlled light source, like a laser, to spoof reflected LiDAR beams [31]. Another study at the University of Cork demonstrated how detailed LiDAR data can be spoofed to trick the car into recognizing pedestrians or vehicles that actually are not there [32]. To mitigate these attacks on LiDAR, the researchers proposed mitigation techniques, such as utilizing V2X connectivity to collaboratively share data and validate results.

Camera systems: Camera systems come in many forms, from mono and stereo-vision to infrared systems. Mono and stereo-vision simply refer to how many cameras are integrated into the system. Mono-vision cameras use a single camera input and require other sensors to extrapolate information, like depth, using a process referred to as *sensor fusion* [33]. Like LiDAR, cameras can also be subjected to adversarial attacks using low-powered lasers [31].

V2X network attacks: The connectivity of connected vehicles enables, both high-speed data transmission to improve vehicle autonomy in addition to inherent security flaws and vulnerabilities presented by this externally facing access point [34]. With V2X, vehicles can connect to one another, and infrastructure as well as unknown edge devices, such as mobile phones and laptops, that might pose a serious threat. At every level of V2X, adversaries can launch impersonation [35], FDI [25], and time delay attacks [36]. Since V2X communications fall under the purview of network security, the focus will be on the vehicle and subsystems rather than the network channels.

Hardware attacks and exploits: In this section, the attacks and exploits that exist within the vehicle, itself, will be discussed. These include vulnerabilities that allow an adversary to gain access to the vehicle's CAN bus, whether through physical means or remote access. For example, the OBD port is present on most vehicles manufactured since 2008 and allows for collecting diagnostic information. As the OBD port is connected directly to the CAN bus, this is an often targeted access point [37]. The downside of this exploit is that an adversary must first gain physical access to the vehicle's port.

Modern vehicles, however, are often equipped with Wi-Fi, Bluetooth, and cellular connectivity, each opening a new, potential access point for a would-be adversary. While the systems that utilize these protocols are usually for in-vehicle infotainment (IVI) and are not *directly* connected to the CAN-bus, once one subsystem is accessed, it is only a matter of time before the adversary is able to freely traverse the entire vehicle. With widespread access, the adversary could inject and monitor CAN traffic as well as tamper with the firmware and updates of components on the vehicle.

7.4.3 Validating the readiness of autonomous vehicles

Validating the readiness of full self-driving (FSD) systems is the crucial, final step before confident adoption of this technology. To accomplish this, an AV's ability to demonstrate crash avoidance is tested and verified using predefined scenarios. In the realm of AV testing and verification, there are several approaches for accomplishing this arduous task. Each of the approaches considers either an individual vehicle subsystem or a combination of subsystems and occurs either in the real world or in simulated environments. Within simulated testing, there exist offline and real-time methods. Offline simulations aim to maximize the computation speed of testing while real-time simulations prioritize testing accuracy in a bounded response time [38]. Furthermore, real-time simulations ensure accessibility to testing data and accelerate the validation phase of testing, maximizing certainty during the development process [39]. Under the umbrella of real-time simulations are the Vehicle-in-the-Loop (ViL) and Hardware-in-the-Loop (HiL) methods.

Software-in-the-loop (SiL): SiL simulation is a powerful technique, often employed in situations not feasibly explored using traditional, real-world testing methods [40]. SiL testing, however, is simply unable to match the accuracy and resolution of other testing methods that integrate the real, physical system.

Hardware-in-the-loop (HiL): HiL is used for exploring how a specific controller responds to virtual stimuli, in real time with physical hardware [41]. Due to the inclusion of the actual, physical hardware being tested, this leads to high-resolution testing results. Furthermore, using the physical systems in question enables more streamlined testing of complex systems with minimal overhead and risk. Improving the testing phase of development has the benefit of ensuring the safety, reliability, and regulatory satisfaction of products while still meeting budgetary and time constraints. When applied to both conventional vehicles and CAEVs, however, HiL testing comes at the cost of accuracy and realism due to not incorporating actual, real-time vehicle dynamics into the simulation.

Vehicle-in-the-loop (ViL): ViL testing improves upon the idea of HiL testing by placing the entire vehicle and all systems into a simulated testing environment. Real-time simulation software, such as IPG CarMaker, Ansys, dSpace, and rFpro, allows for the generation of tailored environments and testing scenarios. However, there does not currently exist an approach or framework for applying ViL testing to the unique case of CAV testing and verification. Applying and refining this approach to target a specific CAV's weaknesses could lead to accelerated performance gain can be achieved at a significantly reduced cost and in a shorter time [42]. ViL testing, however, can only consider events and variables that occur within and only within the simulation. This comes at the cost of accuracy and resolution as no simulated environment can fully replicate the real world.

Real-world testing: Real-world testing is regarded as the most accurate and reliable approach for CAV testing and verification as the entirety of the vehicle is incorporated, including its unique dynamics. Even so, CAV testing and real-world performance demonstrate that these vehicles are still far from exhibiting adequate readiness [43]. The situation is further exacerbated due to the high cost, increased risk, and low closure velocity associated with real-world testing [44–46]. Furthermore, there exists a myriad of scenarios that are unsafe or simply infeasible to test using real-world methods. Edge cases that involve humans, particularly pedestrians, are unsafe to explore using real vehicles and participants. Additionally, testing the security and resiliency of CAEVs is not possible due to cost, safety, and an inability to inject randomized attacks against the system.

7.4.4 Challenges facing AV deployment

In addition to the challenges facing AV testing, there is also the daunting goal of creating the perfect CAV, one that is infallible and guarantees crash avoidance [47]. While this pursuit of perfection is understandable, it may be unnecessary and could even result in more lost life due to a comparatively minuscule margin of error [48,49]. Annually, the US loses at least 38,000 lives due to automobile crashes [50]. Can society afford to wait another decade for perfection and are they willing to lose hundreds of thousands more of their friends and loved ones? Even now, as imperfect as CAEVs are, it is considered that these vehicles could prevent as much as a third of current fatalities [51].

The primary obstacle in the way of CAV perfection is simply the infinite number of possible scenarios in which to test a CAV [52]. This problem is further exacerbated by the critical differences between each CAV model and system, rendering the goal of a finite series of tests that completely address any given CAV impossible.

Another hurdle for CAEVs is their notoriously poor performance in less than ideal weather conditions [53–56]. In addition to weather, CAV sensors are often obscured by dead insects during driving [57]. Ford has taken measures to incorporate insects as a variable in their testing and was even seeking to mitigate this issue to keep sensors clean [58].

Another major issue facing both CAEVs and conventional vehicles is the condition of the US's infrastructure [59]. In 2021, the American Society of Civil Engineers (ASCE) awarded American roads with a "D" and overall infrastructure with a "C-" [60,61]. This remains consistent with their 2017 report, finding the state of roadways to be as severe then as it still is today [62]. In particular, the ASCE's analysis reveals that pedestrian fatalities are on the rise, calling attention to the importance of testing CAV-to-human interactions.

According to AAA's testing of mass-market pedestrian detection systems in 2019, they found that the systems performed abysmally [63]. Specifically, their results showed that vehicles were unable to detect a pedestrian in more than half of the tests—in the best of cases. Furthermore, when tested at night, none of the systems were able to detect the pedestrian. This reveals a major area of improvement for CAEVs as the NHTSA found that roughly 75% of pedestrian fatalities occur during the night.

Currently, there exist no realistic methods for truly training and testing a CAV's performance with pedestrians due to the dangers associated with testing this. Furthermore, while human behavior is mostly predictable, pedestrians have never been exposed to self-driving vehicles in the real world thus there is no pre-existing "training" on how to react [64]. The development of a method for investigating human-computer interactions with regard to CAEVs is critical to minimizing the risk to pedestrians and even helping to improve public trust in this technology.

Public trust in CAEVs has been a constant, uphill battle for the industry, particularly due to an Uber research vehicle's fatal incident with a bicyclist in Arizona [65–68]. The results of a Partners for Automated Vehicle Education (PAVE) survey showed that nearly 75% of respondents felt that the technology is far from ready with nearly half stating that they had no future plans of using any self-driving services [69]. Unfortunately, these results are consistent with the trend of mistrust that has followed several high-profile, CAV-related incidents [70,71].

Public uncertainty is but one of the many hurdles facing CAEVs, followed closely by concerns regarding security and privacy. Traditionally, the automobile had been reliant on mechanical systems that the human pilot used to operate the vehicle [72]. In recent years, technology from the humble GPS to cutting-edge ADAS have pervaded the automotive industry. This integration of technology has also had the unintended side effect of creating vast attack surfaces for would-be adversaries [1]. As early as 2016, major media outlets started bringing attention to the data privacy issues related to CAEVs [2,3].

7.5 Electric vehicle charge station (EVCS) security

The electric vehicle has long held the interest of both the public and scientific community, before that of even internal combustion options [73,74]. On their own, electric vehicles (EVs) are positioned to have a major, positive impact on climate change, energy independence, and future electrification innovations [75–80]. However, when integrated with CAV technology, the full potential of, both, EVs and CAEVs can be taken advantage of (?). EV drivetrains are more linear and predictable than that of their internal combustion counterparts [81]. This enables smoother, more precise control of the drivetrain which directly translates to improved efficiency when using features, like adaptive cruise control. Just like CAEVs, however, EV charge stations (EVCS) are vulnerable to adversarial attack, creating an additional hurdle to a clean, autonomous future [82].

EVCS, just like all edge devices, possess externally facing access points that an adversary will likely seek to exploit. Once an attacker gains access, they could engage in attacks that target the CIA triad—like DoS—or seek to exploit vulnerabilities in the underlying charging controller [83]. For example, work by Bhusal, et al. analyzed the susceptibility of EVCS to data manipulation attacks, finding a concerning lack of physical-level safety mechanisms [84]. Later research by M. Basnet and M. Ali demonstrated—both—the vulnerability of a 5G connected EVCS to FDI and time delay attacks and how to mitigate these attacks with near 100% detection [85]. In early 2022, "Brokenwire," a novel DoS attack against common DC charging systems, was unveiled [86]. While the results of this attack are merely inconvenient, the possible victims of this attack number over 12 million [87]. Improving the state of EVCS security is paramount to the success of the EV as an insecure, unreliable, and unsafe charging infrastructure will tarnish public perception, just as premature adoption of self-driving technology has done for the CAV [88,89]. Furthermore, as EV charging begins to dominate energy consumption, an unstable charging infrastructure could lead to damage across the power grid [90].

7.6 Generalized mathematical model under attacks

A CAV consists of a vehicle having sensor and actuator nodes, a control system having input and output nodes, an intermediate network system having interconnected nodes, and wireless communication links for the data transfer between the different nodes of the plant, controller, and network systems as shown in Fig. 5.

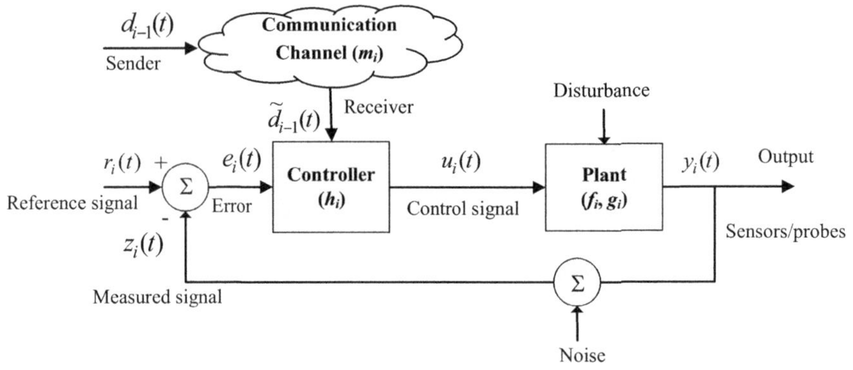

FIG. 5 Block diagram of CAEVs.

Let's consider a nonlinear dynamic model as

$$X(t) = F(t, U(t), X(t)), \tag{1}$$

and

$$Y(t) = G(X(t)), \tag{2}$$

where $F = \{f_1, f_2, \ldots, f_i\}$ and $G = \{g_1, g_2, \ldots, g_i\}$ are the state transit function and measurement function respectively. $X(t) = [x_1(t), x_2(t), \ldots, x_i(t)]^T$, $U(t) = [u_1(t), u_2(t), \ldots, u_i(t)]^T$, and $Y(t) = [y_1(t), y_2(t), \ldots, y_i(t)]^T$ are the state, input, and output vectors. $x_i(t) \in R^a$, $u_i(t) \in R^b$, and $y_i(t) \in R^c$, where a, b, and c are the dimensions of the state, input, and output vectors of the ith CAV, respectively.

Cooperation with the surrounding environment can be modeled as

$$\tilde{d}(t) = M(d, t), \tag{3}$$

where $M = \{m_1, m_2, \ldots, m_i\}$ is the model of communication and $d(t) = [d_0(t), d_1(t), \ldots\ldots, d_{i-1}(t)]^T$ is the transmitted data (e.g., position, velocity, and acceleration). In Eq. (3), $\tilde{d}(t) = d(t)$ when there is no cyber-attack, failure, or fault; Hence, the communication channel is ideal. The controller output can be represented as

$$U(t) = H\left(t, Z(t), r(t), \tilde{d}(t)\right), \tag{4}$$

where $Z(t)$, $r(t)$, and $H = \{h_1, h_2, \ldots\ldots, h_i\}$ are the measured signal, reference signal, and controller function, respectively.

Cyber-attacks, failures, and faults convert the dynamic model of CAEVs to the switched dynamical model. Therefore, the attacks and faults can be modeled as

$$\left(\tilde{x}_i, \tilde{y}_i, \tilde{f}_i, \tilde{h}_i, \tilde{m}_i, \tilde{g}_i, \tilde{z}_i, \tilde{r}_i, \tilde{u}_i, \tilde{t}, \tilde{d}_{i-1}\right) = \pi\left(x_i, y_i, f_i, g_i, z_i, u_i, r_i, t, d_{i-1}\right). \tag{5}$$

The attack/fault function π can be uniquely defined as a linear or nonlinear function. For instance, $\pi(\gamma_i(t)) = \tilde{\gamma}_i(t) = \gamma_i(t - \tau_i(t))$, where τ is a time-varying delay can represent a time delay switch (TDS) attack [36,91–94]. For a FDI attack which has been injected to the signal γ_i, the function π can be defined as $\pi(\gamma_i(t)) = \tilde{\gamma}_i(t) = \gamma_i(t) + \alpha_i(t)$, where $\alpha_i(t)$ is a random variable, can represent a FDI attacks [95–97].

7.7 Case studies

Now, equipped with an understanding of the security challenges facing the next era of transit, let us discuss three case studies targeting connected driving algorithms (CDA) under attack. Our scenarios highlight the vulnerabilities of classical controllers through experimental analysis. The focus of these discussions is on hardware-level security rather than traditional, network-focused security. The reasoning for this is due to the lack of attention that low-level components and systems have received in recent decades. As such, when discussing attacks that exploit V2X communications, it is with the assumption that these protocols are either insecure or have been successfully penetrated.

7.7.1 Vulnerability demonstrations

The experiments are performed as part of our ongoing CAEVs testing and verification efforts. To demonstrate the impacts of insecure CDA in a tangible way, we have developed a real-time testing platform that enables ViL. Currently, our work is focused on the development of a next-generation testing framework that will incorporate ViL methodology into a unified framework to enable the complete integration of a physical vehicle as well as its real-time motion and dynamics into a highly detailed, dynamics simulation environment. The uniquely developed ViL environment enables us to perform the vulnerability analysis in a safe environment.

7.7.1.1 Automatic emergency brake under false data injection attack

In the first scenario, an automatic emergency braking (AEB) algorithm is subjected to an FDI attack on the input signal of the forward-facing radar, resulting in a perceived delay time to collision. We tested the performance of the AEB algorithm using our experimental setup shown in Fig. 6.

In this setup, we used IPG Carmaker software as the simulation environment where the AEB for the passenger vehicle (2017 Ford fusion) is tested when the vehicle under the test gets closer to a simulated vehicle. To demonstrate the vulnerability of an existing AEB system, we injected an adversarial signal into the forward-facing radar input data of our testing platform. Fig. 7A and B illustrate how an adversary might carry this out in the real world, using approaches similar to how LiDAR might be exploited. The plot in Figs. 8–10 displays the data obtained from this test, demonstrating that even small injections of 1–2 m/s^2 to the approach rate were enough to drastically impact the stopping speed and, in turn, the stopping distance from the lead vehicle (shown in Fig. 8).

7.7.1.2 Cooperative adaptive cruise control under FDI attack

The second study is of cooperative adaptive cruise control (CACC) that was exposed to an FDI attack to the lead vehicle acceleration used as a reference by the follower. ViL testing is accomplished using a dSPACE MicroAutobox II which works with MATLAB Simulink (shown in Fig. 11). Using our passenger vehicle and ViL setup, we demonstrated vulnerabilities that exist for traditional CACC algorithms. Due to the connected nature of CACC, an attacker simply can inject a malicious packet into the follower vehicle's leader acceleration reference. This can be accomplished by employing a network-level attack strategy, like "man-in-the-middle" (shown in Fig. 12).

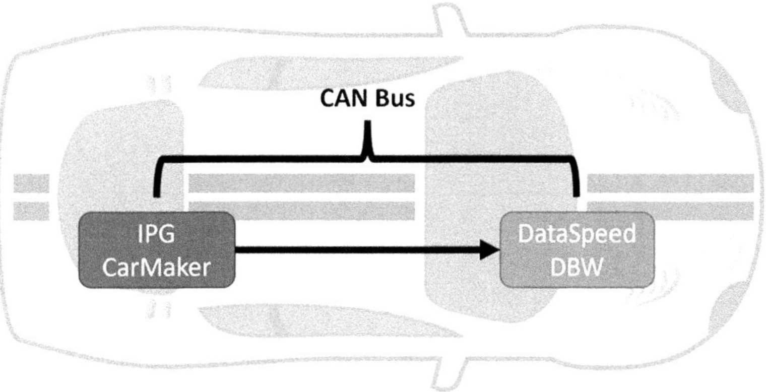

FIG. 6 High-level component and connection diagram: Automatic emergency braking FDI demonstration.

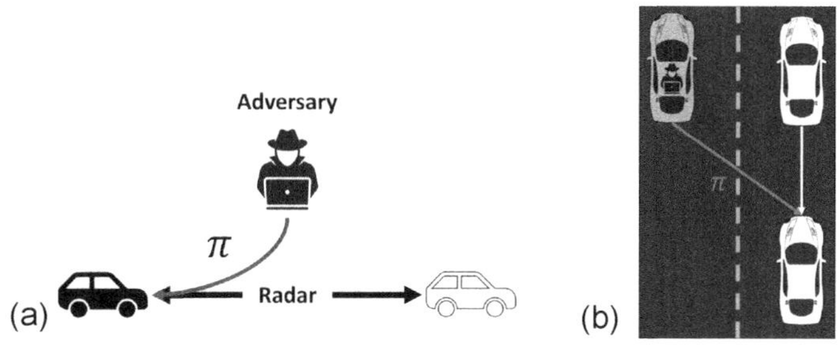

FIG. 7 AEB FDI attack scenario: (A) automatic emergency braking (AEB) attack scenario and (B) radar attack scenario feasibility.

FIG. 8 Resulting AEB stopping distances during FDI testing: (A) baseline stopping location of AEB system on test vehicle and (B) resulting stopping location of AEB system under FDI attack.

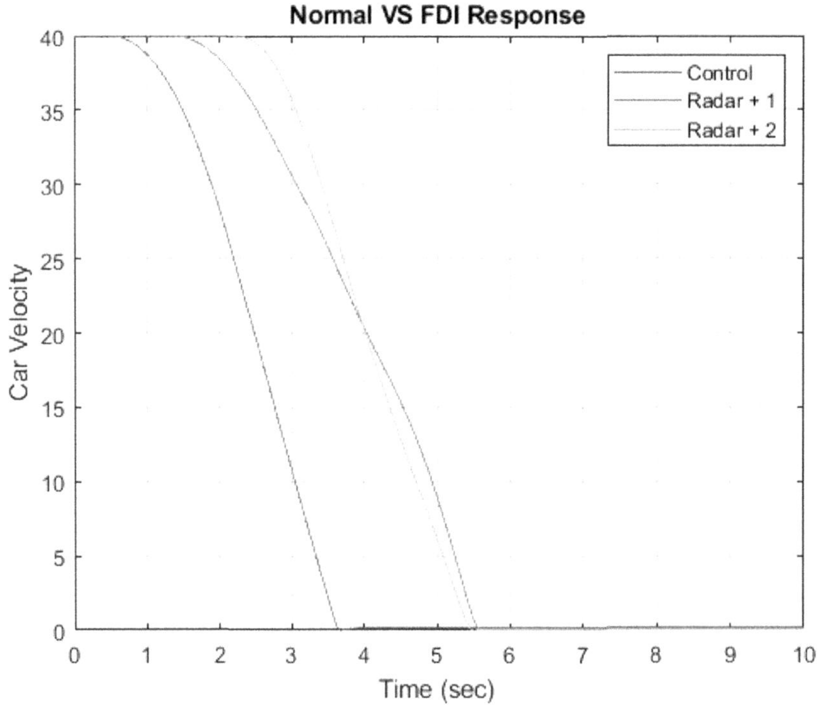

FIG. 9 Stopping location of vehicle under normal and FDI conditions.

The traditional controller was a proportional-integral-derivative (PID) using leader acceleration as the reference signal. By monitoring leader acceleration, the CACC would increase or decrease follower speed to maintain a safe following distance between vehicles. The FDI signal, while only being an increase of $1\,m/s^2$, resulted in the appearance of acceleration when the leader was actually at a constant speed and decreased the perceived rate of deceleration when under braking.

In this scenario, both the test vehicle and a simulated lead vehicle will accelerate to 3 m/s. Fifteen seconds into testing, both vehicles will accelerate to 6 m/s. Five seconds later, the lead vehicle decelerates to 3 m/s, and after another 5 s, a simulated FDI attack is injected. The FDI attack increases the leader's acceleration reference signal, leading the follower vehicle to act as if the lead vehicle is traveling faster than it actually is.

Testing was performed by uploading the CACC model to a MicroAutobox II that interfaces with the test vehicle's CAN bus to receive sensor measurements and send actuator commands. The model was developed in MATLAB/Simulink and uses the dSPACE RTI CAN blockset to interface with our vehicle actuators. By using a DBC file, our model is able to translate the control signal into CAN messages.

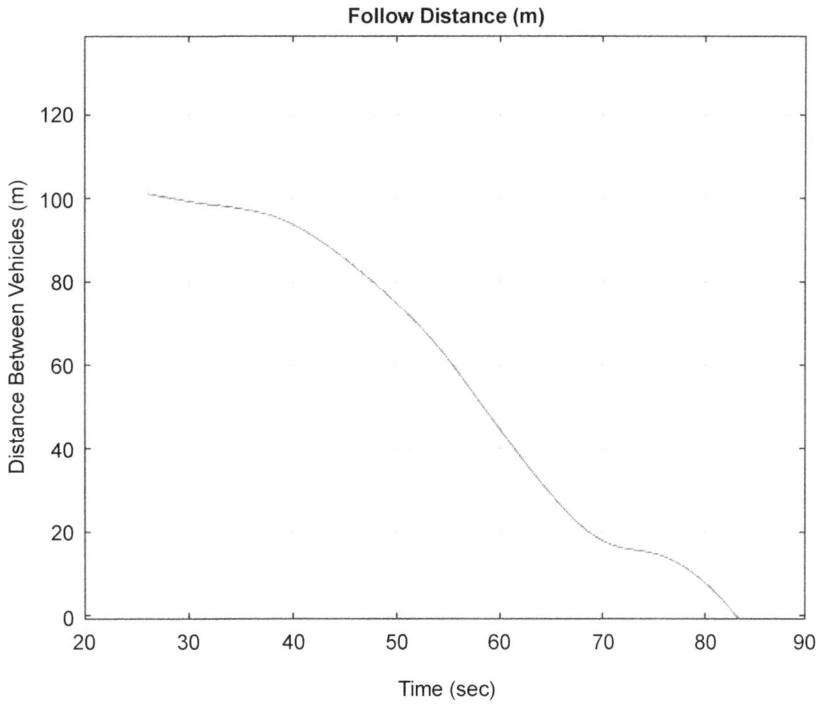

FIG. 10 Distance between leader and follower vehicles.

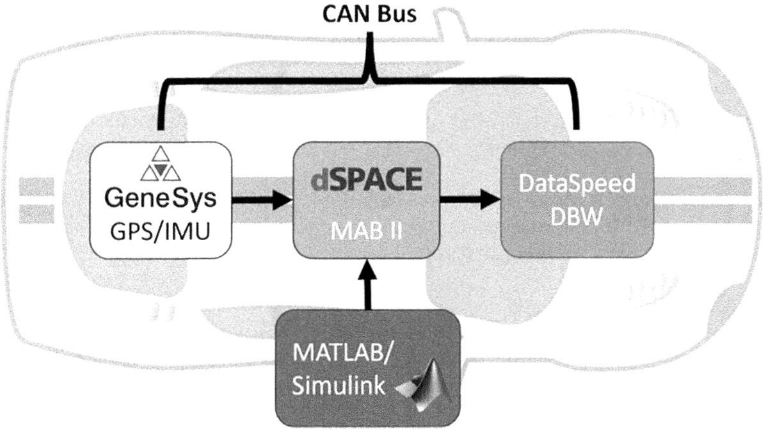

FIG. 11 High-level component and connection diagram: vulnerable cooperative adaptive cruise control.

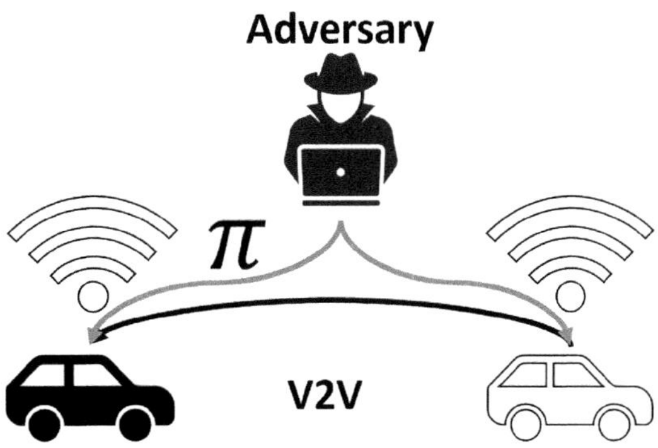

FIG. 12 FDI attack on CACC via leader acceleration reference transmitted via V2X.

When under attack, the traditional controller perceives the lead vehicle to be accelerating, in turn, producing a control signal that increases the follower's speed. The effects of this attack are made apparent in the following distance measurements. The traditional controller, predictably, collides with the lead vehicle as shown by crossing the X-axis.

7.8 Summary and conclusion

The future for autonomous transportation is bright with tens of billions of dollars already invested with billions more raised each year. Even once this technology has demonstrated its efficacy, it will still have not demonstrated its security and resiliency against random disruptions or adversarial attacks. For decades, hardware security has been left ignored in favor of security at higher layers, like the network and application. While these security measures ensure the CIA triangle of *data* security is maintained, these mechanisms do little to ensure *access* to services. In applications, like critical infrastructure, ensuring that the *lights stay on and the water keeps running* is vastly more important than protecting the service's end user's data. Furthermore, as the majority of CPS implemented in critical infrastructure *are* insecure, the importance of this issue should be abundantly clear. Only once CAEVs and the technology that enables these machines have demonstrated complete readiness can society widely and confidently accept them into our everyday lives.

Acknowledgment

Partial support of this research was provided by the National Science Foundation under Grant No. ECCS-EPCN-2241718 and CNS-1919855. Any opinions, findings, and conclusions, or recommendations expressed in this material are those of the author(s) and do not necessarily reflect the views of the sponsoring agency.

References

[1] The attack surface of a connected vehicle. Available at https://argus-sec.com/attack-surface/.
[2] Available at http://www.theguardian.com/technology/2016/jun/08/self-driving-car-legislation-drones-data-security, abstractNote=US states are scrambling to regulate self-driving cars, but the focus on physical safety ignores the more complex issues of data privacy and security, June 2016.
[3] A. LaFrance, The creepy thing about self-driving cars, 2016, Available at https://www.theatlantic.com/technology/archive/2016/03/self-driving-cars-and-the-looming-privacy-apocalypse/474600/. abstractNote=Automated vehicles will learn everything about you—and influence your behavior in ways you might not even realize.
[4] Federal Trade Commission, Connected cars: Privacy, security issues related to connected, automated vehicles, 2017, Available at https://www.ftc.gov/news-events/events-calendar/2017/06/connected-cars-privacy-security-issues-related-connected.
[5] Upstream security's 2020 global automotive cybersecurity report. Available at https://upstream.auto/upstream-security-global-automotive-cybersecurity-report-2020/.
[6] P&S Intelligence, "Automotive cybersecurity market to witness robust growth ahead: P&S intelligence." Available at https://www.prnewswire.com/news-releases/automotive-cybersecurity-market-to-witness-robust-growth-ahead-ps-intelligence-301190123.html.
[7] M. Bubbers, "What kind of data is my new car collecting about me? Nearly everything it can, apparently", The Globe and Mail (Jan. 2020).
[8] G.A. Fowler, What Does Your Car Know About You? We Hacked a Chevy to Find Out, vol. 17, The Washington Post, WP Company LLC, Dec. 2019.
[9] R.E. Latta, Self drive act, 2017, Available at https://www.congress.gov/bill/115th-congress/house-bill/3388.
[10] Late to scene, "safely ensuring lives future deployment and research in vehicle evolution act" ("self drive act") seeks to set course for future development. Available at https://www.natlawreview.com/article/late-to-scene-safely-ensuring-lives-future-deployment-and-research-vehicle-evolution.
[11] B.H. Law, What you need to know about driverless cars and privacy, 2019, Available at https://medium.com/@baumhedlund/what-you-need-to-know-about-driverless-cars-and-privacy-8720d46e8877.
[12] U.S. Department of State, Presidential decision directive 63: protecting America's critical infrastructures [fact sheet], in: Homeland Security Digital Library, United States Department of State, 1998.
[13] J. Monteagudo, The Need to Change the Paradigm of Control System Cyber Security - Part 2: Lack of Control System Cyber Incident Information Sharing, 2022.
[14] N. Kalra, S.M. Paddock, Driving to safety: how many miles of driving would it take to demonstrate autonomous vehicle reliability? Transp. Res. A Policy Pract. 94 (2016) 182–193.
[15] S. Oladimeji, S.M. Kerner, SolarWinds hack explained: everything you need to know, TechTarget Network (Nov. 2023). https://www.techtarget.com/.
[16] J. Bunge, JBS paid $11 million to resolve ransomware attack, The Wall Street Journal (June 2021), www.wsj.com/articles/jbs-paid-11-million-to-resolve-ransomware-attack.
[17] S.M. Kerner, Colonial pipeline hack explained: everything you need to know, TechTarget Network (Apr. 2022).
[18] B. Ballard, Cybercrime apparently cost the world over $1 trillion in 2020, TechRadar, Future US, Inc. (Feb. 2021).
[19] Business in the crosshairs: analyzing attack scenarios, Positive Technologies (Dec. 2021).
[20] ThoughtLab, Cybersecurity solutions for a riskier world, ThoughtLab (2021). https://www.thoughtlabgroup.com/.

[21] D.J. Brin, Data Science Versus Data Analytics: What's the Difference and Why Does It Matter for Your Business? UpCity, 2024.
[22] FieldComm Group, HART Protocol Specifications, FieldComm Group, 2024, https://www.fieldcommgroup.org/ (Accessed 16 May 2024).
[23] PI - Standards for Interoperability, PROFIBUS & PROFINET International (PI), PROFIBUS Nutzerorganisation e.V., 2024, https://www.pi-organization.com/.
[24] Beckhoff Automation GmbH & Co. KG, New Automation Technology, Beckhoff, 2024, https://www.beckhoff.com/en-en/ (Accessed 16 May 2024).
[25] A. Sargolzaei, A secure control design for networked control system with nonlinear dynamics under false-data-injection attacks, in: 2021 American Control Conference (ACC), 2021, pp. 2693–2699.
[26] M.L. Psiaki, T.E. Humphreys, Gnss spoofing and detection, Proc. IEEE 104 (6) (2016) 1258–1270.
[27] X.-C. Zheng, H.-M. Sun, Hijacking unmanned aerial vehicle by exploiting civil gps vulnerabilities using software-defined radio, Sens. Mater. 32 (2020) 2729.
[28] S.Z. Khan, M. Mohsin, W. Iqbal, On gps spoofing of aerial platforms: a review of threats, challenges, methodologies, and future research directions, PeerJ Comput. Sci. 7 (2021) e507.
[29] Todd Humphreys, Statement on the Vulnerability of Civil Unmanned Aerial Vehicles and Other Systems to Civil GPS, The University of Texas at Austin, 2012.
[30] S. Parkinson, P. Ward, K. Wilson, J. Miller, Cyber threats facing autonomous and connected vehicles: future challenges, IEEE Trans. Intell. Transp. Syst. 18 (2017) 1–18.
[31] B.G.B. Stottelaar, Practical Cyber-Attacks on Autonomous Vehicles, University of Twente, 2015.
[32] M. Harris, Researcher Hacks Self-driving Car Sensors, IEEE, Sep. 2015, p. 4.
[33] I. Mahajan, R. Bhosale, P. Kulkarni, Obstacle detection using mono vision camera and laser scanner, Int. J. Res. Eng. Technol. 2 (12) (2013) 684–690.
[34] A.D. Kumar, K.N.R. Chebrolu, R. Vinayakumar, K.P. Soman, A Brief Survey on Autonomous Vehicle Possible Attacks, Exploits and Vulnerabilities, arXiv, 2018.
[35] J. Khan, Vehicle network security testing, in: 2017 Third International Conference on Sensing, Signal Processing and Security (ICSSS), 2017, pp. 119–123.
[36] A. Sargolzaei, F. Zegers, A. Abbaspour, C. Crane, W.E. Dixon, Secure control design for networked control systems with nonlinear dynamics under time-delay-switch attacks, IEEE Trans. Autom. Control 68 (2022) 798–811.
[37] W. Yan, A two-year survey on security challenges in automotive threat landscape, in: 2015 International Conference on Connected Vehicles and Expo (ICCVE), 2015, pp. 185–189.
[38] J. Bélanger, P. Venne, J.-N. Paquin, The What, Where and Why of Real-Time Simulation, Opal-RT Technologies, Montréal, QC, 2010.
[39] O. Gietelink, J. Ploeg, B.D. Schutter, M. Verhaegen, Development of advanced driver assistance systems with vehicle hardware-in-the-loop simulations, Veh. Syst. Dyn. 44 (2006) 569–590.
[40] Advantages and disadvantages of simulation. Available at https://www.concentricmarket.com/blog/advantages-and-disadvantages-of-simulation.
[41] D. Bullock, B. Johnson, R. Wells, M. Kyte, Z. Li, Hardware-in-the-loop simulation, Transp. Res. Part C Emerging Technol. 12 (2004) 73–89.
[42] D. Zhao and H. Peng, "From the lab to the street:solving the challenge of accelerating automated vehicle testing." https://mcity.umich.edu/wp-content/uploads/2017/05/Mcity-White-Paperaccelerated-AV-Testing.pdf.
[43] S.S. Banerjee, S. Jha, J. Cyriac, Z.T. Kalbarczyk, R.K. Iyer, Hands off the wheel in autonomous vehicles?: a systems perspective on over a million miles of field data, in: 2018 48th Annual IEEE/IFIP International Conference on Dependable Systems and Networks (DSN), 2018, pp. 586–597.
[44] A.J. Hawkins, "Toyota will test self-driving car 'edge cases' at new proving ground in Michigan." Available at https://www.theverge.com/2018/5/3/17314778/toyota-self-driving-car-test-proving-ground-michigan.
[45] A.J. Hawkins, Waymo gave me a ride in a car with no driver. Available at https://www.theverge.com/2017/10/31/16579180/waymo-self-driving-test-facility-castle-google.
[46] D. Muoio, Uber built a fake city in Pittsburgh with roaming mannequins to test its self-driving cars. Available at https://www.businessinsider.com/ubers-fake-city-pittsburgh-self-driving-cars-2017-10.
[47] E. Komendantskaya, L. Arnaboldi, and M. Daggitt, Perfecting self-driving cars – can it be done?. Available at http://theconversation.com/perfecting-self-driving-cars-can-it-be-done-157483.
[48] Monica, Why waiting for perfect autonomous vehicles may cost lives, 2017, Available at https://www.rand.org/blog/articles/2017/11/why-waiting-for-perfect-autonomous-vehicles-may-cost-lives.html.
[49] L. Fraade-Blanar, M.S. Blumenthal, J.M. Anderson, N. Kalra, Measuring Automated Vehicle Safety: Forging a Framework, RAND Corporation, 2018. Available at https://www.rand.org/pubs/researchreports/RR2662.html. https://doi.org/10.7249/RR2662.
[50] Annual global road crash statistics. Available at https://www.asirt.org/safe-travel/road-safety-facts/.
[51] Available at https://cal.streetsblog.org/2020/06/09/autonomous-vehicles-really-arent-perfect-report/, June 2020.
[52] P. Koopman, The heavy tail safety ceiling, Automated and Connected Vehicle Systems Testing Symposium, Carnegie Mellon University, 2018.
[53] W. Knight, Snow and Ice Pose a Vexing Obstacle for Self-Driving Cars, Wired, 2020.
[54] S. Zang, M. Ding, D. Smith, P. Tyler, T. Rakotoarivelo, M.A. Kaafar, The impact of adverse weather conditions on autonomous vehicles: Examining how rain, snow, fog, and hail affect the performance of a self-driving car, IEEE Veh. Technol. Mag. PP (2019) 1.
[55] EurekAlert, Heavy rain affects object detection by autonomous vehicle LiDAR sensors, 2021, Available at https://www.eurekalert.org/pubreleases/2021-02/uow-hra022521.php.
[56] How will self-driving cars fare in bad weather? Available at https://www.endurancewarranty.com/learning-center/tech/self-driving-cars-bad-weather/.
[57] S. Abuelsamid, Dead bugs are getting in the way of fully autonomous self-driving cars, 2020, Available at https://debugger.medium.com/dead-bugs-are-getting-in-the-way-of-fully-autonomous-self-driving-cars-abf5bdf932fb.
[58] S. Abuelsamid, Ford launches bugs at sensors because keeping them clean is crucial for self-driving cars. Available at https://www.forbes.com/sites/samabuelsamid/2019/09/04/ford-launches-bugs-at-sensors-keeping-it-clean-will-be-crucial-for-automated-vehicles/.
[59] N. Oliver, K. Potočnik, T. Calvard, To Make Self-Driving Cars Safe, We Also Need Better Roads and Infrastructure, Harvard Business Review, 2018.

[60] Available at https://infrastructurereportcard.org/, 2021.
[61] Available at https://infrastructurereportcard.org/cat-item/roads/, 2021.
[62] Asce's 2017 American infrastructure report card. Available at https://www.infrastructurereportcard.org/wp-content/uploads/2016/10/2017-Infrastructure-Report-Card.pdf. Accessed 6/4/2021.
[63] AAA Warns Pedestrian Detection Systems Don't Work When Needed Most, 2019.
[64] Human Behavior is 93% Predictable, Research Shows.
[65] A. Walker, Are self-driving cars safe for our cities? 2016, Available at https://archive.curbed.com/2016/9/21/12991696/driverless-cars-safety-pros-cons.
[66] A.J. Hawkins, Americans still deeply skeptical about driverless cars: poll. Available at https://www.theverge.com/2018/1/12/16883510/self-driving-car-poll-congress-bill-safety.
[67] F. Richter, Consumer concerns about self-driving cars, 2018, Available at https://www.statista.com/chart/5950/concerns-about-self-driving-cars/.
[68] D. Wakabayashi, Self-driving uber car kills pedestrian in Arizona, where robots roam, The New York Times (2018).
[69] Partners for Automated Vehicle Education (PAVE), Americans wary of AVS but say education and experience with technology can build trust, 2020, Available at https://pavecampaign.org/pave-poll-americans-wary-of-avs-but-say-education-and-experience-with-technology-can-build-trust/.
[70] AAA, Three-quarters of Americans "afraid" to ride in a self-driving vehicle, 2016, Available at https://newsroom.aaa.com/2016/03/three-quarters-of-americans-afraid-to-ride-in-a-self-driving-vehicle/.
[71] Caravan public opinion poll: Driverless cars, 2018. https://saferoads.org/wp-content/uploads/2018/01/AV-Poll-Report-January-2018-FINAL.pdf.
[72] History.com Editors, n.d. Automobile history. Available at https://www.history.com/topics/inventions/automobiles, abstractNote=The automobile was first invented and perfected in Germany and France in the late 1800s, though Americans quickly came to dominate the automotive industry in.
[73] Department of Energy, The History of the Electric Car, 2015.
[74] P. Valdes-Dapena, I. Sherman, Electric cars have been around since before the US Civil War, Cable News Network (2019).
[75] Environmental Protection Agency, Electric Vehicle Myths, 2022.
[76] Electricity Basics, Alternative Fuels Data Center, US Department of Energy's Vehicle Technologies Office, www.afdc.energy.gov/fuels/electricity.html (Accessed 16 May 2024).
[77] Electricity Basics Benefits & Considerations, Alternative Fuels Data Center, US Department of Energy, www.afdc.energy.gov/fuels/electricity_basics.html (Accessed 16 May 2024).
[78] J. Choksey, The Future of EVs, J. D. Power, 2022.
[79] How Electric Vehicles Will Shape the Future, McKinsey & Company, Apr. 2022. https://www.mckinsey.com/featured-insights/themes/how-electric-vehicles-will-shape-the-future.
[80] P.J. Fontaine, Shortening the path to energy independence: a policy agenda to commercialize battery–electric vehicles, Electr. J. (2008).
[81] J. Whitehead, Here's Why Electric Cars Have Plenty of Grunt, Oomph and Torque, The Conversation, 2022.
[82] Panda Security, Electric Vehicle Charging Stations Are Vulnerable to Hacker Attacks, 2022.
[83] R. Metere, M. Neaimeh, C. Morisset, C. Maple, X. Bellekens, R.M. Czekster, Securing the Electric Vehicle Charging Infrastructure, arXiv, 2022.
[84] N. Bhusal, M. Gautam, M. Benidris, Cybersecurity of electric vehicle smart charging management systems, in: 2020 52nd North American Power Symposium (NAPS), 2021, pp. 1–6.
[85] M. Basnet, M.H. Ali, Exploring cybersecurity issues in 5g enabled electric vehicle charging station with deep learning, IET Gener. Transm. Distrib. 15 (24) (2021) 3435–3449.
[86] S. Köhler, R. Baker, M. Strohmeier, I. Martinovic, Brokenwire: Wireless Disruption of CCS Electric Vehicle Charging, arXiv, 2022.
[87] R. Lakshmanan, Brokenwire hack could let remote attackers disrupt charging for electric vehicles, The Hacker News (2022).
[88] ISO 15118-1:2019 Road Vehicles—Vehicle to Grid Communication Interface—Part 1: General Information and Use-Case Definition, second ed., International Organization for Standardization, 2019.
[89] M. Brunner, Safeguarding electric vehicle charging key to secure e-mobility, Fierce Electronics (2020).
[90] M.A. Sayed, R. Atallah, C. Assi, M. Debbabi, Electric Vehicle Attack Impact on Power Grid Operation, Elsevier, 2021.
[91] A. Sargolzaei, Time-Delay Switch Attack on Networked Control Systems, Effects and Countermeasures (FIU Electronic theses and dissertations), Florida International University, 2015.
[92] A. Sargolzaei, K.K. Yen, M.N. Abdelghani, Preventing time-delay switch attack on load frequency control in distributed power systems, IEEE Trans. Smart Grid 7 (2) (2015) 1176–1185.
[93] A. Sargolzaei, K.K. Yen, M. Abdelghani, S. Sargolzaei, B. Carbunar, Resilient design of networked control systems under time delay switch attacks, application in smart grid, IEEE Access 5 (2017) 15901–15912.
[94] A. Sargolzaei, M. Abdelghani, et al., Detection of and responses to time delays in networked control systems, 2018. US Patent 9,946,231.
[95] A. Sargolzaei, B.C. Allen, C.D. Crane, W.E. Dixon, Lyapunov-based control of a nonlinear multiagent system with a time-varying input delay under false-data-injection attacks, IEEE Trans. Industr. Inform. 18 (4) (2021) 2693–2703.
[96] A. Sargolzaei, C.D. Crane, A. Abbaspour, S. Noei, A machine learning approach for fault detection in vehicular cyber-physical systems, in: 2016 15th IEEE International Conference on Machine Learning and Applications (ICMLA), IEEE, 2016, pp. 636–640.
[97] A. Sargolzaei, K. Yazdani, A. Abbaspour, C.D. Crane III, W.E. Dixon, Detection and mitigation of false data injection attacks in networked control systems, IEEE Trans. Industr. Inform. 16 (6) (2019) 4281–4292.

CHAPTER

8

Intelligent energy management system for autonomous vehicles

Vahid Sohrabi Tabar

Faculty of Electrical and Computer Engineering, University of Tabriz, Tabriz, Iran

List of symbols

Parameters	Definitions
MAX	maximum flow rate through the throttle
TC	throttle angle
RPI	pressure ratio
T	temperature
R	gas constant for air
V_m	manifold volume
m_{cyl}	mass flow rate of air entering the cylinder
V_{dis}	volumetric displacement of the engine
η_{vol}	engine's volumetric efficiency
P_m	manifold pressure
ω_{eng}	engine speed
CT	Torque constant
AFI(λ)	function of air-to-fuel ratio
SPI(δ)	ignition time effect on the engine torque
m	mass of the vehicle
θ	road inclination
T_{room}	temperature
M	mass of the air within the cabin
$A(\phi)$	effective surface area of the vehicle
V_w	absolute wind speed
t	time
V_t	vehicle speed
C_{rolling}	road's friction coefficient
g	gravity acceleration
V	vehicle speed
V_w	wind speed
m	mass (vehicle + equivalent rotating parts + passengers)
C_{drag}	drag coefficient (constant)
A	cross-sectional area
ρ	air density
θ	slope of the road
q_{comb}	combustion energy
$\eta_{\text{mechanical}}$	mechanical efficiency

Handbook of Power Electronics in Autonomous and Electric Vehicles
https://doi.org/10.1016/B978-0-323-99545-0.00007-5

η_{engine}	engine efficiency
r	wheel radius
d_r	differential ratio
g_r	gear ratio
τ_{eng}	engine torque
C_{room}	specific heat at constant pressure
$\hat{Q}$	learning function
$r_{(t+1)}$	reward received for changing the state
$s_{(t+1)}$	initial state of s_t
α	learning rate ($0 < \alpha \leq 1$)
γ	discount factor
Δ	maximum number of training cycles of state
μ	self-adaptive exploration factor
ε_t	exploration factor
t	present number of iteration cycles
χ	constant parameter
ε_f	final exploration value from the first step
Δ_1	final episodes with minimum ε
ε_m	minimum value set for this algorithm
ρ	expected probability of the agent success rate
a'	action calculated in the Bellman equation
η_{engine}	control variable's efficiency
P	road power demand
N	driving cycle
m_{fuel}	mass fuel rate consumption of the vehicle
r_t	reward
εt	error (the difference between the actual torque and the optimal torque)
$^{\bullet}\varepsilon t$	derivative of the error
$u_{(t-1)}$	integration of the error until time t
$\varepsilon\varepsilon_m$	exploration probability (initial value)
NoI	number of iteration
$\chi\rho$	constant parameter

8.1 Introduction

The notion of autonomous vehicles finds its roots in the early 20th century, where these visionary ideas inspired generations of scientists and engineers to explore the feasibility of autonomous transportation. The advancement of sensor technology has been paramount in enabling autonomous vehicles to perceive their surroundings. The light detection and ranging, radio detection and ranging, cameras, and ultrasonic sensors have all contributed to creating a robust sensing system that allows vehicles to make real-time decisions. In parallel, the rise of ride-sharing and mobility services recognized the potential of autonomous vehicles in offering cost-effective and efficient transportation. The main benefits related to the penetration of autonomous vehicles into the transportation system can be classified as follows:

- *Enhancing safety and reducing accidents:* The most compelling argument for the adoption of autonomous vehicles is their potential to revolutionize road safety. Human error is a leading reason for accidents on the road, accounting for a vast majority of collisions. Autonomous vehicles, equipped with many sensors and machine learning algorithms, have the ability to process real-time data, making split-second decisions that prioritize safety. They are not prone to the distractions, fatigue, or impaired judgment that can afflict human drivers. This intrinsic safety improvement could save countless lives, reducing traffic fatalities to historically low levels.
- *Alleviating traffic congestion:* The scourge of traffic congestion plagues cities worldwide, resulting in wasted time, increased pollution, and economic inefficiency. Autonomous vehicles, when integrated into a comprehensive transportation network, significantly reduce congestion. They can communicate with each other and traffic management systems to optimize traffic flow, reduce stop-and-go patterns, and increase road capacity. In urban centers, this technology could ease the gridlock that hampers productivity and quality of life.

- *Economic growth and job creation:* The development and deployment of autonomous vehicles represent a significant economic opportunity. A burgeoning industry, with roots in advanced technology, manufacturing, and transportation, is expected to drive job creation and economic growth. From engineers and software developers to fleet managers and maintenance personnel, a wide array of professions will benefit from the autonomous vehicle ecosystem. Moreover, the benefits of autonomous vehicles extend to industries such as delivery, logistics, and e-commerce, which stand to gain from more efficient transportation systems.
- *Environmental sustainability:* Autonomous vehicles can contribute to a more sustainable future by optimizing energy use and reducing emissions. Electric autonomous vehicles, in particular, have the potential to make a substantial impact on reducing greenhouse gas emissions when charged with clean energy sources. Additionally, the efficiency improvements by autonomous driving can reduce fuel consumption and further contribute to environmental conservation.
- *Accessibility and mobility solutions:* For individuals with limited mobility, including the elderly and people with disabilities, autonomous vehicles offer newfound freedom and independence. These individuals can regain their ability to travel independently, fostering inclusivity, and improving their quality of life. Furthermore, autonomous vehicles can offer new mobility solutions for underserved areas, connecting people to jobs, health care, and education.
- *Urban planning and land use:* The advent of autonomous vehicles presents an opportunity to reimagine urban planning and land use. As parking lots and structures become less necessary due to efficient ride-sharing and parking optimization, urban spaces can be repurposed for parks, housing, and commercial development. This shift can lead to more sustainable and liveable cities.

Although penetration of autonomous vehicles has many advantages, they may create significant challenges for users and all other agents, which must be addressed for their successful integration into the transportation systems. Here are the key challenges and considerations associated with autonomous vehicles.

- *Safety and reliability:* Technology reliability: Reliability of the technology is a critical challenge. Autonomous vehicles must function flawlessly in various weather conditions and traffic situations to guarantee safety. Cybersecurity: Autonomous vehicles are susceptible to cyberattacks. Cybersecurity of these vehicles is paramount to prevent unauthorized access and malicious interference. Sensor limitations: Sensors can be affected by adverse weather conditions (e.g., heavy rain or snow) and face challenges in interpreting complex traffic scenarios.
- *Ethical and legal considerations:* Liability issues: Specifying liability in accidents involving an autonomous vehicle is complex. It raises questions about whether the vehicle owner, the software developer, or another party is responsible. Ethical decisions: Autonomous vehicles may need to make ethical decisions in emergency situations. Deciding how vehicles should prioritize different outcomes (e.g., protecting the occupants vs pedestrians) raises ethical dilemmas.
- *Regulatory and legal frameworks:* Lack of regulations: The regulatory framework for autonomous vehicles is still evolving in many regions. Establishing clear and comprehensive regulations that ensure safety and compliance is a significant challenge. Interstate and international coordination: Autonomous vehicles need to operate across state and national borders. Coordinating regulations and standards across various jurisdictions is essential.
- *Infrastructure and connectivity:* Infrastructure compatibility: Existing road infrastructure may not be fully compatible with autonomous vehicles. Investments in infrastructure upgrades, such as road markings and smart traffic management systems, are necessary. Network connectivity: Autonomous vehicles rely on consistent and high-speed network connections for communication with other vehicles and traffic infrastructure. Expanding and securing this connectivity is essential.
- *Data privacy and ownership:* Data collection and ownership: Autonomous vehicles generate a lot of data about their surroundings and occupants. Determining who owns these data and how they are used raises privacy concerns. Data security: Safeguarding the collected data from breaches and misuse is crucial for ensuring user privacy and security.
- *Socioeconomic impacts:* Job displacement: The widespread adoption of autonomous vehicles leads to job displacement in sectors like trucking and taxi services. Preparing for these labor market changes is essential. Economic inequality: autonomous vehicles could exacerbate economic inequality if their benefits are not distributed equitably. Ensuring accessibility and affordability for all is a key consideration.
- *Human-autonomous vehicle interaction:* Human trust and understanding: Users need to trust and understand how autonomous vehicles function. Ensuring clear communication between the vehicle and the user is crucial for a smooth transition to autonomous driving. Mixed traffic environments: Autonomous and human-driven vehicles will coexist for some time. Managing the interactions and behaviors between these two types of vehicles poses challenges.

- *Environmental impact:* Electric vehicle adoption: Autonomous vehicles are often paired with electric power trains. Widespread adoption of electric autonomous vehicles depends on the availability and sustainability of electric charging infrastructure. Induced demand: There is a concern that the convenience of autonomous vehicles could lead to increased demand for personal transportation, potentially offsetting environmental benefits.

8.2 Components and subsystems

An autonomous vehicle is typically constructed from various interconnected sections, each of which plays a specific role in enabling the vehicle to operate autonomously. Here is a breakdown of the key components and sections that make up an autonomous vehicle.

- *Sensors:* Sensors are critical components of autonomous vehicles. They provide data about the vehicles' surroundings, enabling them to perceive the environment.
- *Control systems:* Control systems manage the vehicles' movements and operation. They interpret data from sensors and make real-time decisions to control acceleration, steering, and braking.
- *Computer hardware:* Autonomous vehicles are equipped with powerful onboard computers to process data from sensors and run complex algorithms for decision-making and control.
- *Artificial intelligence and machine learning units:* Artificial intelligence (AI) and machine learning algorithms are used to process sensor data, make decisions, and learn from past experiences. They are crucial for tasks such as object detection, path planning, and decision-making.
- *Connectivity:* Autonomous vehicles often rely on high-speed network connectivity to exchange data with other vehicles, infrastructure, and cloud-based services. This connectivity can be crucial for real-time updates and enhanced situational awareness.
- *Mapping and localization:* Autonomous vehicles use high-definition maps and localization techniques to specify their precise location. These maps include detailed information about road features, traffic signs, and lane markings.
- *Actuators:* Actuators are responsible for physically controlling the vehicles' movements. This includes components like the throttle, brake, and steering systems.
- *Safety systems:* Autonomous vehicles utilize advanced safety systems, such as collision detection and avoidance systems to enhance passenger safety and prevent accidents.
- *Human-machine interface:* A human-machine interface (HMI) provides a way for passengers to interact with the vehicles' autonomous features. This may include touchscreen displays, voice recognition systems, and controls for manual override.
- *Redundancy and fail-safes:* Autonomous vehicles often incorporate redundant systems and fail-safe mechanisms to ensure safety in the event of component failures or unexpected situations.
- *Regulatory compliance:* To meet legal and safety standards, autonomous vehicles need to include compliance mechanisms that adhere to local regulations and ensure safe operation.
- *Power supply:* Electric or hybrid power trains are common in autonomous vehicles. They provide the necessary energy to operate all systems, including propulsion and onboard electronics.
- *External communication:* Autonomous vehicles often need to communicate with other vehicles, traffic infrastructure, and central traffic management systems to optimize traffic flow and safety.

These components and sections work in tandem to enable autonomous vehicles to perceive their environment, make real-time decisions, and safely navigate through various situations. The integration of these systems, along with ongoing advancements in technology and algorithms, is what allows autonomous vehicles to function and improve over time.

8.3 Intelligent energy management system

According to the structure of autonomous vehicles, a robust control unit is required to make efficient and reliable connections between different elements. In the past decade, numerous computer-assisted tools and technologies have been introduced with the aim of efficiently optimizing energy consumption in our daily lives. The primary components of an intelligent energy management system are illustrated in Fig. 1 [1], which includes several main parts as presented in the following:

- sensing and measurement infrastructure
- actuation mechanisms
- processing engine
- user interfaces

As depicted in Fig. 2, each intelligent energy management system (IEMS) necessitates the inclusion of at least one variant of each of these components to function optimally. Data collected from sensors should be stored and processed. This subsystem comprises units responsible for processing all obtained data, executing optimization tasks based on end-user preferences, and learning to recognize occupants' activity patterns. The processing engine also interfaces with the actuation mechanisms and manages anomalies or outliers to adapt the settings of individual appliances or devices according to the context. Additionally, the processing engine occasionally provides recommendations to agents by the user interface, encouraging behavioral changes that can impact overall energy consumption. Such adjustments are aimed at promoting a more sustainable energy footprint for society, while also considering economic implications. Furthermore, users gain access to graphical representations of their daily energy consumption by user interface unit. This element is most commonly found in the form of a computer or smartphone application.

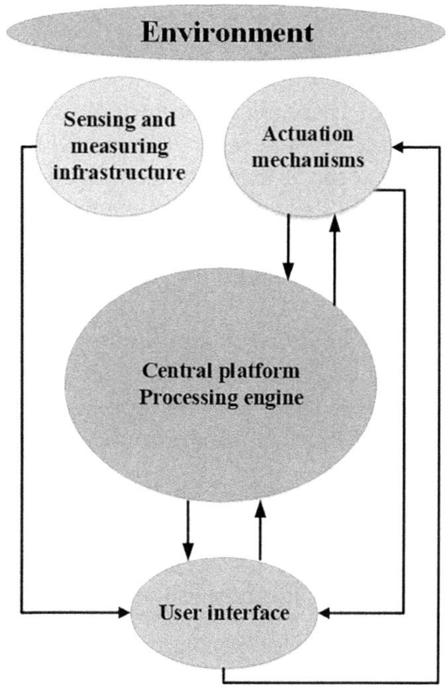

FIG. 1 The base structure of an intelligent energy management system.

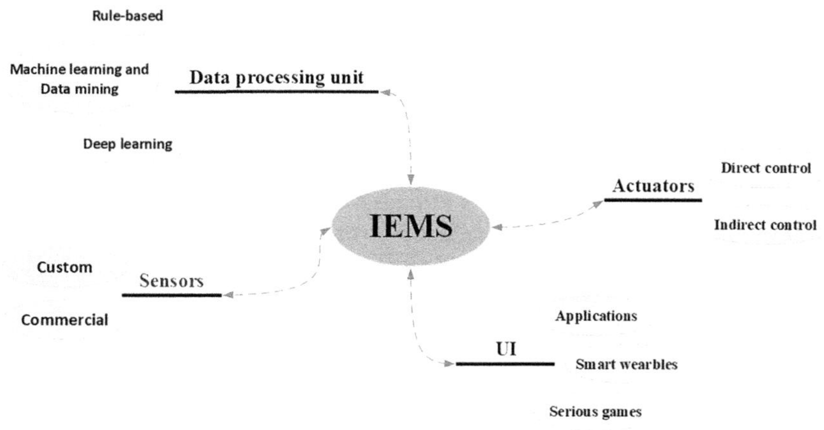

FIG. 2 The main components of an intelligent energy management system.

An IEMS for autonomous vehicles is a crucial component in optimizing the energy consumption and efficiency of these vehicles. It plays a vital role in ensuring that autonomous vehicles can operate efficiently, safely, and without the need for frequent recharging or refueling for long periods. In order to design an optimal IEMS, the operation and role of the following elements and units should be considered:

- *Energy sources*: Autonomous vehicles can use various energy sources, including electric batteries, fuel cells, or hybrid power trains. The IEMS must be designed to manage the available energy source efficiently.
- *Data and sensing*: The IEMS relies on a wide range of data inputs and sensors to make real-time decisions about energy usage. These data include information about the vehicle's state, driving conditions, traffic, weather, and the energy source's current status.
- *Route planning and optimization*: One of the primary functions of the IEMS is to plan and optimize the vehicle's route. It takes into account factors such as traffic congestion, road conditions, elevation changes, and the locations of charging or refueling stations. By optimizing the route, the system can reduce energy consumption and increase range.
- *Driving behavior analysis*: The IEMS can analyze the vehicle's driving behavior and adapt it to be more energy efficient. For example, it can optimize acceleration, deceleration, and speed to maximize energy efficiency.
- *Predictive analytics*: Machine learning and predictive algorithms are often used to forecast energy needs based on historical data and current driving conditions. These allow the system to make proactive decisions, such as when to switch to a different power source or when to engage regenerative braking.
- *User preferences*: Some IEMSs may take into account user preferences and priorities. For example, a user might prioritize a shorter travel time over energy efficiency, and the system can adjust accordingly.
- *Data connectivity*: Autonomous vehicles are typically connected to cloud-based systems that provide real-time updates on traffic, weather, and charging station availability. This connectivity is essential for effective energy management.
- *Safety considerations*: The IEMS must prioritize safety while optimizing energy usage. It ensures that there is always enough energy reserved for essential vehicle functions and emergency situations.
- *Regulatory compliance*: Depending on the region, there may be regulations and standards governing energy management in autonomous vehicles. The IEMS must adhere to these regulations.

In the following section, some of the critical components in the structure of an IEMS including tasks of autonomous vehicles, level of autonomy, sensors, and algorithms will be explained [2].

8.3.1 Vehicle tasks in intelligent energy management system

The principal operation of an autonomous vehicle depends on its task, where the most critical types are described as follows:

- *Dynamic driving task:* Dynamic driving task (DDT) encompasses the tactical functions needed to operate autonomous vehicles safely in real-world conditions. These include vehicle control aspects such as steering (lateral control) and controlling acceleration and deceleration (longitudinal control), along with environmental monitoring and maneuver planning. DDT focuses on real-time, on-road operation, excluding functions like trip planning and destination selection.
- *Automated driving system:* Automated driving system (ADS) refers to both software and hardware units required to continuously perform the DDT without considering its operational design domains.
- *Operational design domain:* Operational design domain (ODD) represents the specific conditions under which driving automation systems or features function. These conditions are specified by geographical, environmental, and time factors. Additionally, they rely on traffic and road characteristics. If an autonomous vehicle can operate on highways and in urban environments, its ODD incorporates data to define the abilities and limitations of the ADS according to the road type, geographical area, and weather conditions.
- *Object and event detection and response:* Object and event detection and response (OEDR) is a subtask of DDT and involves environmental monitoring including the detection, recognition, and classification of objects. It also provides appropriate responses to events and operates within an ODD and enables the ADS to differentiate between various entities such as vehicles, pedestrians, and objects.
- *Active safety systems:* Active safety systems (ASSs) are responsible for monitoring both external and internal conditions within autonomous vehicles to identify potential and existing dangers to passengers, pedestrians, and other vehicles.

They automatically intervene to reduce the likelihood of potential collisions. These interventions can include alerting the driver, adjusting the vehicle's systems, and controlling subsystems like throttle, brakes, and suspensions.

8.3.2 Automation level in intelligent energy management system

The Society of Automotive Engineers introduced a classification system to categorize levels of driving automation, as explained below and illustrated in Fig. 3.

- *Level zero:* There is no automation and the driver is entirely responsible for the DDT. In addition, ASSs provide assistance.
- *Level one:* The ADS handles longitudinal or lateral vehicle control, but not the complete OEDR task. The driver is always in the vehicle and serves as a fallback in case the ADS encounters challenges. Systems such as the electronic stability section, antilock braking unit, and adaptive cruise control fall under this level.
- *Level two:* The ADS manages longitudinal and lateral vehicle control. However, it remains similar to level one in terms of its inability to perform the complete OEDR. The driver should be in the vehicle the whole time and ready to take control if necessary. Automatic emergency braking and collision prevention systems are additional features at this level.
- *Level three:* Level three signifies that the ADS can perform the entire DDT with the driver as the fallback. The driver is notified when the ODD limitations are approached and they receive notifications of system failures.
- *Level four:* This level does not require human intervention in most cases. The ADS at level four of automation conducts the entire DDT within specific ODDs, typically areas with well-mapped and structured roads. The driver can take control if desired, but the ADS minimizes risk significantly.
- *Level five:* Level five represents full automation, where the ADS can handle the entire ODD without restrictions. The ADS manages the DDT fallback and the driver only takes control when they choose to do so.

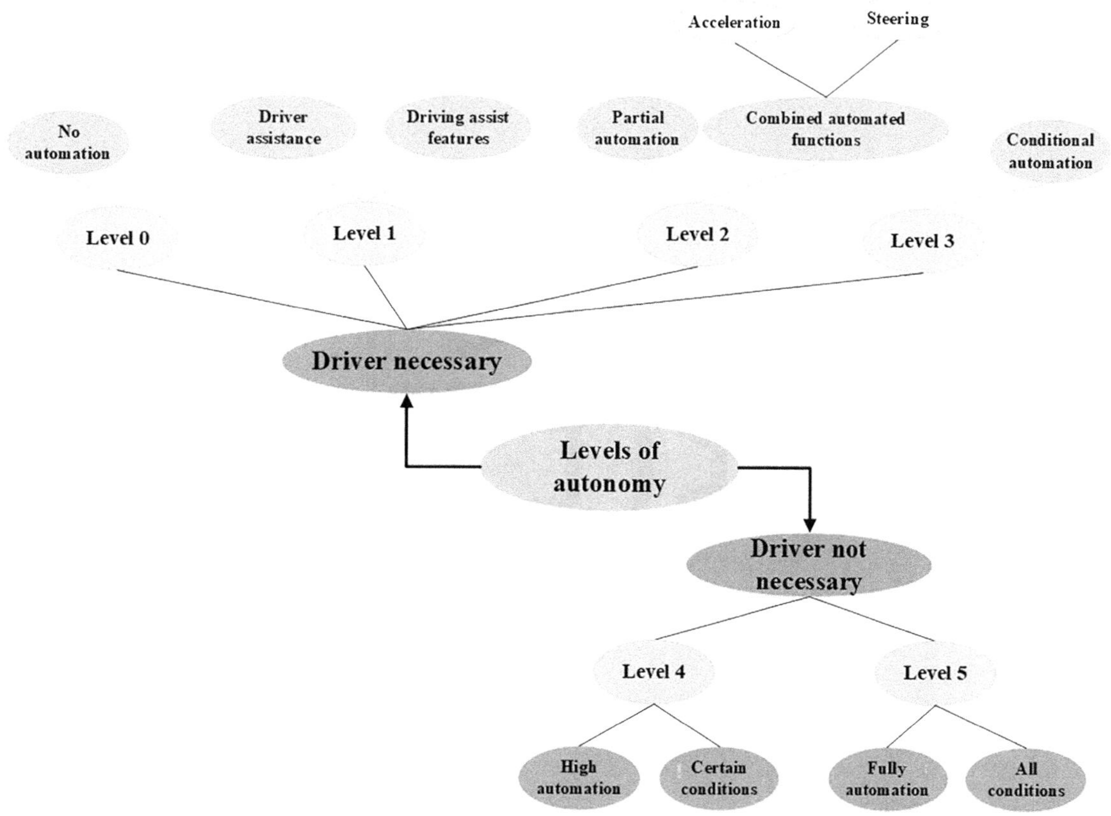

FIG. 3 Level of autonomy based on the society of automotive engineers.

8.3.3 Sensors in intelligent energy management system

The most important and usable sensors in autonomous vehicles, which also can be used in IEMSs are illustrated in Fig. 4 and introduced as follows:

- *Camera:* Autonomous vehicles are equipped with visible light cameras that offer a comprehensive 360-degree view of their surroundings. These cameras excel in detecting and recognizing objects, providing valuable data for AI-driven algorithms. However, their performance is compromised under low-light conditions, and they generate substantial data volumes, necessitating intricate processing. In situations with reduced visibility, infrared cameras are employed to enhance performance.
- *RADAR:* Radio detection and ranging (RADAR) sensor utilizes radio waves to measure critical factors including distance, velocity, and angle. Autonomous vehicles employ transmitters to emit radio waves and receivers to capture reflected waves. RADAR operates effectively across various weather conditions and over long distances, although there is a possibility of false object identification.
- *LiDAR:* Light detection and ranging (LiDAR) sensor employs light to gauge distances. It calculates these distances by measuring the time it takes for emitted laser beams to bounce off the environment and return to a photodetector. The collected beams are combined into a point cloud, creating a detailed three-dimensional (3D) image of the surroundings. While LiDAR offers exceptional sensing capabilities, but it is relatively costly.
- *GNSS:* Global navigation satellite system (GNSS) encompasses the arrangement and alignment of satellites that furnish essential data, including navigation, positioning, and traffic information. GNSS relies on triangulation to determine the location of receivers and computes the distances between satellites in geostationary orbits and the autonomous vehicle. It should be noted that such signals are susceptible to interference due to their inherent weakness.
- *Ultrasonic sensors:* These sensors are short-range and cost-effective, making them suitable for low-speed scenarios. They measure the distance to objects by employing ultrasonic waves, which are then converted into electrical signals.
- *Global positioning system:* Global positioning system (GPS) provides vital spatial information, encompassing navigation, positioning, and timing data. This service is freely accessible to civilians and plays a crucial role in path planning for autonomous vehicles.

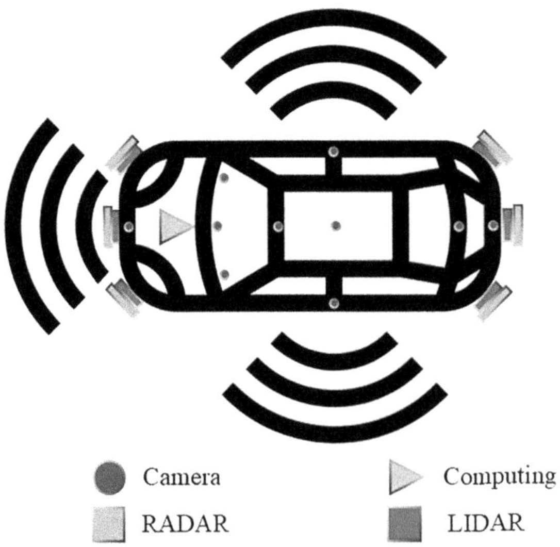

FIG. 4 Type of sensors in autonomous vehicles.

8.3.4 Algorithms and architectures in intelligent energy management system

In order to construct an intelligent energy management system, appropriate processing algorithms should be used. As a result, some of the well-known algorithms are introduced in this section.

- *Convolutional neural networks:* Convolutional neural networks (CNNs) are a type of neural network primarily used for image processing. They excel at extracting distinctive features from images with high accuracy by employing a convolution function. These networks take two-dimensional (2D) image inputs and utilize multiple hidden layers to obtain high-level features. Without the need for extensive preprocessing, CNNs are easily deployable. In the context of autonomous vehicles, CNNs find application in tasks such as path planning and pedestrian detection.
- *Region-based convolutional neural networks:* Region-based convolutional neural networks (RCNNs) are neural networks specifically designed for object detection. They are preferred over conventional CNNs for this task, as using CNNs to detect objects within an image can be resource-intensive, particularly when the number of object occurrences varies. RCNNs employ a selective searching process to identify object boundaries and labels in an image, subsequently creating bounding boxes around them. Linear regression models are then used to refine the bounding box coordinates. In the AV context, RCNNs are employed for detecting pedestrians, objects, and traffic signs.
- *Simultaneous localization and mapping:* Simultaneous localization and mapping (SLAM) is a technique to estimate the relative positions of static objects using LiDAR or RADAR sensors. RADAR-SLAM offers additional capabilities, as it can provide velocity information, functioning as an odometer, and even perform localization based on existing map data. In contrast, LiDAR-SLAM and visual SLAM rely on monocular or stereo cameras to track features in consecutive images, while estimating relative orientation and translation. In the autonomous vehicle context, SLAM is instrumental for tasks such as motion control, path planning, and occasionally, pedestrian detection.
- *K-means clustering:* K-means clustering (KMC) is an unsupervised algorithm designed to cluster unlabeled or unclassified datasets into predefined groups. Each cluster is associated with a centroid, and the algorithm iteratively seeks to minimize the total sum of distances between cluster centroids and their respective data points.
- *You only look once:* You only look once (YOLO) is a developed CNN-based algorithm within the darknet framework, renowned for its efficiency in real-time object detection. Subsequent iterations such as YOLOv2, YOLOv3, and YOLOv4 have emerged in the field of computer vision, building upon the original YOLO concept. YOLO operates as a single-stage detector, handling object identification and classification in a single network pass. Its efficiency and ease of deployment make it a popular choice in computer vision applications.

8.4 Recent studies and developments

The automotive sector faces several critical challenges including traffic congestion, accidents, high emissions, and excessive energy usage [3]. Autonomous vehicles have emerged as a potential solution to improve traffic safety [4] and address energy consumption in transportation [5]. Although battery-powered electric and hybrid electric autonomous vehicles demonstrate better energy efficiency compared to conventional vehicles, reports show that conventional vehicles will persist on roadways in the future [6], despite their low efficiency. For instance, modern internal combustion engines (ICEs) use less than 34% of the injected energy. Emissions remain a persistent concern in ICE-based platforms. Therefore, a primary concern within the automotive industry is the development of new technologies to optimize energy usage and decrease vehicle emissions.

Energy management strategy (EMS) in vehicles controls the energy flow between different energy sources or suppliers. This strategy requires an understanding of vehicle operation and its environment. It is noteworthy that uncertainties also should be considered as a critical factor for optimizing the controller performance, where effective control encompasses the consideration of uncertainties. Such a strategy also attempts to optimize fuel consumption, while minimizing potential adverse impacts on vehicle performance [7]. An ideal EMS should address the uncertainties, nonlinearities, and complexities in the models. The EMS for conventional vehicles can be divided into two main groups: rule-based and optimization-based approaches [8]. Figure 5 illustrates the details regarding these classifications.

Deterministic rule-based strategy involves specific methods for power consumption reduction in conventional vehicles. For instance, in Ref. [9] a deterministic rule-based strategy through a thermal barrier coating is introduced to

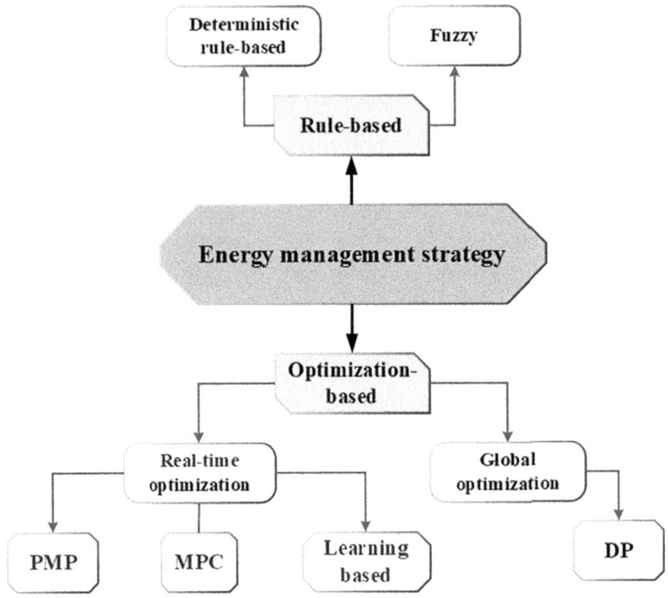

FIG. 5 Classification of energy management strategy in autonomous vehicles.

decrease fuel consumption in conventional vehicles. In a similar study, in Ref. [10] braking protocols are utilized to manage engine torque in optimal bound based on the vehicle's operational regime and reduced fuel consumption. However, such approaches cannot be implemented on uncertain models as they lack an adaptation framework to accommodate model changes, which is a typical attribute of actual models. Furthermore, they require an additional algorithm or method to optimize energy usage and cannot inherently optimize outcomes by itself.

Another method for energy management of vehicles is fuzzy rule-based control strategies, a process that translates human expertise into a computational algorithm. These methodologies are more adaptable and suitable for conventional vehicles compared to deterministic methods. In Ref. [11] the desired engine torque is optimized based on the vehicle's road power demand and uses a proportional-integral-derivative (PID) controller to regulate the air/fuel ratio by adjusting the throttle angle for an autonomous operation mode. In a similar study, an adaptive neuro-fuzzy system is employed to produce optimal torque as a reference signal for a PID controller to manage the throttle using the same procedure as in the earlier research [12]. Nonetheless, all rule-based methods present a challenge when applied as an EMS for autonomous conventional vehicles. They necessitate an additional component to optimize vehicle energy consumption and are unable to serve as a management system for autonomous conventional vehicles and simultaneously minimize fuel usage.

Another approach in vehicle energy management is the optimization-based control strategies. In Ref. [13] Pontryagin's minimum principle is applied to decrease fuel usage in a conventional vehicle; however, this method cannot operate as an online controller since it requires precise prior knowledge and yields a local solution. In Ref. [14] a model predictive control (MPC) with notable prediction capabilities is introduced to optimize fuel demand of a conventional vehicle. This algorithm also has been utilized for vehicles operating in urban traffic [15]. Although MPC optimization is known for its quick computational time, it demands additional processes and knowledge to handle the uncertainty associated with autonomous vehicles.

Learning-based energy management encompasses a spectrum of supervised, unsupervised, and reinforcement learning algorithms that are significantly applied in hybrid electric vehicles (HEVs). These methods have been well documented in various studies [16–20]. To enhance the real-time control capabilities, adapting traditional learning-based algorithms have emerged as an effective strategy. For instance, in Ref. [21] the learning rate index in the deep Q-network algorithm is adjusted to enhance fuel economy in plug-in hybrid electric vehicles (PHEV). This chapter introduces a self-adaptive strategy employing a genetic algorithm (GA) and a deep Q-network to generate an optimal learning rate. In another study, in Ref. [22] an improved Q-learning algorithm consisting of deterministic rules for real-time cooperation between the fuel cell and the supercapacitor in HEVs is proposed. This algorithm aims to improve the adaptability of HEVs across various driving cycles. In Ref. [23] a self-adaptive reinforcement-exploration Q-learning algorithm is presented, aiming to reduce the algorithm's operation time and improve exploration efficiency. The results exhibit notable benefits in the average number of turning times and success rates in route planning.

Although learning-based control systems have been primarily utilized for vehicle movement, the literature review indicates limited utilization in autonomous conventional vehicles.

Dynamic programming (DP) is a robust technique employed to attain optimal fuel economy over specific drive cycles [24]. However, this method operates offline and its results serve mainly as a benchmark to assess and refine existing online algorithms. For managing energy in uncertain conditions, one must employ adaptive controllers that adapt to uncertainty and changing environmental and vehicle operational conditions. Recent studies propose various strategies to reduce fuel consumption, such as traffic signal optimization to streamline urban traffic, leveraging connected vehicles to manage active lanes, and incorporating automated vehicles into mixed traffic scenarios [25,26]. These strategies integrate deep reinforcement learning, DP, and fuzzy logic systems (FLSs) in traffic control strategies. Additionally, redesigning conventional vehicle engines and gearboxes is proposed as another avenue to minimize fuel consumption [27–29].

This newly developed controller presents a novel reinforcement learning algorithm capable of addressing the complexity of vehicle problems and uncertainties under environmental conditions. It significantly reduces fuel consumption compared to optimization- and adaptive-based controllers, while also decreasing the controller's operational time to enable online functionality. Notably, this learning-based approach does not necessitate modification in the face of changing environmental conditions, as it can learn from the training data and apply this knowledge to address unknown conditions.

8.5 Mathematical model of an intelligent energy management system

In order to design an intelligent energy management system for autonomous vehicles, various information related to the operation principle and structure of vehicles is required. In addition, there are different uncertainties that their effects also should be investigated to improve the model's accuracy. The environmental conditions including road specifications, wind speed, air-conditioning, and driving cycles are some of the most critical parameters for the energy management systems to optimize the whole required energy. In order to have a better standpoint, a specific energy management formulation is modeled by Eqs. (1)–(9), where more descriptions can be found in Ref. [20]. In this model, appropriate probabilistic approaches and test cycles have been applied to simulate the behavior of uncertain parameters. In this regard, the mathematical model that is based on reinforcement learning (i.e., Q-learning) is introduced and then, its operation compared to other methods is investigated [20]. It is noteworthy that the autonomous vehicle under study is a conventional type that consumes gasoline as the input source.

$$\dot{m}_{\text{throttle}} = \text{MAX.TC.RPI} \tag{1}$$

$$\dot{P}_m = \frac{\dot{m}_{\text{manifold}}.R.T}{V_m} = \frac{(\dot{m}_{\text{throttle}} - \dot{m}_{\text{cyl}}).R.T}{V_m} \tag{2}$$

$$\dot{m}_{\text{cyl}} = \frac{1}{4\pi} V_{\text{disp}}.\eta_{\text{vol}}.\omega_{\text{eng}}.\frac{P_m}{RT} \tag{3}$$

$$\tau = \frac{C_T.AFI(\lambda).SPI(\delta).V_{\text{disp}}.P_m.\eta_{\text{vol}}}{4\pi.R.T} \tag{4}$$

$$\dot{\gamma}_{\text{combustion}} = (\dot{\gamma}_{\text{fuel}} + \dot{\gamma}_{\text{air}} + \dot{\gamma}_{\text{exhust}}) = P_{\text{RPD}} + \dot{\gamma}_{\text{water/oil}} + \dot{\gamma}_{\text{heatloss}} \tag{5}$$

$$(\dot{\gamma}_{\text{combustion}}).\eta_{\text{otto}}.\eta_{\text{fuel-air}}.\eta_{\text{mechanical}}.\eta_{\text{heatloss}} = P_{\text{RPD}} = \text{P}_{\text{road-friction}} + P_{\text{drag}} + P_{\text{slope}} + P_{\text{accessory}} + P_{\text{driving}} \tag{6}$$

$$\begin{aligned} P_{\text{RPD}} &= \text{P}_{\text{road-friction}} + P_{\text{drag}} + P_{\text{slope}} + P_{\text{accessry}} + P_{\text{driving}} \\ &= C_{\text{rolling}} mg\cos\theta.Vt + C_{\text{drag}}(\phi).\frac{1}{2}\rho(V_m - V_t)^2 A(\phi).V_t + \\ & mg\sin\theta.V_t + MC_{\text{room}}\frac{dT_{\text{room}}}{dt} + mVt\frac{dV_t}{dt} \end{aligned} \tag{7}$$

$$\dot{m}_{\text{fuel}} = \frac{\text{P}_{\text{RPD}}}{(q_{\text{combustion}}).\eta_{\text{mechanical}}.\eta_{\text{engine}}} \tag{8}$$

$$m_{\text{fuel}} = \int_0^t \dot{m}_{\text{fuel}}\, dt \tag{9}$$

8.5.1 Q-learning algorithm

The basic expression for the Q-learning update rule is as follows:

$$\hat{Q}(s_t, a_t) = \hat{Q}(s_t, a_t) + a\left[r_{t+1} + \gamma \max{}_a \hat{Q}(s_{t+1}, a) - \hat{Q}(s_t, a_t)\right] \tag{10}$$

In the Q-learning algorithm, the determination of the optimal action-value correspondence occurs independently of the policy throughout the learning process. It is essential to visit and update all state-action pairs for fast convergence. Properly specifying the state, action, and reward functions plays a critical role in achieving convergence. Once the function is trained, the best action for each state is selected. The Q-learning algorithm, as a reinforcement learning method, trains the agent to estimate rewards.

One challenge encountered in this algorithm is finding the right balance between exploration and exploitation. Exploration involves the agent randomly selecting actions based on previously obtained reward values, while exploitation entails choosing the action with the highest average rewards. These two processes cannot occur simultaneously, requiring the agent to decide when to engage in each. The ε-greedy method offers a simple way to strike a balance, wherein the agent employs a small static probability value, to randomly select an action among others. The ε-greedy method is not well suited for real-time control strategies, especially in online control systems for autonomous vehicles. To address this limitation, the self-adaptive Q-learning algorithm has been developed.

8.5.2 Self-adaptive Q-learning algorithm

To enhance the ε-greedy algorithm, a specific time-dependent exploration factor is inserted into the main equations. The entire exploration process is segmented into two distinct steps. In the initial step that involves agent training, the agent lacks information about its environment and, therefore, needs to explore the environment extensively. Consequently, the exploration factor decay should be gradual to allow the agent to explore most states and compute Eq. (10) for a broader set of states. The proposed model for this phase is presented in Eq. (11).

$$\varepsilon_t = \chi - e^{\left(\frac{t}{\Delta}\right)^2} t \leq \mu\Delta \tag{11}$$

The commencement of the second step occurs when the transition of the agent's emphasis from exploration to exploitation or the use of Q-table elements. In this stage, the agent possesses knowledge about the environment; however, it requires data about future states. The rate of attenuation for the exploration factor is illustrated in Eq. (12).

$$\varepsilon_t = \begin{cases} \varepsilon_f + \left(\varepsilon_m - \varepsilon_f\right) \times \frac{t-\mu\Delta}{\Delta l-\mu\Delta} \times \rho if \varepsilon_f + \left(\varepsilon_m - \varepsilon_f\right) \times \frac{t-\mu\Delta}{\Delta_l-\mu\Delta} \times \rho \geq \varepsilon m \\ \varepsilon_m else \end{cases} \tag{12}$$

As a result, Eq. (10) should be

$$\hat{Q}(s_t, a_t) = \hat{Q}(s_t, a_t) + a\left[r_{t+1} + \gamma \max{}_a \hat{Q}(s_{t+1}, a) - \hat{Q}(s_t, a_t)\right] \tag{13}$$

8.5.3 Dynamic programming

This method decomposes one problem into several subproblems to reduce the simulation time and increase convergence probability. The constraints are

$$\eta_{\min} \leq \eta \leq \eta_{\max} P_{\min} \leq P \leq P_{\max} \tag{14}$$

It should be noted that determining the control sequence to minimize the cost function is the presented control problem:

$$J = \sum_{k=0}^{N} \dot{m_{fuel}}\left(\eta_k, P_k\right) \tag{15}$$

8.5.4 Reinforcement learning controller

To optimize the energy management systems of conventional autonomous vehicles, a reinforcement learning (RL) controller is applied due to its efficiency and adaptability in unfamiliar environments. This controller employs a Q-learning algorithm for overseeing the throttle angle based on an optimal torque.

8.5.5 Q-learning controller algorithm

The Q-learning algorithm consists of two distinct phases: training and testing. In the training phase, a Q-table is constructed using states derived from the real power demand and actions corresponding to various throttle angles. Specifically, the real power demand is discretized into 282 states (representing 282 power consumption values), while there are a total of 766 states for all power consumption values. Additionally, 15 actions are chosen, representing different throttle angles ranging from 0 to 80 degrees. The Q matrix is trained with the 282 states during this phase. Subsequently, in the testing phase, the controller encounters unknown states but selects the appropriate action based on the learning specifications (Eq. 10). Throughout the training step, all elements of the Q-table should be nonzero.

In the testing phase, the learned table is utilized to determine the best action for each state. The error is also incorporated into the reward function considering the system dynamics and requirements of the algorithm. Establishing an appropriate reward function is critical for the design of the reward block. As a result, the employed reward function is created based on a polynomial function that is presented as follows:

$$r_t = -\left(\varepsilon_t^2 + \dot{\varepsilon}_t^2 + 0.001u_{t-1}^2\right) + 5 \tag{16}$$

The mentioned function computes rewards each time, and the Q-function utilizes this outcome as a reward for Q-table training. To regulate the vehicle's fuel consumption, there are limitations on revolutions per minute (RPM), actual, and optimal torques.

8.5.6 Self-adaptive Q-learning controller

To reduce the operation time and increase the reward value of the Q-learning algorithm, the self-adaptive Q (SAQ)-learning algorithm can be used that only has one extra loop in the training part for updating the exploration rate at the end of each interval. The algorithm begins with discrete states and actions according to the vehicle specifications and environmental conditions. Then, the value of ε_m, Δ and Δ_1 is determined to present a time-dependent exploration and exploitation rate based on Eqs. (11) and (12). When the time-dependent rates are generated, Eq. (13) is considered to calculate Q-table, where its values will be checked after each iteration.

8.5.7 Convergence and robustness of model

The convergence and robustness of the control algorithm should be investigated under different conditions. The convergence refers to the number of required iterations to reach a specific value, while the robustness shows the algorithm's response to the fluctuations.

8.6 Simulations and results

The outcomes encompass the results obtained through three methodologies including the FLS-PID model, Q-learning approach, and DP algorithm. It is noteworthy that the results have been presented for 100 s and more analyses can be found in Ref. [20]. The FLS-PID serves as the baseline method for comparison, where this model yields a fuel consumption of 6.7 L/100 km and a vehicle efficiency of 2.7.

The state of Q-learning algorithm is represented by the power demand in each second, while the action corresponds to the throttle angle, leading to the computation of the actual torque. As a result, the fuel consumption per 100 km can be estimated. The Q-learning control system involves two steps consisting of creating and training the Q-table and identifying the highest chance in the mentioned table.

Figure 6 illustrates the road power demand (RPD), calculated from Eq. (7), which remains consistent across all evaluations. This consistency implies identical vehicle specifications, driver behavior, and environmental conditions for all

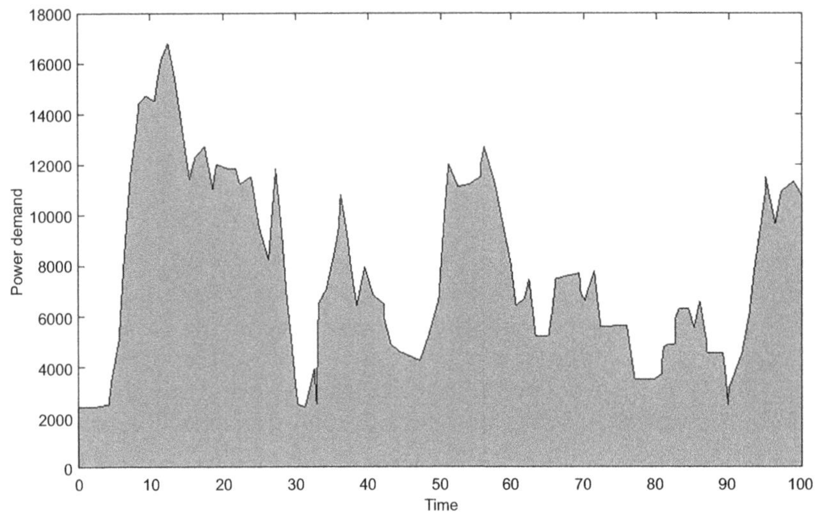

FIG. 6 Power demand obtained from Eq. (7).

simulations. Engine RPMs controlled by FLS-PID, Q-learning controller, and DP are depicted in Fig. 7. FLS-PID exhibits a smoother engine operation compared to the others. The Q-learning controller operates with fluctuations between 1500 and 6300 RPM, averaging 2200 RPM, closely aligning with the optimal RPM range. DP shows less fluctuation with an average of 3400 RPM.

The results in Fig. 8 display the engine torque for different approaches. As evident, FLS-PID operates the engine in the range of 40–75 Nm, and the Q-learning maintains the engine within the same range, averaging 48.5 Nm. Engine energy efficiency, as shown in Fig. 9, shows that FLS-PID achieves an efficiency of 22%–32%, with an average of 28.3%. In contrast, Q-learning results in engine efficiency ranging from 19% to 32%, averaging 29.8%, indicating superior efficiency. DP exhibits an efficiency fluctuating between 30% and 34%, with an average of 33.01%. Figure 10 depicts the fuel demand of FLS-PID and Q-learning, averaging 6.71 L/100 km and 6.69 L/100 km, respectively. Accordingly, Q-learning demonstrates the best performance among controllers, while DP illustrates the fuel consumption of 6.12 L/100 km.

The control signals for FLS-PID and Q-learning controllers are presented in Fig. 11. As seen, the average control signal of the mentioned methods are 0.27 and 0.28, respectively. This higher control signal is a trade-off for improving engine efficiency and fuel consumption.

The Q-learning enhances engine efficiency to 29.81%, while the SAQ-learning further increases efficiency to 30.01%, which is relatively high for a conventional vehicle. Additionally, the Q-learning and SAQ-learning algorithms decrease

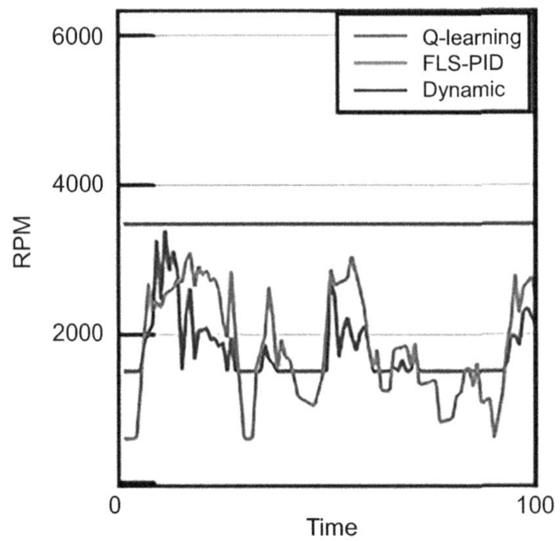

FIG. 7 The comparison of engine RPM for different controllers.

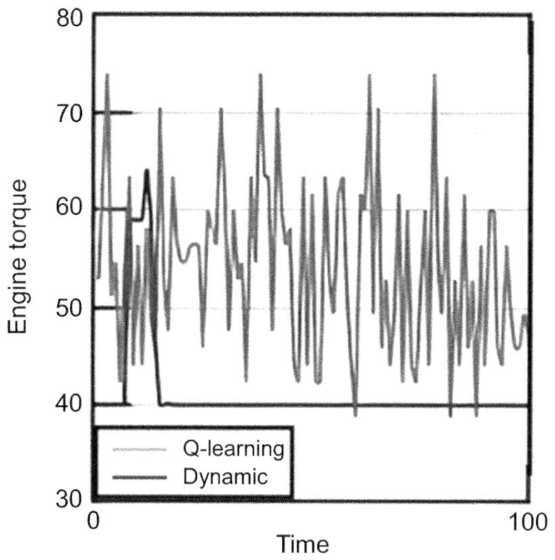

FIG. 8 The comparison of engine torque for different controllers.

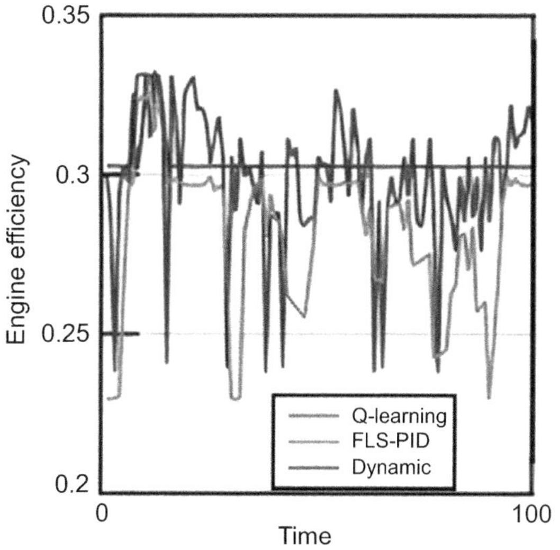

FIG. 9 The comparison of engine efficiency for different controllers.

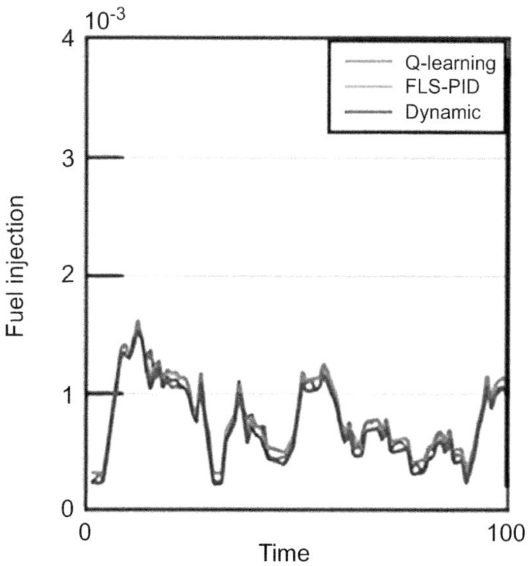

FIG. 10 The comparison of fuel injection for different methods.

fuel demand to 6.64 and 6.56 L/100 km, respectively, compared to the FLS-PID controller. The operation times for Q-learning and SAQ-learning algorithms are 0.0265 and 0.0199 s. These results underscore the significant fuel consumption reduction achieved by the Q-learning control algorithm, making it suitable for real-time control in autonomous vehicles. Although DP yields the best performance, it lacks adaptability and is less effective in uncertain systems. Furthermore, it falls short as a real-time controller for autonomous vehicles.

Figure 12 portrays the fuel demand for the Q-learning and SAQ-learning controllers. The average fuel consumption for the engine with the Q-learning controller is 6.64 L/km, and this value decreases to 6.56 L/km with the SAQ-learning controller. The SAQ-learning controller emerges as a more suitable option for conventional autonomous vehicles due to its operation time and efficiency, making it feasible for implementation in autonomous vehicles as a real-time controller.

The incorporation of new training data in the algorithm ensures the stability of SAQ-learning and its adaptability to uncertainty. These new data, as illustrated in Fig. 13, are used for an additional simulation. These new training data generate a new power demand, as depicted in Fig. 14. Figure 15 demonstrates that the SAQ-learning algorithm exhibits long-lasting results according to the training data, and the vehicle's fuel demand remains within an acceptable range. In contrast, DP fails to achieve the same level of success, with fuel consumption using SAQ-learning and DP measured at 6.65 and 8.88 L/km, respectively.

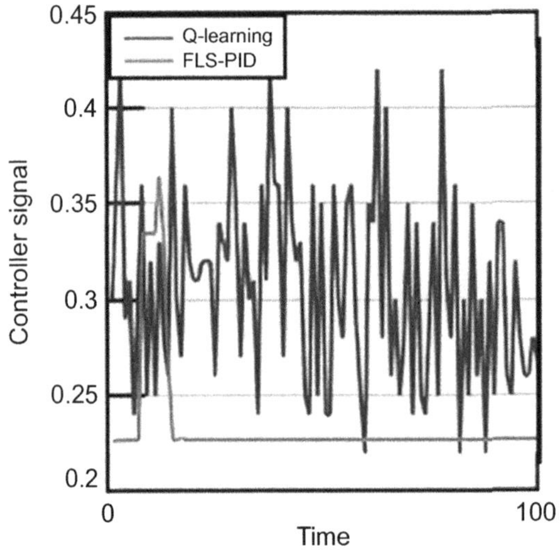

FIG. 11 The comparison of output signals for different controllers.

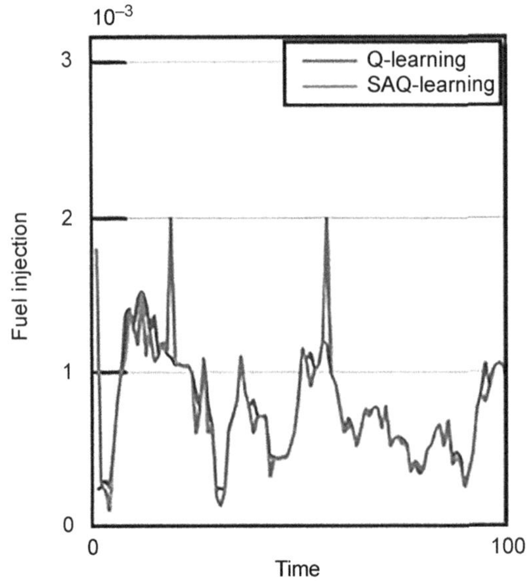

FIG. 12 The power demand of engine for different learning-based controllers.

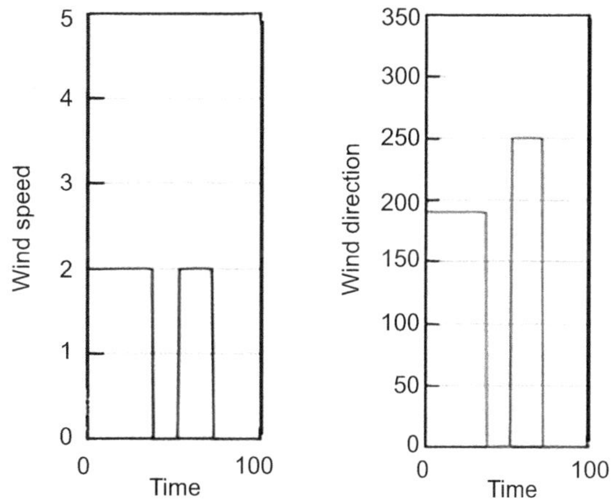

FIG. 13 The training data for wind speed and wind direction.

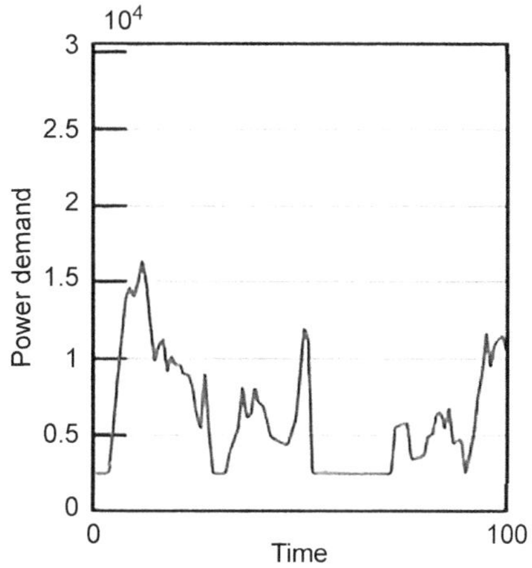

FIG. 14 The power consumption considering the uncertainty.

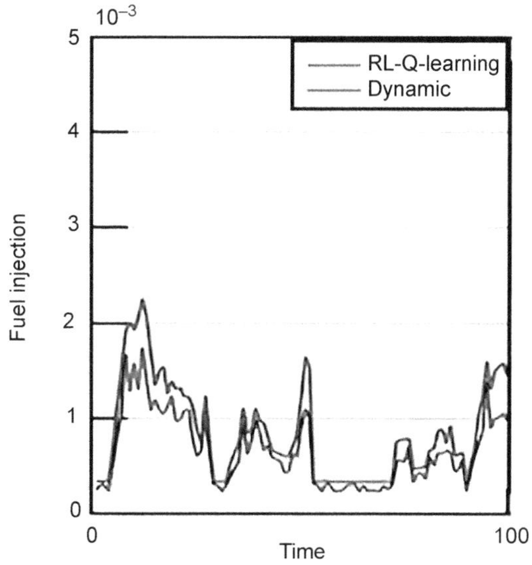

FIG. 15 The comparison of fuel demand for different controllers considering the uncertainty.

8.7 Summary

This chapter assesses the significance of IEMSs in the context of autonomous vehicles. The analysis underscores that the development of an optimal IEMS necessitates the consideration of various economic and technical criteria. According to the presence of numerous sensors and units constituting in such systems, achieving optimal performance requires precise integration among them, a challenge that becomes more pronounced with increasing levels of vehicle autonomy. Vehicles with higher autonomy levels demand a more accurate real-time management unit. The development of a real-time controller capable of handling uncertainty and complexity in automotive control poses a formidable challenge. For instance, the operational performance of a Q-learning controller falls short of meeting real-time requirements. In return, the SAQ-learning controller emerges as a solution, effectively reducing both operation time and fuel consumption. This positions the SAQ-learning controller as the most proficient real-time controller for conventional autonomous vehicles.

References

[1] S. Mischos, E. Dalagdi, V. Dimitrios, Intelligent energy management systems: a review, Artif. Intell. Rev. 56 (2023) 11635–11674, https://doi.org/10.1007/s10462-023-10441-3.
[2] D. Parekh, N. Poddar, A. Rajpurkar, M. Chahal, N. Kumar, G. Prasad Joshi, W. Cho, A review on autonomous vehicles: progress, methods and challenges, Electronics 11 (2022) 2162, https://doi.org/10.3390/electronics11142162.
[3] H. Khayyam, B. Javadi, M. Jalili, R.N. Jazar, Artificial intelligence and internet of things for autonomous vehicles, in: R.N. Jazar, L. Dai (Eds.), Nonlinear Approaches in Engineering Applications: Automotive Applications of Engineering Problems, Springer International Publishing, Cham, 2020, pp. 39–68.
[4] H. Marzbani, H. Khayyam, C.N. To, Đ.V. Quoc, R.N. Jazar, Autonomous vehicles: autodriver algorithm and vehicle dynamics, IEEE Trans. Veh. Technol. 68 (4) (2019) 3201–3211, https://doi.org/10.1109/TVT.2019.2895297.
[5] H. Miao, H. Jia, J. Li, T.Z. Qiu, Autonomous connected electric vehicle (ACEV)-based car-sharing system modeling and optimal planning: a unified two-stage multi-objective optimization methodology, Energy 169 (2019) 797–818, https://doi.org/10.1016/j.energy.2018.12.066.
[6] A. Carvalho, S. Lefévre, G. Schildbach, J. Kong, F. Borrelli, Automated driving: the role of forecasts and uncertainty—a control perspective, Eur. J. Control. 24 (2015) 14–32, https://doi.org/10.1016/j.ejcon.2015.04.007.
[7] L. Olatomiwa, S. Mekhilef, M.S. Ismail, M. Moghavvemi, Energy management strategies in hybrid renewable energy systems: a review, Renew. Sust. Energ. Rev. 62 (2016) 821–835, https://doi.org/10.1016/j.rser.2016.05.040.
[8] F.R. Salmasi, Control strategies for hybrid electric vehicles: evolution, classification, comparison, and future trends, IEEE Trans. Veh. Technol. 56 (5) (2007) 2393–2404, https://doi.org/10.1109/TVT.2007.899933.
[9] A. Roberts, R. Brooks, P. Shipway, Internal combustion engine cold-start efficiency: a review of the problem, causes and potential solutions, Energy Convers. Manag. 82 (2014) 327–350, https://doi.org/10.1016/j.enconman.2014.03.002.
[10] M. Koot, J. Kessels, B. De Jager, P. Van Den Bosch, Fuel reduction potential of energy management for vehicular electric power systems, Int. J. Altern. Propuls. 1 (1) (2006) 112–131, https://doi.org/10.1504/IJAP.2006.010761.
[11] D. Phan, et al., Neuro-fuzzy system for energy management of conventional autonomous vehicles, Energies 13 (7) (2020), https://doi.org/10.3390/en13071745.
[12] D. Phan, et al., Intelligent energy management system for conventional autonomous vehicles, Energy 191 (2020) 116476.
[13] J. Wang, H.A. Rakha, Fuel consumption model for conventional diesel buses, Appl. Energy 170 (2016) 394–402, https://doi.org/10.1016/j.apenergy.2016.02.124.
[14] O. Gomozov, J.P.F. Trovão, X. Kestelyn, M.R. Dubois, Adaptive energy management system based on a real-time model predictive control with nonuniform sampling time for multiple energy storage electric vehicle, IEEE Trans. Veh. Technol. 66 (7) (2017) 5520–5530, https://doi.org/10.1109/TVT.2016.2638912.
[15] M.A.S. Kamal, M. Mukai, J. Murata, T. Kawabe, Model predictive control of vehicles on urban roads for improved fuel economy, IEEE Trans. Control Syst. Technol. 21 (3) (2013) 831–841, https://doi.org/10.1109/TCST.2012.2198478.
[16] X. Hu, T. Liu, X. Qi, M. Barth, Reinforcement learning for hybrid and plug-in hybrid electric vehicle energy management: recent advances and prospects, IEEE Ind. Electron. Mag. 13 (3) (2019) 16–25, https://doi.org/10.1109/MIE.2019.2913015.
[17] Y. Hu, W. Li, K. Xu, T. Zahid, F. Qin, and C. Li, "Energy management strategy for a hybrid electric vehicle based on deep reinforcement learning," Appl. Sci., vol. 8, no. 2, https://doi.org/10.3390/app8020187.
[18] R. Liessner, C. Schroer, A.M. Dietermann, B. Bäker, Deep reinforcement learning for advanced energy management of hybrid electric vehicles, in: ICAART (2), 2018, pp. 61–72.
[19] T. Liu, X. Hu, W. Hu, Y. Zou, A heuristic planning reinforcement learning-based energy management for power-split plug-in hybrid electric vehicles, IEEE Trans. Industr. Inform. 15 (12) (2019) 6436–6445, https://doi.org/10.1109/TII.2019.2903098.
[20] M. Fayyazi, et al., Real-time self-adaptive Q-learning controller for energy management of conventional autonomous vehicles, Expert Syst. Appl. 222 (2023) 119770, https://doi.org/10.1016/j.eswa.2023.119770.
[21] R. Zou, Y. Zou, Y. Dong, L. Fan, A self-adaptive energy management strategy for plug-in hybrid electric vehicle based on deep Q learning, J. Phys. Conf. Ser. 1576 (1) (2020) 012037. IOP Publishing.
[22] Y. Li, J. Tao, L. Xie, R. Zhang, L. Ma, Z. Qiao, Enhanced Q-learning for real-time hybrid electric vehicle energy management with deterministic rule, Meas. Control 53 (7–8) (2020) 1493–1503, https://doi.org/10.1177/0020294020944952.
[23] L. Zhang, L. Tang, S. Zhang, Z. Wang, X. Shen, Z. Zhang, A self-adaptive reinforcement-exploration Q-learning algorithm, Symmetry 13 (6) (2021), https://doi.org/10.3390/sym13061057.

[24] J. Liu, Y. Chen, J. Zhan, F. Shang, Heuristic dynamic programming based online energy management strategy for plug-in hybrid electric vehicles, IEEE Trans. Veh. Technol. 68 (5) (2019) 4479–4493, https://doi.org/10.1109/TVT.2019.2903119.

[25] Z.H. Khattak, B.L. Smith, M.D. Fontaine, J. Ma, A.J. Khattak, Active lane management and control using connected and automated vehicles in a mixed traffic environment, Transp. Res. Part C Emerging Technol. 139 (2022) 103648, https://doi.org/10.1016/j.trc.2022.103648.

[26] I. Mahdinia, A. Mohammadnazar, R. Arvin, A.J. Khattak, Integration of automated vehicles in mixed traffic: evaluating changes in performance of following human-driven vehicles, Accid. Anal. Prev. 152 (2021) 106006, https://doi.org/10.1016/j.aap.2021.106006.

[27] S.F.Z. Abidin, et al., The effect of 48V mild hybrid technology on fuel consumption of a passenger car by using simulation cycle, Case Stud. Therm. Eng. 28 (2021) 101492, https://doi.org/10.1016/j.csite.2021.101492.

[28] J.J. Eckert, S.F. da Silva, F.M. Santiciolli, Á.C. de Carvalho, F.G. Dedini, Multi-speed gearbox design and shifting control optimization to minimize fuel consumption, exhaust emissions and drivetrain mechanical losses, Mech. Mach. Theory 169 (2022) 104644, https://doi.org/10.1016/j.mechmachtheory.2021.104644.

[29] Ş. Saragea, G. Frăţilă, G. Badea, Fuel consumption and emission improvements obtained using a mild hybrid electrical system on a vehicle equipped with a manual gearbox, IOP Conf. Ser.: Mater. Sci. Eng. 1037 (1) (2021) 012053. IOP Publishing.

CHAPTER

9

Hardware security of autonomous vehicles

Faiq Khalid[a] *and Syed Rafay Hasan*[b]
[a]Vienna University of Technology, Vienna, Austria
[b]Tennessee Technological University, Cookeville, TN, United States

9.1 Hardware security

Hardware security of electronic systems has gained attention in the recent past from researchers in academia and industry. One of the main reasons for this concern is the requirement of several untrusted players in developing any electronic system. For example, the development of autonomous vehicles needs to attain several sensors, integrated circuits (ICs), dedicated embedded systems, etc., which are woven into a complex electronic system to enable automation features in autonomous vehicles. It is prohibitively expensive for any business model to ascertain the trust of a complete supply chain of these components. To provide a perspective, each intelligent IC (one of the major components in an autonomous vehicle's electronic system) contains several soft and hard IPs. These IPs are typically designed at separate facilities, with system integration done on a different site, and ICs are fabricated mostly in foundries in a foreign country. *Hence, so many supply chain locations are beyond the control of a single IC vendor, which leads to several issues in terms of hardware security and hardware trust.*

This section provides a general description of hardware security, its conventional definition, and the different types of threats and defenses available in the literature. Since an autonomous vehicle utilizes all sorts of modern electronic systems as a commercial-off-the-shelf (COTS) component, it is vulnerable to all the security threats discussed in this section. Researchers have divided hardware vulnerabilities into two interrelated issues: *hardware security issues* and *hardware trust issues.*

9.1.1 What is hardware security issue?

Hardware security issues of electronic components, in general, can be defined as any hardware-oriented action that *can damage the intellectual and economic well-being of the electronic component manufacturer.* For example, these hardware-oriented actions can cause functionality alteration, parametric information sniffing, performance degradation, sabotaging the device, reverse engineering, and other similar acts. From now on, in this chapter, we will use the term hardware security only for hardware security related to electronic components.

9.1.2 What are hardware trust issues?

Hardware trust issues arise from untrusted players in the supply chain of the electronic hardware system. As discussed earlier, these untrusted players can be anyone from the IP design house for the IC manufacturers to the system-on-chip (SoC) design integrator, to the foundry used for IC fabrication, to the untrusted end users sniffing the information using side channels.

Hardware security and trust issues are interrelated since what sort of hardware security challenges can be inserted is dependent on at which level of the supply chain the untrusted player has access to the electronic hardware system. For example, a third-party intellectual property (3PIP) vendor can insert backdoor channels, the SoC design integrator can clone the IP, the foundry may be responsible for implanting interlayer Trojan, and end users can do reverse engineering or side-channel attacks. Hence, all the untrusted players can cause hardware security issues in their own capacity.

Handbook of Power Electronics in Autonomous and Electric Vehicles
https://doi.org/10.1016/B978-0-323-99545-0.00012-9

9.1.3 Hardware security models

To understand the feasibility of an attack and defense mechanism, one needs to know what information the attacker and defender have. A hardware security model is a combination of threat and trust models [1]. The trust model describes which level of the supply chain is untrusted, e.g., foundry, SoC integrator, and 3PIP vendor. On the other hand, the threat model provides details on the attacker's motive, like whether the adversary is leaking the information or triggering a payload to malfunction the device. The threat model also details the mechanisms of the attack, such as the insertion of hardware Trojans during the 3PIP design phase to introduce a trigger on a rarely occurring event.

9.1.4 Hardware attacks

Hardware attacks on electronic components can be of multiple types. These include hardware Trojan-based attacks, side-channel attacks, IP theft, counterfeit attacks, physical attacks, etc. Autonomous vehicles may also be subject to a new type of hardware-oriented attack on the environment of their sensors. Examples of such attacks include evasion attacks by subtle changes in the objects of interest that can adversely affect the camera [2] and LiDAR sensors [3,4]. Some background information about these mentioned attacks is provided below.

9.1.4.1 Hardware Trojan-based attacks

What is hardware Trojan?—Hardware Trojans are malicious circuits added during design time in any electronic component. Most of the time, it can be added within an IC or SoC [5]. Some hardware Trojans in PCB modules have recently been investigated [6].

Security model—Hardware Trojans can alter functionality, deteriorate performance, and destroy the electronic component. 3PIP providers, SoC integrators, and PCB module designers can insert hardware Trojans. Anyone of the supply chain players can introduce this attack.

General mechanism—Hardware Trojan usually contains a trigger circuit that waits for a particular event to trigger the payload of hardware Trojan. The adversary tries to make these triggers and payload stealthy by ensuring a small footprint, minuscule performance loss, and power degradation.

Why is this attack a threat to autonomous vehicles?—Automakers use several COTS board-level electronic components. Hardware Trojans can be responsible for simple performance degradation to catastrophic AV disasters.

9.1.4.2 Side-channel attacks (SCA)

What is side-channel attack (SCA)?—Any attack that can be done by sniffing the electrical characteristics of the electronic component to reveal the circuit operations for potentially adverse effect fall under side-channel attack. Power, timing, and electromagnetic signals are used for such attacks. Many adversaries utilize machine learning to steal information or introduce fault injection attacks [7–9].

Security model—Such attacks reveal critical system information, especially at the deployment stage. This can be used to invade privacy, reveal secret keys, understand the hardware architectures, etc. As alluded to earlier, such attacks occur after deployment in the field.

General mechanism—Attackers sniff the power profile, timing characteristics, electromagnetic spectrum, or their combination over a period of time to steal the required information. Some more sophisticated attacks also use machine learning techniques [9].

Why is this attack a threat to autonomous vehicles?—Such attacks can be used to break into the secret keys of encrypted data and reveal proprietary secret design choices.

9.1.4.3 IP theft attacks

What are IP theft attacks?—This type of attack occurs when 3PIPs are misused.

Security model—SoC Integrator is considered a malicious party, and 3PIP is the victim.

General mechanism—When 3PIP is provided to the SoC integrator, it either steals the design approach or uses the IP illicitly, e.g., over usage of IPs in other SoCs.

Why is this attack a threat to autonomous vehicles?—This type of attack may not directly affect the operation of the AVs, but IP theft can lead to legal consequences, which will be detrimental to the reputation of the AV market.

9.1.4.4 Counterfeit attacks

What is counterfeit attack?—Counterfeit attacks are when a compromised component is used in the electronic system, for example, if an old component is sold as a new component or a standard component is sold as a military-standard component [10].

Security model—The supply chain is considered untrusted, where some vendors sell compromised components (fake or used) as genuine components.

General mechanism—Attackers sell the used or fake electronic components as genuine and new after refurbishing and crafty relabeling them.

Why is this attack a threat to autonomous vehicles?—If such components become part of AV's critical intelligence decision, e.g., timely braking system, LiDAR's distance measurement, or video processing before feeding to the intelligent system, etc., it can lead to catastrophic consequences.

9.1.4.5 Hardware attacks due to adversarial environment to the sensors

What is this new type of attack?—In some of the recent literature, it is shown that image classification can be fooled by the very subtle embellishment of the physical object of interest [2,4,11]. Hence, the sensors become subject to an adverse environment, and this can also be true for other types of sensors like LiDAR and RADAR.

Security model—It is assumed that all the hardware system is honest and the user is honest, but an intruder can change the environment of AV to fool its intelligence, such as image classification, voice recognition, and distance identification.

General mechanism—An intruder can get hold of the environment, such as maliciously changing the object of interest, e.g., subtle embellishment on STOP signs, and disturbing LiDAR sensors' capability.

Why is this attack a threat to autonomous vehicles?—Some of the manifestations of these attacks include but are not limited to, misreading the road sign, being unable to properly measure the distance between the car and objects, etc. Such a scenario may lead to AV cars being at the mercy of such attackers, who can manufacture accidents that can cause severe damage.

9.1.5 Countermeasures against hardware attacks

Several countermeasure techniques have been proposed to counter hardware attacks. This section summarizes the state-of-the-art countermeasure and mitigation against hardware attacks.

9.1.5.1 Hardware Trojan detection and mitigation techniques

- *Vector sustained* technique to increase the differential power between genuine and Trojan circuits [12]. This can be implemented at the electronic system integration level.
- *The timing paths* of the circuits are also used as an HT detection metric. Conventionally, timing delay is measured as the golden model and checked against the runtime delay measurements to detect Trojans [13]. A relative path ordering technique is proposed to avoid the golden model and make the technique robust to process variation [14]. These techniques can be applied during electronic design under test by measuring the circuit timing through test vector injection.
- *Hardware obfuscation*—Another hardware Trojan mitigation approach is hardware obfuscation, which can be at the layout level, usually called *split manufacturing*. Other techniques include gate camouflaging and logic locking [15]. In split manufacturing, different layers of the IC design are sent out to the foundry as independent modules while combining the less expensive fabrication steps in the trusted facility [16]. Gate camouflaging is also done at the layout level, where an attempt is made to keep the layout of all the different logic gates to be the same [17]. In comparison, logic locking is obtained by adding extra gates, which prevents the circuit from correct operation unless these additional gates get the right combination of *locked* key vectors [18]. These techniques are strictly design time techniques. In most, if not all, cases, these ICs are outsourced by AV manufacturers. Hence, these techniques must be prebuilt once the IC is shipped to the AV manufacturer. It is possible that an AV manufacturers request the IC design to include one of these mitigation techniques and provide the details of their approach to the AV manufacturer. This can be helpful for AV's system integrator to test the system against possible threats. However, to the best of our knowledge, such an approach is not considered standard practice at this time.

9.1.5.2 *Side-channel mitigation techniques*

Many design time interventions have been proposed that help SoC integrators or end users perform runtime side-channel mitigation approaches. The basic idea is to sniff the electrical parameters, such as power usage, latency, and electromagnetic spectrum, and analyze whether the electronic component is compromised or not. Some modern side-channel approaches use machine learning techniques where the golden model is used to train the machine learning model [19]. Another class of machine learning approach also works without a golden model, using clustering approaches or inserting trusted components in the design under test [20,21].

9.1.5.3 *Physically unclonable functions (PUF)*

Each integrated circuit component has some unique identification due to process and environmental variations in die-to-die and wafer-to-wafer. Exploiting this idea, researchers obtain unique fingerprints of each design using challenge-response pairs. These fingerprints are used in the appropriate statistical analysis, which can help to identify the compromised device from the original implementation [22].

9.1.5.4 *Component for hardware security*

The hardware security review is incomplete without discussing hardware components used in modern processors and electronics to counter possible attacks. Hardware performance counter (HPC) has been inserted into the processors for a very long time to measure the processor's performance. In the last 5 years or so, researchers have proposed to use them to analyze malware attacks in conjunction with machine learning [23–26]. Another approach is to use hardware-assisted security modules to combat traditional attacks in IoTs [27] and FPGAs [28] and against adversarial examples [29].

9.2 Effect of hardware security on autonomous vehicle

AVs are vulnerable to several hardware attacks, and it is critical for the safety and financial feasibility to study hardware security and its impact on AV comprehensively. This section summarizes the critical hardware security issues in AV and provides guidelines for their respective countermeasures.

9.2.1 Hardware security for system architecture of the autonomous vehicle

Different components are required to achieve the above-mentioned levels of automation. These components help the vehicle assess the environment and use it to make effective and efficient decisions for smooth autonomous vehicle control [30,31]. The system architecture based on these components consists of the following layers, as shown in Fig. 1.

9.2.1.1 *Perception layer*

One of the key features of any automation is to sense the surroundings and external factors to get the required relevant information for identifying or determining appropriate decisions [32]. Therefore, the perception layer consists of

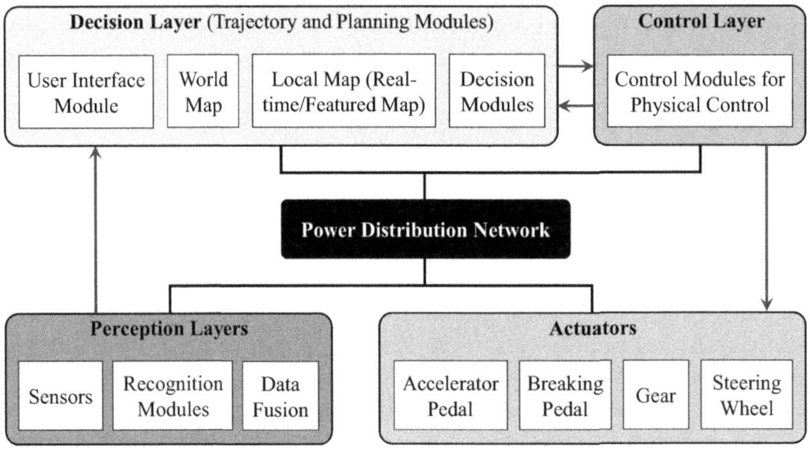

FIG. 1 Overview of the autonomous system architecture.

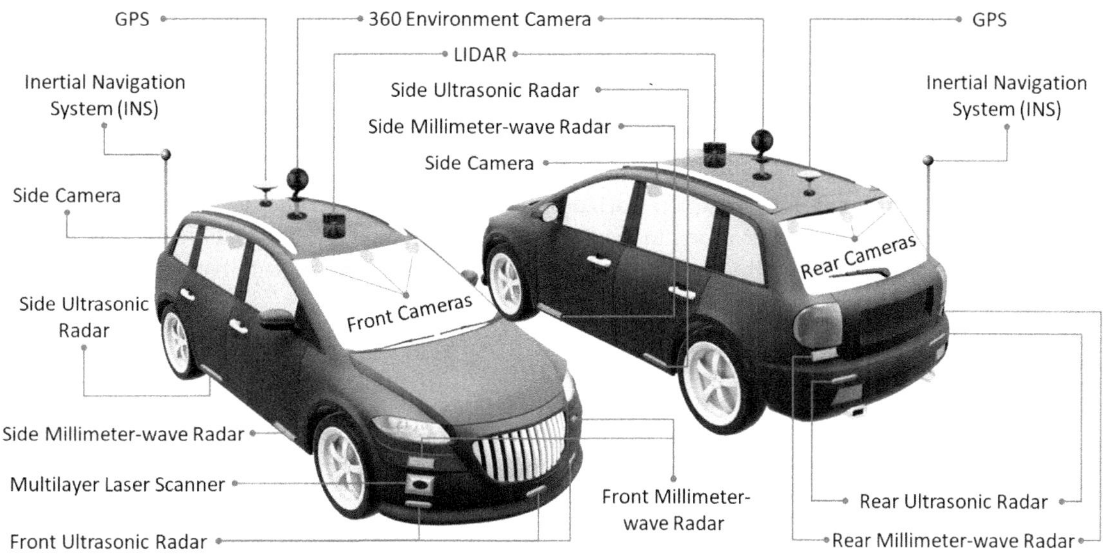

FIG. 2 Illustration of some of an autonomous vehicle's most commonly used sensors. Note that the 3D car model is taken from the built-in library of Microsoft Powerpoint under the educational use license.

three major components, i.e., sensors, recognition modules, and data fusion, as some examples are illustrated in Fig. 2. Some of the most commonly used sensors in autonomous vehicles are LiDAR, camera, GPS, and real-time kinetic, which collect information about different aspects and environmental factors. Recognition modules use this information to perform advanced detection and recognition frameworks like unknown obstacle recognition, vehicle positioning, and localization modules. Finally, data from different recognition modules are fused and transmitted to the decision layer. All these sensors are vulnerable to hardware attacks mentioned in Section 9.1.4.5. For example, adversarial attacks that alter the environment are one of the biggest security threats because the effects of these attacks affect all the layers [2,4]. Moreover, these sensors can also provide the surface for information-stealing attacks and lead to IP theft of the decision module [32].

9.2.1.2 Decision layer

This layer is the brain of the autonomous vehicle, and it utilizes the fused information from the perception layer to perform different key functions like path planning and object/obstacle avoidance. It is important to note that, like the human brain, these modules use the current and information from the past to identify the appropriate course of action, e.g., real-time map information, traffic information, and key information from other vehicles. Moreover, some of these modules also anticipate future failures and prepare the system for an appropriate response. This layer is the most critical but vulnerable to several security attacks, e.g., adversarial attacks on machine learning (ML)-based systems, traditional hardware attacks on ML-based components/systems, and hardware-assisted adversarial attacks.

- *Adversarial attacks:* These attacks manipulate either the training dataset or the ML model, which is the key part of the decision modules, for example, data poisoning [2], backdoor attacks [33,34], evasion attacks [35–37], or model-level Trojans [38,39]. To counter these attacks, several defenses are proposed, which are termed pre-processing-based defenses [40–43], and adversarial training [44]. However, none of these attacks provide a 100% guarantee to counter these attacks.
- *Hardware attacks:* To achieve the best performance, typically, ML-based solutions run on accelerator-based hardware, which is very specific to a certain type of ML model. Therefore, this hardware can be vulnerable to typical hardware attacks, for example, hardware Trojan attacks on ML models [45], row-hammer attacks on the weights [46], and many more. The existing defenses against hardware attacks can be used with certain modifications to counter these attacks. However, developing efficient and effective defenses is one of the greatest challenges because of AVs' limited resources and energy. For example, some of the existing solutions are side-channel-based online anomaly detection [47], hardware isolation [48], and protected memories [49]. These solutions require many modifications in AV's hardware to provide real-time mitigation. For example, hardware isolation requires a certain level of redundancy to existing in the AVs system or AV's control should be designed to work safely with the

isolated hardware component. Hence, context-based solutions for AVs under such hardware attacks require further investigation.

- *Hardware-assisted adversarial attacks:* Recently, several attacks have been proposed that exploit the hardware vulnerabilities to extract the key information for launching adversarial attacks. Typically, these kinds of attacks target the collaborated inference, which is one of the most used inference techniques for larger ML models on resource-constrained devices like many decision modules in AVs. Some of these attacks are StAIn [50], LaBANI [51], FeSHI [52], and modified adversarial attacks on horizontally collaborated capsule networks [53]. To counter these attacks, researchers have found modifications in existing defenses against adversarial are required. Furthermore, investigation is required to study the trade-offs and efficiency of these defenses against such attacks. This is currently an open-research question.

9.2.1.3 Control layer

This layer consists of the traditional open-loop and closed-loop control systems. These components process the information from the decision layer to activate or deactivate the appropriate system. In conventional control systems, it is very difficult to obfuscate the hardware Trojans because these systems are sensitive to a fraction of changes. Therefore, attackers typically use the communication network to manipulate the configuration, which eventually leads to the failure of several associated features, like changing the Programmable Logic Controllers (PLC) [54,55] and tampering with safety controls [56].

9.2.1.4 Actuators

This layer consists of the actuators mounted on the vehicle to control its movements, such as motors for the accelerator pedal, the brake pedal, the steering wheel, and many other actuators. These actuators are controlled by a dedicated control system based on the commands passed by the decision layer. Similar to the sensors, actuators are also vulnerable to several hardware attacks. For example, hardware intervention can cause authentication failure (adding the functional hardware Trojan that disables the authentication output or adds incorrect value to the output [57]). Similarly, deadline-miss (adding the bogus circuit to introduce additional delay in different computations, which eventually generates the mismatch in the computational delay [57]) in the actuators may lead to several consequences. One of the possible solutions to avoid authentication failure is to add redundancy under the assumption that each authentication hardware replica is designed by a different third-party design. Although the redundancy can solve the authentication issue, it is very costly and resource-hungry, which is not possible in all actuators of AVs. Similarly, several delay analysis-based solutions exist in the literature to identify the delay-based Trojan. For example, some of the commonly used techniques are nondestructive logic-based testing methods [58,59], nondestructive side-channel-based testing methods [60], destructive physical inspection techniques [61], and path-delay analysis methods [61,62]. However, the effectiveness of these methods is limited because they all require a golden signature or chip, which in most cases is not available when designing the AV using COTS components.

All the above-mentioned layers are tightly connected to efficiently gather the information from the physical domain and control the vehicle after processing this information. For example, the control layer connects the autonomous vehicle's cyber domain (recognition modules and data fusion from the perception layer and all the components in the decision layer) to the physical environment via actuators.

9.2.2 Hardware security for management modules in autonomous vehicles

The system architecture of autonomous vehicles requires perfect synchronization between different layers and proper management, which is achieved by efficient inter/intracommunication networks and power management modules connected to power distribution networks.

9.2.2.1 Communication network of autonomous vehicle

In a typical AV, two communication networks can be categorized as an in-vehicle communication network and an out-vehicle communication network. These communication networks contain multiple protocols and systems depending upon the information and commands. The communication between different types of systems, components, and other entities makes it vulnerable to several hardware attacks. Therefore, this section introduces these communication networks and discusses the associated hardware security threats.

In-vehicle communication: This term represents all communication networks within the vehicle that are responsible for sharing and transferring information between different modules in AVs, e.g., communication between decision modules to control systems and actuators and between sensors and decision modules [63]. Typically, these networks consist of wired control area networks (CAN). However, CAN-based communication is vulnerable to several hardware attacks, e.g., direct onboard diagnostics (OBD) port attack [64], malicious node insertion attacks [65], and indirect attack that uses multimedia to gain access to CAN [66]. Several methods have been proposed to counter these attacks. For example, network segmentation separates the CAN into multiple independent smaller networks. This approach can reduce the impact of attacks from a whole network to a small subnet. Although it somewhat solves the security threats on CAN, it complicates the network's maintenance and makes it costly. Apart from this approach, several other solutions have been proposed, like encrypting the CAN, introducing authentication, and intrusion detection systems (using delay, physical characteristics, specifications, or features of the CAN) [67]. The existing solution provides a glimpse of things that can be done to secure the CAN, but due to a lack of attack models and resources, the scope of these defenses is very limited and cannot be directly used in real-world AVs.

Out-vehicle communication: This term represents the two types of communication, vehicle-to-infrastructure (V2I) and vehicle-to-vehicle (V2V) communication networks. The main functionality of these two networks includes traffic-status information from traffic management infrastructure or navigation-related information received from other AVs on the road. These communication networks interact with several third-party devices. Therefore, out-vehicle communication opens a huge attack surface related to hardware security attacks. For example, remote attacks using Bluetooth, WiFi, keyless entry, immobilizer, VANET, GPS, and Radio channels (radio data system and traffic message channel) have been proposed and greatly impacted AVs [67]. Several defenses have also been proposed, such as power consumption-based anomaly detection, communication pattern-based anomaly detection, and formal checkers-based anomaly detection. However, all of these defenses somewhat require a reference or baseline dataset that is almost impossible in AVs due to the diversity of components and complicated environmental interaction. Therefore, to ensure the security of AVs, there is a need to design more rigorous and effective defense solutions against all security attacks on V2I and V2V.

9.2.2.2 Hardware security in AV's electrical power distribution system

Hardware security can affect shared resources of AV, such as power systems and control units. Fig. 3 shows a simplified block diagram of several components involved in the overall electrical power distribution within AV. Due to untrusted suppliers of various electric components, hardware attacks are possible. This section elaborates on some of the hardware attacks investigated in the existing literature, other potential possibilities of hardware attacks, and limitations in the current countermeasures.

In the electrical power systems of AV, hardware attacks have not been investigated extensively. Admittedly, it is a difficult problem, but on the other hand, the feasibility of such hardware attacks is also not very well established. Fig. 3 shows a simplistic view of the overall centralized electrical power distribution system for AV. As discussed earlier, there can be several possibilities of hardware attacks, but their practical implementation and at what stage of the design and implementation cycle an intruder can introduce anomaly, trust model, is challenging to establish. In this

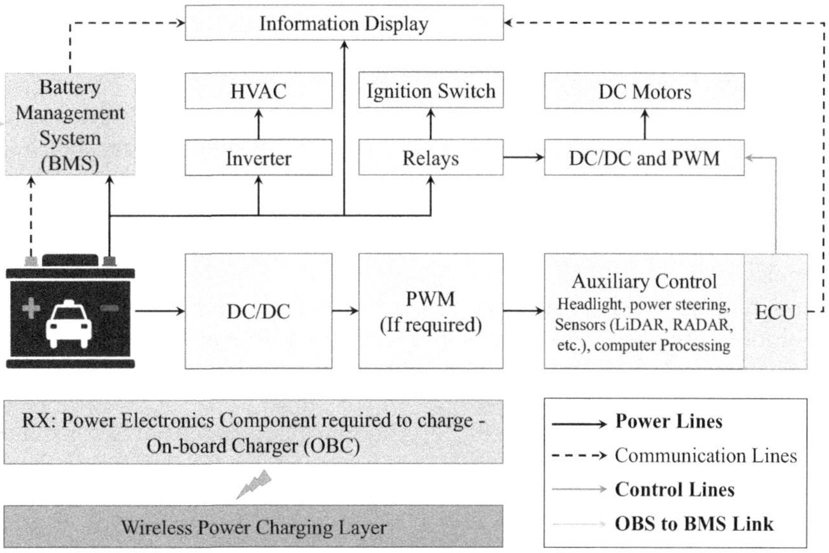

FIG. 3 Overall depiction of a centralized electrical power distribution system for AVs (inspired from Refs. [68,69]).

section, firstly, we will discuss the attacks that have been investigated in the existing literature, and then, we will further elaborate on how accurate the threat is and what can be done to mitigate this.

In literature, several hardware attacks are investigated on AVs' electric power and control systems. The summary of these attacks and associated shortcomings is given as follows:

- *Hardware Trojan in PWM:* Several of the power electronic components utilize pulse width modulators (PWM) to control the speed of motors (see Fig. 3). In [70], it is discussed that hardware Trojan can affect the PWM in unmanned aerial vehicles (UAVs), which can cause failures in the supported flight control system PX4[a] firmware. A machine learning-based intrusion detection system (IDS) can be designed to understand the PWM functionality in this environment [70]. A similar approach can be applied to ground AVs. However, such an approach's threat model must be established in advance. PWMs are certainly critical components. Broadly, malfunctioning can be introduced during design time or during runtime. Theoretically, it can be introduced by the irate employer during design time, but chances are low. However, if the control module is ceased to be the attacker through any conventional hacking mechanisms during run time, then it can certainly be catastrophic. Hence, it is highly recommended that all access to PWM must be secure via encryption. Here, it is to be noted that such attacks are also possible via ECU, which was discussed later in this section.
- *Attacks on motor drives:* Recently, researchers have been exploring possible attacks on electric drive systems [71–73]. It is argued in Ref. [72] that launching a data integrity attack on phase current sensors can cause the electric drive system to become unstable. In Ref. [71], cyberattacks on electric vehicles' power traction motor is discussed. It is stated that device-level information is needed to detect the abnormal state of the motors. Here, they generated a dataset by performing various attacks focusing on untrusted signals from the controller unit. On these datasets, statistical analysis is performed using various binary classifiers. This work is effective in a very short time. However, further investigation on false-positive and false-negative rates is required. Another key investigation is required to suggest required actions to prevent any damage during runtime, especially where human life is at risk.
- *Hardware attacks to detriment the onboard charging (OBC) systems:* Since OBC systems are within the AVs, a malicious hardware-based attack can make the load side or input side relay open, which can result in voltage surge [74]. These surges are detrimental to the power electronic switches used in OBC. Hardware attackers, if gained access to the power factor corrector (PFC) unit in OBC, can introduce short circuits to the grid side, which can be catastrophic to MOSFET switches and may end up malfunctioning. As with the earlier attack, such attacks are more likely to happen during run time. When a malicious user introduces wrong control signals by seizing access to the appropriate microcontroller unit, it is not a common practice to have these power electronics control units in the AV to be secure. Hence, the possibility of such attacks for a knowledgeable attacker does exist. A distributed security approach may mitigate such threats considerably and can be a future study.
- *Attacks on battery management system:* OBC is coupled with battery management systems (BMS). Hence, it is also possible that the communication channel established between OBC and BMS (see Fig. 3) is compromised. Cyberattacks in this domain via cloud network have been studied in various research papers, where it is assumed that attacker illegally hijacked the energy management system (EMS) and can reprogram the EMS [74,75]. This type of attack also falls under the category of hardware attacks. A detection technique requires the knowledge of the battery's existing state of charge (SOC) and compared against the expected SOC [74]. This analysis can arguably be effective against a long-term attack.

  Nevertheless, it cannot prevent the attacker if they intentionally or advertently lead to acute failure of the operation. Cyberattacks on OBC can also lead to the revelation of subsystem sensory data and possible alteration of data, which can lead to a catastrophic impact on the vehicle and interfacing grid, as explained in Ref. [76]. This also warrants securing these communication channels. However, using a standard interface such as Bluetooth is vulnerable to such attacks. Hence, researchers must establish a wrapping protocol to secure such standard communication channels.

  Along the same lines, high-speed charging systems can be forced to malfunction by interfering with the battery current sensor [77]. Solutions are proposed, such as introducing additional logical circuitry between the controller and gate driver (to drive power MOSFETs) [77]. However, such techniques have to be customized based on the individual controller and gate driver circuit.
- *Attacks on electronic control unit (ECU):* As seen in Fig. 3, AVs may have dozens of ECUs that control various critical functions, including, but not limited to, brakes, gears, and engine status [78,79]. Inside the vehicle, controller area network (CAN) bus provides connectivity to ECUs to onboard diagnostics (OBD), sensors, and other external

[a] PX4 is an open flight control software for UAVs—https://px4.io/software/getting-started/.

interfaces (e.g., Bluetooth, WiFi, etc.). Recently, the CAN bus has been replaced with next-generation networks, like high bandwidth CAN flexible data (FD) and third-generation CAN Extra Long (XL). Traditionally, the CAN bus is intended for reduced cost, simple implementation, and efficient real-time communication without much emphasis on malicious adversarial attacks [78,80]. Vulnerabilities in the sensors and CAN bus protocols can lead to attacks on ECU. These severe threats can manifest via compromises in sensors, peripherals of sensors, malicious access through OBD-II (onboard diagnostics—second generation) port, and firmware intervention via remote attacks. In Ref. [81], it is mentioned that sensors, such as cameras and LiDAR in AV, are vulnerable to hardware and cyberattacks. It has been demonstrated that by having physical access to the OBD-II port, one may inject a CAN packet sniffer to control the ECU. This attack erases any of its evidence by restarting the ECU [81]. Malicious apps on the smartphone may send malicious CAN frames to the vehicles via Bluetooth connection to the OBD-II port [82].

Lack of encryption, authentication, and data integrity checking in CAN bus protocols exposes ECUs to various attacks. For example, some of them are sniffing data through OBD-II port [83], injecting false CAN data frames to misguide ECUs [84,85], increasing the frequency of highest priority messages to cause denial of service (DoS) attacks [86], and spoofing attack [1].

Countermeasures against these attacks: Two types of countermeasures are proposed to prevent CAN buses from attacking ECUs. One type is encryption based, and the other is an intrusion detection system. Secure hardware extension (SHE) protocol through *E*-safety vehicle intrusion protected applications (EVITA) projects propose HSM for automotive embedded systems [87]. These HSMs can be added to ECUs to implement lightweight encryption techniques [81].

With the continuous technological innovation of AV, researchers will have to deal with new attack surfaces. At the moment, to the best of the author's knowledge, there is no standardized approach or consensus among the AV manufacturers on the need to embed proposed hardware security countermeasures. Most modern AVs use CAN FD (CAN with flexible data rate) to accommodate the needs of varying data payload size and bandwidth consumption. However, it has been demonstrated that several attacks, such as eavesdropping and replay attack, are still possible [32].

Shortcomings in existing security scenarios for the hardware attacks discussed in Section 9.2.2.2: Overall, these attacks have been studied on an ad hoc basis, and we believe that integrating the knowledge of the hardware security model into these studies requires further investigation. For example, for wireless energy security, maximum efficiency band tracking (MEBT) is a traditional method against energy theft against nonsecure operating frequency [88]. Encryption is proposed in Ref. [88], which requires transmitting the 2D security keys to the receiver to implement this security measure. A hardware attack, such as hardware Trojan or side-channel techniques, has been demonstrated to steal such keys through either information leakage or differential power analysis [89]. Along the same lines, solutions proposed in Ref. [77] for possible threats on fast OBC need customization based on the individual controller and gate driver circuit. Similarly, CAN-based solutions also require further investigation as it has been demonstrated that several attacks, such as eavesdropping and replay attack, are still possible [32] on CAN FD. We suggest that a holistic approach should be used to score vulnerable hardware security models, then mitigation techniques should be devised for a few of these components, and lastly, this mitigation effect needs to be again studied and justified from the overall AV systems' perspective.

9.3 Roadmap of autonomous vehicles and associated security risks

The autonomy in autonomous vehicles requires multiple hardware and software components working in perfect synchronization. The complex interconnectivity and interdependability between these components make the design of autonomous vehicles challenging. Although researchers are exploring different solutions to achieve autonomy in autonomous vehicles, achieving complete and efficient automation is still an open-research challenge. Depending on the automation, the Society of Automotive Engineers (SAE)[b] defines the six levels of automation in autonomous vehicles. Several companies and institutes are working toward achieving level 5 autonomous vehicles, but none are available for public testing [90]. Table 1 briefly overviews the key features, associated systems, current development stage, and starting year of the respective technology from each automation level. It also provides some real-world examples of AVs from each level. Table 2 shows the distinguishing features of AV and summarizes the role of humans and machines at each level. Although automation is providing a lot of features and moving toward

[b] The next automation level contains the distinguishing features of all previous automation levels.

TABLE 1 A list of the example features and examples of existing cars in the respective automation level of autonomous vehicles.

	Level 0	Level 1	Level 2	Level 3	Level 4	Level 5
	No automation	Driver assistance	Partial driving	Confitional driving	High driving	Full driving
Automated emergency brakes (AEB)	✓	✓	✓	✓	✓	✓
Blind spot warning (BSW)	✓	✓	✓	✓	✓	✓
Adaptive cruise control (ACC)	–	✓	✓	✓	✓	✓
Parking helper (PH)	–	PH 1.0	PH 2.0	PH 2.0	PH 2.0	PH 2.0
Active lane centering (ALC)	–	✓	✓	✓	✓	✓
Highway pilot (HP)	–		HP 2.0	HP 2.0	HP 2.0	HP 2.0
Traffic jam pilot (TJP)	–			✓	✓	✓
Automated driving system (ADS)	–			ADS 3.0	ADS 4.0	ADS 4.0
Design/development cycle	P	P	P	T/CP	D/T	D/T
Starting year of the technology	1939	1958–1960	2009, 2015	2023...	2025...	2040...
Real world examples	All model without basic ACC	Any model with basic ACC and lane centering	GE Super Cruise, Mercedes-Benz Distronic Plus, Nissan ProPilot Assist, Tesla Autopilot	Audi A8 Sedan (waiting for level 3 approval)	Test cars from Google, Hyundai, and Waymo	

Note the term used in the table is defined as P (approved for public use), T/CP (conditional/awaiting approval or testing phase), and D/T (design or testing phase). The timeline for AVs is taken from Ref. [97].

TABLE 2 Distinguishing features and the role of humans and machines in each level.

	Level 0	Level 1	Level 2	Level 3	Level 4	Level 5
Driving						
Supervision						
Control override						
Safety						
Examples of distinguishing features (the next level contains the distinguishing features of the previous level)	Automatic emergency braking, blind spot warning	Lane centering OR adaptive cruise control	Lane centering AND adaptive cruise control at the same time	Traffic chauffer	Independent driving in specific conditions	It can drive independently in all conditions

In this table, : human is solely responsible, : machine provides support to human, : machine operates only if there is an emergency, : machine give only safety warning, : machine provide safety warning and possible solutions, : machine is solely responsible, and : human involves only if machine requests. It also provides some of distinguishing features of each level.

driverless cars, the associated hardware security threats can impact the timeline of autonomous vehicles, as discussed next (see Table 3):

- *Level 0 (no automation):* At this level, the driver is solely responsible for driving the vehicle. It is possible to have assistance in emergencies like the emergency brake, but the system is not responsible for driving the car. In some cases, it is possible for automation features to override the controls.

  Associated security hardware threat and associated impact: The automation features at this level are not sophisticated but still can be a source of a security attack. For example, if the attacker inserts a malicious control switch or a malicious circuit in the control hardware of the automatic braking system [91] or blind spot warning [92]. Apart from traditional hardware attacks, the risk of an adversarial attack is another major threat. For example, the pass-by vehicle can attach a rod or extension that can go undetected by humans and blind spot warning systems [91,92]. Hence, all these attacks can lead to catastrophic outcomes that can be fatal to the driver and passenger and financially expensive.
- *Level 1 (driver assistance):* At this level, the driver has basic level assistance by controlling the steering wheel (power steering wheel) and maintaining a constant speed (cruise control) individually. Apart from the mentioned assistance, the adaptive cruise control also comes under level 1 automation. It is important to note that in this level of automation, each assistance system works individually.

TABLE 3 The list of key components in each automation level that are vulnerable to hardware security threats.

<table>
<tr><th></th><th></th><th>Level 0</th><th>Level 1</th><th>Level 2</th><th>Level 3</th><th>Level 4</th><th>Level 5</th></tr>
<tr><td rowspan="4">Hardware security threats</td><td>Hardware Trojan</td><td>Emergency braking system</td><td rowspan="3">Steering and braking ECU, airbag ECU</td><td rowspan="3">Steering and braking ECU, airbag ECU</td><td colspan="3" rowspan="4">ECU (airbag, steering and braking, vehicle access, ADAS, engine and transmission, lighting system), remote link type app, on-board diagnostics (OBD), USB, bluetooth, power management systems, communication networks, remote key, and all sensors (camera, GNN/IMU, ultrasonic sensor, RADA, LIDAR, etc.)</td></tr>
<tr><td>Adversarial attacks</td><td>Blind spot warning system</td></tr>
<tr><td>Remote attacks</td><td>–</td></tr>
<tr><td>Side channel attacks</td><td>–</td><td>–</td><td>–</td></tr>
<tr><td colspan="2">Impact of hardware security threats
(all impacts of the previous level are also applicable to the next levels)</td><td>Late braking, sudden brakes</td><td>Incorrect lane-detection, inappropriate acceleration/ deceleration</td><td>Inappropriate lane-changing</td><td>Incorrect traffic data, IP stealing</td><td>Incorrect detection of driving conditions</td><td>The impacts of security threats are endless</td></tr>
</table>

It also shows the impact of security hardware attacks on different levels of automation. Note the attack surface is expanding and its impact is becoming more prominent with increase in the automation in each level.

Associated security hardware threat and associated impact: In addition to the security risks of level 0, the lane centering and cruise control open a new attack surface to the attacker. For example, malicious circuits in cruise control [93,94] and lane centering module [95] can perform malicious activities, like inappropriate acceleration/deceleration or wrong/premeditated/late lane changing. These activities can lead to catastrophic consequences.

- *Level 2 (partial driving automation)*: At this level, the vehicle simultaneously controls the basic level of assistance. For example, the driving assistance system can control the steering wheel and acceleration/deceleration. Some of the existing examples are Tesla Autopilot [96] and Cadillac (General Motors) Super Cruise systems [97]. It is important to note that this level of automation still requires a human in the driving seat.

  Associated security hardware threat and associated impact: This level of automation is exposed to the same threats as level 1. However, at this level, the impact of attacks becomes more severe. This is because, in level 2, the simultaneous activation of the lane centering and cruise control can be fatal as it can change the vehicle's direction and accelerate it simultaneously.
- *Level 3 (conditional driving automation):* This level of automation involves advanced features like environment detection to control the vehicle effectively. However, it is required that human still has the power to override the control.
- *Level 4 (high driving automation)*: In this level, the vehicle has all the features of level 3, but the vehicle can also intercede if there is a failure in the system. Therefore, the requirement of having human supervision is no longer required.
- *Level 5 (full driving automation):* This automation level does not require human intervention and eliminates the requirement of steering wheels, braking pedals, accelerator pedals, and several other controls that require human intervention.

  Associated security hardware threat and associated impact: In all the automation levels above level 2, the AVs are moving toward driverless driving, which opens a vast attack surface targeting functionality, finances, and privacy. These hardware attacks can be either adding malicious circuits [5,6,10], remote attacks [64,65], side-channel attacks [7–9,61], or adversarial attacks at [12] architecture, component, or management levels [35–38,98,99]. The impacts of these attacks are endless. For example, incorrect traffic data can be costly and fatal, IP stealing of the components or system can be expensive, and false detection of driving conditions can be deadly in level 4.

Based on the above discussion, we conclude that AVs are evolving toward driverless driving. However, apart from the design challenges and safety concerns, security threats are becoming one of the major bottlenecks in the evolution of AVs. Several security incidents were reported related to different levels of automation, e.g., the Hacking of General Motors OnStar [100], speed limit attacks on Tesla autopilot Mobileye EyeQ3 camera system [101], adversarial attacks on Tesla autopilot [96], and Laser attack deletes the pedestrians [102]. In summary, level 5 automation in autonomous vehicles may be achieved sooner, but incorporating existing security threats may take several years of research and still be unable to resolve. Therefore, we can safely conclude that security, especially hardware security, must be incorporated adequately along with safety and other evaluation parameters in the AV evolution timeline.

References

[1] K. Iehira, H. Inoue, K. Ishida, Spoofing attack using bus-off attacks against a specific ECU of the CAN bus, in: 2018 15th IEEE Annual Consumer Communications & Networking Conference (CCNC), IEEE, 2018, pp. 1–4.

[2] W. Jiang, H. Li, S. Liu, X. Luo, L. Rongxing, Poisoning and evasion attacks against deep learning algorithms in autonomous vehicles, IEEE Trans. Veh. Technol. 69 (4) (2020) 4439–4449.

[3] Y. Cao, C. Xiao, B. Cyr, Y. Zhou, W. Park, S. Rampazzi, Q.A. Chen, K. Fu, Z.M. Mao, Adversarial sensor attack on Lidar-based perception in autonomous driving, in: Proceedings of the 2019 ACM SIGSAC Conference on Computer and Communications Security, 2019, pp. 2267–2281.

[4] M. Abdelfattah, K. Yuan, Z.J. Wang, R. Ward, Adversarial attacks on camera-Lidar models for 3d car detection, in: 2021 IEEE/RSJ International Conference on Intelligent Robots and Systems (IROS), IEEE, 2021, pp. 2189–2194.

[5] S. Bhunia, M. Tehranipoor, The Hardware Trojan War, Springer, Cham, Switzerland, 2018.

[6] M. McGuire, U. Ogras, S. Ozev, PCB hardware Trojans: attack modes and detection strategies, in: 2019 IEEE 37th VLSI Test Symposium (VTS), IEEE, 2019, pp. 1–6.

[7] B. Timon, Non-profiled deep learning-based side-channel attacks with sensitivity analysis, in: IACR Transactions on Cryptographic Hardware and Embedded Systems, 2019, pp. 107–131.

[8] D. Chen, Z. Zhao, X. Qin, Y. Luo, M. Cao, X. Hua, A. Liu, MAGLeak: a learning-based side-channel attack for password recognition with multiple sensors in IIoT environment, IEEE Trans. Industr. Inform. 18 (1) (2020) 467–476.

[9] S. Picek, G. Perin, L. Mariot, W. Lichao, L. Batina, SoK: deep learning-based physical side-channel analysis, ACM Comput. Surv. 55 (11) (2023) 1–35.

[10] M. Tehranipoor, J. Villasenor, The hidden dangers of chop-shop electronics, IEEE Spectr. (September 2013).

[11] J. Sun, Y. Cao, Q.A. Chen, Z.M. Mao, Towards robust Lidar-based perception in autonomous driving: general black-box adversarial sensor attack and countermeasures, in: USENIX Security Symposium (Usenix Security'20), 2020.

[12] M. Banga, M.S. Hsiao, A novel sustained vector technique for the detection of hardware Trojans, in: 2009 22nd International Conference on VLSI Design, IEEE, 2009, pp. 327–332.

[13] B. Cha, S.K. Gupta, Trojan detection via delay measurements: a new approach to select paths and vectors to maximize effectiveness and minimize cost, in: 2013 Design, Automation & Test in Europe Conference & Exhibition (DATE), IEEE, 2013, pp. 1265–1270.

[14] X. Cui, E. Koopahi, W. Kaijie, R. Karri, Hardware Trojan detection using the order of path delay, ACM J. Emerg. Technol. Comput. Syst. 14 (3) (2018) 1–23.

[15] L. Alrahis, S. Patnaik, F. Khalid, M.A. Hanif, H. Saleh, M. Shafique, O. Sinanoglu, GNNUnlock: graph neural networks-based oracle-less unlocking scheme for provably secure logic locking, in: 2021 Design, Automation & Test in Europe Conference & Exhibition (DATE), IEEE, 2021, pp. 780–785.

[16] K. Vaidyanathan, B.P. Das, E. Sumbul, R. Liu, L. Pileggi, Building trusted ICs using split fabrication, in: 2014 IEEE International Symposium on Hardware-Oriented Security and Trust (HOST), IEEE, 2014, pp. 1–6.

[17] J. Rajendran, M. Sam, O. Sinanoglu, R. Karri, Security analysis of integrated circuit camouflaging, in: Proceedings of the 2013 ACM SIGSAC Conference on Computer & Communications Security, 2013, pp. 709–720.

[18] J. Rajendran, H. Zhang, C. Zhang, G.S. Rose, Y. Pino, O. Sinanoglu, R. Karri, Fault analysis-based logic encryption, IEEE Trans. Comput. 64 (2) (2013) 410–424.

[19] K.G. Liakos, G.K. Georgakilas, S. Moustakidis, P. Karlsson, F.C. Plessas, Machine learning for hardware Trojan detection: a review, in: 2019 Panhellenic Conference on Electronics & Telecommunications (PACET), IEEE, 2019, pp. 1–6.

[20] F. Khalid, S.R. Hasan, S. Zia, O. Hasan, F. Awwad, M. Shafique, MacLeR: machine learning-based runtime hardware Trojan detection in resource-constrained IoT edge devices, IEEE Trans. Comput. Aided Des. Integr. Circuits Syst. 39 (11) (2020) 3748–3761.

[21] H. Mohammed, F. Khalid, P. Sawyer, G. Cataloni, S.R. Hasan, InTrust-IoT: intelligent ecosystem based on power profiling of trusted device (s) in IoT for hardware Trojan detection, in: Workshop on Hardware and Architectural Support for Security and Privacy, 2021, pp. 1–8.

[22] S. Bhunia, M. Tehranipoor, Hardware Security: A Hands-on Learning Approach, Morgan Kaufmann, 2018.

[23] K. Basu, P. Krishnamurthy, F. Khorrami, R. Karri, A theoretical study of hardware performance counters-based malware detection, IEEE Trans. Inf. Forensics Secur. 15 (2019) 512–525.

[24] P. Cronin, C. Yang, Lowering the barrier to online malware detection through low frequency sampling of HPCs, in: 2018 IEEE International Symposium on Hardware Oriented Security and Trust (HOST), IEEE, 2018, pp. 177–180.

[25] A.P. Kuruvila, X. Meng, S. Kundu, G. Pandey, K. Basu, Explainable machine learning for intrusion detection via hardware performance counters, IEEE Trans. Comput. Aided Des. Integr. Circuits Syst. 41 (2022) 4952–4964.

[26] K. Basu, S.S. Hussain, U. Gupta, R. Karri, COPPTCHA: COPPA tracking by checking hardware-level activity, IEEE Trans. Inf. Forensics Secur. 15 (2020) 3213–3226.

[27] C. Sozio, Z. Jordan, G. Skipper, A. Lukefahr, A. Duncan, Patchable hardware security module (PHaSM) for extending FPGA root-of-trust capabilities, in: 2021 IEEE Physical Assurance and Inspection of Electronics (PAINE), IEEE, 2021, pp. 1–8.

[28] A.A. Ayoub, M.D. Aagaard, Application-specific instruction set architecture for an ultralight hardware security module, in: 2020 IEEE International Symposium on Hardware Oriented Security and Trust (HOST), IEEE, 2020, pp. 69–79.

[29] J. Zhang, S. Peng, H. Yupeng, F. Peng, H. Wei, J. Lai, J. Ye, X. Wang, HRAE: hardware-assisted randomization against adversarial example attacks, in: 2020 IEEE 29th Asian Test Symposium (ATS), IEEE, 2020, pp. 1–6.

[30] F. Khalid, S. Rehman, M. Shafique, Overview of security for smart cyber-physical systems, in: Security of Cyber-Physical Systems: Vulnerability and Impact, Springer, Cham, 2020, pp. 5–24.

[31] D. Ratasich, F. Khalid, F. Geissler, R. Grosu, M. Shafique, E. Bartocci, A roadmap toward the resilient internet of things for cyber-physical systems, IEEE Access 7 (2019) 13260–13283.

[32] C. Gao, G. Wang, W. Shi, Z. Wang, Y. Chen, Autonomous driving security: state of the art and challenges, IEEE Internet Things J. 9 (10) (2021) 7572–7595.

[33] F. Khalid, M.A. Hanif, S. Rehman, M. Shafique, Security for machine learning-based systems: attacks and challenges during training and inference, in: 2018 International Conference on Frontiers of Information Technology (FIT), IEEE, 2018, pp. 327–332.

[34] J. Dai, C. Chen, Y. Li, A backdoor attack against LSTM-based text classification systems, IEEE Access 7 (2019) 138872–138878.

[35] F. Khalid, H. Ali, M.A. Hanif, S. Rehman, R. Ahmed, M. Shafique, Fadec: a fast decision-based attack for adversarial machine learning, in: 2020 International Joint Conference on Neural Networks (IJCNN), IEEE, 2020, pp. 1–8.

[36] F. Khalid, M.A. Hanif, M. Shafique, Exploiting vulnerabilities in deep neural networks: adversarial and fault-injection attacks, arXiv preprint arXiv:2105.03251 (2021).

[37] F. Khalid, M.A. Hanif, S. Rehman, R. Ahmed, M. Shafique, TrISec: training data-unaware imperceptible security attacks on deep neural networks, in: 2019 IEEE 25th International Symposium on Online Testing and Robust System Design (IOLTS), IEEE, 2019, pp. 188–193.

[38] A.S. Rakin, Z. He, D. Fan, Tbt: targeted neural network attack with bit Trojan, in: Proceedings of the IEEE/CVF Conference on Computer Vision and Pattern Recognition, 2020, pp. 13198–13207.

[39] R. Tang, D. Mengnan, N. Liu, F. Yang, H. Xia, An embarrassingly simple approach for trojan attack in deep neural networks, in: Proceedings of the 26th ACM SIGKDD International Conference on Knowledge Discovery & Data Mining, 2020, pp. 218–228.

[40] F. Khalid, M.A. Hanif, S. Rehman, J. Qadir, M. Shafique, Fademl: understanding the impact of pre-processing noise filtering on adversarial machine learning, in: 2019 Design, Automation & Test in Europe Conference & Exhibition (DATE), IEEE, 2019, pp. 902–907.

[41] H. Ali, F. Khalid, H.A. Tariq, M.A. Hanif, R. Ahmed, S. Rehman, SSCNets: robustifying DNNs using secure selective convolutional filters, IEEE Des. Test 37 (2) (2019) 58–65.

[42] F. Khalid, H. Ali, H. Tariq, M.A. Hanif, S. Rehman, R. Ahmed, M. Shafique, QuSecNets: quantization-based defense mechanism for securing deep neural network against adversarial attacks, in: 2019 IEEE 25th International Symposium on Online Testing and Robust System Design (IOLTS), IEEE, 2019, pp. 182–187.

[43] K.D. Doan, Y. Lao, P. Yang, P. Li, Defending backdoor attacks on vision transformer via patch processing, arXiv preprint arXiv:2206.12381 (2022).

[44] T. Bai, J. Luo, J. Zhao, B. Wen, Q. Wang, Recent advances in adversarial training for adversarial robustness, arXiv preprint arXiv:2102.01356 (2021).

[45] Y. Liu, Y. Xie, A. Srivastava, Neural Trojans, in: 2017 IEEE International Conference on Computer Design (ICCD), IEEE, 2017, pp. 45–48.

[46] Y. Zhao, X. Hu, S. Li, J. Ye, L. Deng, Y. Ji, J. Xu, W. Dong, Y. Xie, Memory Trojan attack on neural network accelerators, in: 2019 Design, Automation & Test in Europe Conference & Exhibition (DATE), IEEE, 2019, pp. 1415–1420.

[47] Y. Han, M. Chan, Z. Aref, N.O. Tippenhauer, S. Zonouz, Hiding in plain sight? On the efficacy of power side {channel-based} control flow monitoring, in: 31st USENIX Security Symposium (USENIX Security 22), 2022, pp. 661–678.

[48] H. Omar, O. Khan, Ironhide: a secure multicore that efficiently mitigates microarchitecture state attacks for interactive applications, in: 2020 IEEE International Symposium on High Performance Computer Architecture (HPCA), IEEE, 2020, pp. 111–122.

[49] S. Mittal, H. Gupta, S. Srivastava, A survey on hardware security of DNN models and accelerators, J. Syst. Archit. 117 (2021) 102163.

[50] A.A. Adeyemo, J.J. Sanderson, T.A. Odetola, F. Khalid, S.R. Hasan, StAIn: stealthy avenues of attacks on horizontally collaborated convolutional neural network inference and their mitigation, IEEE Access 11 (2023) 10520–10534.

[51] T.A. Odetola, F. Khalid, S.R. Hasan, LaBaNI: layer-based noise injection attack on convolutional neural networks, in: Proceedings of the Great Lakes Symposium on VLSI 2022, 2022, pp. 143–146.

[52] T.A. Odetola, F. Khalid, H. Mohammed, T.C. Sandefur, S.R. Hasan, FeSHI: feature map-based stealthy hardware intrinsic attack, IEEE Access 9 (2021) 115370–115387.

[53] A. Adeyemo, F. Khalid, T. Odetola, S.R. Hasan, Security analysis of capsule network inference using horizontal collaboration, in: 2021 IEEE International Midwest Symposium on Circuits and Systems (MWSCAS), IEEE, 2021, pp. 1074–1077.

[54] T. Alladi, V. Chamola, S. Zeadally, Industrial control systems: cyberattack trends and countermeasures, Comput. Commun. 155 (2020) 1–8.

[55] W. Duo, M.C. Zhou, A. Abusorrah, A survey of cyber attacks on cyber physical systems: recent advances and challenges, IEEE/CAA J. Autom. Sin. 9 (5) (2022) 784–800.

[56] Z. Pang, F. Yuan, H. Guo, J. Sun, Analysis of stealthy false data injection attacks against networked control systems: three case studies, J. Syst. Sci. Complex. 36 (2023) 1407–1422.

[57] J. Plusquellic, F. Saqib, Detecting hardware Trojans using delay analysis, in: The Hardware Trojan War: Attacks, Myths, and Defenses, Springer, Cham, 2018, pp. 219–267.

[58] M. Banga, M. Chandrasekar, L. Fang, M.S. Hsiao, Guided test generation for isolation and detection of embedded Trojans in ICs, in: Proceedings of the 18th ACM Great Lakes Symposium on VLSI, 2008, pp. 363–366.

[59] H. Salmani, M. Tehranipoor, J. Plusquellic, A novel technique for improving hardware Trojan detection and reducing Trojan activation time, IEEE Trans. Very Large Scale Integr. VLSI Syst. 20 (1) (2011) 112–125.

[60] I. Giechaskiel, K. Rasmussen, Taxonomy and challenges of out-of-band signal injection attacks and defenses, IEEE Commun. Surv. Tutorials 22 (1) (2019) 645–670.

[61] F. Farahmandi, Y. Huang, P. Mishra, F. Farahmandi, Y. Huang, P. Mishra, Hardware Trojan detection schemes using path delay and side-channel analysis, in: System- on-Chip Security: Validation and Verification, Springer, Cham, 2020, pp. 221–271.

[62] S. Giridharan, S.R. Ramesh, M. Nirmala Devi, A MUX based latch technique for the detection of HardwareTrojan using path delay analysis, in: 2021 IEEE International Conference on Electronics, Computing and Communication Technologies (CONECCT), IEEE, 2021, pp. 1–4.

[63] J. Liu, S. Zhang, W. Sun, Y. Shi, In-vehicle network attacks and countermeasures: challenges and future directions, IEEE Netw. 31 (5) (2017) 50–58.

[64] A. Palanca, E. Evenchick, F. Maggi, S. Zanero, A stealth, selective, link-layer denial-of-service attack against automotive networks, in: Detection of Intrusions and Malware, and Vulnerability Assessment: 14th International Conference, DIMVA 2017, Bonn, Germany, July 6–7, 2017, Proceedings 14, Springer International Publishing, 2017, pp. 185–206.

[65] C. Miller, C. Valasek, A survey of remote automotive attack surfaces, Black Hat USA 2014 (2014) 94.

[66] Z. Ju, H. Zhang, X. Li, X. Chen, J. Han, M. Yang, A survey on attack detection and resilience for connected and automated vehicles: from vehicle dynamics and control perspective, IEEE Trans. Intell. Veh. 7 (2022) 815–837.

[67] F. Fakhfakh, M. Tounsi, M. Mosbah, Cybersecurity attacks on CAN bus based vehicles: a review and open challenges, Library Hi Tech 40 (5) (2022) 1179–1203.

[68] R.W. Wall, J. Bennett, G. Eis, Creating a low-cost autonomous vehicle, in: IEEE 2002 28th Annual Conference of the Industrial Electronics Society. IECON 02, Vol. 4, IEEE, 2002, pp. 3112–3116.
[69] J.A. Baxter, D.A. Merced, D.J. Costinett, L.M. Tolbert, B. Ozpineci, Review of electrical architectures and power requirements for automated vehicles, in: 2018 IEEE Transportation Electrification Conference and Expo (ITEC), IEEE, 2018, pp. 944–949.
[70] M.A. Rahman, M.T. Rahman, M. Kisacikoglu, K. Akkaya, Intrusion detection systems-enabled power electronics for unmanned aerial vehicles, in: 2020 IEEE CyberPELS (CyberPELS), IEEE, 2020, pp. 1–5.
[71] B. Yang, J. Ye, L. Guo, Fast detection for cyber threats in electric vehicle traction motor drives, IEEE Trans. Transp. Electrif. 8 (1) (2022) 767–777.
[72] B. Yang, L. Guo, F. Li, J. Ye, W. Song, Impact analysis of data integrity attacks on power electronics and electric drives, in: 2019 IEEE Transportation Electrification Conference and Expo (ITEC), IEEE, 2019, pp. 1–6.
[73] L. Guo, J. Ye, Cyber-physical security of electric vehicles with four motor drives, IEEE Trans. Power Electron. 36 (4) (2020) 4463–4477.
[74] A. Chandwani, S. Dey, A. Mallik, Cybersecurity of onboard charging systems for electric vehicles—review, challenges and countermeasures, IEEE Access 8 (2020) 226982–226998.
[75] T. Kim, J. Ochoa, T. Faika, A. Mantooth, J. Di, Q. Li, Y. Lee, An overview of cyber-physical security of battery management systems and adoption of blockchain technology, IEEE J. Emerging Sel. Top. Power Electron. 10 (2020) 1270–1281.
[76] Z.-Y. Hou, P.-Y. Lou, C.-C. Wang, State of charge, state of health, and state of function monitoring for EV BMS, in: 2017 IEEE International Conference on Consumer Electronics (ICCE), IEEE, 2017, pp. 310–311.
[77] Y. Park, O.C. Onar, B. Ozpineci, Potential cybersecurity issues of fast charging stations with quantitative severity analysis, in: 2019 IEEE CyberPELS (CyberPELS), IEEE, 2019, pp. 1–7.
[78] E. Aliwa, O. Rana, C. Perera, P. Burnap, Cyberattacks and countermeasures for in-vehicle networks, ACM Comput. Surv. 54 (1) (2021) 1–37.
[79] International Organization for Standardization, ISO 11898-1:2015 Road vehicles—Controller Area Network (CAN)—Part 1: Data Link Layer and Physical Signalling, 2015. https://www.iso.org/standard/63648.html. (Last accessed 11/2022).
[80] L. Apvrille, L.W. Li, A. Bracquemond, Design and verification of secure autonomous vehicles, in: 12th European ITS Congress, Strasbourg, France, 2017.
[81] P. Cui, U. Guin, M. Tehranipoor, Trillion sensors security, in: Emerging Topics in Hardware Security, Springer, Cham, 2021, pp. 61–93.
[82] S. Woo, H.J. Jo, D.H. Lee, A practical wireless attack on the connected car and security protocol for in-vehicle CAN, IEEE Trans. Intell. Transp. Syst. 16 (2) (2014) 993–1006.
[83] R. Currie, Developments in Car Hacking, 2016. https://www.sans.org/white-papers/36607/. (last accessed: 11/22).
[84] J. Khan, Vehicle network security testing, in: 2017 Third International Conference on Sensing, Signal Processing and Security (ICSSS), IEEE, 2017, pp. 119–123.
[85] M. Marchetti, D. Stabili, Anomaly detection of CAN bus messages through analysis of ID sequences, in: 2017 IEEE Intelligent Vehicles Symposium (IV), IEEE, 2017, pp. 1577–1583.
[86] C. Miller, C. Valasek, Remote exploitation of an unaltered passenger vehicle, Black Hat USA 2015 (S) (2015) 91.
[87] O. Henniger, H. Seudié, EVITA-project.org: E-safety vehicle intrusion protected applications, in: 7th Escar (Embedded Security in Cars) Conference, Düsseldorf, Germany, November 2009 (invited talk – Slides).
[88] W. Liu, K.T. Chau, C.H.T. Lee, W. Han, X. Tian, W.H. Lam, Full-range soft-switching pulse frequency modulated wireless power transfer, IEEE Trans. Power Electron. 35 (6) (2019) 6533–6547.
[89] J. Chen, J.-S. Ng, K.-S. Chong, Z. Lin, B.-H. Gwee, A novel normalized variance-based differential power analysis against masking countermeasures, IEEE Trans. Inf. Forensics Secur. 16 (2021) 3767–3779.
[90] Synopsys, The 6 Levels of Vehicle Autonomy Explained, 2023, Online access: April 16 https://www.synopsys.com/automotive/autonomous-driving-levels.html.
[91] Z. El-Rewini, K. Sadatsharan, N. Sugunaraj, D.F. Selvaraj, S.J. Plathottam, P. Ranganathan, Cybersecurity attacks in vehicular sensors, IEEE Sensors J. 20 (22) (2020) 13752–13767.
[92] B. Nassi, Y. Mirsky, J. Shams, R. Ben-Netanel, D. Nassi, Y. Elovici, Protecting autonomous cars from phantom attacks, Commun. ACM 66 (4) (2023) 56–69.
[93] A. Chowdhury, G. Karmakar, J. Kamruzzaman, A. Jolfaei, R. Das, Attacks on self-driving cars and their countermeasures: a survey, IEEE Access 8 (2020) 207308–207342.
[94] P. Zhang, H. Sun, C. Peng, C. Tan, Sigmoid-like event-triggered security cruise control under stochastic false data injection attacks, Processes 10 (7) (2022) 1326.
[95] T. Sato, Q.A. Chen, Towards driving-oriented metric for lane detection models, in: Proceedings of the IEEE/CVF Conference on Computer Vision and Pattern Recognition, 2022, pp. 17153–17162.
[96] Research Team at Tencent Keen Security Lab, Experimental Security Research of Tesla Autopilot, Tencent Keen Security Lab White Papers, 2019. Online access: 2023, April 16 https://keenlab.tencent.com/en/whitepapers/Experimental_Security_Research_of_Tesla_Autopilot.pdf.
[97] GlobalData Thematic Research, Autonomous Vehicles: Timeline, 2021, Online access: 2023, April 15 https://www.roadtraffic-technology.com/comment/autonomous-vehicles-timeline/.
[98] T.A. Odetola, A. Adeyemo, S.R. Hasan, Hardening hardware accelerator based CNN inference phase against adversarial noises, in: 2022 IEEE International Symposium on Hardware Oriented Security and Trust (HOST), IEEE, 2022, pp. 141–144.
[99] A. Adeyemo, T. Sandefur, T.A. Odetola, S.R. Hasan, Towards enabling dynamic convolution neural network inference for edge intelligence, in: 2022 IEEE International Symposium on Circuits and Systems (ISCAS), IEEE, 2022, pp. 1833–1837.
[100] A. Greenberg, GM Took 5 Years to Fix a Full-Takeover Hack in Millions of OnStar Cars, WIRED, 2015. Online access: 2023, April 16 https://www.wired.com/2015/09/gm-took-5-years-fix-full-takeover-hack-millions-onstar-cars/.
[101] S. Povolny, Model Hacking ADAS to Pave Safer Roads for Autonomous Vehicles | McAfee Blog, McAfee Labs, 2020. Online access: 2023, April 16 https://www.mcafee.com/blogs/other-blogs/mcafee-labs/model-hacking-adas-to-pave-safer-roads-for-autonomous-vehicles/.
[102] Y. Cao, S.H. Bhupathiraju, P. Naghavi, T. Sugawara, Z.M. Mao, S. Rampazzi, You can't see me: physical removal attacks on LiDAR-based autonomous vehicles driving frameworks, arXiv preprint archive (2022).

CHAPTER

10

Propulsion drives and control algorithms of electrical vehicles

Mahmoud Ibrahim, Anton Rassõlkin, Toomas Vaimann, and Ants Kallaste
Tallinn University of Technology, Tallinn, Estonia

10.1 Introduction

In recent times, electric and hybrid vehicle EV/HEV has highly gained the interest of manufacturers and governments, as it is considered the next generation of transportation systems.

Several associations, such as the European Association for Electromobility AVERE, the United States Council for Automotive Research (USCAR), and the United Nations Economic and Social Commission for Asia (UN ESCAP), have defined a set of targets to be met by the EV/HEV propulsion drive systems [1]. Those targets concern system efficiency, reliability, power losses, power density, and economic costs. To achieve all these market objectives, it becomes clear that it is necessary to provoke a change in society and make the EV an attractive road transport alternative [2]. As a result, the EV/HEV propulsion drive system must overcome several technological challenges to comply with the technical objectives.

From the end-user point of view, the main technical aspects considered to decide whether to purchase an EV or HEV are vehicle power, efficiency, maximum speed, dynamic response, reliability, autonomy, and economic costs. All these items are related to the vehicle propulsion drive system, including electric motors and power electronics. From a technical point of view, significant advances will be required to achieve the required cost reductions for the popularization of EV/HEV technologies.

10.2 EV propulsion drive system architectures

Generally, the main electric vehicle architectures are battery-powered electric vehicles (EV), hybrid electric vehicles (HEV), fuel cell vehicles (FC), and fuel cell hybrid electric vehicles (FCHEV). The main difference between those architectures is the energy source. For example, EVs and HEVs rely on batteries, while FCs rely on hydrogen fuel cell stacks (zero-emission vehicle only emits water and heat). Major applications of FC are slow-speed vehicles (buses, trams, etc.). Although FC supplies constant power, it does not adapt properly to a rapid change of power demand [3,4]. However, an FCHEV configuration comprises a combination of FC as the main energy source and battery as an additional storage system, FCHEV can get an increased operating range and speed range at the expense of modifying the powertrain. In this setting, manufacturers, such as Toyota, Honda, and Hyundai, made high-performance FCHEVs [5–7].

10.2.1 Topologies overview

The internal combustion engine's efficiency (ICE) commonly ranges from 20% to 40% and in a very small area of the engine map. Usually, the best performance ICE is revealed during highway cruising; thus, it is worse for urban cruising. Many ways are available to combine an ICE with an electric machine (EM) [7]. The electric motor is designed to handle transient power variations and helps the engine to operate more constantly so that higher efficiency and lower tailpipe emissions can be achieved [8,9]. The variety of ICE/EM connection topologies and designs of the power train

Handbook of Power Electronics in Autonomous and Electric Vehicles
https://doi.org/10.1016/B978-0-323-99545-0.00009-9

can be varied by the traction effort proportion, operating time, and totally provided power. Most widely, EV topologies are addressed in detail next.

10.2.1.1 *Battery/full EVs*

EVs use a chemical battery as primary energy storage and an EM only for traction. It means that they have zero tailpipe emissions and are more friendly to the environment than other types of vehicles.

The EV has many advantages over the conventional ICEV, such as zero emissions, high efficiency, independence of fossil fuels, and quiet and smooth operation [10]. Usually, EV simplifies the mechanical motor-to-wheel transmission part that is like the series HEV or the in-wheel HEV. The design and parameters of the traction system of the EV depend mostly on the speed-torque and speed-power characteristics [11,12]. The proportional schematic of an EV is shown in Fig. 1E.

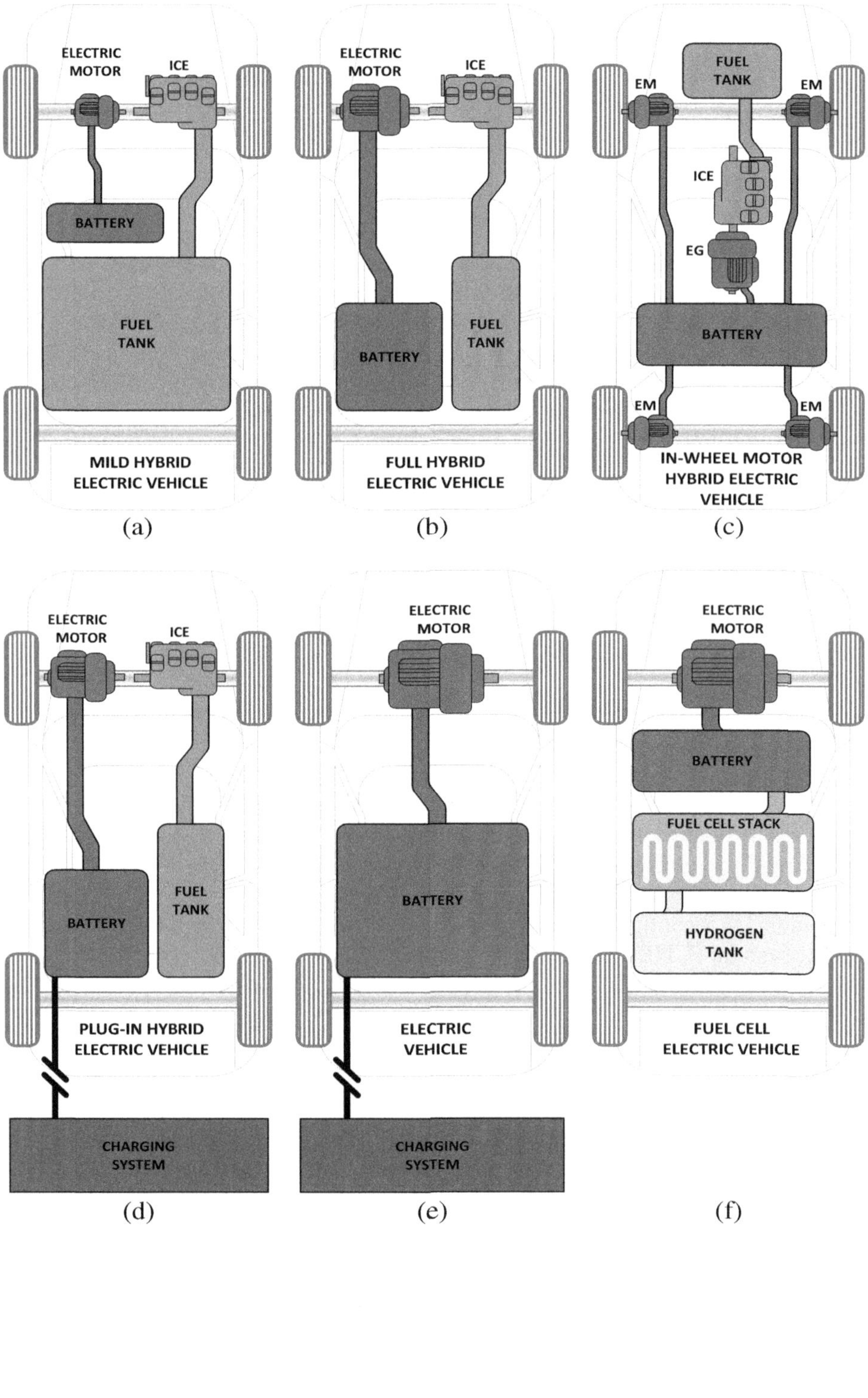

FIG. 1 Hybrid and electric vehicle topologies: (A) mild hybrid electric vehicle; (B) full hybrid electric vehicle; (C) in-wheel motor hybrid electric vehicle; (D) plug-in hybrid electric vehicle; (E) battery electric vehicle; and (F) fuel cell electric vehicle.

All the major automotive manufacturers produce EVs, many of which are available for sale or lease to the public. While the EV is a novel technology, many challenges have emerged in its development and research areas [13]. New standards and requirements should be prepared before the mass implementation of EVs. Moreover, public recharging infrastructure is required and is being used nowadays, and electrical grids need to be renewed or even rebuilt to withstand the increasing consumption load. Due to European Union directives, the governments support the idea of the EV/HEV with taxes cuts and other benefits for EV/HEV owners. However, it seems that the main problem is the driver's habits on vehicle use and refueling.

10.2.1.2 Assist HEV

The electric motor is essentially a very large motor that operates when the engine needs to be turned over and when the driver presses the throttle pedal and requires extra power [14]. Because of the small electric drive range, the assist hybrid cannot operate in pure electric mode, but it can be used as an assist motor in a wide range of controlled torques and speeds. In addition, part of kinetic energy could be converted into electric energy during regenerative braking while the vehicle decelerates. The size of the electric battery used in assist hybrids has extended compared to the ICEV, which increases the vehicle's total weight.

10.2.1.3 Mild HEV

The advanced version of the assist HEV is called a mild HEV. The mild hybrid has a small electric motor that is mainly used as a starter and a charging generator; sometimes, it does operate as a traction motor as well. Usually, the power rating of the electric motor may be in the range of about 10% of the engine power rating [15]. The mild hybrid could operate as a pure EV (the combustion engine is switched off in that mode) only in a small speed range, up to 10 km/h. The design of the traction part of the mild hybrid is like the IECV traction system. The main difference is the extended battery storage system, like in assisted hybrids. The proportional schematic of the mild HEV is shown in Fig. 1A.

10.2.1.4 Plug-in hybrid electric vehicles

The only difference between the plug-in hybrid electric vehicles (PHEVs) from the HEV is the presence of the connection with an electric grid. Depending on the energy flow direction, plug-in systems could be divided into several parts: grid-to-vehicle (G2V) if the batteries are charged from the grid; vehicle-to-grid (V2G) if the energy flow goes to the grid from the vehicle, some solutions are called vehicle-to-home (V2H), if the vehicle is connected to the Smart Grid (SG). G2V allows recharging the batteries without using the ICE, making daily used lightweight PHEV more environmentally friendly [16]. The proportional schematic of the PHEV is shown in Fig. 1D.

10.2.1.5 Fuel cell electric vehicle

A fuel cell (FC) is a galvanic cell in which the chemical energy of a fuel is converted directly into electrical energy utilizing electrochemical processes [17]. There are special tanks required to store the fuel, which usually is hydrogen. It makes the fuel cell electric vehicle (FCEV) more like the series hybrid but with fewer energy conversion stages. A FCEV is schematically shown in Fig. 1F.

The FC presents a novel technology that can be used for ground vehicle technology and for airplane solutions [18]. The main benefits and drawbacks of the FCEV relate to the FC, but vehicles powered solely by the FC have some other drawbacks, such as a heavy and bulky power unit caused by the low power density of the fuel cell system, long start-up time, and slow power response [19]. Some hybrid solutions are presented to overcome these drawbacks, where the FCEVs are combined with some other vehicle technologies, such as supercapacitors (SC).

10.2.1.6 Full HEV

Full hybrids are a type of vehicle that can use pure electric traction mode proportionally EM and ICE are equal, which notably reduces CO_2 emission. The hybrid-drive concept appears in many forms depending on the mix of energy sources and propulsion systems used on the vehicle [13,20]. The energy required for traction could be received from two separate energy sources: EM and ICE; several combinations are available. Schematically, the full HEV is shown in Fig. 1B.

10.2.1.7 In-wheel motor HEV

Some variations of the series HEV have separate small electric in-wheel motors (sometimes called hub motors) installed independently at each wheel. That allows separate control of the power delivery to each wheel, and therefore, it simplifies the vehicle's traction control [21]. The acceleration/deceleration torque and speed values are regulated independently for each wheel. The proportional schematic of the in-wheel motor HEV is shown in Fig. 1C. The mechanical part could be reduced even more than in the series HEV, i.e., additional batteries can be installed on the free space occupied by the transmission, which helps increase the driving range per charge [22]. Most conventional EMs (such as ac excited or brushed dc motors) are not suitable for applications of in-wheel motor drive because of their poor torque density and overload capability. For that reason, in-wheel hybrids require specially designed motors like axial-flux ironless permanent magnet motors or synchronous permanent magnet outer rotor motors.

10.2.2 Overview of HEV topologies

Sometimes, series HEVs are referred to as extended-range electric vehicles (EREVs). There are several variables in a combination of ICE and electrical drive propulsion systems that are presented next.

10.2.2.1 Series HEV

Series HEV uses the only electric drive for traction. An ICE is decoupled from the mechanical transmission and used to produce kinetic energy for the generator that converts it into electrical energy. That means the traction system requires double energy conversion: mechanical-electrical-mechanical. A weakness of a series hybrid system is that series hybrids require a separate motor and generator portions, which can be combined in some parallel hybrid engines; the combined efficiency of the motor and generator will be lower than that of a conventional transmission, thereby offsetting the efficiency gains that might otherwise be realized [23]. Schematically, the configuration of a series HEV is shown in Fig. 2. Some part of energy disengaged during the braking could be regenerated and returned to the battery, only a small amount of braking energy, up to 35%, could be returned back to the battery [24].

10.2.2.2 Parallel HEV

A parallel HEV is only a hybrid vehicle that can switch between two types of drives and could be used separately or simultaneously, depending on the traction mode. Schematically, the configuration of the parallel HEV is shown in Fig. 3, and the conventional signs are the same as in Fig. 3. Parallel HEVs have the same features as series HEVs during regenerative braking. Moreover, an EM can also be used as a generator to recharge the batteries when the engine

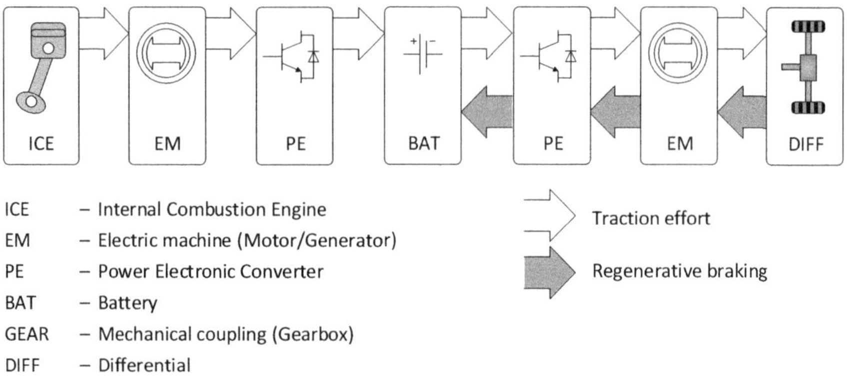

FIG. 2 Functional diagram of series hybrid electric vehicle.

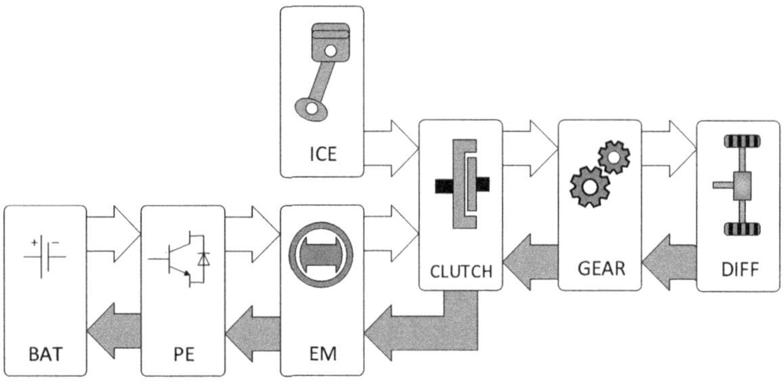

FIG. 3 Functional diagram of parallel hybrid electric vehicle.

produces more power than is needed to propel the vehicle [25]. An ICE and an EM are coupled mechanically by a special gear construction and could be unplugged by re-switching of a clutch. A parallel hybrid uses mostly the electric drive-in urban areas because of the specific qualities of urban driving: fast acceleration/deceleration and short-time duty cycles. The ICE is commonly used during highway cruising, while it has its best performance without speed and torque variations. The main drawbacks of the parallel HEV are a doubled traction system, which increases the price of the vehicle as well [26]. Flexibility and the number of combinations of the working modes of the parallel HEV make it the most used technology for today's lightweight hybrid vehicles.

10.2.2.3 Series-parallel HEV

Series-parallel HEV has two couplings: mechanical and electrical, which allows overcoming the drawbacks of booth technologies and using their benefits with the best performance. Schematically, the configuration of the series-parallel HEV is shown in Fig. 4, and the conventional signs are the same as in Fig. 4. The main aim of such systems is to reduce fuel consumption. In addition, drivability can be optimized based on the vehicle's operating condition [27]. The main drawback of the series-parallel hybrid is higher vehicle prices than other hybrids.

10.3 EV/HEV propulsion drive system

EV propulsion drive system consists of two main parts: electric motor and power electronics (converter, inverter, motor control unit, and on-board charger).

10.3.1 Electric motor

The electric motor is considered the base stone of an EV. It is used for converting the energy from electrical to mechanical and vice versa. It provides power and torque to the transaxle for propulsion. An electric motor's energy conversion efficiency is higher than the conventional internal combustion engine ICE, in between 80% and 95% [28]. EV motor must meet different vehicle requirements. The most important characteristics of the EV motor are wide-speed constant power capability, low-speed high-torque hill-climbing, high-torque density (Nm/kg), high-power density (kW/kg), minimum size, and mainly high operational efficiencies [1,29].

Propulsion motors very harsh environmental conditions, such as a wide temperature range (−30°C to 60°C), severe vibration and shock, high electromagnetic noise, size and weight constraints, stringent safety, and reliability requirements; as a result, there are many unique aspects in the design, development, analysis, manufacturing, and research of electric motors and drives for propulsion applications, which are all important aspects [30]. While the type of the propulsion motor determines the parameters of the whole drive system, the motors could be studied apart. Several types of direct current (DC) and alternating current (AC) electric motors are used in propulsion motor drives, and it is necessary to address their classification and characteristics.

10.3.1.1 Brushed DC motor

The brushed DC motors were used in the last century in electric transportation due to their developed status and suppleness of speed control. The series excited DC motor speed-torque characteristics are an excellent match with the

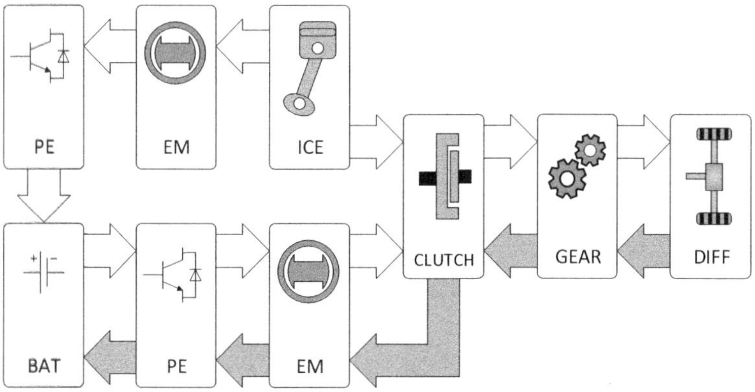

FIG. 4 Functional diagram of series-parallel hybrid electric vehicle.

road load characteristics. Therefore, they were widely used in public transport: tramway cars, trolleybuses, electric trains, and diesel-electric locomotives. Several early prototypes of BEV cars in the 1980s and earlier were built based on the DC series motors as well. However, DC motors' size and maintenance requirements make their use obsolete, not just in the automotive industry but in all motor drive applications [31].

The main drawback of the DC motor is a brushed collector that is required to change the direction of the current inside the rotating part of the motor (armature). The high friction between the brush contacts and the collector results in high friction losses, temperature rise in the machine, low reliability, lower lifecycle, and low efficiency.

According to Eq. (1), the rotational speed of a DC motor can be adjusted in three ways: by changing the supply voltage U, by changing the magnetic flux F (which, due to saturation, can only be reduced), and by changing the resistance of the armature circuit (this can be done by connecting an additional resistance R_{add} to the armature in series).

$$\omega = \frac{U}{k_E F} - \frac{T r_a}{k_E k_T F^2}. \tag{1}$$

Fig. 5 shows the adjustment of speed by changing the supply voltage of the armature circuit.

For start-up, a low voltage is selected for the motor (0.1 times the rated voltage in point A in Fig. 5). At start-up, the rotor is standing still and the electromotive force of the machine is equal to zero. Thus, the intensity of the starting current is determined only by the resistance of the armature circuit: $I_{\text{st}} = U/r_{\text{a}}$. The resistance of the armature circuit is so low that if the rated voltage U_{n} were used for start-up, the starting current would exceed the rated current by more than 20 times, which would destroy the motor. At the proposed low voltage, however, the current will remain within the acceptable range and the machine will start to rotate slowly (point B). Thereafter, the voltage is slowly increased until the desired rotational speed is reached (point C). Changing the voltage rapidly would cause the current to increase, and care should be taken to ensure that the current does not get too high. For example, if we were to rapidly shift from a voltage of 0.75 U_{n} to 0.62 U_{n} (reducing the voltage), we would move abruptly from point C to point D, where braking would occur at the rated torque in generator mode. At this point, the current is equal to the rated current, but opposite in direction to the initial current. Due to the braking torque, the speed would drop and we would end up in point E.

Fig. 6 shows the adjustment of speed by reducing the excitation current. The formula (1) indicates that when the magnetic flux F of the poles decreases, the rotational speed increases. In Fig. 6, we can see that as the pole flux drops,

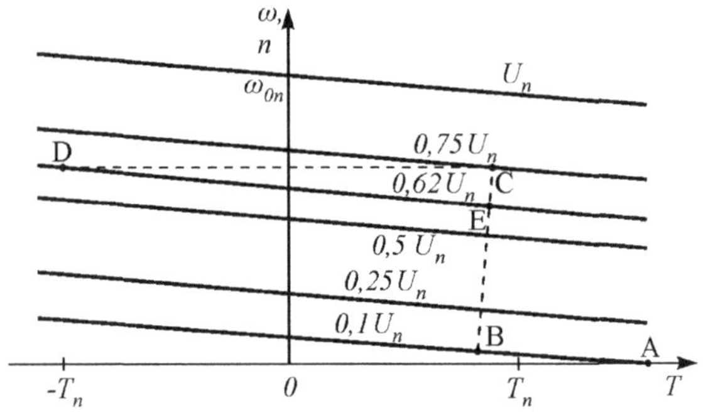

FIG. 5 Mechanical characteristics of a DC machine when reducing the supply voltage.

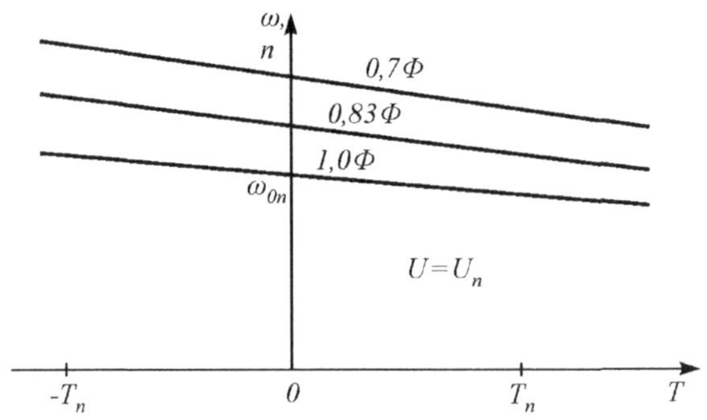

FIG. 6 Mechanical characteristics of a DC machine when reducing the excitation flux of the poles.

the speed becomes higher. Nevertheless, there is a limit to how much the speed can be increased. According to Eq. (2), as the magnetic flux decreases, so, too, does the torque developed by the motor.

$$T = k_I F\, I. \tag{2}$$

However, since the resistive torque of the appliance does not change, then in order to balance out this resistive torque, the motor current needs to increase. To prevent the motor from being overloaded with current, when the excitation field is reduced, the load torque needs to be lower than the rated torque. The more the speed is increased, the less the motor may be loaded.

An additional resistance does not cause the no-load speed to change, but when the machine is loaded with torque, the speed will begin to decrease more. The greater the additional resistance, the more the speed will drop (in Fig. 7, $R_{add2} > R_{add1}$). An additional resistance can also be used when starting a motor. If, at start-up, a resistor R_{add2} is present in the rotor circuit, this resistor will limit the starting current, putting us in point A in Fig. 7. As the speed begins to increase, we will reach point B, where the torque of the motor becomes equal to the load torque. To increase the speed further, we can reduce the resistance of the armature circuit by switching over to the resistance R_{add1}. This first puts us in point C, whereafter the speed increases and we get to point D.

Using an additional resistance for controlling the speed is not as efficient as changing the supply voltage. Additional resistances were used before the invention of semiconductor converters.

There are also DC motors where the excitation current and armature current can only be adjusted together, as well as motors where excitation is achieved by means of permanent magnets. If excitation is achieved by means of permanent magnets mounted on the stator, the magnitude of the excitation field cannot be changed and the motor speed can only be adjusted by changing the armature voltage. In separately excited motors, the armature circuit and excitation circuit have different voltage sources, in which the voltage can be adjusted separately. The effect of this adjustment was explored above. Motors where the armature winding and excitation winding are connected in parallel and supplied from the same voltage source are called shunt motors. Shunt motors can be started and controlled by means of variable resistors that are connected in series either with the armature only or with both the armature and the excitation winding. Shunt motors behave similarly to separately excited motors. The breaking of the excitation circuit due to a fault during the operation of the motor is dangerous for a shunt motor. If this happens, the armature current will increase abruptly and the speed will begin to increase unfettered, as when the excitation current is lost, there will remain only a weak residual magnetic field. This can be prevented with an appropriate protective circuit.

10.3.1.2 Brushless DC motor

Brushless DC motor (BLDCM) is a motor with an independently excited magnet. The speed-torque diagram of a BLDCM is close to the parallel excited DC motor. This type of motor usually has high efficiency, compactness, ease of control and cooling, low maintenance, great longevity, reliability, and low noise emission. Because of the limited PM technologies, the power range of the BLDCM is quite small. These drawbacks make the PM DC motor unsuitable for high-power traction applications. Also, the price of the BLDCM is quite high, the PM can be demagnetized, and the construction is dangerous. Moreover, because of the PMs on the rotor, the BLDCM presents major risks in the case of short-circuit failures of the inverter [32].

In BLDC motors, instead of mechanical commutation of the current, commutation is electronic—performed by an inverter controlled by the position sensor. The mechanism for producing the excitation field in a servomotor is located on the rotor, instead of the stator, and usually consists of permanent magnets. The commutated winding, conversely, is located on the stator, instead of the rotor, so that current would not need to be transmitted from the rotating part to the

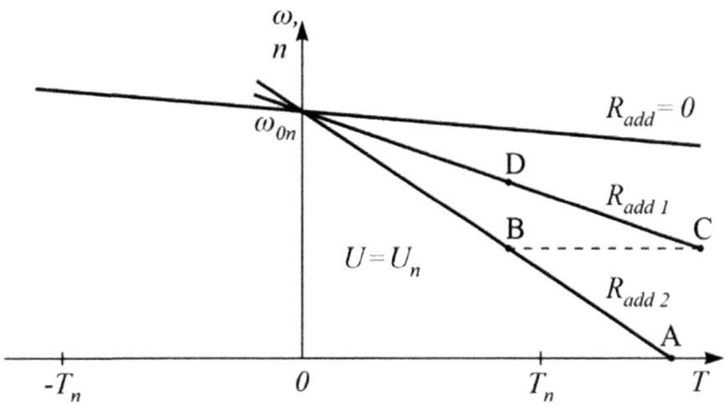

FIG. 7 Mechanical characteristics of a DC machine when an additional resistor is connected in series to the armature.

stationary part. This is known as an inverted design. To make the electronic commutator simpler, only three winding sections are used. In a conventional DC machine, such a small number of sections would result in an unacceptably high voltage between the commutator segments. In an electronic commutator, however, excessive voltage is prevented by means of an appropriate circuit. The integrated position sensor makes it easy to use these machines in route control systems and tracking systems. It is also why they are called AC servomotors.

BLDCs are supplied with adjustable voltage from a DC power supply via an inverter. In higher-power applications, a thyristor inverter with a DC choke, which forms a current inverter, is used. In lower-power applications, a voltage inverter with power transistors is used. When the DC voltage supplied to the inverter is adjusted, the motor's speed changes in a similar manner as when the armature voltage is adjusted in a DC motor.

10.3.1.3 Induction motors

The relatively mature AC induction motor (IM) technology has advanced fast with the progress of power electronics in the last 30 years. A squirrel cage IM has a very simple rotor construction, which increases the reliability and performance of the motor. The absence of the friction parts (except bearings) increases the efficiency of the IM [33]. In contrast to the SRM, the rotor winding of the IM is made from copper, which increases the price of a single device. Vector control of the IM can decouple its torque control from the frequency control. This allows the motor to behave similarly as a separately excited DC motor [34].

Because of low stability and overheating at low speed, the IM is mostly applied in high-speed vehicles; otherwise, some additional mechanical transmission is required for speed reduction. The efficiency of the IM in no-load or light-load operation modes is very small—ca 20:30%, but it increases gradually with the rated load and reaches ca 75:92%. Drawbacks of the IM are limited torque capabilities at low speeds, lower torque density, and lower efficiency and noise due to stator/rotor slot combinations [33].

The rotational speed of the rotor of an IM can be changed based on the formula

$$n = \frac{60 f_1}{p}(1 - s). \tag{3}$$

This formula indicates that the rotational speed changes when the supply current frequency f_1, number of pole pairs p, or slip s changes.

Changing the stator voltage frequency f_1 enables the rotational speed of an IM to be changed seamlessly within a relatively wide range. Seamless frequency changing can be achieved by means of an inverter. Today, inverters are the most common means for controlling the rotational speed of an IM.

At the rated frequency, the rated voltage is applied to the motor and a rotating magnetic flux Φ_r will be produced in the air gap of the machine, the magnitude of which can be found from the electromotive force equation for asynchronous machines

$$F_p = \frac{U_1}{f_1} \cdot 4,44 w_1 k_{w1}, \tag{4}$$

where U_1, stator voltage;
f_1, frequency of stator voltage;
w_1, number of turns in the stator winding phase;
k_{w1}, winding factor of the stator winding.

The increasing of the magnetic flux Φ_p during speed adjustment is undesirable, as this would cause the magnetic circuit to become saturated and the no-load current to increase significantly. In order to prevent the magnetic flux Φ_r from increasing when the frequency f_1 decreases, the ratio U_1/f_1 needs to be kept constant. This can be done by decreasing the voltage by the same amount as the frequency.

For the abovementioned reason, below the rated frequency, the magnetic flux is kept constant, as a result of which the maximum torque of the motor will also remain constant. This can be seen in the mechanical characteristic curves for the different frequencies shown in Fig. 8.

In the figure, the torque and speed values are shown as per-unit quantities with respect to the rated torque and rated synchronous speed. A positive torque corresponds to motor mode, while a negative one corresponds to generator mode. If, during adjustment, the torque is kept constant and the speed is reduced, the power decreases with the speed.

When the speed is raised above the rated speed, the frequency is raised above the rated frequency (in Fig. 8 as an example above 50 Hz). The formula (4) shows that an increase in the frequency f_1 is accompanied by a decrease in the magnetic flux Φ_p, and thus, there is no risk of saturation of the magnetic circuit. Due to this, the voltage is usually not

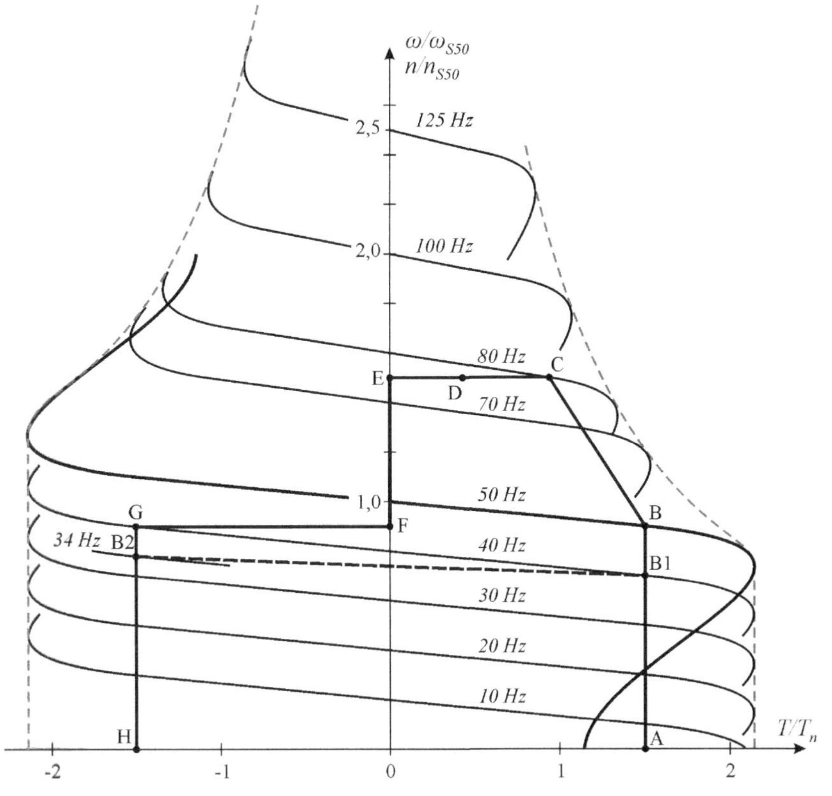

FIG. 8 Mechanical characteristics of an asynchronous machine at different adjusted frequencies.

increased when the frequency is raised above the rated frequency. However, the decrease in Φ_r will result in a decrease in the maximum torque of the machine. This is usually accepted, and, at higher speeds, the mechanical characteristics will look like those shown in Fig. 8. At speeds above the rated speed, the power will not decrease, because the speed will increase in proportion to the decrease in torque, which means that their product remains constant.

If the electric vehicle needs to get from point A to point B and there are no unexpected obstacles on the road, the operation of the powertrain of the vehicle may be divided into five stages, which are shown in Fig. 9.

The speed of the vehicle (which is directly proportional to the rotational speed of the motor) and the torque and power of the motor are shown as time-dependent quantities. The figure shows five stages: start-up and acceleration at a constant torque, acceleration at a constant power, driving at a steady speed, traveling with the engine off, and braking until the speed drops to zero. Let us consider these stages together with the IM mechanical characteristics shown in Fig. 8.

Stage 1. Traveling starts from standstill. A constant torque is applied to the traction motor via a converter. The constant torque produces a constant tractive force. The tractive force causes the vehicle to accelerate. The acceleration is directly proportional to the applied force and inversely proportional to the mass of the vehicle. During acceleration, the speed of the vehicle and the rotational speed of the motor in the powertrain increase. As the rotational speed of the IM increases, so do the frequency and voltage of the current supplied to the motor by the converter. During the first stage, the frequency rises to the rated frequency. In Fig. 8, the first stage occurs along the line segment AB, where the torque is 1.5 times the rated torque and the frequency rises to 50 Hz.

Stage 2. During the second stage, the frequency of the frequency converter rises above the rated frequency and, simultaneously, the torque starts decreasing, because the voltage is no longer boosted. The power of the motor remains constant. In Fig. 8, this stage occurs along the line segment BC. The frequency rises to 80 Hz.

Stage 3. During the third stage, the speed of the vehicle no longer changes. This is possible if the torque of the motor balances out the counter-torque produced by the motion resistance. On a uniform horizontal road, the motion resistance torque is constant. The torque developed by the motor drops exponentially during the transition from the second stage to the third stage (transition from point C to point D in Fig. 8), because the vehicle stops accelerating. The third stage occurs in point D in Fig. 8, where torque, speed, and power are all constant.

Stage 4. The powertrain is switched off. The torque of the motor drops to zero (line segment DE in Fig. 8). The vehicle continues traveling under the effect of inertia, and the speed drops (line segment EF).

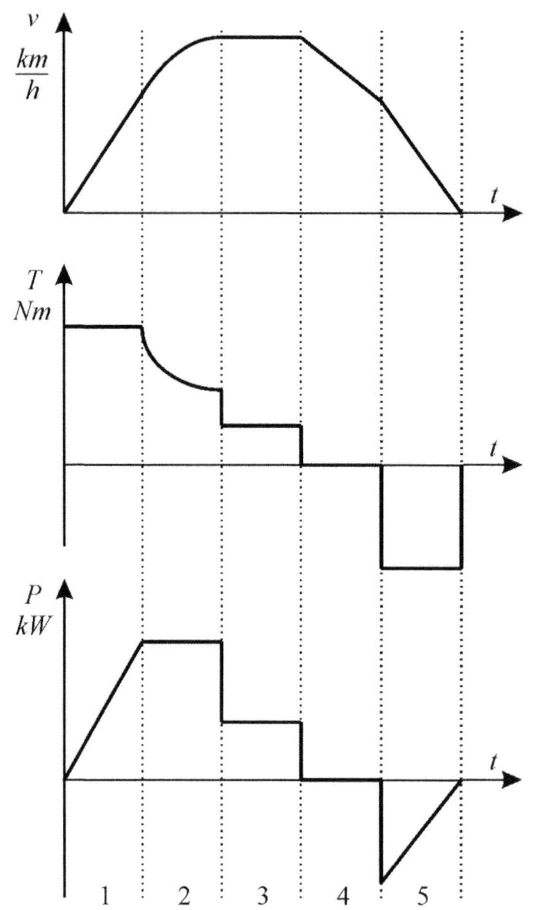

FIG. 9 Typical change in electric vehicle speed, motor torque, and power during a journey from point A to point B on a uniform horizontal road.

The frequency is selected by the program in the control circuit of the frequency converter according to the rotational speed of the motor, the converter frequency, and the position of the pedals. Most frequency converters are equipped with a programmable control circuit and can be used in motors with different speed and power ratings (within certain limits) utilized for performing different tasks.

Stage 5. consists in braking. Braking is used to stop at the intended destination. The direction of the torque during braking is the opposite of that during starting. In Fig. 8, the braking torque arises along the line segment FG, where, at point G, the torque is negative. The necessary braking torque occurs when the frequency converter shifts the frequency to 40 Hz. During braking, the traction motor is in generator mode. The energy from the generator passes through the inverter and enters the battery.

During braking, the torque is kept constant and the frequency of the frequency converter is lowered. In Fig. 8, this occurs along the line segment GH.

When driving in real life, unexpected obstacles may be encountered on the road, in which case the vehicle will need to brake. For example, when picking up speed (stage 1), point B1 may have been reached on the mechanical characteristic curve, when, suddenly, it may be necessary to switch over from the acceleration torque to a braking torque. To do this, the frequency may be reduced from 40 to 34 Hz, putting as at point B2.

Of course, drivers will never actually adjust the frequency directly. Instead, they operate the vehicle by means of gas and brake pedals. The frequency is selected by the program in the control circuit of the frequency converter according to the rotational speed of the motor, the converter frequency, and the position of the pedals. Most inverters are equipped with a programmable control circuit and can be used in motors with different speed and power ratings (within certain limits) that are employed in different applications.

10.3.1.4 Permanent magnet synchronous motor

Usually, the permanent magnet synchronous motor (PMSM) motor has three-phase (rarely one phase) windings, and the permanent magnet excite the rotor. For EV applications, motor size is relatively large compared to PM motors' other smaller power applications, which amplifies the cost problem [35]. However, the PM AC motors could be used in the assist and mild HEV. The benefit of this motor is that currents do not need to be induced in the rotor (like in IMs), making them somewhat more efficient and giving slightly greater specific power, at the same time, the drawback is that it is more expensive due to the presence of a PM. PM motors face the possibility of demagnetization at extremely high temperatures, limited speed range, and difficulty in protecting the powertrain during a fault condition [36].

A comparison of PMSM and IM would reveal that the former have a higher energy conversion efficiency. The rotor windings of IM, unlike PMSM, are subject to relatively high losses. Nevertheless, this advantage of PMSM only becomes apparent if the rotational speed of both types of motors needs to be adjusted and they are assisted by n inverter. If a single steady speed is required (what is very rare for EVs), the efficiency of a powertrain with an IM is higher.

10.3.1.5 Permanent magnet-assisted synchronous reluctance motors

This motor started gaining manufacturers and researchers' interest recently as it does not rely on rare-earth-based magnets [37]. Permanent magnet-assisted synchronous reluctance motors (PM-SynRM) rotor is manufactured in a way that the asymmetry between the *d*- and *q*-axes is maximized. As a result, its reluctant torque is maximized, allowing to achieve power densities of around 75% of an interior PMSM for the same size [38]. As it is made of a ferrite PM, the main drawback is that it may suffer from demagnetization at low temperatures, which can be avoided or minimized by proper design or preheating the magnets before the machine starts.

In a synchronous motors (SM), booth PMSM and PM-SynRM, the rotor always moves synchronously with the rotating magnetic field. As such, the rotational speed of the motor can be changed only by changing the rotational speed of the stator field. The stator field's rotational speed n_S is dependent on the number of pole pairs p in the motor and the supply voltage frequency f.

$$n_S = 60f/p. \tag{5}$$

Changing the rotational speed of a SM requires changing the frequency of the supply voltage. This can be achieved by means of an inverter with a similar power component as that utilized in inverters for IMs. Fig. 10 shows the mechanical characteristics for frequency adjustment in a SM.

The motor's speed is shown with respect to the rated speed n_n, while the torque is shown with respect to the rated torque T_n. The motor's speed can be adjusted to be higher or lower than the rated speed. Since the stator of a SM is similar to that of an IM, when reducing the frequency with a frequency converter, it is also necessary to reduce the voltage supplied to the motor from the frequency converter (otherwise the stator's magnetic circuit will become saturated).

The rotational speed of a SM, unlike that of an IM, is not dependent on the load torque (the characteristics shown in Fig. 10 are straight horizontal lines). That is, until the breakdown torque is reached, where the motor's operation becomes unstable. Reaching the breakdown torque is generally unacceptable in synchronous machines, because the unstable operation manifests in strong vibrations and dangerously high currents.

Lowering the speed below the rated speed ($\omega/\omega_n < 1$) does not cause the breakdown torque nor the slightly lower maximum permissible load torque to change. However, the maximum power decreases in proportion to the decrease

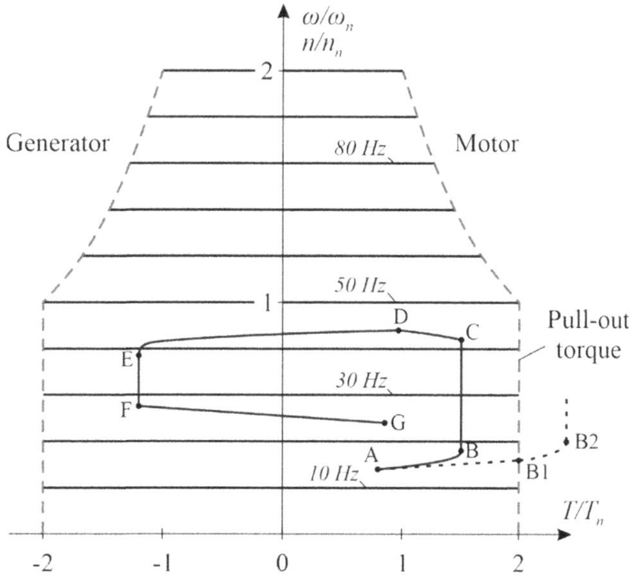

FIG. 10 Mechanical characteristics of a synchronous motor at different adjusted frequencies.

in speed, because $P = \omega T$. At speeds greater than the rated speed, however, further increases in speed are accompanied by the decreasing of the maximum torque (ωT = const), while the maximum power remains unchanged.

In steady-state operation (point A in Fig. 10), the torque T_M developed by the motor is equal to the load torque of the appliance T_{load}. In point A, the motor's per-unit speed is 0.28. Given a rated speed of, for example, 3000 rpm, the speed in this point would be 840 rpm, or 87.96 rad/s. The motor's per-unit torque in this point is 0.8. The load angle then is arcsin 0.4 = 23.58 degree, because the per-unit breakdown torque is 2 and the relationship between the load angle and the load torque is given by the following equation:

$$T_{em} = \frac{3p}{\omega}\frac{EU}{X_S}\sin\theta. \tag{6}$$

Let us see what happens if we increase the angular velocity of the stator's rotating field by 0.1 rad/s and assume that the rotor's speed does not change. Then, after 1 s, the angle between the space vector of the stator field and the rotor pole will have increased by 0.1 rad, or 5.7 degree. This angle is the load angle. Thus, if the speed of the stator's rotating field increases by a small amount, the load angle θ starts to increase slowly. This load angle could not change if there did not exist a small difference between the rotational speed of the stator field and that of the rotor for a short while. Therefore, during transient responses to a changeover, the rotor of a synchronous machine may not rotate in perfect synchronism with the stator field.

If we were to increase the angular velocity ω_B of the stator field by means of a frequency converter in such a way that $d\omega_B/dt$ = const, the load angle θ would start to increase in an accelerating manner. According to formula (6), when the load angle increases, so does the motor's torque T_M, which becomes higher than the load torque T_{load}. A dynamic torque T_{dyn} arises, which starts to accelerate the rotor,

$$T_{d\ddot{u}n} = T_M - T_{load}. \tag{7}$$

The higher this dynamic torque, the faster the increase in the rotor's angular velocity ω_R:

$$d\omega_R/dt = T_{dyn}/J. \tag{8}$$

where J is the moment of inertia of the rotating parts of the motor and the appliance.

Eventually, the increase in rotor speed becomes equal to the increase in the rotational speed of the stator field:

$$d\omega_R/dt = d\omega_B/dt. \tag{9}$$

This is the case in point B in Fig. 10. When this happens, the load angle stops increasing (as does the motor's torque). Thereafter, the increase in the speed of the rotor occurs synchronously with that of the stator field. The motor's torque remains constant and exceeds the load torque by the magnitude of the dynamic torque. The motor's speed will increase for as long as the frequency converter is made to increase the frequency. In Fig. 10, this occurs up to point C. When point C is reached, the rate of increasing the frequency is gradually reduced, and in point D, the frequency becomes constant. Then the acceleration of the rotation stops, the dynamic torque drops to zero, and the torque of the motor again becomes equal to the load torque. All of this, however, occurs at a higher per-unit rotational speed of 0.88.

The higher the chosen acceleration $d\omega_B/dt$ of the rotating field, the higher the resulting acceleration $d\omega_R/dt$ of the rotor. Note that, according to formula (8), this must also result in a higher dynamic torque. Here, it is necessary to ensure that the motor's torque $T_M = T_{load} + T_{dyn}$ does not exceed the breakdown torque T_{max}. The motor's torque must be lower than the breakdown torque, i.e.,

$$T_{load} + J(d\omega_R/dt) < T_{max}. \tag{10}$$

Rearranging Eq. (10) and taking into account that, in this case, the condition $d\omega_R/dt = d\omega_B/dt$ is also met, we get the maximum acceleration of the rotation of the stator field, which must not be exceeded:

$$d\omega_B/dt < (T_{max} - T_{load})/J. \tag{11}$$

In order for the operation of the motor to remain stable when the speed is increased (so that there is no breakdown), the frequency of the frequency converter must not be increased too rapidly. The higher the load torque and the moment of inertia of the rotating parts, the slower the frequency must be increased.

Points B1 and B2 in Fig. 10 illustrate what happens if the frequency is increased too fast. In order for the rotor to receive the same acceleration as the stator field, it is necessary to apply a torque that is higher than the breakdown torque (point B2). Reaching point B2, however, is impossible, because the motor stops operating in a stable manner when point B1 is reached, which never should be.

Reducing the speed, too, must be done sufficiently slowly as not to reach the maximum torque for generator mode. Fig. 10 shows how the load point changes along the mechanical characteristic curves when the rotational speed is decreased. When moving from point D to point E, the speed of the stator field decreases by a greater amount than the rotor speed and, as a result, the load angle decreases. Between point E and point F, braking occurs at a constant torque. The magnitude of the braking torque depends on how rapidly the frequency converter lowers the frequency. Between point F and point G, the lowering of the frequency is gradually stopped.

Frequency converters are equipped with a control program that selects the suitable acceleration or deceleration of the rotational speed according to the parameters of the connected motor and the instantaneous value of the torque. Some frequency converters are equipped with programs that can be used to control both synchronous and IMs. In such converters, the type of motor is defined for the converter during setup.

10.3.1.6 Switched reluctance motor

Switched reluctance motors (SRMs) are also known as doubly salient machines that have a simple construction and excellent fault tolerance characteristics. The stator and the rotor of the SRM are made of iron, which is magnetized by the current through the stator winding. These motors have no windings, magnets, or cages on the rotor, which helps to increase the torque and inertia rating. The design of SRMs is similar to that of SMs. What sets them apart is that SRMs also contain a rotor position sensor. The stator winding is three-phase like that of conventional SM, but is powered by an inverter controlled by signals received from the rotor position sensor.

The motor speed-torque characteristics are an excellent match with the road load characteristics, and the performance of switched reluctance motors for BEV applications has been found to be excellent [39].

The main drawback of SRMs is that the timing of the turning-on and turning-off of the stator currents must be much more carefully controlled, and a more complex control device is required. In addition, SRMs have such challenges as torque ripple [40]. The motor types of some EV available in the market are presented in Table 1.

TABLE 1 Motor types of some EV examples already in the market.

Manufacturer	Year of release	Model	Configuration	Motor type
Mercedes	2010	Class B F-Cell	FCEV	PMSM
Mercedes	2014	SLS AMG ED	EV	PMSM
Mercedes	2014	Class B ED	EV	PMSM
Mercedes	2019	Benz EQC 400	EV	PMSM
Kia	2014	*E*-Soul	EV	PMSM
Kia	2020	Sorento	HEV	PMSM
Kia	2022	Niro EV	EV	PMSM
Nissan	2010	Leaf	EV	PMSM
Mitsubishi	2022	Outlander	HEV	PMSM
Renault	2020	Twingo	EV	PMSM
Renault	2021	Master E-Tech	EV	PMSM
Tesla	2012	Model S	EV	IM
Tesla	2015	Model X	EV	IM
Tesla	2017	Model 3	EV	PM-SynRM
Tesla	2021	Cyber Truck	EV	PM-SynRM
Peugeot	2014	Partner	EV	PMSM
Peugeot	2019	e-208	EV	PMSM
Toyota	2012	RAV4	EV	IM
Toyota	2015	Mirai	FCEV	PMSM
Land Rover	2015	Defender	EV	SRM

The key research items that will improve the features of electric motors can be summarized as:

- Reducing cost, weight, and volume of the motor: (a) usage of new magnet materials without rare earth and heavy rare earth (b) simplification of the cooling system (air cooling is considered among other solutions).
- Improving performance, efficiency, and reliability: (a) development of high-performance alloys, such as magnetic steels and copper alloys, and (b) the design of machines with low copper and iron losses.
- Developing innovative modular and scalable designs.
- Improve the manufacturability and accelerate commercialization.

10.3.2 Power electronics

Portable energy storage systems ESSs are mostly DC energy sources, so some power electronic converters are required to provide the electric energy conversion between the ESS and the propulsion system of the EV. Inverters are required to provide a connection between AC propulsion motors and the batteries. Rectifiers are used to apply AC generators as kinetic-to-electrical energy converters. If more than one is used, DC/DC converters equal the DC link voltage from all energy sources. Moreover, the charging systems require special attention and have specific requirements and standards [41].

Historically, EV power electronics development has passed different phases by the time. From 1995 to 2005, most of the research focused on developing basic automotive components (electric motor drives and power modules with maximum junction temperatures of 125°C and switching frequencies between 2 kHz and 10 kHz) [42]. From 2005 to 2015, huge efforts were made to achieve higher junction temperatures (up to 150°C) and higher switching frequency operation (up to 12 kHz) for the power electronics [42]. Currently, IGBT devices are being used in almost all the commercially available EVs, HEVs, and PHEVs. The IGBTs will continue to be the technology soon until the silicon carbide (SiC) and gallium nitride (GaN)-based devices are commercially available at a reasonable cost as silicon IGBTs. Lately, noticeable progress has been made in these devices' technology for both automotive and other power applications [20,22]. Silicon carbide (SiC) is considered the next-generation power semiconductor device. The main features of SiC devices are inherent radiation resistance, high-temperature operating capacity, high voltage, and power handling capacity, high-power efficiency. SiC has great properties such as wider bandgap, larger critical electric field, and higher thermal conductivity that enable its operation at higher temperatures and higher voltages, which offer higher power and current density than the Si-based devices. Gallium nitride GaN devices are predicted to have higher performance over Si-based devices and much better performance than SiC-based devices due to the excellent GaN material properties such as high electron mobility, high breakdown field, and high electron velocity. GaN-based devices are characterized by low on-resistance and fast switching, leading to a higher reduction in both conduction and switching losses. GaN devices can also be produced in high ratings that allow superior performance and affordable manufacturing. Compared to Si, GaN is certainly higher in cost, but it costs less than SiC because GaN is compatible with silicon substrates. As power devices' dc voltage rating requirement in almost propulsion drives is <1000 V, GaN will be more suitable for EVs applications than the SiC devices. In the following context, different EV power electronics circuits will be addressed.

10.3.2.1 Rectifiers

In series and series-parallel HEVs, a synchronous generator is frequently used to convert mechanical energy produced by the ICE into electrical one. Rectifiers are required to convert the output AC of a synchronous generator into DC for a battery. There are four common rectifier circuits in series HEV: pulse-width modulation (PWM) voltage-source current-controlled rectifier, uncontrolled full-bridge diode rectifier, uncontrolled full-bridge diode rectifier with a DC/DC boost converter, and controlled full-bridge thyristor rectifier [43]. Topologies of 3-phase rectifiers are shown in Fig. 11. Single-phase rectifiers are rarer than 3-phase, used only in auxiliary circuits.

The rectifier presented in Fig. 11A is bidirectional, i.e., the energy flow could be directed from the ESS to the propulsion motor or in the opposite direction. That feature could be used during regenerative braking while the propulsion motor works as a generator. The rectifier presented in Fig. 11B is an uncontrolled unidirectional rectifier, which does not allow the bidirectional energy flow. Rectifiers presented in Fig. 11C, D are controlled rectifiers, they allow output voltage adjustment. To apply a fully controlled thyristor bridge rectifier (Fig. 11D) as a bidirectional rectifier, special additional circuits are required to close thyristors under DC current (Fig. 11).

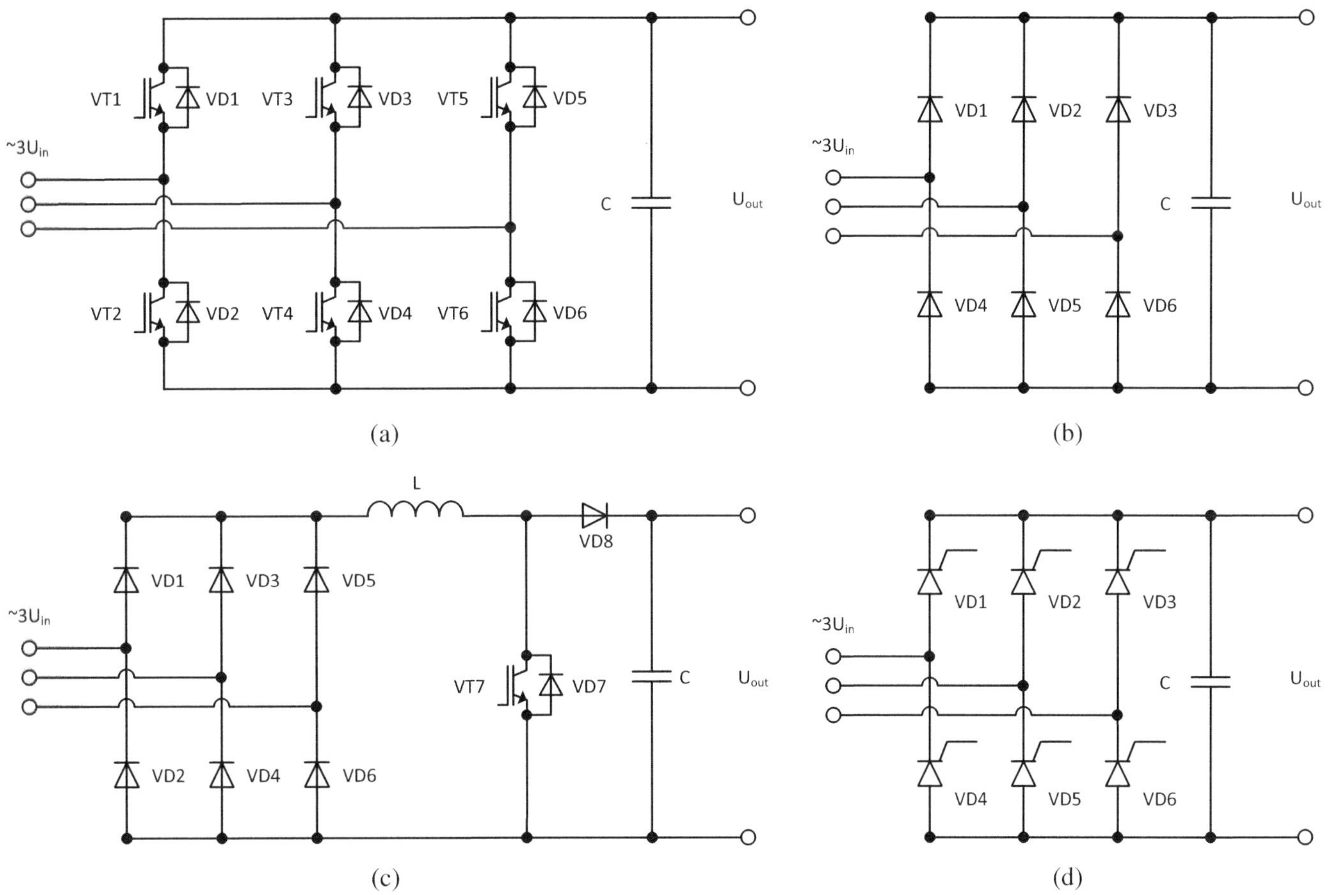

FIG. 11 Rectifiers used in a BEV: (A) PWM voltage-source current-controlled rectifier; (B) uncontrolled diode bridge rectifier; (C) diode rectifier with boost DC/DC converter; (D) fully controlled thyristor bridge rectifier.

10.3.2.2 Inverters

While most energy storage devices are voltage sources, the voltage source inverter (VSI) is the most common in EV applications, but it presents difficult hurdles for the automakers to address the important issues on the system cost and weight volume, and reliability [41]. VSIs are used to control the speed of IMs and PM AC motors. The switches are usually IGBTs for high voltage, high-power hybrid configurations or MOSFETs for low voltage designs [36]. Voltage surges caused by rapid voltage transitions made by the PWM control technique can cause motor insulation degradation, bearing failure due to the resulting shaft leakage current, and unacceptable electromagnetic interference effects upon the control circuits as acoustic noises in the motor [44]. Nowadays, many different torque control techniques are used for AC motors (IM, SRM, and PMSM). The main benefit of such techniques is the direct control of the motor flux that directly influences the torque developed by the motor that increases the motor controllability.

The current source inverter (CSI) uses an inductor for energy-boosting as distinct from the VSI that uses a capacitor. It has several inherent advantages that include its voltage boosting capabilities, its natural shoot-through and short-circuit protection capability, and a sinusoidal output voltage due to the effect of the output AC filter capacitors, which are much smaller than the VSI's DC capacitor [45]. Inverter topologies used in the BEV are shown in Fig. 12.

10.3.2.3 DC/DC converters

The variety of DC/DC converter topologies used in the BEV is wider than the rectifier or inverter topologies, and some circuits are shown in Fig. 13.

The typical application of a buck converter in a BEV is to step down the high voltage battery voltage (typically 200–400 V) to charge the auxiliary battery (14 V) [46]. At such a high step-down voltage rate, the buck converter's duty cycle is small, which affects the design of the inductor and the capacitor, as well the current and voltage ripples. Moreover, controlling and regulating the buck converter output voltage becomes more difficult.

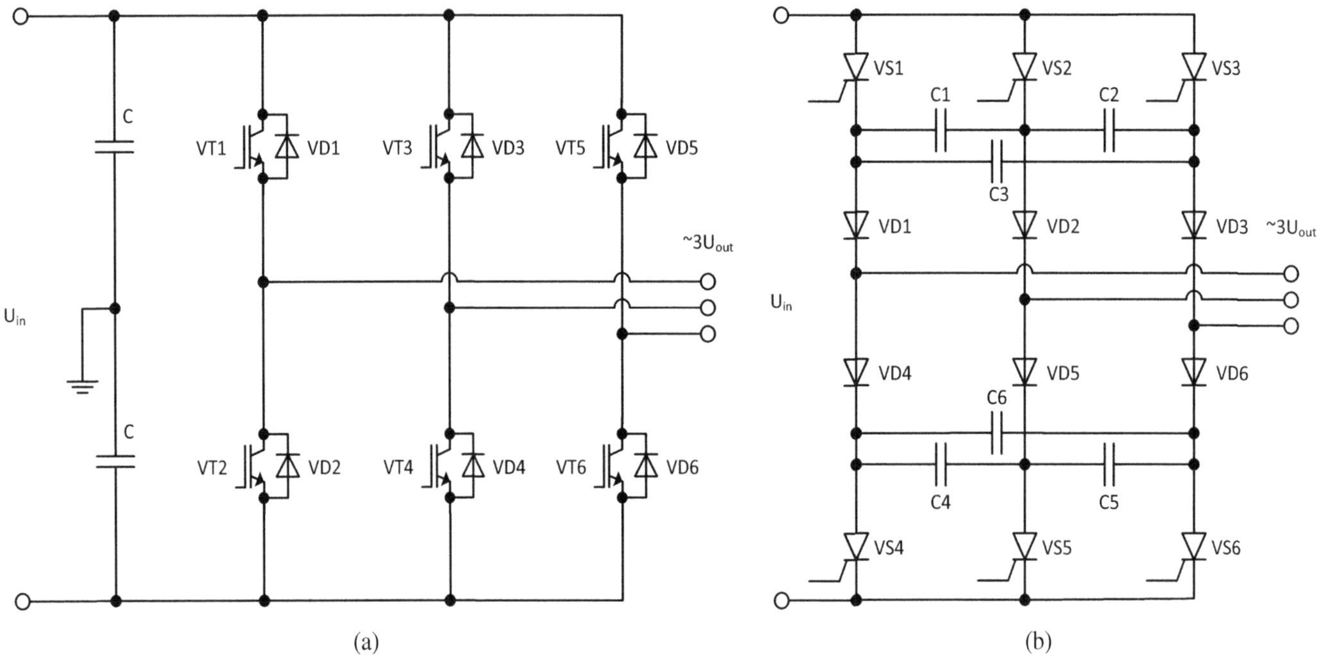

FIG. 12 Inverters used in a BEV: (A) voltage source inverter; (B) current source inverter.

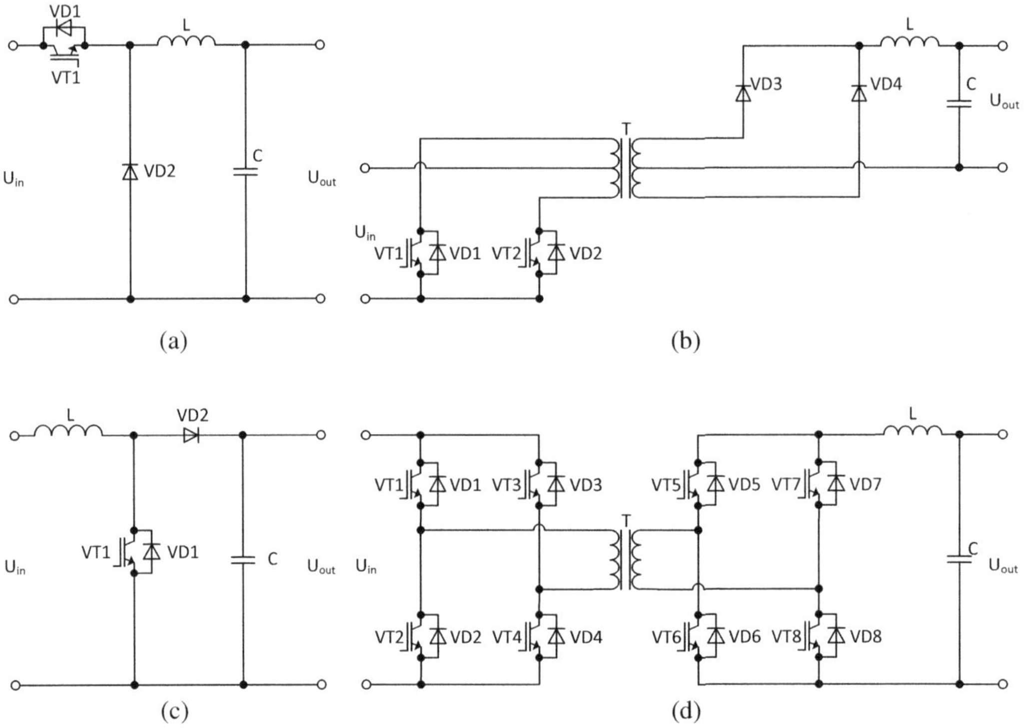

FIG. 13 DC/DC Converters used in a BEV: (A) buck converter; (B) boost converter; (C) isolated unidirectional half-bridge DC/DC converter; (D) isolated bidirectional full-bridge DC/DC converter.

A boost converter or a nonisolated bidirectional DC/DC converter is used to step up the high voltage battery up to the DC bus. The nonisolated converter type is generally used where the voltage needs to be stepped up or down by a relatively small ratio (less than 4:1) [47]. The best operating voltage for a VSI that controls the traction motor is around 600 V. In addition, that converter allows the backward energy flow from the DC link back to the battery during the regenerative braking.

In some cases, nongalvanic isolation between the battery and the DC link is required. That is, isolated unidirectional or bidirectional DC/DC converters with a high-frequency AC link are used. Usually, the ratio of stepping down or stepping up the voltage of the bidirectional DC/DC converters is high.

Current research items concerning EV power electronics focus on:

- High efficiency targets include implementing wide bandgap (WBG) semiconductor technology with low switching losses.
- High-power density targets, (a) by introducing WBG technology including advanced packaging, and (b) by implementing optimum thermal management Improvements in thermal management and reliability.
- The step-up function of the converter.
- Low electromagnetic interference.

10.4 Control algorithms of EV propulsion drives

The torque-speed profile of a well-controlled EV motor drive is shown in Fig. 14A. It includes two distinguishable regions—constant torque (CT) and constant power (CP). The constant torque region maximum speed is usually called base speed. The lower than rated speed (base speed) results in a larger maximum torque and a wider operating speed range at a given power rating and maximum speed [48]. A term, speed ratio (α), given in Fig. 14B, is defined as the ratio of the maximum speed to the base speed. The higher α, the lower the power motor is, with reduced size, cost, and high efficiency.

The propulsion motor controller plays a key role in controlling motor speed and torque during acceleration and deceleration. As the driver impulse, the accelerator, commanded torque signal is processed through the controller, and the motor input is varied to meet the commanded torque or speed. Several control techniques can be used depending on the EV-motor type.

10.4.1 Field-oriented control

Also known as vector control, it controls separately the torque producing and magnetizing the flux components. This control technique is based on projections that transform a three-phase (time and speed-dependent system) into a

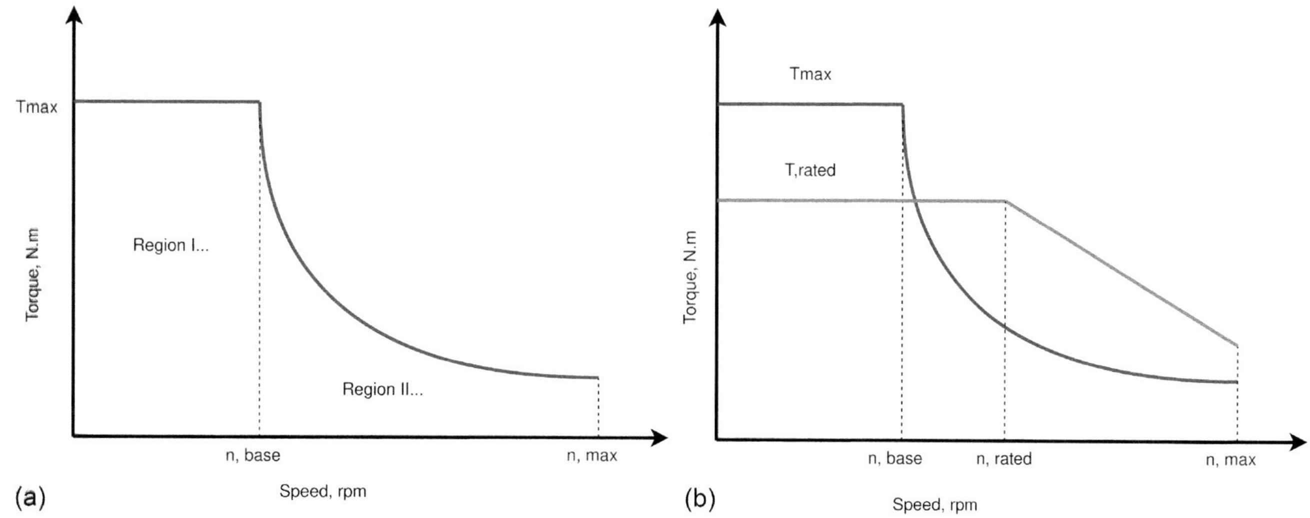

FIG. 14 (A) Ideal output characteristics of EV-motor drive. (B) EV-motor profile with different speed ratio α.

two-coordinate (d and q frame) time-invariant system. These transformations and projections lead to a structure similar to that of a DC machine control. Field-oriented control (FOC) needs two constants as input references: the torque component (aligned with the q coordinate) and the flux component (aligned with d coordinate). It decouples torque and flux components of the stator current to get a fast dynamic torque response, which improves the drive system performance [49].

FOC can be broadly classified into Indirect FOC (IFOC) and Direct FOC (DFOC) schemes. In the DFOC strategy, the rotor flux vector is either measured by means of a flux sensor mounted in the air gap or by using the voltage equations starting from the motor parameters. FOC is widely used for controlling IM, PMSM, and BLDC. Since mounting sensors to the air gap of any machine is not preferable, the IFOC method has been proposed to overcome this difficulty. IFOC algorithm is mainly based on estimating rotor flux according to the measured currents and estimates the speed without any flux sensors [50]. Because of the better control performance of IFOC, it is highly suited for EV drive applications. However, there are some drawbacks of IFOC, such as the transformation requirement for coordinates and sensitivity to motor values, which should be handled in the design stages.

10.4.2 Direct torque control

The direct torque control (DTC) technique is different from the conventional vector control method FOC in the sense that the reference frame in DTC is stator flux instead of rotor flux. DTC control scheme abandons the stator current control strategy, and it directly controls the flux itself. Under the classic DTC scheme, if it is assumed that the rotor flux remains constant over the time interval of interest and the stator resistance drop does not exist, the stator voltage remains constant over the time interval [51]. DTC outperforms FOC in that it offers a fast response of motor speed and torque, as well as it is senseless, so no speed or position encoders are required. It only requires voltage and current measurements. This reduces cost and improves reliability. DTC is applicable for high-speed operation and allows frequent starting/stopping and acceleration. DTC enables robust flux weakening control to be implemented and enables dynamic operation of the motor. However, the conventional DTC strategy has serious electromagnetic torque and flux linkage ripples that result in poor steady-state performance [52].

10.4.3 Space vector modulation direct torque control

To eliminate the previously addressed problems of DTC, an improved scheme based on the control of stator flux, torque angle, and torque was proposed, which used a voltage vector selection strategy and the technology of space vector modulation (SVM) to generate the applied voltage vector instead of switching table. SVM objective is to enable optimal selection of the switching vectors, allowing for reducing torque/flux ripples and harmonic distortion in the current waveform by maintaining a constant switching frequency. DTC-SVM mechanism depends only on the motor's stator parameters [53]. DTC-SVM has several advantages: better DC bus utilization, lower torque ripples, lower total harmonic distortion in the motor current, lower switching loss, and mainly suppress torque ripple and flux [54]. DTC-SVM is mainly used in an EV with IM, PMSM, or BLDC motor drive.

10.4.4 Torque sharing function

It is a control technique especially used for EV SRM drive. As previously mentioned, SRM's main drawback is higher torque ripples than PMSM, resulting in acoustic noise and vibration. Most torque ripple in SRM comes from the total torque pulsation during the commutation between phases [55]. To remain the constant resultant torque during the commutation, the torque sharing function (TSF) technique is presented. In the TSF technique, torque interchanging between two mutual phases is done in such a way that the resultant motor torque remains constant at reference value. The total torque produced by the motor is obtained from the phase torques summation. Generally, the conduction period of each phase is divided into two intervals. One is single-phase conduction, and the other is two phases or commutation between adjacent phases. In the TSF technique, the incoming phase reference torque is considered a specific function so that the results produced by the outgoing and entering phase stay constant [56]. The techniques used for control EV motors are presented in Table 2.

TABLE 2 Some control techniques used in real EV models from the market.

Manufacturer	Model	Type	Motor	Control technique
Tesla	Model S	EV	IM	DTC-SVM
Tesla	Roadster	EV	IM	DTC-SVM
Nissan	Leaf	EV	PMSM	IFOC
Mitsubishi	Outlander	HEV	PMSM	DTC-SVM
Land Rovers	Defender	EV	SRM	TSF

Acknowledgment

The research has been supported by the Eesti Teadusagentuur under grant PSG453 "Digital twin for propulsion drive of autonomous electric vehicle."

References

[1] I. López, E. Ibarra, A. Matallana, J. Andreu, I. Kortabarria, Next generation electric drives for HEV/EV propulsion systems: technology, trends and challenges, Renew. Sust. Energ. Rev. 114 (2019) 109336, https://doi.org/10.1016/J.RSER.2019.109336.
[2] J.A. Sanguesa, V. Torres-Sanz, P. Garrido, F.J. Martinez, J.M. Marquez-Barja, A review on electric vehicles: technologies and challenges, Smart Cities 4 (1) (2021) 372–404, https://doi.org/10.3390/SMARTCITIES4010022.
[3] C. Mahmoudi, A. Flah, L. Sbita, An Overview of Electric Vehicle Concept and Power Management Strategies (Undefined), 2014, https://doi.org/10.1109/CISTEM.2014.7077026.
[4] H.S. Das, C.W. Tan, A.H.M. Yatim, Fuel cell hybrid electric vehicles: a review on power conditioning units and topologies, Renew. Sust. Energ. Rev. 76 (2017) 268–291, https://doi.org/10.1016/J.RSER.2017.03.056.
[5] S. Conti, et al., Solar electric vehicles: state-of-the-art and perspectives, in: 2018 110th AEIT Int. Annu. Conf. AEIT 2018, 2018, https://doi.org/10.23919/AEIT.2018.8577289.
[6] L. Zhang, D. Meng, G. Chen, Noise, Vibration and Harshness of Electric and Hybrid Vehicles, SAE International, 2020.
[7] K.V. Singh, H.O. Bansal, D. Singh, A comprehensive review on hybrid electric vehicles: architectures and components, J. Mod. Transp. 27 (2) (2019) 77–107, https://doi.org/10.1007/S40534-019-0184-3 (2019 271).
[8] F. Magnussen, On Design and Analysis of Synchronous Permanent Magnet Machines for Field-Weakening Operation in Hybrid Electric Vehicles, Elektrotekniska System, 2004.
[9] W. Enang, C. Bannister, Modelling and control of hybrid electric vehicles (a comprehensive review), Renew. Sustain. Energy Rev. 74 (2017) 1210–1239, https://doi.org/10.1016/J.RSER.2017.01.075.
[10] P. Nieuwenhuis, L. Cipcigan, H.B. Sonder, The electric vehicle revolution, in: Futur. Energy Improv. Sustain. Clean Options Our Planet, 2020, pp. 227–243, https://doi.org/10.1016/B978-0-08-102886-5.00011-6.
[11] Q. Tang, X. Shu, G. Zhu, J. Wang, H. Yang, Reliability study of BEV powertrain system and its components—a case study, Processes 9 (5) (2021) 762, https://doi.org/10.3390/PR9050762.
[12] K.T. Chau, Pure electric vehicles, in: Altern. Fuels Adv. Veh. Technol. Improv. Environ. Perform. Towar. Zero Carbon Transp, 2014, pp. 655–684, https://doi.org/10.1533/9780857097422.3.655.
[13] E.A. Nanaki, Electric vehicles, in: Electr. Veh. Smart Cities, 2021, pp. 13–49, https://doi.org/10.1016/B978-0-12-815801-2.00006-X.
[14] B.-H. Lee, S.-I. Kim, J.-J. Lee, J.-P. Hong, C.-S. Park, Design of an interior permanent magnet synchronous in-wheel for electric vehicles, in: 2010 International Conference in Electrical Machines and Systems (ICEMS), 2010, pp. 1226–1229.
[15] W. Zhuang, et al., A survey of powertrain configuration studies on hybrid electric vehicles, Appl. Energy 262 (2020) 114553, https://doi.org/10.1016/J.APENERGY.2020.114553.
[16] F. Zhang, X. Hu, R. Langari, D. Cao, Energy management strategies of connected HEVs and PHEVs: recent progress and outlook, Prog. Energy Combust. Sci. 73 (2019) 235–256, https://doi.org/10.1016/J.PECS.2019.04.002.
[17] B. Tanç, H.T. Arat, E. Baltacıoğlu, K. Aydın, Overview of the next quarter century vision of hydrogen fuel cell electric vehicles, Int. J. Hydrog. Energy 44 (20) (2019) 10120–10128, https://doi.org/10.1016/J.IJHYDENE.2018.10.112.
[18] A. Misra, Energy storage for electrified aircraft: the need for better batteries, fuel cells, and supercapacitors, IEEE Electrif. Mag. 6 (3) (2018) 54–61, https://doi.org/10.1109/MELE.2018.2849922.
[19] H. Fathabadi, Fuel cell hybrid electric vehicle (FCHEV): novel fuel cell/SC hybrid power generation system, Energy Convers. Manag. 156 (2018) 192–201, https://doi.org/10.1016/J.ENCONMAN.2017.11.001.
[20] D.S. Cardoso, P.O. Fael, A. Espírito-Santo, A review of micro and mild hybrid systems, Energy Rep. 6 (2020) 385–390, https://doi.org/10.1016/J.EGYR.2019.08.077.
[21] Y. Tang, J.J.H. Paulides, I.J.M. Besselink, F. Gardner, E.A. Lomonova, Indirect drive in-wheel system for HEV/EV traction, in: 2013 World Electr. Veh. Symp. Exhib. EVS 2014, 2014, https://doi.org/10.1109/EVS.2013.6915011.

[22] D. Lanzarotto, M. Marchesoni, M. Passalacqua, A.P. Prato, M. Repetto, Overview of different hybrid vehicle architectures, IFAC Pap. OnLine 51 (9) (2018) 218–222, https://doi.org/10.1016/J.IFACOL.2018.07.036.

[23] L. Li, S. You, X. Wang, C. Yang, Structure optimization and generalized dynamics control of hybrid electric vehicles, in: Model. Dyn. Control Electrified Veh, 2018, pp. 207–244, https://doi.org/10.1016/B978-0-12-812786-5.00006-9.

[24] M. Ehsani, Conventional fuel/hybrid electric vehicles, in: Altern. Fuels Adv. Veh. Technol. Improv. Environ. Perform. Towar. Zero Carbon Transp, 2014, pp. 632–654, https://doi.org/10.1533/9780857097422.3.632.

[25] A. Pam, A. Bouscayrol, P. Fiani, F. Faval, P. Barrade, Integration of the road slope in the optimization of the energy management strategy of a parallel HEV, IFAC Pap. OnLine 52 (5) (2019) 28–33, https://doi.org/10.1016/J.IFACOL.2019.09.005.

[26] D. Maamria, A. Sciarretta, F. Chaplais, N. Petit, Online energy management system (EMS) including engine and catalyst temperatures for a parallel HEV, IFAC Pap. OnLine 50 (1) (2017) 8913–8920, https://doi.org/10.1016/J.IFACOL.2017.08.1291.

[27] T. El Hajji, B. Kabalan, Y. Cheng, E. Vinot, C. Dumand, Sensitivity analysis on the sizing parameters of a series-parallel HEV, IFAC Pap. OnLine 52 (5) (2019) 405–410, https://doi.org/10.1016/J.IFACOL.2019.09.065.

[28] K. Akatsu, N. Matsui, New trend of motor technology for automobiles—introduction and overview, in: 2013 IEEE ECCE Asia Downunder—5th IEEE Annu. Int. Energy Convers. Congr. Exhib. IEEE ECCE Asia 2013, 2013, pp. 130–135, https://doi.org/10.1109/ECCE-ASIA.2013.6579085.

[29] K.T. Chau, W. Li, Overview of electric machines for electric and hybrid vehicles, Int. J. Veh. Des. 64 (2014) 46–71.

[30] S. Madichetty, S. Mishra, M. Basu, New trends in electric motors and selection for electric vehicle propulsion systems, IET Electr. Syst. Transp. 11 (3) (2021) 186–199, https://doi.org/10.1049/ELS2.12018.

[31] P. Bhatt, H. Mehar, M. Sahajwani, Electrical motors for electric vehicle—a comparative study, SSRN Electron. J. (2019), https://doi.org/10.2139/SSRN.3364887.

[32] H.F. Prasetyo, A.S. Rohman, F.I. Hariadi, H. Hindersah, Controls of BLDC Motors in Electric Vehicle Testing Simulator, 2017, pp. 173–178, https://doi.org/10.1109/ICSENGT.2016.7849645.

[33] P. Lumyong, P. Sarikprueck, A study on induction motor efficiency improvement for implementing in electric vehicle, in: ICEMS 2018—2018 21st Int. Conf. Electr. Mach. Syst, 2018, pp. 616–619, https://doi.org/10.23919/ICEMS.2018.8549478.

[34] T. Rampradesh, S. Harini, C. Christober Asir Rajan, Design and implementation of efficient torque control of induction motor for electric vehicles, in: Proc. Int. Conf. Artif. Intell. Smart Syst. ICAIS 2021, 2021, pp. 1571–1575, https://doi.org/10.1109/ICAIS50930.2021.9395932.

[35] I. Husain, Electric and Hybrid Vehicles, CRC Press, Boca Raton/London/New York/Washington, DC, 2011.

[36] C. Mi, M.A. Masrur, D.W. Gao, Hybrid Electric Vehicles: Principles and Applications with Practical Perspectives, first ed., A John Wiley & Sons, Ltd., United Kingdom, 2011.

[37] P. Li, W. Ding, G. Liu, Sensitivity analysis and design of a high performance permanent-magnet-assisted synchronous reluctance motor for EV application, in: 2018 IEEE Transp. Electrif. Conf. Expo, ITEC 2018, 2018, pp. 477–481, https://doi.org/10.1109/ITEC.2018.8450176.

[38] S. Panda, R.K. Keshri, Evaluation of permanent magnet assisted synchronous reluctance motor for light electric vehicle applications, in: 2020 IEEE Int. Conf. Power Electron. Smart Grid Renew. Energy, PESGRE 2020, 2020, https://doi.org/10.1109/PESGRE45664.2020.9070410.

[39] O. Zayed, M. Omar, M. Bakr, M. Narimani, A. Emadi, B. Bilgin, Switched reluctance motor design for an EV propulsion application, in: IECON Proc. Industrial Electron. Conf., vol. 2021-October, 2021, https://doi.org/10.1109/IECON48115.2021.9589953.

[40] Y. Lan, et al., Switched reluctance motors and drive systems for electric vehicle powertrains: state of the art analysis and future trends, Energies 14 (8) (2021) 2079, https://doi.org/10.3390/EN14082079.

[41] J. Reimers, L. Dorn-Gomba, C. Mak, A. Emadi, Automotive traction inverters: current status and future trends, IEEE Trans. Veh. Technol. 68 (4) (2019) 3337–3350, https://doi.org/10.1109/TVT.2019.2897899.

[42] M. Pahlevani, P.K. Jain, Soft-switching power electronics technology for electric vehicles: a technology review, IEEE J. Emerg. Sel. Top. Ind. Electron. 1 (1) (2020) 80–90, https://doi.org/10.1109/JESTIE.2020.2999590.

[43] I. Aretxabaleta, I.M. De Alegria, J. Andreu, I. Kortabarria, E. Robles, High-voltage stations for electric vehicle fast-charging: trends, standards, charging modes and comparison of Unity power-factor rectifiers, IEEE Access 9 (2021) 102177–102194, https://doi.org/10.1109/ACCESS.2021.3093696.

[44] M. Anwar, M.K. Alam, S.E. Gleason, J. Setting, Traction power inverter design for EV and HEV applications at general motors: a review, in: 2019 IEEE Energy Convers. Congr. Expo. ECCE 2019, 2019, pp. 6346–6351, https://doi.org/10.1109/ECCE.2019.8913098.

[45] Z. Wu, G.-J. Su, High-performance permanent magnet machine drive for electric vehicle applications using a current source inverter, in: 2008 34th Annual Conference of IEEE Industrial Electronics, Nov, 2008, pp. 2812–2817, https://doi.org/10.1109/IECON.2008.4758404.

[46] R. Hariri, F. Sebaaly, H.Y. Kanaan, A review on modular multilevel converters in electric vehicles, in: IECON Proc. Industrial Electron. Conf., vol. 2020-October, 2020, p. 4987, https://doi.org/10.1109/IECON43393.2020.9255037.

[47] M. Al Sakka, J. Van Mierlo, H. Gualous, U. Brussel, S. Soylu (Eds.), DC/DC converters for electric vehicles, in: Electric Vehicles—Modelling and Simulations, vol. c, InTech, 2011, p. 466.

[48] N.A. Elsonbaty, M.A. Enany, M.I. Hassanin, An efficient vector control policy for EV-hybrid excited permanent-magnet synchronous motor, World Electr. Veh. J. 11 (2) (2020) 42, https://doi.org/10.3390/WEVJ11020042.

[49] S.M. Liu, C.H. Tu, C.L. Lin, V.T. Liu, Field-oriented driving/braking control for electric vehicles, Electron 9 (9) (2020) 1484, https://doi.org/10.3390/ELECTRONICS9091484.

[50] M.S. Sepeeh, et al., A comprehensive review of field-oriented control in sensorless control techniques for electric vehicle, Int. Rev. Electr. Eng. 13 (6) (2018) 461–475, https://doi.org/10.15866/IREE.V13I6.15536.

[51] R. Morales-Caporal, M.E. Leal-Lopez, J. De Jesus Rangel-Magdaleno, O. Sandre-Hernandez, I. Cruz-Vega, Direct torque control of a PMSM-drive for electric vehicle applications, in: 2018 28th Int. Conf. Electron. Commun. Comput. CONIELECOMP 2018, vol. 2018-January, 2018, pp. 232–237, https://doi.org/10.1109/CONIELECOMP.2018.8327204.

[52] M. Aktas, K. Awaili, M. Ehsani, A. Arisoy, Direct torque control versus indirect field-oriented control of induction motors for electric vehicle applications, Eng. Sci. Technol. Int. J. 23 (5) (2020) 1134–1143, https://doi.org/10.1016/J.JESTCH.2020.04.002.

[53] F. Ben Salem, DTC-SVM approaches of an induction motor dedicated to position Control applications, in: Direct Torque Control Strateg. Electr. Mach., Nov, 2020, https://doi.org/10.5772/INTECHOPEN.94436.
[54] M. Abassi, A. Khlaief, O. Saadaoui, A. Chaari, M. Boussak, Performance analysis of FOC and DTC for PMSM drives using SVPWM technique, in: 16th Int. Conf. Sci. Tech. Autom. Control Comput. Eng. STA 2015, 2016, pp. 228–233, https://doi.org/10.1109/STA.2015.7505167.
[55] L. Zheng, Y. Ren, Q. Huang, Y. Li, Z. Zhan, The investigation of control strategies of switched reluctance motor to reduce the torque ripple in vehicle, in: SAE Tech. Pap., vol. 2015-April, 2015, https://doi.org/10.4271/2015-01-1218.
[56] K.B. Hunasikatti, R.L. Naik, V.D. Nayak, Critical review of control strategies for switched reluctance motor employed in electric vehicle, in: Proc. 2018 2nd Int. Conf. Adv. Electron. Comput. Commun. ICAECC 2018, 2018, https://doi.org/10.1109/ICAECC.2018.8479510.

C H A P T E R

11

Power electronics controlled electric propulsion systems

Ayse Kocalmis Bilhan and Ersan Kabalci

Electrical and Electronics Engineering Department, Faculty of Engineering and Architecture, Nevsehir Haci Bektas Veli University, Nevsehir, Turkey

List of symbols

B_m	friction coefficients
$\psi_{d,q}$	*d*- and *q*-axis stator flux linkage
ψ_f	permanent magnet flux linkage
$i_{sd,q}$	*d*- and *q*-axis stator currents
J_m	the inertia of rotor
$L_{d,q}$	*d*- and *q*-axis stator inductance
P_{AG}	air gap power
P_I	electrical input power
P_M	mechanical power
P_O	mechanical output power
R_r	rotor resistance
R_s	stator resistance
T_e^*	reference electromagnetic torque
$V_{sd,q}$	*d*- and *q*-axis stator voltages
Vs	space vectors of stator voltage
X_{LS}	stator inductance
ω_e	synchronous frequency
ω_m	the angular velocity
AV	airway vehicles
BEV	battery electric vehicles
EV	electrical vehicles
FCEV	fuel cell electric vehicles
HEV	hybrid electric vehicles
ICE	internal combustion engine
MV	marine vehicles
PHEV	plug-in hybrid electric vehicles
RV	railway vehicles

11.1 Introduction

Electrical propulsion systems for electric vehicles (EVs), marine vehicles (MVs), railway vehicles (RVs), and airway vehicles (AVs) are a combination of different technologies, which include multiple engineering fields such as mechanical, electrical, automotive, chemical engineering, and electronics. By combining different technologies, the overall efficiency and fuel consumption of these vehicles are demanded to be reduced, which makes them more efficient in comparison with conventional vehicles [1,2]. These vehicles typically include three main parts: motors, energy storage and production, and power electronic devices. People are increasingly turning to nonpolluting or renewable energy sources as energy demand continues to rise. Power electronic switches and devices have grown in importance as a result of these methodologies [3]. In addition to converting electrical power to a different level or form, power

https://doi.org/10.1016/B978-0-323-99545-0.00019-1

electronic systems are also used to link the power source to the load or network. The most well-known power electronic devices are rectifiers, inverters, and power converters. A rectifier is a device that transforms AC voltage to DC. Today, it is utilized in practically all household appliances and electrical motor power sources. An inverter is a device that converts DC voltage to AC voltage, while a power converter transforms AC to AC or DC to DC at variable values. The distributed power networks, practically all motor control circuits, and a lot more applications utilize rectifiers. On the other hand, a DC-DC converter clamps a variable or unregulated input voltage to the desired output voltage level.

In order to meet the ever-increasing vehicular load demands, the automotive industry is clearly trending toward the use of more electrical systems. As a result, it is obvious that automotive electrical power systems will undergo significant change during the next couple of decades. The current state of the automotive industry is such that the need for greater fuel efficiency and more electric power is pushing the voltages of advanced vehicular power systems to higher levels. Because of the high voltage levels produced by HEVs, DC/DC converters are required to supply all auxiliary loads on board the vehicle. Although DC/DC converter technology is well developed for low-power, low-cost applications, much work remains to be done for high-power applications. The requirements for packaging, reliability, and electromagnetic compatibility (EMC), as well as all vehicle standards for EMI and EMC, are extremely difficult to meet. Furthermore, power electronic converters control when and how fuel or electricity is used in HEVs. A suitable DC/AC inverter that draws DC power from the batteries powers the electric traction motor, which controls the wheels. Vehicle battery packs are composed of a number of series- and parallel-connected battery cells and can thus be designed for practically any voltage. The majority of production BEVs up to this point have used traction inverters with 600 V rated IGBT modules, which has kept the battery pack voltage to a maximum of about 400 V. As a result, BEVs have typically used around 96 series-connected battery cells, resulting in a peak bus voltage of 403 V when using 4.2 V peak Li-ion cells. Such battery packs have previously worked well for vehicles, but the charging power is limited by the maximum current for DC fast charger cables, which motivates the use of 800-V battery packs [4].

11.2 Electrical vehicles

Over the last decade, intensive use of fossil fuels has already resulted in global environmental issues such as high heat and high hydrocarbon, nitrogen oxide, and carbon monoxide emissions, among other things [5]. Furthermore, the price of oil has risen significantly in recent years as a result of the growing disparity between available fossil fuel supplies and global energy demand. All of these issues, combined with rising demand for more environmentally friendly and fuel-efficient vehicles, have prompted auto manufacturers to search for viable alternatives to internal combustion engine (ICE) vehicles. The alternatives of ICEs are classified as electric vehicles (EVs), hybrid electric vehicles (HEVs), plug-in hybrid electric vehicles (PHEVs), and fuel cell vehicles (FCEVs). These vehicles would also be able to fulfill the rising need for electrical power as a result of the increased usage of electronic systems to improve vehicle performance, fuel economy, emissions, passenger comfort, and safety. The problems in EVs, HEVs, PHEVs, and FCEVs are to achieve high efficiency, ruggedness, compact design, and low costs in power converters and electric machines, as well as accompanying electronics. While different EV driving systems have distinct system designs and charging mechanisms for various energy sources, energy flow in EVs is carried out by flexible electrical lines with no mechanical connections [6]. Fig. 1 depicts a thorough layout of an electric vehicle system and its connections with various components. The fundamental components of an electric vehicle system are the motor, controller, power source, and transmission system.

The battery energy sources and battery chargers are the main supply units of EVs to feed the propulsion and traction systems. The power electronic converter block includes any or both the DC/DC converter and the DC/AC inverter regarding the motor type used in the traction system.

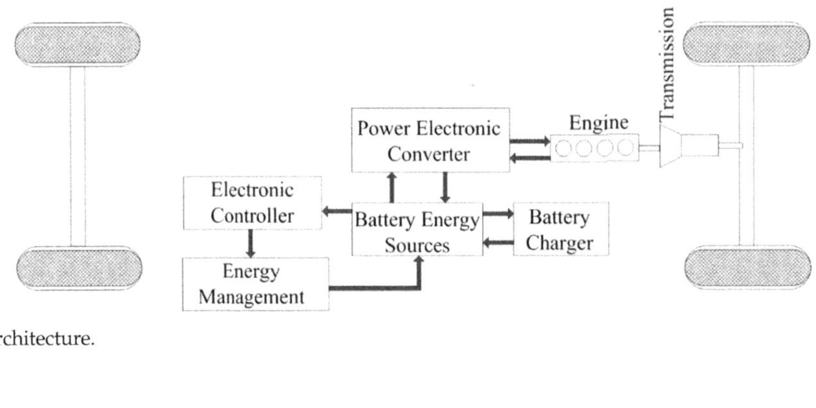

FIG. 1 Vehicle power architecture.

11.2.1 Hybrid electrical vehicles

In the early years of the 20th century, thousands of hybrid and electric vehicles were built. However, it was not in high demand among consumers due to their poor performance in those years. Today, the situation is very different, and electric and hybrid vehicles are becoming increasingly popular. HEVs' propulsion systems include both an electric motor and an ICE. The energy system is powered by a battery/ultracapacitor and a fuel tank. The benefits of HEVs include high fuel efficiency, very low emissions, high-tech components, and a long driving range. Despite these benefits, battery sizing and management issues, as well as control and optimization issues caused by multiple energy sources, are regarded as major issues. HEVs are propelled by connecting an ICE and an electric motor in series, parallel, or in series/parallel configurations. The electric motor regenerates energy during braking and increases efficiency by contributing to fuel economy. All the while, the ICE delivers the vehicle a longer driving range [7].

In general, there are three basic categories of HEVs, namely series hybrids, parallel hybrids, and series/parallel hybrids [8]. Also, PHEVs and battery electric vehicles (BEVs) have a large market share. A typical configuration of a series HEV drive train layout is shown in Fig. 2. The ICE mechanical output is first converted into electricity using a generator. This power is used to charge the batteries or drive an electric motor that provides the final power to the vehicle wheels [9]. The efficiency of series HEVs is very low, because there are three propulsion systems (ICE, generator, and electric motor). Also, multiple conversion stages increase the cost and reduce efficiency. However, the ICE works with maximum efficiency because there is no mechanical connection between the ICE and the gearbox [10].

The basic power electronic components for a series HEV include an AC-DC converter for converting the generator output to DC for charging the batteries or propulsion on wheels, as well as a DC-AC converter (an inverter) for the propulsion motor. A DC-DC converter is also used to charge a 12-V battery, which is used to control other high-tech electronic equipment in the vehicle. There are a couple of operation modes available for series HEVs:

- Battery-fed operation: When the ICE is off, the battery is the only source of power for the car,
- ICE-fed operation: The vehicle is powered by the ICE and generator if the battery is turned off,
- Dual-fed operation: When both the ICE and the battery are turned on, power is generated by both.

Also, regenerative braking mode and stationary charging mode are counted as the operation mode, too. In a parallel HEV, both the ICE and electric motor transmit power in parallel to drive the wheels. The parallel HEV can be considered an electrically assisted ICE to decrease the fuel consumption and carbon emission. There are several configuration options for the ICE and traction motor in a parallel HEV. The most common technique is to utilize the motor alone at low speeds because it is more effective than the ICE, and allow the ICE to operate alone at higher speeds. When only the ICE is used, the traction motor can serve as a generator, to charge the battery. In addition to a fixed-step transmission, a parallel HEV can include a continuously variable transmission (CVT).

In comparison with a series HEV, the parallel HEV architecture has the benefit of having fewer energy conversion steps, which results in reduced energy loss. In addition, when compared to other HEV architectures, the parallel HEV drive train has fewer losses and hence a greater total drive train efficiency [8]. A typical configuration of a parallel HEV drive train layout is shown in Fig. 3.

In contrast to series HEVs, parallel HEVs can offer lower costs and different design capabilities for engines, batteries, and motors. On the other hand, a parallel hybrid vehicle requires a sophisticated control system [11]. In parallel HEVs, the ICE and the electric motor can be used separately or together to propel the vehicle and there are a couple of operation modes similar to series HEVs:

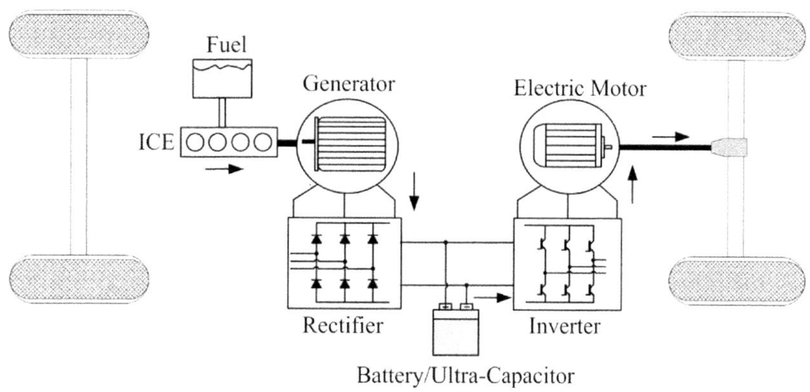

FIG. 2 A series HEV drive train layout.

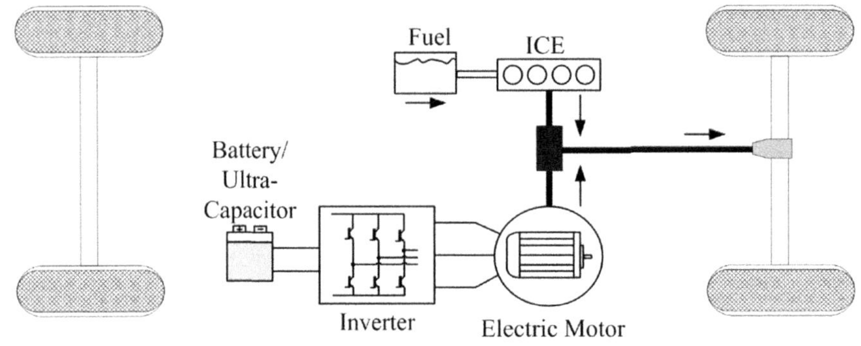

FIG. 3 A parallel HEV drive train layout.

- If ICE is off, then the vehicle is only powered by the electric motor.
- If battery is off, then ICE provides power for the vehicle.
- The power is provided by both ICE and battery to the vehicle.
- The motor transforms into a generator, and the ICE powers the vehicle and charges the battery.

In addition to these modes, the regenerative braking mode and stationary charging mode are also considered operation modes.

A series-parallel HEV is also known as dual-mode HEV or power-split HEV [1]. A typical configuration of a series-parallel HEV drive train layout is shown in Fig. 4. The structure of a series-parallel HEV consists of both series and parallel HEVs but involves additional mechanical components such as the additional link between the generator and electric motor different from series HEVs and the additional generator different from parallel HEVs [8]. The main advantage of the system is that it can operate in both series and parallel modes. However, it is very complicated and expensive. In series-parallel HEVs, there are various operation modes.

The series configuration can be active or the parallel configuration can be active; besides, the electric motor can be more active than the ICE, which is called an electric-intensive configuration for propulsion, or the ICE can be more active, which is called an engine-intensive configuration. During normal driving, the ICE alone propels the vehicle, while the ICE and electric motor propel the vehicle in the electric-intensive configuration. When acceleration is needed, the electric traction motor is used in combination with the ICE to give extra power in both configurations. During braking or deceleration, the electric motor is used as a generator to charge the battery [8,12]. A PHEV can have series, parallel, and series-parallel configurations as well as HEV [13]. A PHEV has different operating modes depending on their configuration which may be one of series, parallel, or series-parallel. In Fig. 5, a plug-in series HEV drive train layout is shown. A plug-in HEV is promising in terms of fuel economy because it has an option to recharge its energy storage system (ESS) with electricity from the grid [14]. From this point of view, while working with the battery, it does not emit exhaust emissions, while various polluting gases are released while generating electricity in the power plant.

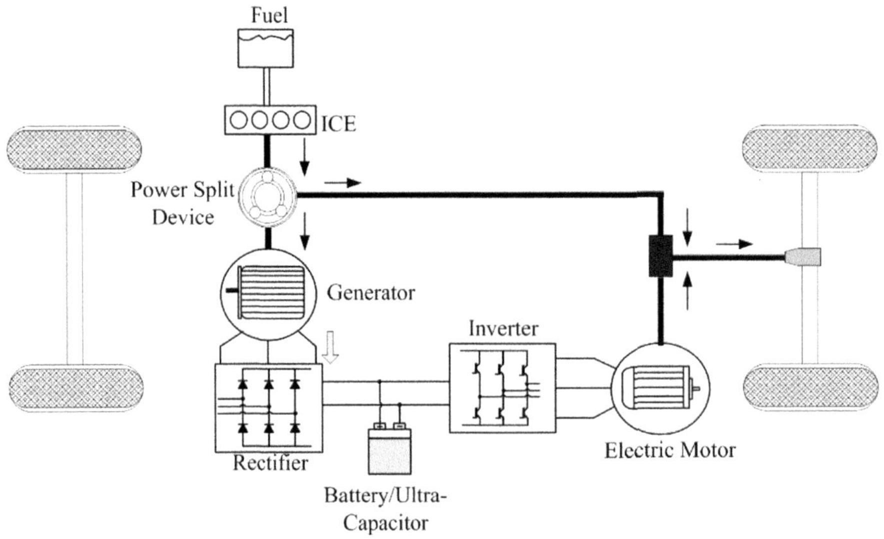

FIG. 4 A series/parallel HEV drive train layout.

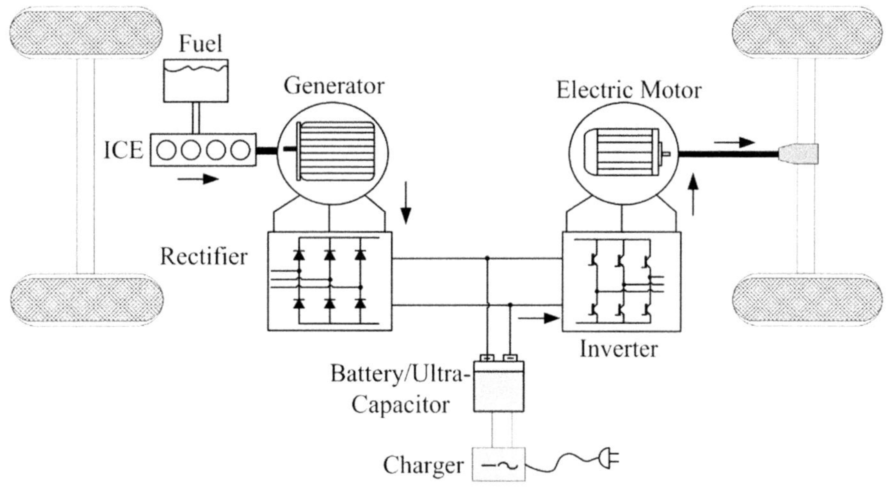

FIG. 5 A plug-in series HEV drive train layout.

The ESSs such as battery can be charged at home using an ordinary plug or at a charging station or at work. PHEV owners often use utility power to recharge their batteries overnight [15]. For short distances, mostly electric motors are used, while long distances also use fuel power when battery/ultracapacitor power is consumed [15]. The driving range can be extended by charging the batteries/ultracapacitors from the ICE. Recently, auto companies and researchers have paid more attention to PHEVs, but ESS size, cost, and battery life turned out to be the main issues to overcome.

In recent years, BEVs become more popular. BEVs have more battery capacity than HEVs and have no ICE. Thus, the largest range with the lowest local emissions can be reached without burning any fuel. It can be considered as one of its biggest advantages because it eliminates one of the environmental concerns. In Fig. 6, a battery electric vehicle (BES) drive train layout is shown.

BEVs are called pure EVs because there is no fuel used and all electricity is provided by the battery [1]. However, despite providing all the electricity needed for driving from batteries, its driving range is less than that of other EVs. The batteries can mainly be charged with a plug-in charger, and the electricity can be obtained by using wind, sun, water, or other renewable energy sources [16]. In addition, charging time for BEVs is longer than for other EVs. For this reason, it must be given attention to charging time in use. Despite the various EVs, fuel cell electric vehicles (FCEVs) have been considered as future EVs [17]. A FC is basically a device that converts chemical energy directly from an oxygen and hydrogen reaction into electrical energy, releasing water and heat. Thus, it is a popular energy source because it enables zero-emission transportation (only water and heat), providing quietness of operation and reliability.

Many automobile manufacturers and researchers consider that BEVs for short-range small vehicles and fuel-cell hybrid electric vehicles (FCHEVs) for long-range large vehicles will be widely used in the future. Although FCEVs are similar to BEVs in their features such as power controllers and electric motors, they have a big difference in terms of main energy sources. FCEVs use fuel cells, which have better features than batteries in many ways. However, FCEVs can also be equipped with batteries or supercapacitors for the storage of regenerative braking energy, which is generated during slowdown and braking or for other electronic equipment in the car. This topology is called as fuel cell hybrid electric vehicle (FCHEV) that is shown with the drive train in Fig. 7 [18].

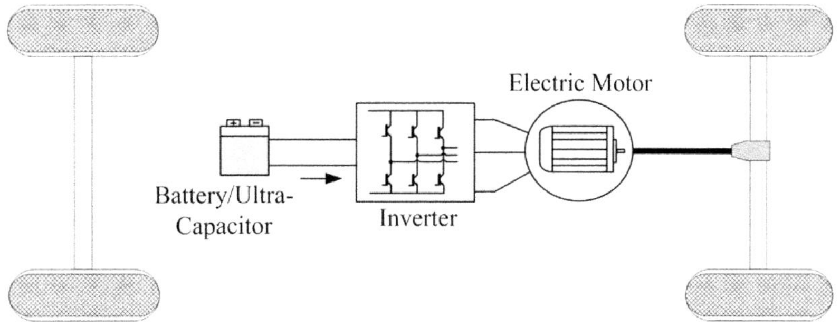

FIG. 6 A battery electric vehicle (BEV) drive train layout.

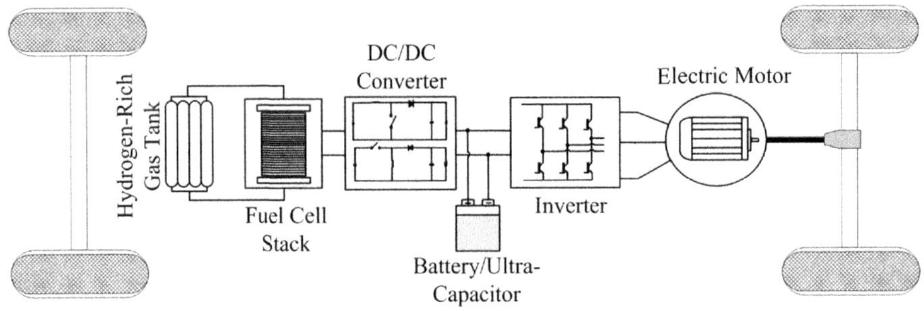

FIG. 7 A fuel cell electric vehicle (FCEV) drive train layout.

Thus, in a FCHEV, FC is the primary energy source to provide average output and the battery/ultra-capacitor recovers regenerative braking energy and provides peak power [19]. According to the connection method of the fuel cell and the battery/ultracapacitor, they can be divided into several topologies such as the parallel connection of dual-energy sources or indirect parallel connection of dual-energy sources. In Fig. 7, the direct parallel connection of the battery is shown.

11.2.2 Summary

The limited fossil fuel resources, the environmental concerns, the increasing population, people's needs, and other factors lie behind the increasing interest in clean and sustainable vehicles, especially in the transportation sector. Due to many reasons, developing countries, automobile manufacturers, and researchers are focusing mainly on EVs and developing new and alternative vehicles to replace the ICE with HEVs. Today, HEV and its topologies, BEV and FCEV, do not have a significant market share even though they are receiving a lot of attention in the automobile industry. However, it is predicted that they will gain popularity around the market due to increasing fuel prices in the world and important developments in EV management systems.

11.3 Power electronics and propulsion systems of electric vehicles

The electrical system in ICE vehicles is designed to meet the needs of lighting, electric motors for fans and pumps, and other internal or external security and safety appliances. According to [8], the 6-V electrical system of ICE vehicles was converted to a 12-V supply system in the mid-1950s. The new supply system is defined as a14-V system due to the distribution level along the electrical devices of the vehicle. However, the 14-V electrical system of any vehicle is not capable to meet the requirements of HEVs and EVs. In an EV, the traction motors perform the most crucial function of the power system due to the dramatically increased power rates and control capability required. The increase in operating voltage leveraged the use of power electronics in EVs since DC and AC power sources and load types are accommodated in the system. The high reliability and security standards defined for EVs leveraged the device topology and control methods of power electronic systems [8].

The advancements in power electronics have improved EV propulsion technologies by improving power control and motor drive methods. The power electronic devices and electric motor drive in an EV enable the bidirectional control of the energy flow between the vehicle and the utility grid. Energy flows from vehicle to grid (V2G) or grid to vehicle (G2V) owing to the power electronic converters managing the exchange. In addition to the conventional electric appliances of an ICE vehicle, EVs include many power electronics systems for power control, multimedia, safety, entertainment, and security purposes [20,21]. The main contribution of power electronics to EVs is the improvement of electric propulsion and battery management systems (BMSs). On the other hand, the electric motor that drives an EV's electric propulsion system operates with a rapidly changing and reacting profile. The frequent start-stop and rapid acceleration demands of an EV increase the requirements of high torque per inertia, higher power density versus mass, high speed, and increased efficiency compared to conventional ICE vehicles.

The electric motors of an EV are mostly one of permanent magnet synchronous motors (PMSMs), AC synchronous motors (SMs), or AC induction motors (ACIMs). The common inverter topology used to drive the electric motor is a two-level inverter while the novel models include multilevel topologies. Although the two-level topology is the most commonly used one, inverter topology of an EV should be improved and controlled to meet the specific requirements of the traction and propulsion systems. The increased switching losses and low efficiency of two-level inverter

TABLE 1 Comparison of the most recent EVs available in the market.

	Kia Niro [26]	Nissan Leaf S [27]	BMW i3 [28]	Tesla Model S [29]	Hyundai Kona [30]
Electric motor type	PMSM	SM	SM	ACIM	PMSM
Electric motor power (kW)	150	110	125	580	204
Electric motor torque (Nm)	395	320	250	1140	395
Top speed (km/h)	167	145	150	250	167
Battery type	Lithium-ion	Lithium-ion	Lithium-ion	Lithium-ion	Lithium-ion
Battery capacity (kWh)	64	40	42	100	64
Voltage (V)	356	360	352	350	356
Battery charging time (h) (% 80)	1.25	0.67	0.7	0.5	1.25
Total range (km)	384.55	239.74	335	560	449

topologies cause several drawbacks in the propulsion control. The difficulties of increased dv/dt ratio, high value of electromagnetic inference (EMI) noise, and low power quality due to a high value of total harmonic distortion (THD) in two-level inverters accelerated the widespread use of multilevel inverters (MLIs). The neutral point clamped (NPC) and its varieties are the most widely used MLI topology in motor drive of EVs [22–24]. Table 1 presents performance details and comparisons including electric motor type, motor power ratings, top speed, battery types, battery voltages, battery charging time up to 80% charge, and range data. The motor types are listed as PMSMs, SMs, and ACIMs while the speed and torque values vary in a range scale [25]. The features of electric motors and power electronic devices used to drive the motors are introduced in the following sections.

11.3.1 Electric motors used in EVs

The electric machine technology has been improved rapidly due to its use in EVs and HEVs. The ACIM used in Tesla Roadster has a peak power density more than two times that used in General Motor EV1, which was produced in the late 1990s. The most important performance indicators of an electric motor used in EVs are high values of efficiency, torque, power density, and ease of field weakening operation. It is noted in [31] that Toyota Prius, Ford Escape, and Chevy Volt are some of the EVs using interior permanent magnet SM (IPM) due to these advantages. Although the IPM provides good opportunities for EVs, the increased cost of high-quality magnets forces researchers to obtain the performance of the IPM without using permanent magnets and benefiting from other motor technologies. The ACIM, switched reluctance machine (SRM), and synchronous reluctance motors (SynRMs) are some candidates to replace the IPM in terms of performance, high efficiency, and low cost issues [31]. The electric motors used in EVs are mostly PMSMs and ACIMs as stated in Table 1. The traction controllers of such motors are based on various inverter topologies that convert the DC voltage of the batteries to AC for speed and torque control purposes.

The regenerative braking operation that enables motors to be operated in generator mode is also performed with the bidirectional power flow capability of an inverter. The energy density of inverters is accepted as a performance indicator in EV applications where a 30 kW continuous rated power inverter is expected to present a power density of 13.4 kW/L and a specific density of 14.1 kW/kg as reported in [32]. It is also noted that traction inverters with 55 kW peak power rates of 15 kW/L and 17 kW/kg are revealed in the field. Although the variable-speed motors draw reliable characteristics in terms of speed and power, the control method of the inverter should ensure the balance between power and lifetime of the battery. The rated power of an EV varies from a few kilowatts to 200 kW for high-performance vehicles. The comparison of motors used in any EV shows that PMSMs present higher efficiency in overload and constant-speed applications. The reluctance motors (RMs) that are also seen in some EVs provide higher performance under high-speed conditions [33]. However, PMSMs benefit from continuous torque, a high-power density, lower energy consumption, and improved efficiency, making it easier to employ EVs on a broad scale. ACIMs may function in any hostile field without encountering any speed constraints due to its high starting torque, improved efficiency, and robust power profile ensured by its simple construction and high reliability. However, the control system of ACIM is quite complicated, and the power density issue persists. The energy efficiency of an ACIM is defined by the overall quantity of losses including iron losses, copper losses, controller losses, and mechanical losses [34].

Another motor type used in EVs is switched reluctance motor (SRM), which has the most basic configuration compared to the aforementioned motors. The SRM is not based on permanent magnets, which makes it low cost compared to PMSMs, but the stator winding creates the electromagnetic force. Although the SRM suffers from acoustic and electromagnetic interference (EMI) noise, high ripple of torque, and featured driver requirement, it is a candidate for EV applications. In the following, the major motor topologies are discussed in terms of rotor and stator topology. The major motor topologies are covered in terms of rotor and stator topology in the following sections [34].

11.3.1.1 AC induction motors

The IMs are widely used in industrial applications due to their low cost, simple structure, high efficiency, and robust control features. In addition to industrial drivers, IMs are also used in the traction system of EVs such as General Motors EV1, Tesla Roadster and Model S, and Toyota RAV4 [31]. The operation regions of an IM used in EVs are defined into three areas such as constant torque, constant power, and reduced power regions as shown in Fig. 8. The characteristic speed-torque curve and power rates are in accordance with the requirements of an electrical propulsion system, and the power electronics and control systems of the machine driver are improved considering the operation regions [35]. The single-phase electrical equivalent circuit of an IM is seen in Fig. 9 where the stator and rotor circuits are depicted with the inductance and resistance of stator and rotor windings. The equivalent stator circuit is seen on the left hand side up to $R_m//X_m$ magnetizing circuit of Fig. 9. The stator circuit is comprised of stator resistance (R_S) and stator inductance (X_{LS}). The core losses of stator and rotor are depicted in the magnetizing circuit. The power flow diagram of an IM model is shown in Fig. 10 where the electrical input power (P_I), air gap power (P_{AG}), mechanical power (P_M), and mechanical output power (P_O) are illustrated together with copper and rotating losses.

The parameter m denotes the number of phases in the IM, while T is torque and ω_m is the angular velocity. The equivalent rotor circuit is comprised of resistance and inductance of the rotor winding and the transformed rotor resistance against the slip factor. The slip of the rotor that IM suffers is determined by the difference between synchronous speed and the rotor speed. The main losses of an IM are listed as stator losses, rotor losses, mechanical losses denoting

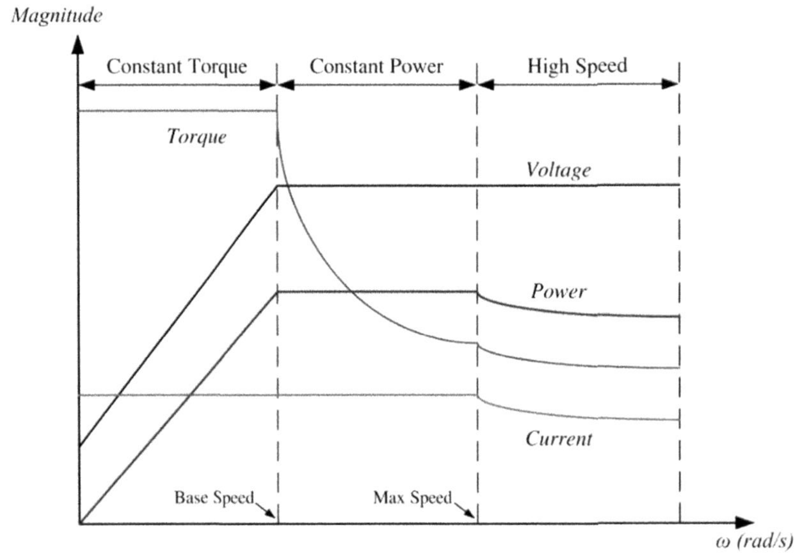

FIG. 8 Operation regions of an IM in an EV.

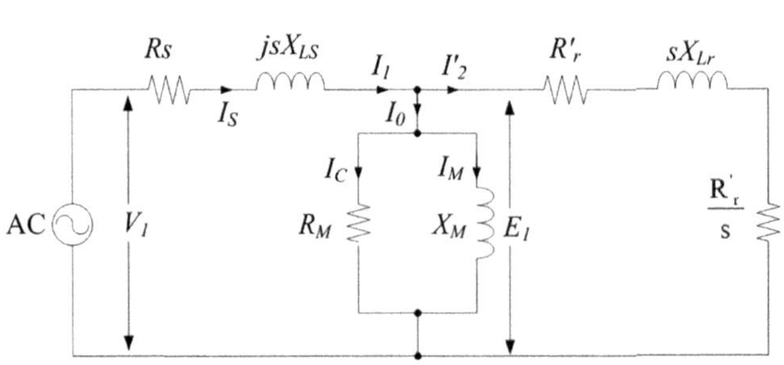

FIG. 9 Electrical equivalent of a single-phase IM in motoring mode.

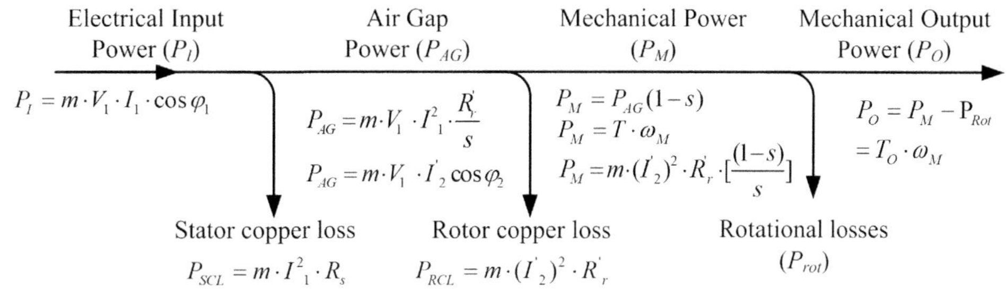

FIG. 10 Power flow chart of an IM.

the rotational losses, and stray losses. The stator losses are comprised of copper losses that are a result of current flowing over stator winding and the resistance of stator winding.

On the other hand, iron losses namely magnetic core losses are caused by eddy current and hysteresis losses. The rotor losses are also comprised of copper and iron losses as well. The mechanical losses are the sum of friction and ventilation losses that depend on the speed and are related to the design structure of the machine. The control methods used in EV motors are based on adjustable speed control (ASD) and direct torque control (DTC) techniques. Although the scalar control method operated with voltage-frequency control is easy to implement, the performance of the controller is not sufficient for many applications. The vector control methods significantly improve the performance and efficiency of IM. The coupling with electromagnetic torque and flux of an IM causes the complexity of improving vector-based controllers for IMs. However, proven controllers such as direct torque control (DTC) and direct and indirect field-oriented control (FOC) techniques are securely used for many years. In the FOC-based controller, the machine parameters are transformed into a coordinate system that synchronously rotates with rotor flux, and it ensures decoupling of flux and torque without the effect of machine nonlinearity. The FOC method decouples stator current into two components of torque and flux. The current measurement required for coordinate transformation is performed in d_s-q_s stationary frame, and the current vectors of I_{sds} and I_{sqs} are converted to d-q rotating frame. The motor model is represented with the following equation in d_s-q_s stationary frame:

$$V_{sds} = R_s I_{sd_s} + \frac{d\psi_{sd_s}}{dt} \tag{1}$$

$$V_{sq_s} = R_s I_{sq_s} + \frac{d\psi_{sq_s}}{dt} \tag{2}$$

$$0 = R_r I_{rd_s} + \frac{d\psi_{sd_s}}{dt} + p\omega_m \psi_{rq_s} \tag{3}$$

$$0 = R_r I_{rq_s} + \frac{d\psi_{sq_s}}{dt} - p\omega_m \psi_{rd_s} \tag{4}$$

$$\psi_{sd_s} = L_s I_{sd_s} + L_M I_{rd_s} \tag{5}$$

$$\psi_{sq_s} = L_s I_{sq_s} + L_M I_{rq_s} \tag{6}$$

$$\psi_{rd_s} = L_r I_{rd_s} + L_M I_{sd_s} \tag{7}$$

$$\psi_{rq_s} = L_r I_{rq_s} + L_M I_{sq_s} \tag{8}$$

$$\frac{d\omega_m}{dt} = \frac{1}{J}\left[p\frac{m_s}{2} I_m\left(\psi_{sd_s} I_{sq_s} - \psi_{sq_s} I_{sd_s}\right) - T_L\right] \tag{9}$$

where V_s denotes space vectors of stator voltage, I_s and I_r are space vectors of stator and rotor currents in d and q frames, R_s and R_r are resistances of stator and rotor, L_s and L_s are self-inductances of stator and rotor, ω_m is the mechanical rotor speed, and ψ_s and ψ_{sr} are space vectors of stator and rotor flux linkage in d and q frames. The adoption of the FOC algorithm ensures that motor parameters are obtained with high reliability in both steady-state and transient operations. The FOC technique, which is based on inheriting rotor flux data, acquires all rotating frame characteristics and converts them to rotor flux. The rotational flux frame parameters at synchronous speed are converted to DC components. When the rotor flux is allocated to the direct axis of reference frame, the *d* parameter denoting flux of stator current and the *q* parameter denoting torque component of stator current are employed. The conducted transformation allows for the decoupling of controlled ACIM flux and torque measurements to a two-axis frame [36,37]. Vector controllers with suitable feedback signals from encoders can be used to operate an induction motor. This technique uses a

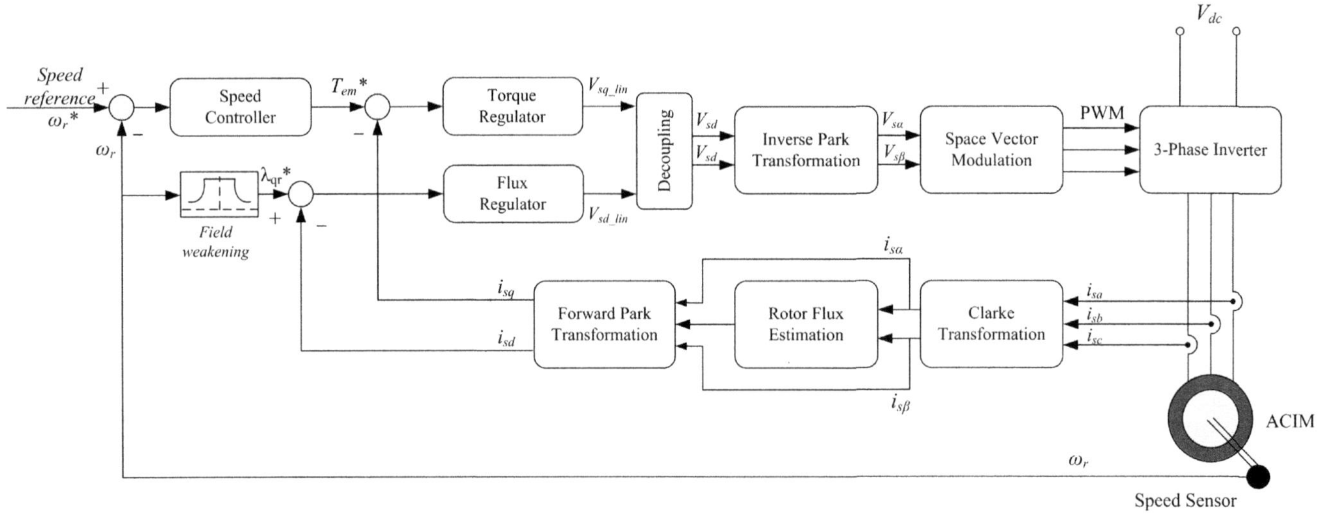

FIG. 11 Block diagram of field-oriented control (FOC).

PWM-synthesized sinusoidal waveform to control speed. High efficiency winding, an effective lamination design, and high-temperature insulating materials are frequently used while designing vector-controlled motors. Vector control using scalar control methods can achieve tighter speed regulation of less than 0.01% of the fixed speed. A vector controller enables configurable speed ranges ranging from zero to five times the fundamental speed. The rotor flux estimator receives input from the stator currents as shown in Fig. 11 where the block diagram of the FOC controller is illustrated. In the estimator or observer application, the rotating coordinate reference currents I_{sdr} and I_{sqr} are calculated using the reference torque and flux values.

The speed controller is based on a PI regulator that regulates the motor slip ratio, with the value computed by the regulator added to the motor speed to generate the demand frequency. The latter frequency is also employed to provide the required voltage in order to maintain the V/f ratio of the motor consistently. A three-phase generator, low-pass filter, α-β transformation, α-β sector, ramp generator, and switching time calculator are all included in the space vector PWM modulator. The modulating signals are provided by a three-phase generator with a 120 degree phase shift, and the α-β transformer creates a three-phase to two-phase conversion for use in sector preparation. The low-pass bus filter is used to reduce sudden transients from the measurement of DC bus voltage [38].

Although the FOC methods are convenient in the control of IMs, they are quite sensitive to parameter fluctuations caused by the inaccurate estimation of flux or speed values. The DTC method is improved to overcome the reference frame transformation requirement of the FOC controller. The reference frame transformation requirement of the FOC controller has been overcome by improving the DTC method.

11.3.1.2 Permanent magnet synchronous motor

PMSMs provide constant magnetization at the rotor with high magnetic field induced in the air gap. The excitation current that holds almost half of the machine losses does not exist in PMSM, and thus it represents very high efficiency due to lack of exciting currents. The disadvantage of this situation is overcome by complex control methods required to regulate the excitation field. Although there are many permanent magnet allocation and geometry possibilities, the most of these machines are implemented in radial or axial flux configurations. The magnets are mounted on the surface of the rotor in an axial-flux machine while they are either surface mounted or internal mounted in radial-flux PMSMs [33]. The magnetomotive force (mmf) of PMSM is sinusoidal and the voltage and current waveforms are as well. The sinusoidal distribution of the air-gap flux and stator winding enables the machine to operate synchronously. The rare earth materials used as magnets are ferrites, samarium cobalt (SmCo), and neodymium-iron-boron (NdFeB), which help increase the flux density in the air gap, the motor power density, and torque-to-inertia and thus can be operated over a wide constant power speed range [6].

The interior PMSM (IPMSM) provides the most efficient configuration, and it is the most preferred motor type in EVs. In addition to the large torque ripple and complex radial force introduced in PMSM due to high air gap density, the difference in d- and q-axis inductances of IPMSM creates reluctance torque and distorts the air gap flux distribution. Therefore, decreasing the torque ripple and vibration of the IPMSM which is to be used in an EV traction system should be ensured with precise design of the motor. The electromagnetic oscillation of IPMSM that is produced by the

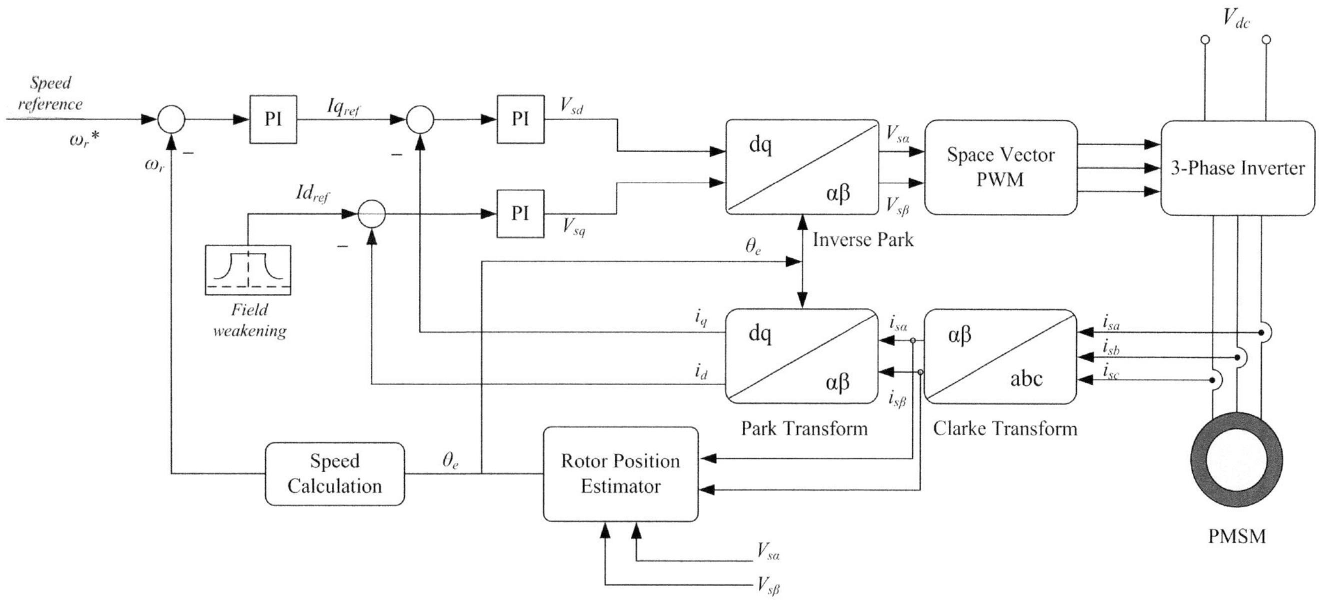

FIG. 12 Block diagram of FOC-controlled PMSM.

excitation force is one of the main reasons for the vibration in the drive system. The torque ripple, power converter switching, and torsional force increase the vibration of the traction system in the EV application of IPMSM [39]. In the rotor flux-oriented control-based PMSM driving system, the stator voltage equation in the synchronous rotating frame can be expressed as follows [40]:

$$\begin{bmatrix} V_{sd} \\ V_{sq} \end{bmatrix} = R_s \begin{bmatrix} i_{sd} \\ i_{sq} \end{bmatrix} + p \begin{bmatrix} \psi_d \\ \psi_q \end{bmatrix} + \omega_e \begin{bmatrix} -\psi_q \\ \psi_d \end{bmatrix} \tag{10}$$

$$\begin{bmatrix} \psi_d \\ \psi_q \end{bmatrix} = \begin{bmatrix} L_d & 0 \\ 0 & L_q \end{bmatrix} \begin{bmatrix} i_{sd} \\ i_{sq} \end{bmatrix} + \begin{bmatrix} \psi_f \\ 0 \end{bmatrix} \tag{11}$$

where $V_{sd,q}$ are *d*- and *q*-axis stator voltages, R_s is the stator resistance, $i_{sd,q}$ are *d*- and *q*-axis stator currents, *p* is the differential operator, $\psi_{d,q}$ are the *d*- and *q*-axis stator flux linkage, ω_e is the synchronous frequency, $L_{d,q}$ are the *d* and *q*-axis stator inductance, and ψ_f is the permanent magnet flux linkage.

The PMSM control methods are mostly based on FOC and DTC methods as introduced for IMs. The block diagram of a FOC-based variable frequency drive of a PMSM is shown in Fig. 12. The controller is implemented with sensorless operation where the motor currents are measured with current sensors and the rotor flux angle is estimated by determining the back EMF instead of using a speed sensor. The measured current signals are transformed from the three-axis reference frame (*abc*) to the two-axis orthogonal reference frame (*αβ*), which is then transformed to the two-axis orthogonal rotating reference frame (*dq*) in the Park Transform block. The Park transformation is driven by the rotor flux estimation.

The measured and estimated *d*- and *q*-axis currents are compared to the desired current values to generate the error signal. The *d*-axis current is used to control the flux, while the *q*-axis reference current is used to control the torque. The PI regulators provide the control mechanism for determining the correction voltage reference regarding the error signal rates. The correction voltages of the *d*- and *q*-axis are transformed to the three-axis reference frame (*abc*) voltages that are applied to the motor terminals by switching modulators.

11.3.1.3 Synchronous reluctance motors

The synchronous reluctance motor (SynRM) is designed with high and low reluctance areas in its magnetic structure, and the operation principle complies with SMs as its name implies. The excited stator windings generate a magnetic field, and the rotor is aligned with its highest magnetically conducting axis, which is *d*-axis. In this situation, the applied field minimized the reluctance of the magnetic circuit. Although the configuration of SynRM is like IM, the rotor of SynRM is not symmetrical that is defined as anisotropic. The simplicity of the construction provides advantages for simple production as in IM. Although the configuration is like PMSMs, the absence of permanent magnets decreases the cost of the machine. Therefore, SynRM provides an efficient and cost-effective alternative to the PMSM and IMs [6,41]. The voltage and flux equations of an SynRM in a d- and q-axis rotor reference frame are given as follows:

$$V_d = R_s I_d + \frac{d\psi_d}{dt} - \omega_r \psi_q \tag{12}$$

$$V_q = R_s I_q + \frac{d\psi_q}{dt} + \omega_r \psi_d \tag{13}$$

$$\psi_d = L_d I_d \tag{14}$$

$$\psi_q = L_q I_q \tag{15}$$

V_d and V_q denote d- and q-axis stator voltages, R_s is the stator resistance, I_d and I_q are d- and q-axis stator currents, ψ_d, and ψ_q are the d- and q-axis stator flux linkage, ω_r is the rotor speed, and L_d, and L_q represent the d- and q-axis stator inductance. The electromagnetic torque T_e and dynamic equations of the SynRM are defined as follows:

$$Te = \frac{3}{2} p_m \left(\psi_d I_q - \psi_q I_d\right) = \frac{3}{2} p_m (L_d - L_q) I_d I_q \tag{16}$$

$$Te - T_L = J_m \frac{d\omega_r}{dt} + B_m \omega_r \tag{17}$$

where p_m denotes the number of pole pairs, T_L is the load torque, and J_m and B_m are the inertia of the rotor and friction coefficients, respectively. The electromagnetic torque of a SynRM is robustly controlled depending on the accurate values of d- and q-axis stator currents [42]. Although the electromechanical configuration and control methods of SynRM seem like PMSM with distributed windings, the FOC-based controller has some differences as seen in Fig. 13.

The FOC controller is comprised of three loops where the external loop is responsible for controlling the rotor speed and providing the reference electromagnetic torque T_e^*. The internal loops regulate the d- and q-axis stator currents for controlling the torque and flux of the motor. Both internal control loops are enhanced with PI controllers to compare the actual d- and q-axis stator currents to reference values and to generate the reference frame voltages. The d- and q-axis voltages generated are transformed to stationary reference frame and supplied to the space vector modulator [31,42].

11.3.2 Power electronics system

The EV and HEV propulsion systems are enhanced with the contribution of power electronic devices and systems. The power electronics of a propulsion system is comprised of a power converter that is built with switching devices and the closed-loop control system for regulating the operation conditions. The power semiconductors, converter topologies, control and switching methodology, and cooling methods used in the design of propulsion system power

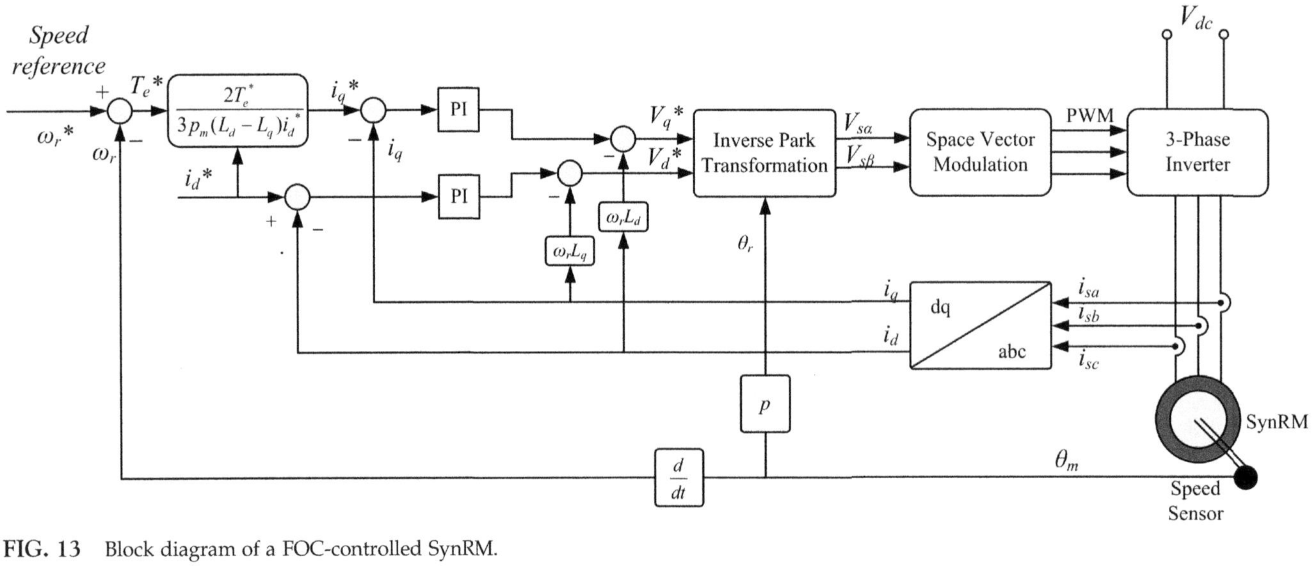

FIG. 13 Block diagram of a FOC-controlled SynRM.

electronics determine the overall efficiency and performance of the EV. The propulsion mechanism of an EV is comprised of three-phase electric machines introduced in the recent section. The majority of EV and HEV propulsion systems are based on a two-level three-phase inverter topology to convert the battery voltage for controlling the motor that requires variable voltage and variable frequency for driving needs. The two-level inverter topology is operated with hard switching-based control methods due to its reliable configuration that complies with the electromagnetic structure of electric machines. Although the IGBT-based two-level inverter topology is widely used in EVs, the increased switching losses of these traction drives limit the switching frequency up to 20 kHz for suppressing the losses. Another drawback of two-level inverter topology is the high-frequency ripples on the current that may increase up to 60% of the rated current at the source side. This situation increases the requirement of high-value DC-link capacitors to get rid of the current ripples. The three-level multilevel inverter (MLI) topologies such as NPC and T-type NPC are proposed as alternative drivers to overcome two-level topologies [31,43]. The advantages of MLI topologies over conventional two-level inverters can be described as decreased switching losses, lower output dv/dt ratios, lower current ripple on the source side, and low capacitor requirements. The MLI topologies also increase the power quality by decreasing the THD ratios and voltage disturbances.

The developments in semiconductor technology enabled to have high-power IGBT modules with low costs and facilitated the improvement of various inverter topologies with an increased number of IGBTs. Moreover, the high-electron mobility devices such as silicon carbide (SiC) and gallium nitride (GaN)-based devices are much more available. Both of these semiconductors replace the conventional silicon-based IGBT and MOSFETs due to their higher efficiency, higher power capacity, and lower switching losses that are very useful in the propulsion drivers of EVs and HEVs [31,43]. The improvements in MLI inverters are the key progress on EV propulsion systems. The novel topologies increase the inverter efficiency by more than 95% while increasing the power and energy density compared to regular topologies. Besides, novel SiC and GaN devices allow to increase the switching frequency much more than 10–20 kHz limits of conventional Si IGBT devices. Therefore, the slew rate of the dv/dt ratio exceeds 1000 V/μs, and the value of the reactive components used at the input and output of the inverter is decreased [44]. The battery voltages of EVs are set to 400 V as a standard for many years, but now there are some studies to increase and to standardize the 800 V battery voltage for ultrafast charging applications. The 800 V batteries decrease the copper and switching losses and ensure a decreased size of the motor and wiring in addition to the faster charging opportunity. The power converters are also becoming adapted to the increased battery voltage [4]. The topologies of EV propulsion systems comprised of two-level and multilevel inverters are introduced in the following subheadings.

11.3.2.1 Conventional two-level inverter

The EV and HEV traction systems require compact, reliable, and efficient power controllers. The conventional two-level inverter topology that is widely used for the IM drive of Toyota, Honda, Mercedes, BMW, and Nissan EVs has become a standard in EV industry. The power block or brick structure power modules of six-switch configuration facilitate the use of a three-phase two-level inverter topology with robust and reliable designs [45]. The control method of the two-level inverter that is mostly comprised of IGBTs is based on either hard-switching or soft-switching algorithms. In some applications, the battery is directly connected to the DC supply bus of the inverter, while some others interface the battery with bi-directional intermediate DC-DC converters as shown in Fig. 14A and B [32,45]. Another two-level voltage source inverter (VSI) used in the traction drive of EVs is based on an open-end winding connection that is shown in Fig. 14C. In addition to the DC-DC converter-interfaced and open-end winding configurations, there are many VSI topologies that are improved with multiphase structure, star connection, and four-leg topologies. These variations of conventional two-level inverters aim to decrease the common mode voltage and leakage current disturbances to minimize motor bearing failures. Another inverter configuration is defined as the current source inverter (CSI) that is supplied by a DC current source to produce AC voltage at the output. A DC voltage source with a large inductor connected in series provides the current source requirement of this topology. Therefore, the battery of an EV can be used as a current source in this configuration by replacing shunt capacitors with series inductors in Fig. 14A.

One of the advantages of CSI inverter is its booster feature that allows having a higher output voltage level compared to the input DC voltage rate. Although this situation eliminates the boost converter requirement and increases the constant power range of the electrical machine, the lower efficiency and higher cost over VSI topologies limit the use of CSI in EV traction control [32].

11.3.2.2 Multilevel inverters

The NPC inverter topology that generates three-level output voltage proposed by Nabae et al. [46] may be addressed as the first comprehensive topology of MLIs. The main motivation of the first proposed NPC topology was to eliminate the losses caused by harmonics and the pulsating torques in an AC motor drive system. The NPC

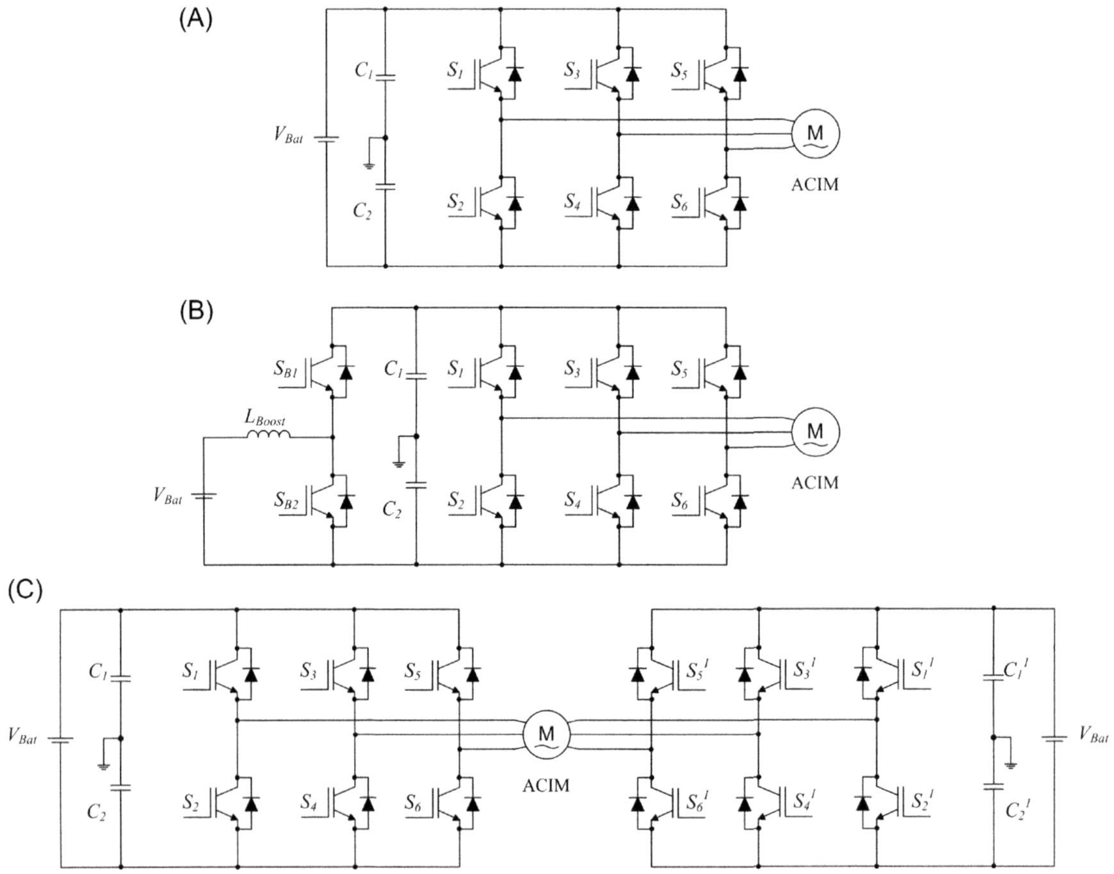

FIG. 14 Two-level inverter topologies, (A) conventional, (B) DC-DC converter interfaced, and (C) open-end winding IM driver.

topology overcomes low efficiency and a higher magnitude of harmonics in a motor drive system controlled with a two-level inverter. In the following years after Nabae's proposal, MLI topologies and application areas are rapidly improved owing to developments in semiconductor and microcontroller technologies. The cascaded H-bridge and flying capacitor topologies have been proposed as alternative MLI configurations to the NPC topology for various application areas. The MLI topologies provide remarkable advantages due to lower common mode voltage, decreased dv/dt ratio, low voltage stress on switching devices, and low harmonic distortion ratio [47]. Although there are many MLI topologies in different application areas, the NPC, active NPC (ANPC), and T-type NPC configurations are the most widely used inverters in EV traction drives.

The three-level NPC inverter is comprised of two additional clamping diodes per phase arm, and each switching device in a two-level inverter is series connected with another switch. The common point of the clamping diodes in the NPC topology shown in Fig. 15 creates the neutral point of the inverter that enables the connection of each phase

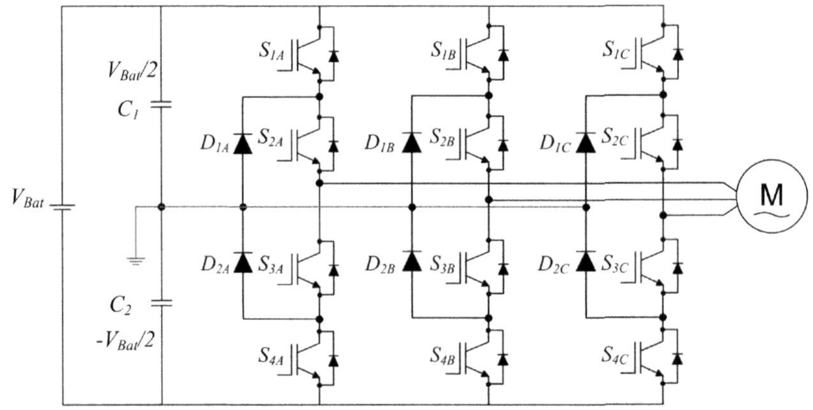

FIG. 15 Three-level neutral point clamped (NPC) inverter for EVs.

TABLE 2 The switching states of an NPC inverter.

	Switching status (*Phase n; a/b/c*)				
Switching state S_n	**S_{1n}**	**S_{2n}**	**S_{3n}**	**S_{4n}**	**Output voltage level**
P	1	1	0	0	$V_{dc}/2$
O	0	1	1	0	0
N	0	0	1	1	$-V_{dc}/2$

output to the neutral of the inverter. The DC source capacitors share the input voltage equally to produce positive and negative voltage levels [25,48].

The output voltage level of an NPC inverter can be higher than three level, but this causes the use of a more complex controller to make sure the equality of the capacitor voltages and to generate the required switching states. The three-level NPC configuration is widely accepted as an industrial standard for low- and medium-voltage drivers due to its reliable and efficient operation with space vector-based control methods. Many EV traction control systems are also improved with a three-level NPC topology as an alternative to the conventional two-level inverter topology [25,49]. The switching states of the three-level NPC MLI are shown in Table 2 where the output voltage levels are generated regarding the presented switching states.

The switching states of the inverter are defined by $S_n = (S_a, S_b, S_c)$ where each switching state generates the output voltage levels at $V_{dc}/2$, 0, and $-V_{dc}/2$ referring to the P, O, and N states, respectively. The space vector modulation requires 27 switching states for three-level NPC where the state number is calculated as $L^3 = 27$ where L is the number of voltage level.

The switching states of a space vector (V_S) are calculated referring to the switching order in the complex α-β frame as given by

$$\begin{aligned} V_S &= \frac{2}{3}\left[V_{AN} + aV_{BN} + a^2V_{CN}\right] \\ &= \frac{2}{3}\left[S_A\frac{V_{DC}}{2} + aS_B\frac{V_{DC}}{2} + a^2S_C\frac{V_{DC}}{2}\right] \end{aligned} \tag{18}$$

where $a = e^{j2\pi/3}$.

The disadvantage of the conventional NPC topology is the problem of dc-bus voltage balancing. The efficiency and reliability of an NPC inverter are dependent on the modulator control of the neutral point voltage. To address the issue of dc-bus balancing in regular NPC configurations, the ANPC topology shown in Fig. 16 has been proposed. The ANPC topology has solved the problem by replacing clamping diodes with active switches as its name implies. The power switches located on the upper and lower sections of the neutral point remove the inequality of neutral point voltages, and they help to reduce the switching losses [27,28]. The upper and lower neutral point paths of the NPC topology are determined referring to the direction of phase current. The switching devices S_{2n} and S_{3n} of NPC topology are always turned on during the zero (O) switching state as shown in Table 2.

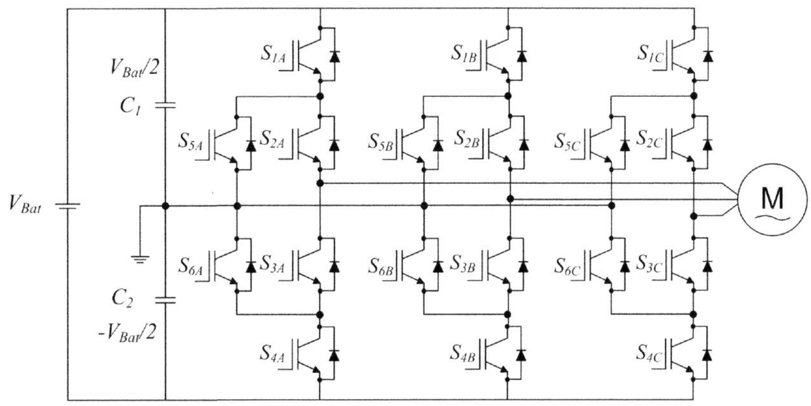

FIG. 16 Three-level ANPC inverter for EVs.

TABLE 3 The switching states of an ANPC inverter.

Switching state S_n	Switching status (*phase n; a/b/c*)						Output voltage level
	S_{1n}	S_{2n}	S_{3n}	S_{4n}	S_{5n}	S_{6n}	
P	1	1	0	0	0	1	$V_{dc}/2$
OU2	0	1	0	0	1	0	0
OU1	0	1	0	1	1	0	0
OL1	1	0	1	0	0	1	0
OL2	0	0	1	0	0	1	0
N	0	0	1	1	1	0	$-V_{dc}/2$

In case the active switching devices are applied in NPC topology, the phase current is conducted on the upper path of the neutral point regarding the switching state of S_{2n} and S_{5n} devices. This turns on the controller that enables the bidirectional current flow between the battery and the electrical machine of an EV. In case the S_{3n} and S_{6n} switching devices are turned on, the bidirectional current flow is allowed in the lower path of the neutral point. It should be noted that only one of both switching pairs of $S_{2n}-S_{5n}$ or $S_{3n}-S_{6n}$ can be turned on at any switching interval, and the current flow is determined by the characteristic feature of the upper or lower path switching order. The switching order of an ANPC inverter is given in Table 3 where S_{6n} should be switched on to provide the voltage equality between S_{3n} and S_{4n} in the P state. Similarly, S_{5n} should be turned on for the same voltage equality during the N state. The switching losses can be controlled by defining the upper or lower current paths during zero states that are shown as OU2, OU1, OL1, and OL2 states in Table 3 [50].

In [41], a study comparing the switching losses of NPC and ANPC is presented where additional switching states distribute the switching losses among active switches at each interval. According to the study's proposed loss balancing approach, the output power of ANPC MLI has been increased by 20%, while the switching frequency can be increased by up to 85% higher than in conventional NPC. Although the surveyed paper represents ANPC results in medium voltage (MV) applications, it demonstrates increased efficiency and ease of control for EV applications when compared to conventional NPC topology [25]. Another NPC topology that provides higher efficiency over the NPC and ANPC inverters is defined as the T-type NPC owing to its configuration. The switching and conduction losses of a T-type NPC inverter are remarkably decreased, and it is noted to be efficiently operated in a medium frequency of up to 30 kHz [51]. The schematic diagrams of common-emitter (CE) and common-collector (CC) T-type NPC inverter topologies are presented in Fig. 17A and B, respectively. This configuration removes the clamping diodes that are used to clamp neutral voltage to positive or negative DC voltage in NPC, and it enables bidirectional power flow due to active switching devices replacing the clamping diodes. The CE configuration needs to commutate each active switch with an isolated gate driver, while the active switches in the CC connection are commutated by a single isolated drive since emitter connections of S_{2n} switches share the common emitter connection at the neutral point.

The switches S_{1n} and S_{4n} block the total battery voltage as seen in Fig. 17, while the switches S_{2n} and S_{3n} are subjected to half of the battery voltage due to capacitor voltages at the neutral point. The decreased number of devices and the elimination of clamping diodes decrease the switching losses in low switching frequency operation. The switching status of the T-type NPC inverter is same as that of the NPC inverter in terms of P, O, and N states as presented in Table 2. The commutation order presented in Table 2 denotes that the switches S_{1n} and S_{3n} are controlled in complementary states, while S_{2n} and S_{4n} are complementary to each other.

The current equation of an T-type NPC inverter is derived as follows:

$$i_{ON} = S_{AO}i_A + S_{BO}i_B + S_{CO}i_C \tag{19}$$

where i_{ON} denotes the current flowing from neutral point to the load end, while S_{NO} represents the switching status of [0,1] applied to any active switch on phase arms. The voltage equality of the input capacitors should be obtained in the T-type inverter depending on the PWM-based control methods [52].

11.3.3 Summary

This section dealt with electric machines, control methods, and power electronic devices used in the propulsion systems of EVs. ACIM and PMSMs have widespread use in traction drive systems, and the control methods are mostly

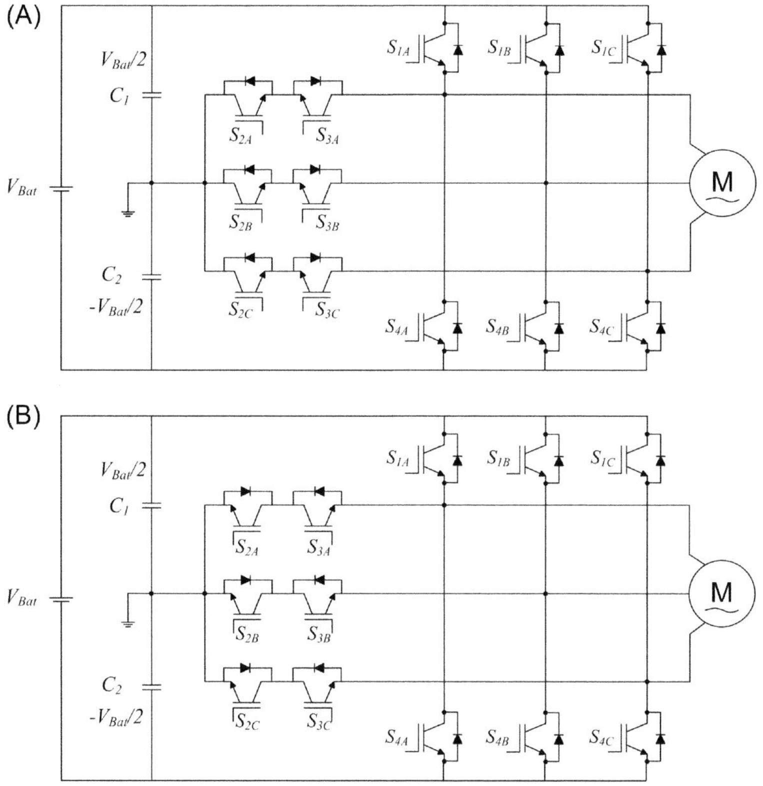

FIG. 17 Three-level T-type NPC inverter: (A) common-emitter configuration and (B) common-collector configuration.

based on vector-oriented algorithms such as FOC, DTC, and space vector PWM methods. The electric motors are controlled either with direct measurements inherited from sensors or with estimation algorithms without any measurement of speed. The most recent inverter topologies that are NPC, ANPC, and T-type NPC used in the control of propulsion systems have been introduced in this section.

11.4 Marine vehicles and electric propulsion systems

Due to its long history as the primary facilitator of international trade, maritime transportation has been one of the factors that supports countries' economic progress. Specifically, marine transportation represents about 80%–90% of international commerce. According to the Second International Maritime Organization's (IMO) Greenhouse Study 2009, carbon dioxide (CO_2) emissions from the shipping industry are estimated to be responsible for 12%–18% of global CO_2 emissions by 2050. The hybrid propulsion system that is proposed to overcome these emission rates is a brand-new design generally consisting of a combination of ICEs and an electric motor driven by an ESS. To ensure that ships have an effective and sustainable propulsion system, shipping firms are trying to achieve these new design and technical criteria [53,54]. One of the most important reasons for this is that transportation is carried out between continents by marine shipping. Today marine shipping is not just about the successful movement of the ship but also ensuring a better safety standard for the marine ecosystem, cost efficiency, and low emission values. Marine propulsion is a system used to generate thrust to move a boat or ship across the water [55]. In the early times, it was provided by paddles or human power; now in most modern ships, a propeller is turned by an electric motor or engine, or in pump jets, an impeller is propelled by mechanical systems [56]. In Fig. 18, a basic scheme of a propeller and rudder layout is shown. The generated power is transmitted to the propeller via a shaft and reduction gear, which converts this power into thrust used in moving a vessel forward. The propulsion unit is not only responsible for forward movement, but it is also responsible for stopping and maneuvering the marine vehicle when required. For this reason, it is a critical issue to consider the marine vehicle propulsion system selection. While in the past limited propulsion systems were used for ships, today there are several ship propulsion systems such as wind propulsion, steam turbine propulsion, diesel propulsion, gas turbine propulsion, nuclear propulsion, fuel cell propulsion, and solar propulsion,

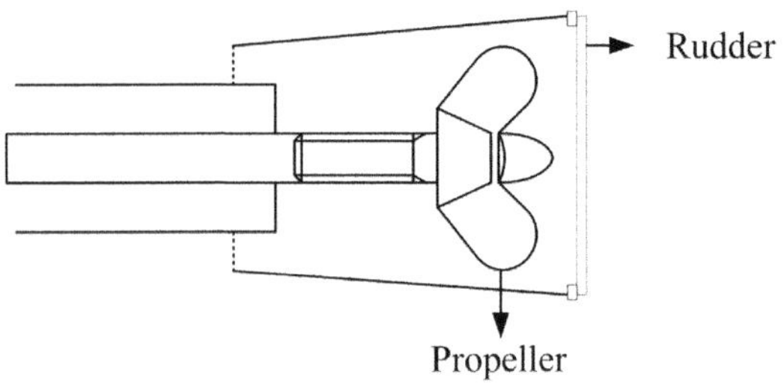

FIG. 18 A basic scheme of a propeller.

depending on the intended use of ships [57]. With the improvement of power electronic switches and devices, electrical propulsion systems became more important, and more popular. The electrical propulsion system that is supported by converter and inverter circuits continues to operate against changes that may occur in the value of power supplies. Also, the speed and direction of the ship can be easily controlled by power electronic systems, and it allows hybrid operation in which different power sources are used together.

Operational safety is the primary requirement for propulsion systems used in maritime vehicles. The IMO has established strict requirements for the safe movement of passengers, crew, and cargo aboard ships. As a result, all equipment and major machinery employed on the maritime vessel must match these requirements. There are also local emission rules that maritime vehicles must follow [58]. The economics is the most important factor for a commercial ship, but high speed/maneuverability, volume, and weight parameters are more important for warships. In Fig. 19, the schematic of the electrical propulsion system of a ship is represented where the diesel engine, fuel-based motor, or renewable energy source can be considered as a prime mover that rotates the generator. This rotating generator produces electricity and supplies it to the transformer that supplies the input voltage of the rectifier. The DC/DC converter (buck, boost, or buck/boost) is used to regulate the rectified DC voltage and feeds the inverter that drives the electric motor. In this scheme, the pulse width modulation technique is applied to the converter and inverters for controlling the power switches. The speed of the propellers and the currents of the rotor and stator windings are controlled and measured in the controller block.

Until the change in the 19th century, shipping relied heavily on wind power. However, due to the unpredictability of wind power, low efficiency, unreliability, and the emergence of steam turbines, it has been pushed aside for nearly two centuries, especially in cargo and passenger transportation. Today, wind propulsion for marine vehicles again gains popularity because IMO launched its first emission reduction strategy in 2018 to align with the emissions reduction targets set out in the 2015 Paris Agreement. Various wind-assisted ship propulsion (WASP) products have been developed and tested to utilize wind energy in modern ships. WASP is a promising solution because of its fuel and emissions savings. It is considered an important renewable energy source for the future of the shipping industry. In Fig. 20, a wind-propelled ship is schematically shown.

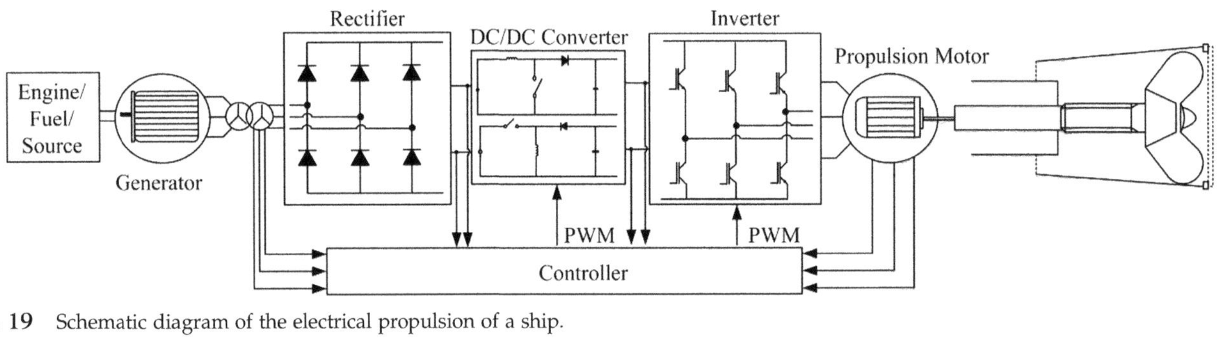

FIG. 19 Schematic diagram of the electrical propulsion of a ship.

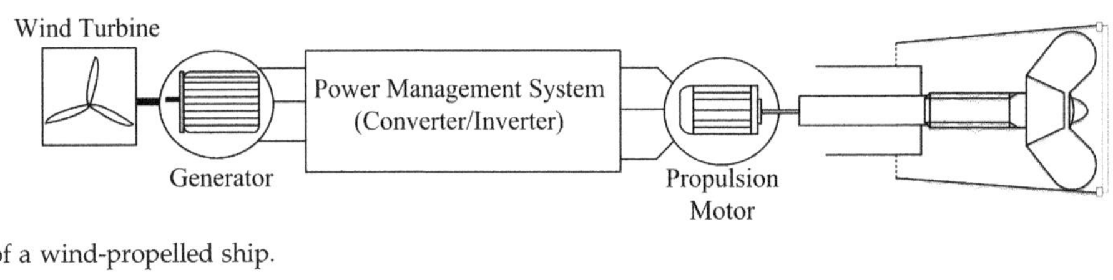

FIG. 20 A scheme of a wind-propelled ship.

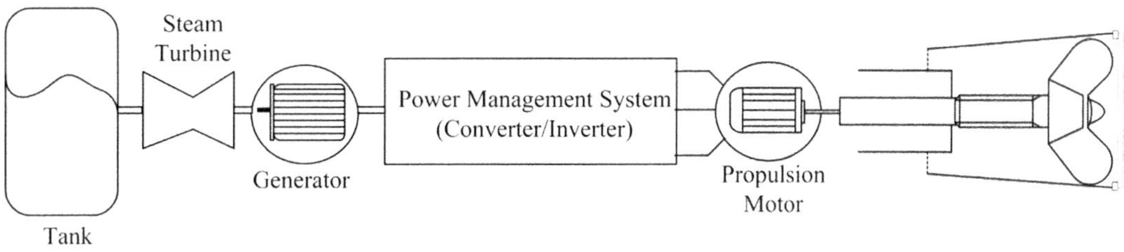

FIG. 21 Simplified schematic of steam-propelled ship.

When the first steam engines started to provide controllable and reliable amounts of energy, it was considered that this opportunity could be used efficiently in land vehicles and marine vehicles [59]. With the use of steam engines in passenger ships and transportation, the power, speed, and reliability of these ships increased over time and the steam engine began to be used in the navy. By the end of the 19th century, naval fleets of the steam engine had already formed. The steam engine has a piston and cylinder. Steam is obtained from fossil fuels (petroleum, natural gas, or coal). In this system, steam is directly injected into the rotating turbine blades. Energy is obtained by the rotation of the steam by striking the turbine blades. It is the power unit that converts the potential energy of steam into kinetic energy and kinetic energy into mechanical energy. In Fig. 21, a basic scheme of the steam-propelled ship is schematically shown.

The diesel propulsion system converts heat forces into mechanical energy. The extensive usage of petroleum and petroleum derivatives at the end of the 19th century resulted in the adoption of steam boiler propulsion units that used petroleum instead of coal [60,61]. In later years, as the demand for distilled fuel and automotive fuel increased, the development of boilers and diesel engines that could burn cheaper fuels from refined residues accelerated. Today, however, this kind of propulsion system is commonly used in marine vehicles such as small boats, ships, or navy and also, they are used for pumps, compressors, and land vehicles, too. In Fig. 22, a diesel-propelled ship is schematically shown.

Although the use of gas turbines started with submarines in the 1937s, the first ship using gas turbines was launched in 1947. In this first application, gas turbine design and operation were experienced, and after 1950, gas turbines began to be widely used in marine vehicles. The gas turbine consists of a compressor, a combustion chamber, and a turbine. The compressor pressurizes the air it draws by filtering it from the atmosphere and sends it to the combustion chamber. The fuel mixed with the air in the combustion chamber burns, and the resulting burned gases are sent to the turbine. When gases with a high temperature and pressure coming from the combustion chamber to the turbine pass through the turbine, the heat energy is transformed into mechanical energy. Gas turbines are a good alternative to high-speed diesel engines because of their low weight/power ratio in cruise ships, ferries, and high-speed small boats. In Fig. 23, a gas turbine-propelled ship is schematically shown.

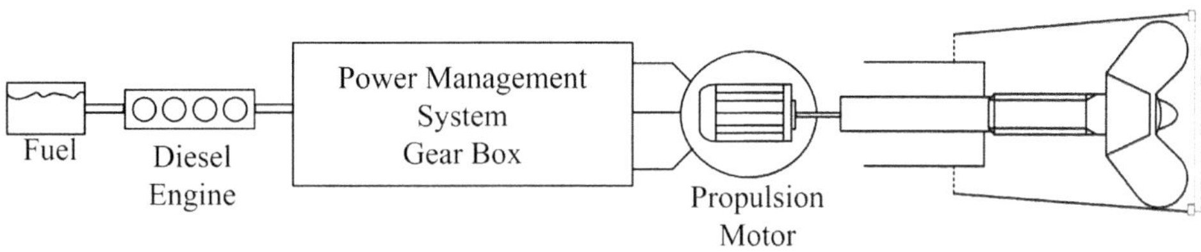

FIG. 22 A scheme of a diesel-propelled ship.

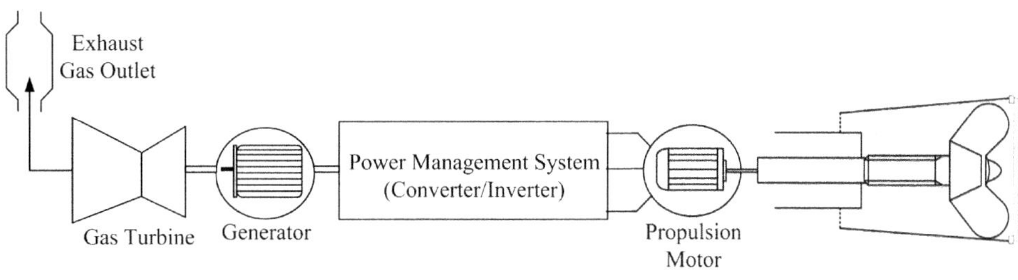

FIG. 23 A scheme of a gas turbine-propelled ship.

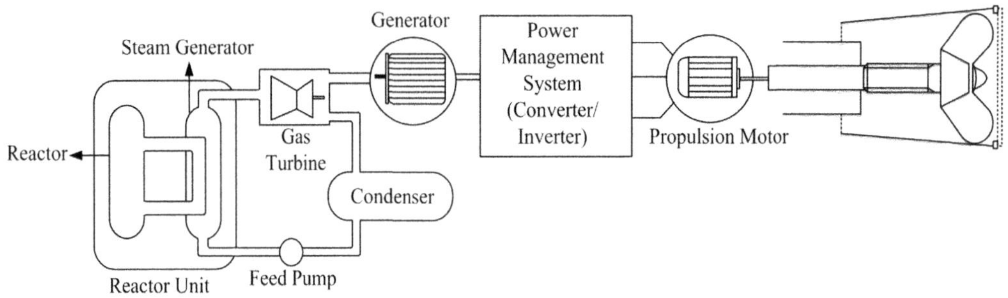

FIG. 24 A scheme of a nuclear propelled ship.

In the last decades, it has been shown that it is important in terms of preventing both global warming and climate change in studies under the ocean as well as long-term maritime transport. Also, the longer time spent at sea showed that more equipment and food were needed. However, a strong and long-lasting power supply was needed for it. At that time, in the 1940s, the idea of using nuclear power in ships emerged, and now, nuclear-powered ships have been used in navies and icebreakers for about 50 years. These kinds of ships can stay away from home for months. It should also be highlighted that interest in nuclear-powered ships has increased due to lower fuel prices and the dependence on fossil fuels. Nuclear power enhances resistance to shipboard power outages. However, nuclear propulsion is a highly complicated system that uses a variety of equipment to feed fuel to the ship, including a reactor, pumps, steam generators, and so on. As a result, it takes a long time and a team of professionals to create a nuclear propulsion marine vehicle. In Fig. 24, a nuclear-propelled ship is schematically shown.

In fuel cell propulsion systems, hydrogen is the main fuel component. Hydrogen is converted into electricity by fuel cells, and the electricity is used to move the electric motors. The proton exchange membrane fuel cells (PEMFCs) that use hydrogen as a fuel have been popular in marine vehicles due to their high energy efficiency, low noise, low maintenance cost, flexibility, and eco-friendly structure. However, hydrogen storage, safety, operating costs, etc., can be listed as disadvantages of fuel cell propulsion marine vehicles. This technology is still new, it needs improvement, and it is considered a promising technology. In Fig. 25, a fuel cell-propelled ship is schematically shown.

Solar energy is one of the most popular renewable energy sources since it is easily obtained from nature. It is now used practically anywhere that requires clean energy. Photovoltaic cells turn the heat and light of the sun into electricity. Solar panels, batteries, electric motors, and power electronic converters are all required in solar-powered maritime vehicles. Solar energy is used to provide electrical propulsion during the day, but at night or on overcast, rainy days, a battery or another energy source is used to supply electrical energy to the marine vehicle. In Fig. 26, a solar-propelled ship is schematically shown.

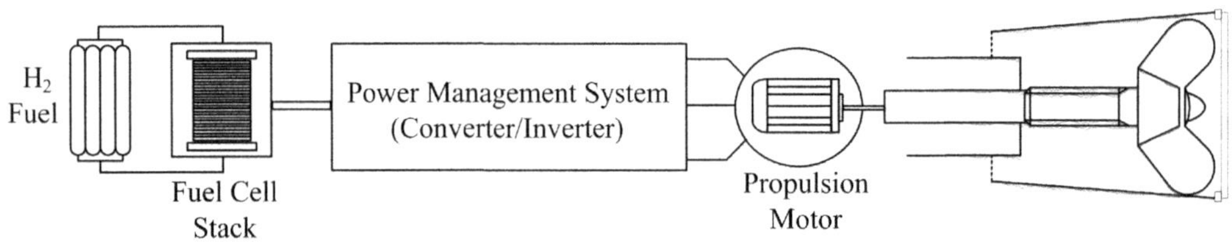

FIG. 25 Simplified schematic of a fuel cell-propelled ship.

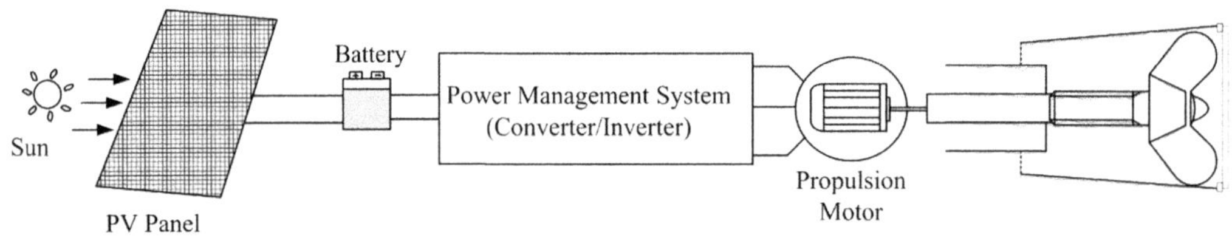

FIG. 26 Simplified schematic of a solar-propelled ship.

11.4.1 Summary

In the last few decades, electrically propelled ships that are supported by power electronic devices gained popularity, especially for medium-capacity container ships and submarines. In earlier times, the working mechanism for the propulsion of sea vessels was mechanical. However, today, with the development of power electronic devices, these mechanically powered ships have started to leave their place for electric ships with an electric propulsion system. In this way, longer range travels come with more comfortable and newly added features.

11.5 Electric propulsion systems for railway vehicles

The advancement of electrical railway power systems (ERPSs) has always been closely related to the technological advancement available at the time. Utility grids were smaller and weaker at the start of railway electrification, and using large motors at industrial frequency presented several challenges. These factors triggered the development of dedicated DC and single-phase low-frequency AC systems as the best solutions for railway systems. With the subsequent advancement of the utility grid and technological advancements in the field of electric motors, the AC railway systems that are directly fed by power transformers became possible. As a result of this evolution, the following electrification schemes are currently available: 750 V DC, 1500 V DC, 3000 V DC, 15/11 kV 162/25 Hz, and 25 kV 50/60 Hz. Power supply systems of the electrical railway can be divided into three main systems in terms of electrical power engineering: the electrical system, the traffic system, and the control system.

11.5.1 Electrical railway power systems

The main component of ERPSs is the railroad, which is further divided into the traction substations, the railway transmission grid, the generators and storage systems, and the catenary. It is worth noting that the term catenary refers to the circuit as well as the positive and negative feeders, rails, and potential return lines. The ERPSs can be divided into two types of topologies as centralized and decentralized. In centralized systems shown in Fig. 27, the catenary is connected to a power system via an intermediary railway transmission grid that is supplied by dedicated railway power plants and/or the utility grid, whereas in decentralized systems, the catenary is directly connected to the utility grid. The centralized systems were developed at the beginning of railway electrification due to the requirement of devoted lower frequency generation. However, decentralized systems have become the dominant topology today with the recent advancement of power converters. It is important to express that decentralized systems may also include a parallel railway transmission grid, as well as generation or storage systems that are either directly or indirectly connected to the catenary [62–64].

Most of the railway system is powered by 1500 V DC and 25 kV/50 Hz single-phase electrified lines. The substations use transformers to decrease the voltage in the high-voltage energy distribution system and diode rectifiers to convert AC to DC in order to distribute the current to the rolling stock via the catenary and the pantograph. The supply of single-phase catenaries is achieved by converting the high voltage drawn between two phases of a three-phase transmission network (63 kV, 90 kV, 225 kV, and 400 kV) to a voltage of 25 kV [65]. The block diagram of a DC railway system is illustrated in Fig. 27 where the three-phase transmission network of 63 kV is stepped down to 25 kV to comprise the railway distribution network. The substations are equipped with step-down transformers and 12-pulse rectifiers that produce a 1500 V output voltage used to feed the catenaries. The regenerative inverters may also be accommodated in stations to inject the dump regenerative energy inherited from rails, rolling stocks, and rectifiers to supply the substation bus bar.

The bi-directional regenerative inverter operates in three modes: rectifying, circulating current, and inverter modes. The circulating current mode is essentially used to control the exchange of an inverter from rectifying mode to inverter mode by controlling the small circulation current. The inverter mode is operated in both constant current and constant voltage in order to efficiently manage power exchange between upstream and the catenary line [66]. The transformer-based AC railway systems are configured with two 25 kV single-phase transformers at each traction substations as seen in Fig. 28. The transformers are commonly connected in different phase angles to decrease the power imbalance of the grid, and they are used to connect the catenary and the utility grid. The neutral network installation builds the insulation sections along the catenary grid, and all substations have independent neutral sections. Another AC network configuration is defined as a bi-voltage system where two single-phase 50 kV transformers are center tapped to feed +25 kV and −25 kV at the secondary side and connected to autotransformers at the rail side in the catenary line side [63].

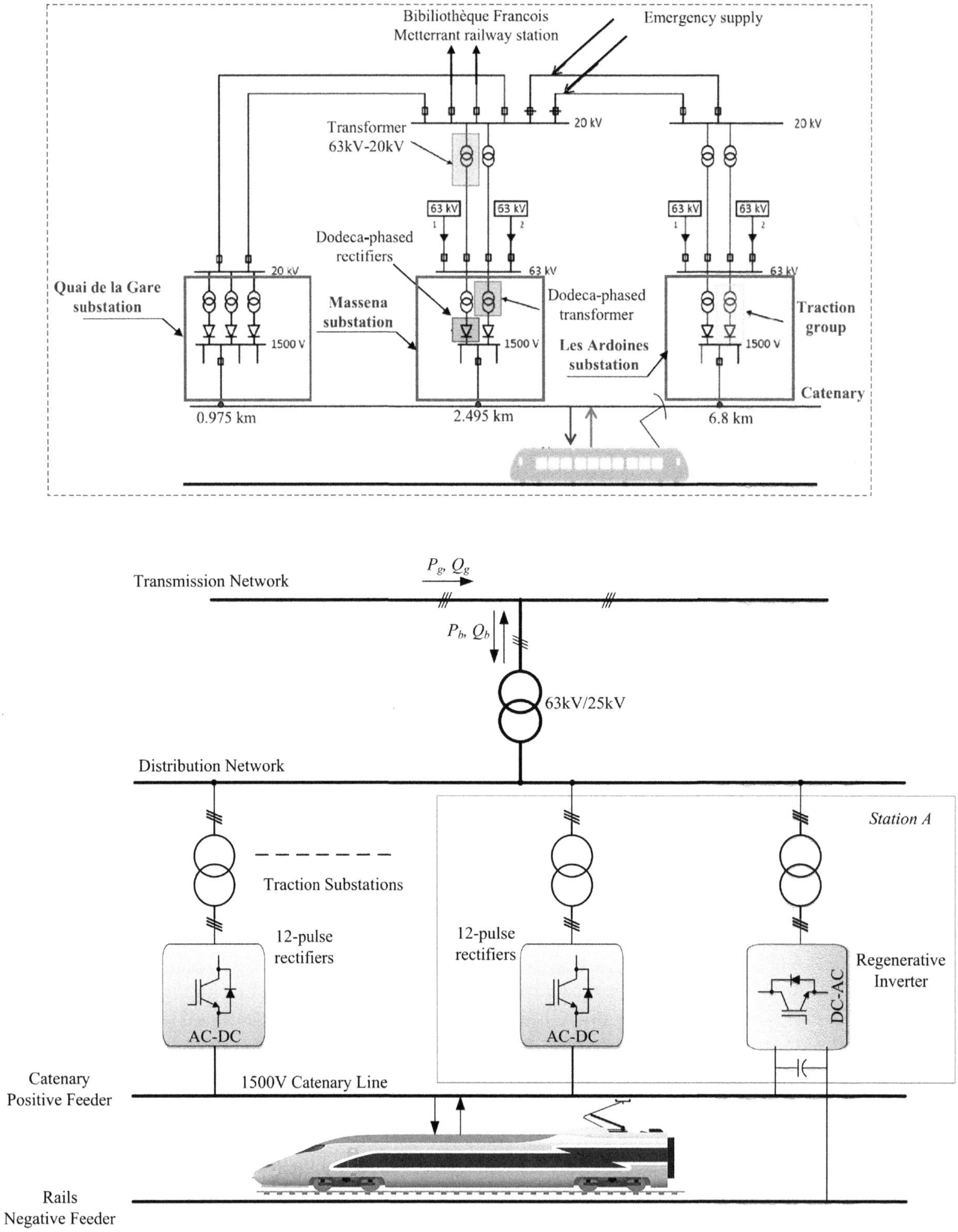

FIG. 27 Single-wire schematic and block diagram of a DC railway power system.

In AC railway systems, static VAR compensators (SVCs) and STATCOMs are widely used to regulate power quality along the railway power system. Static VAR compensators are used to provide adjustable impedance in relation to the reactor or capacitors. Thyristor-controlled reactor (TCR) SVC devices control inductance, whereas thyristor-switched capacitors (TSCs) control impedance through capacitance adjustment. SVC devices can be installed on the grid or at the railway substation to compensate for the current components. Fig. 29A shows a schematic diagram of a TSC, which is

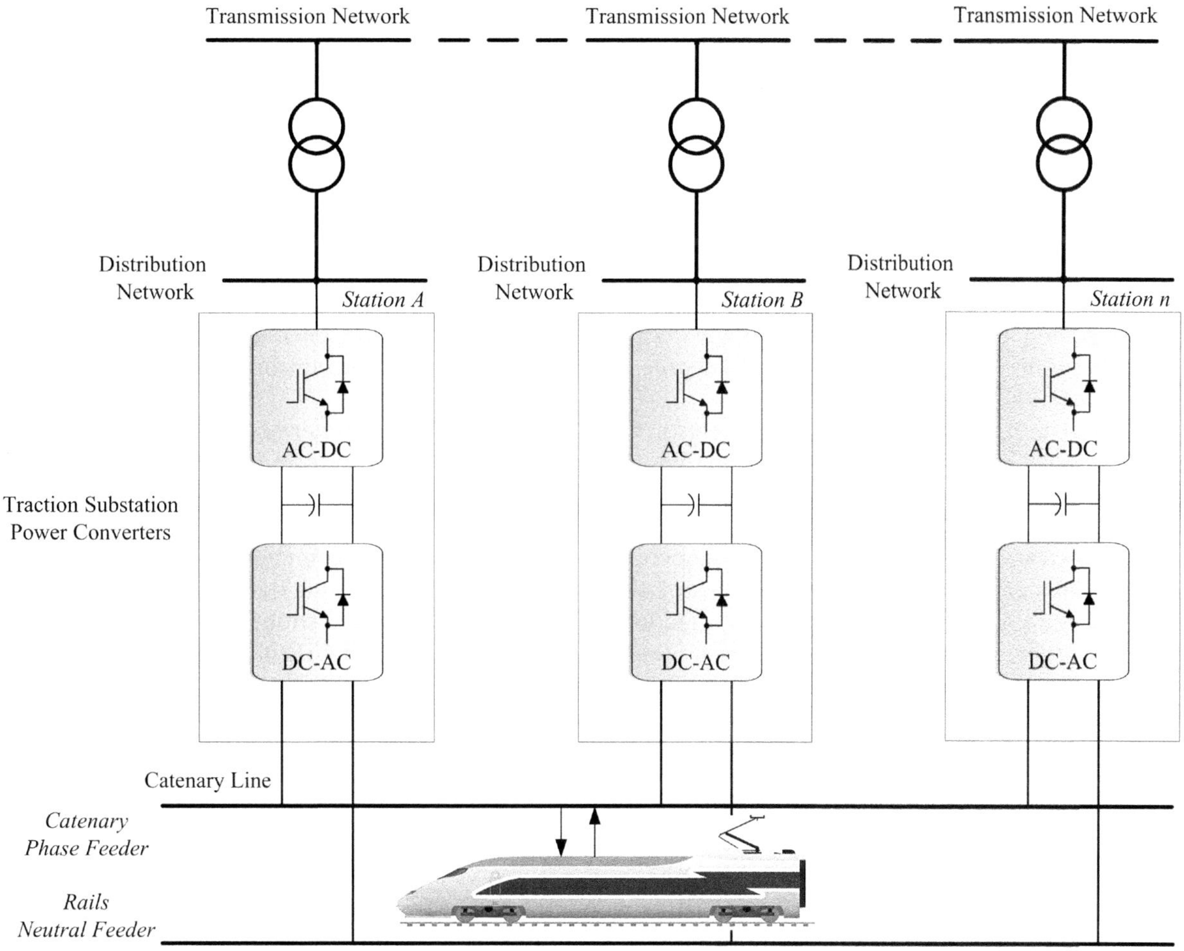

FIG. 28 Single-wire schematic and block diagram of AC railway power system.

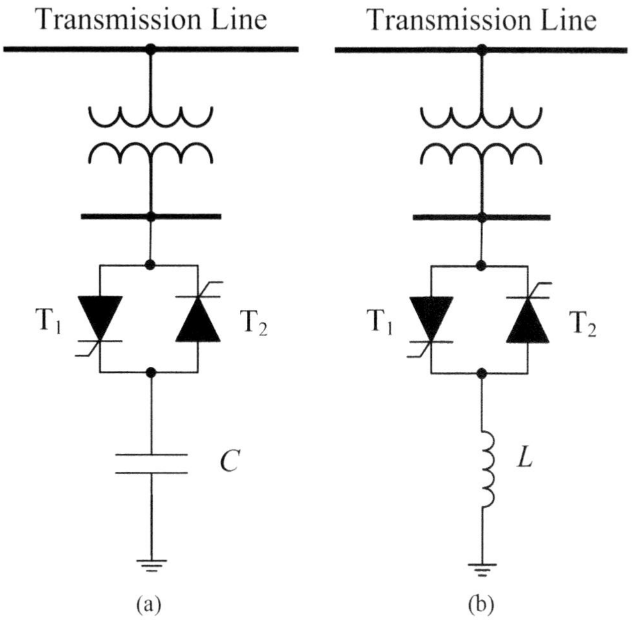

FIG. 29 SVC devices: (A) thyristor-switched capacitor and (B) Thyristor-controlled reactor.

implemented with a line transformer, two-way thyristor valve, and a capacitor to which a small inductance can be added in series to limit the current. The thyristors are turned on at precise angles to adjust the required capacitance to compensate for the power factor of the load. In ideal conditions, the thyristor voltage that is controlled at turn-on and turn-off transitions is forced to zero. However, in practice, the required positive amplitude across anode to cathode for switching thyristors is provided, and the thyristor valve provides capacitor connection and disconnection to the line. When using a small reactor, it should be placed between the transformer and the thyristor valve to prevent inrush currents under critical conditions [67].

The TCR controls the inductive reactance of the line owing to its shunt connection and controlled phase angle as shown in Fig. 29B. The voltage and current waveforms of the compensator are controlled at switching angle α, which increases or decreases the fundamental magnitude of reactor current I_L. Starting with a voltage crossing interval of zero, the switching angle can be adjusted to any degree between 90 degree and 180 degree. At 90 degree, the reactor is completely connected to the line, and at 180 degree, it is completely disconnected from the line, with a partial value connection available between both cases.

SVC-based solutions have less satisfactory performance when compared to other power electronic-based systems, limiting their use significantly. However, it is still a viable option when low costs and simple controls are required. Static synchronous compensators (STATCOMs) are an evolution of SVCs that include self-commutated switches. Due to the additional grade of freedom the self-commutated devices introduce, STATCOMs can simultaneously address the issues of harmonics, negative sequence current, and reactive compensation, in contrast to SVCs. A STATCOM that is shunt connected to the transmission line acts as a compensating device based on a voltage source converter structure. STATCOMs that are integrated into the transmission line and shunt to generators operate by generating or absorbing reactive load to compensate the transmission line. The STATCOM improves power quality by implementing a number of compensations, including active and reactive power (P-Q) control in transmission systems, oscillation damping of power lines, flicker and sag-swell controls, and dynamic voltage control. Furthermore, STATCOMs serve as SVC replacements in distribution systems.

11.5.2 Power electronics of electric railways

Due to their high torque and ease of control, the first electric railway traction motors were DC commutated with diode rectifiers. The disadvantages of DC motor commutators and brushes accelerated the adoption of AC motors such as IM, PMSM, and SynRM in railway traction systems. The IM-based AC traction systems introduced in the early 1990s are still used in many electric railway applications today. The safe and low-cost structure of the squirrel-cage design increased the popularity of IMs in railways. Another advantage provided by IMs is the parallel connection capability of motors that enables the use of a controller for driving many motors at the same time. This drive control is named after 1C2M implying "one controller two motors." The PMSM, SRM without permanent magnet, and SynRM motors have been used in traction control of railway systems. However, there is not any remarkable achievement reported to replace the use of IMs [68]. The use of power electronic converters has increased in railway system as much as in other industries in recent years. The rapid and precise control of electric motors is ensured with appropriate power electronic topologies and control methods that facilitate the design of lightweight and highly efficient devices. The IM-based electric rolling stocks dominate the railway industry, and the single-phase inverters interfacing the three-phase IM drive systems are one of the most widely seen device topologies [62]. The 25 kV single-phase station distribution line shown in Fig. 28 is connected to a single-phase converter in this configuration to produce the DC bus voltage, which is 3000 V in high-speed railways (HSRs). The DC bus voltage is converted to three-phase AC voltages, and IM drive algorithms are executed under the control of this inverter, which is known as the motor drive inverter. The schematic diagram of power converters used in Shinkansen N700A series rolling stock is presented in Fig. 30 [69]. The transformer supplied by the distribution line over the pantograph has a secondary voltage of 1520 V. The three-level PWM converter and three-level three-phase PWM inverter are comprised of 3.3 kV 1.2 kA IGBTs to drive 305 kW traction motors as seen in the figure. The motors are IMs running at 3260 rpm rated rotational speed. The switching frequencies of the converter and inverter are relatively low compared to any other motor drive application due to their power rate.

Power electronic locomotives may interact negatively with each other or with power system components such as rotating synchronous frequency converters. This situation causes the power system to become unstable. The low-frequency stability studies by time-domain simulations and linear analysis can be performed in fundamental frequency RMS mode rather than time-domain simulations with instantaneous voltage and current values in traditional three-phase power systems that are highly equipped with electric machines and passive loads. These conventional tools might not be entirely valid any longer in a power system with predominating nonlinear components, and in

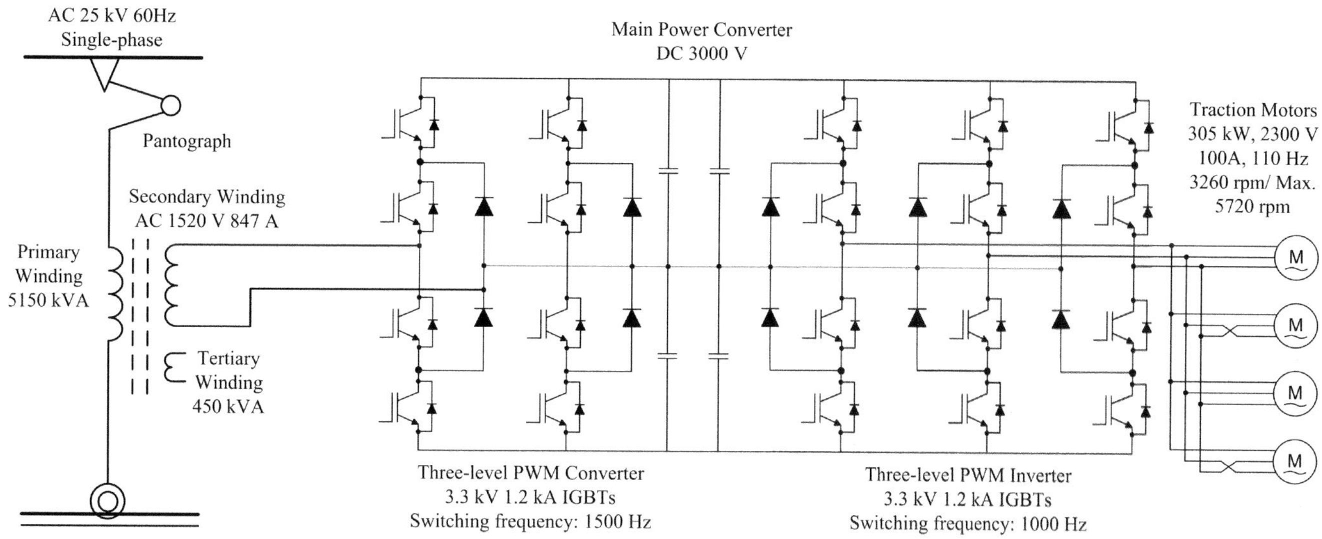

FIG. 30 Schematic diagram of power converters in Shinkansen series N700A [69].

single-phase systems in particular. If established and traditional methods for power system stability studies could be applied to systems with power electronic components as well, that would be interesting and advantageous [62]. Rolling stock is the primary load on the railway power supply system, and due to repeated acceleration and braking in short time spans, the substation load increases from 0 to several MWs in less than 30 s. Another issue with the power electronic drives met in electric railways is low voltage and high current operation. The traction drives except for the HSR systems that are fed with a 1500 V catenary line. The current of each feeder line increases to thousands of amperes since the load is supplied with power on the order of MWs at this voltage.

Another widely used power electronics device is based on a phase-shifted full-bridge (PSFB) topology as seen in Fig. 31. This topology is commonly used as an auxiliary power unit (APU) to feed the onboard loads such as lightings, air conditioners, compressors, and power systems of control signals. The three-phase AC supply voltage is converted

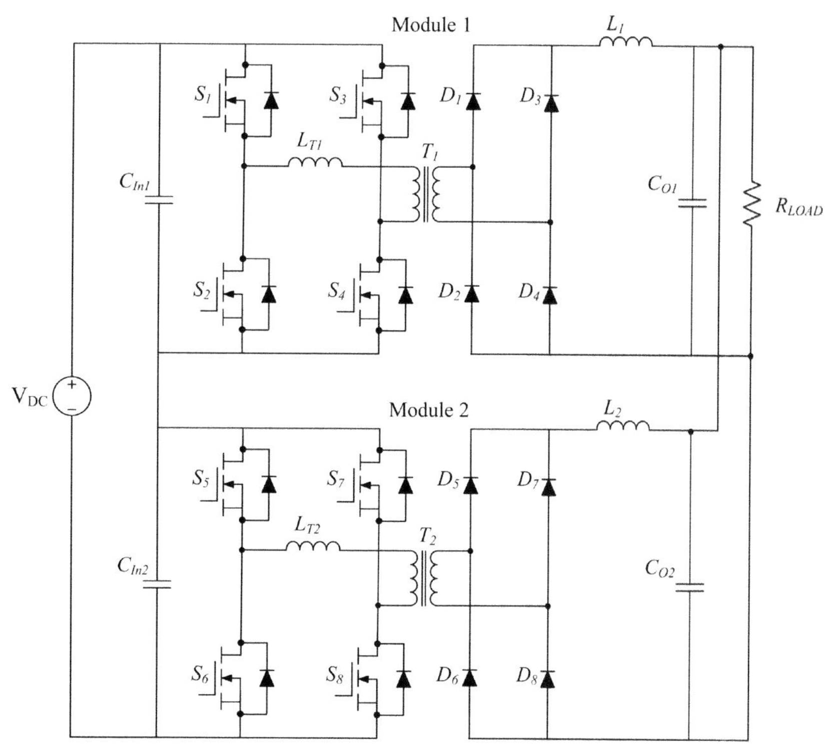

FIG. 31 Schematic diagram of SiC MOSFET input-series output-parallel PSFB converter.

to 1500 V DC voltages and used to feed PSFB converters that are operated uninterruptedly and with high efficiency. The two modules of PSFB converters seen in Fig. 31 that are series connected at the input and parallel connected at the output. This configuration is titled as input-series output parallel (ISOP). The main module of the converter is comprised in H-bridge topology with switching devices $S_1 - S_4$ with antiparallel diodes $D_1 - D_4$ that are the rectifying diodes, L_{T1} is the resonant inductor, including the leakage inductance of transformer, and L_{O1} and C_{O1} are the components of the output filter. The full-bridge rectifier is used at the output side of the phase-shifting full-bridge converter due to the high input voltage of the APU, which can meet the requirements of high output voltage and high power [70–72].

SiC MOSFET power converters with lower losses are preferable for downsizing and the higher efficiency of APU. In case the DC system is operated at 1.5 kV, 3.3 kV MOSFETs are required to form simple circuit configurations. However, the 3.3 kV SiC MOSFETs are quite expensive and the production capacity is very limited. Therefore, circuit configurations allowing the use of lower voltage range devices, such as 1.7 kV MOSFETs, are proposed by many researchers. The DC-DC converter presented in [71] isolates the input side from the output side while converting DC 1.5 kV to DC 700 V. Two full-bridge inverters constituted by 1.7 kV SiC MOSFETs are connected in series on the input side to share the DC 1.5 kV input. The isolation transformer operates at a 40 kHz frequency. Two-diode full-bridge rectifiers are connected in parallel on the output side. Assuming 700 VDC on the output side of the DC-DC converter, the transformer's output voltage is around AC 600 V. It is noted that the mass of the transformer is reduced by 1/1.4 due to the 40 kHz switching frequency when compared to conventional 10 kHz switching by the same type of IGBT power converter. The next generation of high-voltage power semiconductor switching devices is anticipated to employ SiC power semiconductor devices. One of the successful applications of these devices is railway car traction. Nevertheless, technological challenges remain in the high-power and dependable traction application. Designing the parasitic inductance is one of the most crucial technical design difficulties. It is also challenging to design the parasitic inductance. Another technological concern is how much the inductance should be. Less parasitic inductance is desired in order to prevent overvoltage caused by the larger current change ratio.

The traction systems, motor types, and rated speed of Japanese, European, and Chinese HSR systems are listed in Table 4 [68]. The traction systems of HSR are configured in two alternatives: concentrated and distributed tractions. The train in concentrated traction (CT)-HSR utilizes two powered cars one at each end of the train, including vehicles without a tractive force known as trailer cars in the middle. In the CT-HSR configuration, a powered vehicle typically creates the thrust power utilizing a large motor of the megawatt class. On the other hand, the distributed traction (DT)-HSR utilizes motors in the kilowatt range, and the tractive force is shared by a dispersed set of motors placed on the wheels of passenger vehicles. The traction units are also known as electric multiple units (EMUs).

The benefits of a distributed traction system include improved acceleration and deceleration performance, efficient use of regenerative brakes, and a lower axle load. All of this has a direct influence on the performance of energy consumption and reduces the impact on the track.

TABLE 4 Traction system and motor types of HSR applications.

	Motor type	Traction mode	Rated speed (km/h)	Maximum speed (km/h)
Siemens Velaro E	IM	Distributed	250	403
Talgo 350	IM	Concentrated	300	365
ETR 500	IM	Concentrated	300	–
Thalys	IM	Concentrated	300	–
AGV Italo	PMSM	Distributed	360	574.8
Eurostar	PMSM	Distributed	320	574.8
Shinkansen Series 700	IM	Distributed	285	–
Shinkansen Series N700A	IM	Distributed	300	–
Shinkansen Series N700S	IM	Distributed	300	–
China Star	IM	Concentrated	270	321.5
China Railway High Speed	IM	Distributed	250	380
Korea KTX-Sancheon	IM	Distributed	300	330

11.5.3 Summary

The ERPS is comprised of a wide variety of line voltages comprising 1500 V or 3000 V station power buses. This section deals with the common configuration of ERPS and the power electronics converting the ERPS voltage levels to catenary and rolling stock connection.

11.6 Electric propulsion systems for aircraft vehicles

Globally, aviation is responsible for 3% of all greenhouse gas emissions and 2% of CO_2 emissions; over time, this percentage is predicted to increase by 4%–5% yearly. By 2050, the emissions are projected to approximately quadruple, making up 25% of the world's carbon emissions if no action is taken to reduce them. The aviation sector has been able to minimize noise and pollution levels because of advancements in technology. Regulatory organizations have changed certification requirements based on technological maturity [73]. Also, due to high oil prices, developments in battery technologies, and environmental concerns, electrically propelled aircraft have become popular. However, switching from aircraft fueled by fossil fuels to those propelled by electricity requires significant design adjustments. Improvements are especially needed in electricity generation, distribution, and conversion. For instance, certain converter types are employed to power low-power electrical devices, while others are included in propulsion systems where the power need might exceed megawatts [74]. One of them is the more-electric aircraft (MEA) in which progressive electrical systems are activated, and the other is the electric propulsion aircraft (EPA), which fundamentally changes the electric propulsion architecture. Table 5 compares traditional airplane and more-electric aircraft topologies based on the electrical networks and power electronic converters used. In this list, the two largest civil aircraft planes and a military plane are given. As it can be seen from Table 5, the generated power for all aircraft has different values. Furthermore, the architecture of an AC-AC converter is required for varied frequencies like 360–800 Hz. The DC-AC converter is used to start the engine, protect the wing engine, and manage the system [75–77]. In conventional civil aircraft, generators produce a constant-frequency AC bus at 115 V_{AC}/400 Hz, while AC/DC converters transmit a part of the power to the 28 V_{DC} bus. In case of an emergency, the 28 V_{DC} battery is also used to power the loads. Typically, the total electrical power output is less than 300 kVA. The Airbus A320, for example, has two generators that provide a total of 180 kVA. The elimination of the constant speed drive between the engine variable speed shaft and the generators was one of Airbus' major advancements in the A380. This results in bulk and weight reductions, as well as better dependability.

Some military aircraft have a particular architecture that is separate from regular airplanes. This is applicable for the Lockheed Martin F-22 and F-35, which utilize switched reluctance generators to power a 270 V_{DC} bus through AC/DC converters. As a result, the main bus is a DC network, with DC/AC and DC/DC converters supplying the electric loads that require 115 V_{AC} and 28 V_{DC} buses. Furthermore, a 270 V_{DC} Li-ion battery has been installed in the F-35 electrical network to power the APUs during starting operations and to serve as an emergency power supply for flight control systems. This architecture is intriguing since it establishes the primary DC bus for future electric aircraft [75].

An electric propulsion system for aircraft consists of four essential components: the electric power source, the control module, the electric motor, and the fan or propeller that controls the energy flow. In general, there are numerous electric propulsion systems that have been proposed for aircraft. However, hybrid electric, turboelectric, and all electric

TABLE 5 The Comparative list of conventional aircraft and more-electric aircraft topologies.

	Airbus A320	F-35 (military)	Airbus A380 MEA	Boeing 787 MEA
Generated power	2 × 90 kVA	2 × 80 kVA	4 × 150 kVA	4 × 250 kVA
AC bus	115 V/400 Hz	115 V/400 Hz	115 V/400 Hz 115 V/360–800 Hz	115 V/400 Hz 230/360–800 Hz
HC DC bus	–	270 V	±135 V	±250 V
LV DC bus	28 V	28 V	28 V	28 V
Power electronic converters	AC/DC	AC/DC DC/AC DC/DC	AC/DC DC/AC DC/DC	AC/DC DC/AC DC/DC

can be considered as main electric propulsion architectures for aircraft [73,78–80]. In Fig. 32A and B, parallel-hybrid propulsion and series-hybrid propulsion have been shown.

In a hybrid electric propulsion architecture, the energy is obtained by jet engine fuel and/or electric energy storage devices such as battery packs. In a parallel hybrid configuration, a propulsor fan is powered by an electric motor and a jet engine separately or together to achieve higher efficiency. In series hybrid configuration, the propulsor fan is powered by an electric motor.

In Fig. 33, the turboelectric propulsion of aircraft is shown. In this propulsion architecture, the motor is powered by the electric energy provided by the onboard electric generator, which is driven by a jet engine. The propulsor fan is powered by a turbine engine, which is the source of kinetic energy for the turbine. This architecture helps to design more efficient aircraft with higher performance propulsion systems. In Fig. 34, an all-electric aircraft propulsion topology has been given. In this scheme, a battery pack is the only power supply for the aircraft thrust. The main advantage of this architecture is that zero emission can be reached with this propulsion topology. However, because of the energy storage devices, the implementation of this topology is limited.

In turboelectric and hybrid aircraft propulsion, since there are many generators connected to the shafts of the engines, rectifiers known as AC-DC converters are used to convert AC power to DC power, but since the power generation and storage systems are DC in all-electric aircraft, there is no need for an AC-DC converter.

To improve power quality in electric aircraft, several rectifier topologies for multipulse have been developed, and also transformer rectifier units (TRUs) have been used to reach higher voltage ratio conversion. To establish the various frequencies between 360 and 800 Hz, AC-AC converters are needed. Hence, back-to-back converter or matrix converter

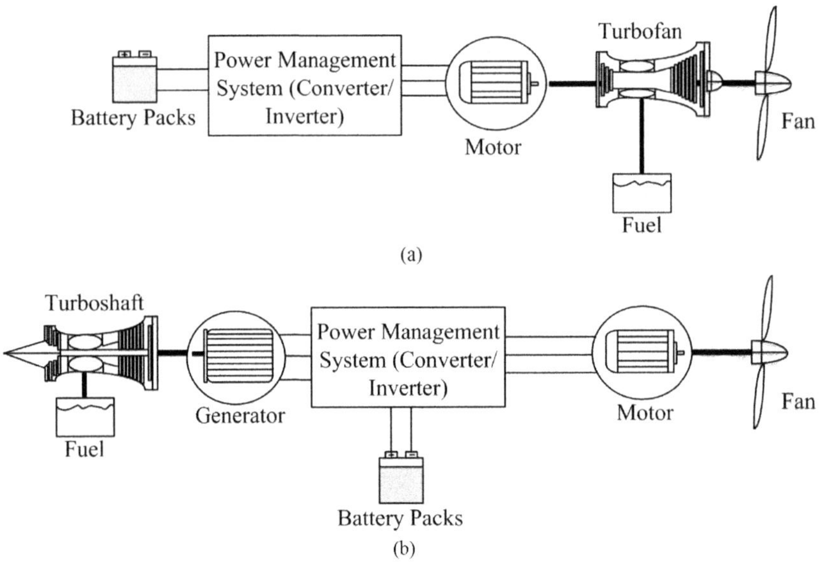

FIG. 32 A scheme of (A) parallel hybrid aircraft propulsion and (B) series hybrid aircraft propulsion.

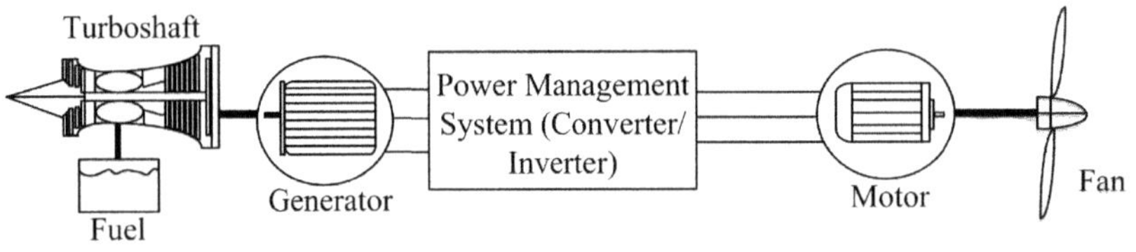

FIG. 33 A scheme of turboelectric aircraft propulsion.

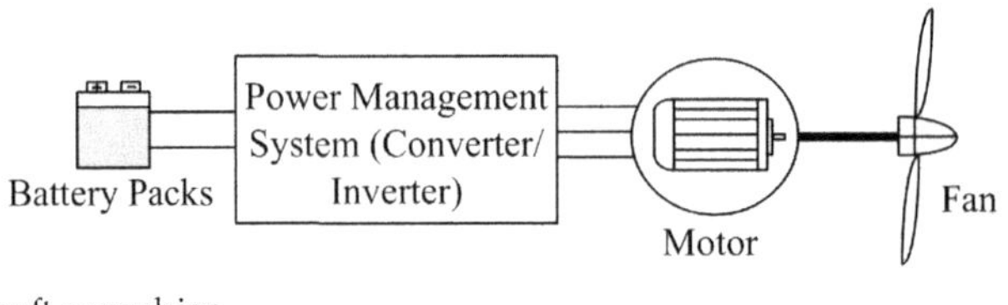

FIG. 34 Schematic of all-electric aircraft propulsion.

topologies have been used and developed [81,82]. A DC-DC converter is another important topology for electric aircraft because it is used to support emergency load. Although buck, boost, and buck-boost converters are the simplest topologies for DC/DC conversion, nowadays topologies with high power density, high efficiency, high reliability, and high flexibility have also been developed. A DA-AC converter that is also known as an inverter is used to supply AC loads. It is used to convert DC to AC under variable voltage and frequency. In the literature, many DC-AC converter topologies such as multilevel inverter, multiphase inverter, the NPC inverter, the T-type inverter, and the Z-source inverter have been proposed and are still under development. Today, all power electronic components and topologies have become essential to designing aircraft because of safety, efficiency, power density, and power quality.

In electric aircraft propulsion systems, power management systems can be considered a key point. With the conversion of AC-DC, AC-AC, DC-DC, and DC-AC, it can control speed, current, and voltage and give stability to all topologies. In the design of power electronic circuits, packing, reliability, EMI shielding, weight, thermal design, cooling requirements, power component requirements, power converter topology type, modulation, control techniques, and so on have to be considered.

11.6.1 Summary

Recent technological developments, environmental concerns, support of governments, and some other factors show that electrical propulsion for large aircraft in long distance flight will be developed in the near future. In these all-technological developments, power electronics can be considered as a key to efficient and effective design because of its voltage, current, and frequency control abilities. In a power management system (PMS), the storage of electrical energy, electrical machines, distribution, and many other factors can be controlled and observed.

References

[1] S. Verma, S. Mishra, A. Gaur, S. Chowdhury, G. Dwivedi, P. Verma, A comprehensive review on energy storage in hybrid electric vehicle, J. Traffic Transp. Eng. (Engl. Ed.) 8 (5) (2021) 621–637, https://doi.org/10.1016/j.jtte.2021.09.001.

[2] A. Emadi, Advanced Electric Drive Vehicles, first ed., CRC Press, Boca Raton, 2014.

[3] F. Barrero, J. Rodas, Control of power electronics converters and electric motor drives, Energies 14 (4591) (2021).

[4] I. Aghabali, J. Bauman, P.J. Kollmeyer, Y. Wang, B. Bilgin, A. Emadi, 800-V electric vehicle powertrains: review and analysis of benefits, challenges, and future trends, IEEE Trans. Transp. Electrif. 7 (3) (2021) 927–948, https://doi.org/10.1109/TTE.2020.3044938.

[5] L. Lu, X. Han, J. Li, J. Hua, M. Ouyang, A review on the key issues for lithium-ion battery management in electric vehicles, J. Power Sources 226 (2013) 272–288, https://doi.org/10.1016/j.jpowsour.2012.10.060.

[6] A. Karki, S. Phuyal, D. Tuladhar, S. Basnet, B.P. Shrestha, Status of pure electric vehicle power train technology and future prospects, Appl. Syst. Innov. 3 (3) (2020), https://doi.org/10.3390/asi3030035 (Art. no. 3).

[7] C.C. Chan, The state of the art of electric, hybrid, and fuel cell vehicles, Proc. IEEE 95 (4) (2007) 704–718, https://doi.org/10.1109/JPROC.2007.892489.

[8] A. Emadi, S.S. Williamson, A. Khaligh, Power electronics intensive solutions for advanced electric, hybrid electric, and fuel cell vehicular power systems, IEEE Trans. Power Electron. 21 (3) (2006) 567–577, https://doi.org/10.1109/TPEL.2006.872378.

[9] A. García, J. Monsalve-Serrano, Analysis of a series hybrid vehicle concept that combines low temperature combustion and biofuels as power source, Results Eng. no. 1 (2019), https://doi.org/10.1016/j.rineng.2019.01.001 (Elsevier Enhanced Reader).

[10] M. Sivertsson, L. Eriksson, Optimal powertrain lock-up transients for a heavy duty series hybrid electric vehicle, FAC Pap. OnLine 50 (1) (2017) 7842–7848, https://doi.org/10.1016/j.ifacol.2017.08.1062.

[11] S.F. Tie, C.W. Tan, A review of energy sources and energy management system in electric vehicles, Renew. Sust. Energ. Rev. 20 (2013) 82–102.

[12] D.W. Gao, C. Mi, A. Emadi, Modeling and simulation of electric and hybrid vehicles, Proc. IEEE 95 (4) (2007) 729–745, https://doi.org/10.1109/JPROC.2006.890127.

[13] D. Huo, P. Meckl, Power management of a plug-in hybrid electric vehicle using neural networks with comparison to other approaches, Energies 15 (5735) (2022), https://doi.org/10.3390/en15155735.

[14] P.G. Anselma, Computationally efficient evaluation of fuel and electrical energy economy of plug-in hybrid electric vehicles with smooth driving constraints, Appl. Energy 307 (2022) 118247, https://doi.org/10.1016/j.apenergy.2021.118247.

[15] R.A. Waraich, K.W. Axhausen, Plug-in hybrid electric vehicles and smart grids: investigations based on a microsimulation, Transp. Res. C 28 (2013), https://doi.org/10.1016/j.trc.2012.10.011 (Elsevier Enhanced Reader).

[16] W. Li, R. Long, H. Chen, J. Geng, The batteries can mainly be charged with a plug-in charger, Renew. Sust. Energ. Rev. 78 (2017) 318–328.

[17] K. Wang, C.Y. Lin, L. Zhu, D. Qu, F.C. Lee, J.S. Lai, Bi-directional DC to DC converters for fuel cell systems, in: Power Electronics in Transportation (Cat. No.98TH8349), 1998, pp. 47–51, https://doi.org/10.1109/PET.1998.731056.

[18] X. Zhao, X. Yan, Energy management strategies for fuel cell hybrid electric vehicles: classification, comparison, and outlook, Energy Convers. Manag. 270 (2022), https://doi.org/10.1016/j.enconman.2022.116179.

[19] D. Gao, Z. Jin, L. Qingchun, Energy management strategy based on fuzzy logic for a fuel cell hybrid bus, J. Power Sources 185 (2008) 311–317.

[20] P.K. Maroti, S. Padmanaban, M.S. Bhaskar, V.K. Ramachandaramurthy, F. Blaabjerg, The state-of-the-art of power electronics converters configurations in electric vehicle technologies, Power Electron. Devices Compon. 1 (2022) 100001, https://doi.org/10.1016/j.pedc.2021.100001.

[21] M. Ehsani, K.V. Singh, H.O. Bansal, R.T. Mehrjardi, State of the art and trends in electric and hybrid electric vehicles, Proc. IEEE 109 (6) (2021) 967–984, https://doi.org/10.1109/JPROC.2021.3072788.

[22] M.A. González, M.F. Escalante, Traction system for an EV based on induction motor and 3-level NPC inverter multilevel converters, in: 12th IEEE International Power Electronics Congress, Aug, 2010, pp. 73–77, https://doi.org/10.1109/CIEP.2010.5598833.

[23] R. Teichmann, S. Bernet, A comparison of three-level converters versus two-level converters for low-voltage drives, traction, and utility applications, IEEE Trans. Ind. Appl. 41 (3) (2005) 855–865, https://doi.org/10.1109/TIA.2005.847285.

[24] A. Choudhury, P. Pillay, S.S. Williamson, DC-link voltage balancing for a three-level electric vehicle traction inverter using an innovative switching sequence control scheme, IEEE J. Emerg. Sel. Top. Power Electron. 2 (2) (2014) 296–307, https://doi.org/10.1109/JESTPE.2013.2296973.

[25] E. Kabalcı, A. Boyar, Multilevel inverter applications for electric vehicle drives, in: Multilevel Inverters, Elsevier, 2021, pp. 185–208, https://doi.org/10.1016/B978-0-323-90217-5.00006-X.

[26] Kia Soul EV. Accessed: Jun. 15, 2020. [Online]. Available: https://www.kiamedia.com/us/en/models/soul-ev/2018/specifications.

[27] Nissan Leaf S. Accessed: Jul. 06, 2020. [Online]. Available: https://www.nissanusa.com/vehicles/electric-cars/leaf.html.

[28] BMW i3. Accessed: Jul. 06, 2020. [Online]. Available: https://www.bmwusa.com/vehicles/bmwi/i3/sedan/pricing-features.html.

[29] Tesla Model S. Accessed: Jun. 15, 2020. [Online]. Available: https://www.tesla.com/models.

[30] Hyundai Kona. Access: Jul. 06, 2020. [Online]. Available: https://www.hyundai.co.uk/new-cars/kona-electric.

[31] K. Rajashekara, Present status and future trends in electric vehicle propulsion technologies, IEEE J. Emerg. Sel. Top. Power Electron. 1 (1) (2013) 3–10, https://doi.org/10.1109/JESTPE.2013.2259614.

[32] J. Reimers, L. Dorn-Gomba, C. Mak, A. Emadi, Automotive traction inverters: current status and future trends, IEEE Trans. Veh. Technol. 68 (4) (2019) 3337–3350, https://doi.org/10.1109/TVT.2019.2897899.

[33] J. de Santiago, et al., Electrical motor drivelines in commercial all-electric vehicles: a review, IEEE Trans. Veh. Technol. 61 (2) (2012) 475–484, https://doi.org/10.1109/TVT.2011.2177873.

[34] M.R. Wahid, B.A. Budiman, E. Joelianto, M. Aziz, A review on drive train technologies for passenger electric vehicles, Energies 14 (20) (2021) 6742, https://doi.org/10.3390/en14206742.

[35] H. El Hadraoui, M. Zegrari, A. Chebak, O. Laayati, N. Guennouni, A multi-criteria analysis and trends of electric motors for electric vehicles, World Electr. Veh. J. 13 (4) (2022) 65, https://doi.org/10.3390/wevj13040065.

[36] C.M.F.S. Reza, M.D. Islam, S. Mekhilef, A review of reliable and energy efficient direct torque controlled induction motor drives, Renew. Sust. Energ. Rev. 37 (2014) 919–932, https://doi.org/10.1016/j.rser.2014.05.067.

[37] E. Kabalci, Multilevel Inverters: Control Methods and Advanced Power Electronic Applications, Academic Press, London, 2021. Accessed: Aug. 02, 2022. [Online]. Available: https://search.ebscohost.com/login.aspx?direct=true&scope=site&db=nlebk&db=nlabk&AN=2712992.

[38] E. Kabalci, Chapter 9—power electronics and drives used in automotive applications, in: Autonomous Vehicles: Intelligent Transport Systems and Smart Technologies, Nova Publishers, New York, 2014, pp. 249–274.

[39] D. Wang, C. Peng, J. Li, C. Wang, Comparison and experimental verification of different approaches to suppress torque ripple and vibrations of interior permanent magnet synchronous motor for EV, IEEE Trans. Ind. Electron. 70 (3) (2023) 2209–2220, https://doi.org/10.1109/TIE.2022.3156034.

[40] X. Zhang, B. Wang, Y. Yu, J. Zhang, D. Xu, Torque adaptive hexagon voltage extension method for PMSM flux-weakening control based on dual PI Cascade structure, IEEE Trans. Power Electron. 38 (1) (2023) 332–345, https://doi.org/10.1109/TPEL.2022.3200976.

[41] D. Pasqualotto, S. Rigon, M. Zigliotto, Sensorless speed control of synchronous reluctance motor drives based on extended Kalman filter and neural magnetic model, IEEE Trans. Ind. Electron. 70 (2) (2023) 1321–1330, https://doi.org/10.1109/TIE.2022.3159962.

[42] H. Hadla, F. Santos, Performance comparison of field-oriented control, direct torque control, and model-predictive control for SynRMs, Chin. J. Electr. Eng. 8 (1) (2022) 24–37, https://doi.org/10.23919/CJEE.2022.000003.

[43] S. Bhattacharya, D. Mascarella, G. Joos, J.-M. Cyr, J. Xu, A dual three-level T-NPC inverter for high-power traction applications, IEEE J. Emerg. Sel. Top. Power Electron. 4 (2) (2016) 668–678, https://doi.org/10.1109/JESTPE.2016.2517819.

[44] K. Sayed, A. Almutairi, N. Albagami, O. Alrumayh, A.G. Abo-Khalil, H. Saleeb, A review of DC-AC converters for electric vehicle applications, Energies 15 (3) (2022) 1241, https://doi.org/10.3390/en15031241.

[45] A. Matallana, et al., Power module electronics in HEV/EV applications: new trends in wide-bandgap semiconductor technologies and design aspects, Renew. Sust. Energ. Rev. 113 (2019) 109264, https://doi.org/10.1016/j.rser.2019.109264.

[46] A. Nabae, I. Takahashi, H. Akagi, A new neutral-point-clamped PWM inverter, IEEE Trans. Ind. Appl. IA-17 (5) (1981) 518–523, https://doi.org/10.1109/TIA.1981.4503992.

[47] I. Colak, E. Kabalci, R. Bayindir, Review of multilevel voltage source inverter topologies and control schemes, Energy Convers. Manag. 52 (2) (2011) 1114–1128, https://doi.org/10.1016/j.enconman.2010.09.006.

[48] J. Rodriguez, S. Bernet, B. Wu, J.O. Pontt, S. Kouro, Multilevel voltage-source-converter topologies for industrial medium-voltage drives, IEEE Trans. Ind. Electron. 54 (6) (2007) 2930–2945, https://doi.org/10.1109/TIE.2007.907044.

[49] B. Chikondra, U.R. Muduli, R. Kumar Behera, Fault-tolerant DTC technique for five-phase three-level NPC inverter-fed induction motor drive with an open-phase fault, in: 2019 IEEE Energy Conversion Congress and Exposition (ECCE), 2019, pp. 5281–5287, https://doi.org/10.1109/ECCE.2019.8912513.

[50] T. Bruckner, S. Bernet, H. Guldner, The active NPC converter and its loss-balancing control, IEEE Trans. Ind. Electron. 52 (3) (2005) 855–868, https://doi.org/10.1109/TIE.2005.847586.

[51] E. Kabalci, Multilevel Inverters for Emergent Topologies and Advanced Power Electronics Applications, Elsevier, Waltham, 2021.

[52] J. Chen, C. Zhang, A. Chen, X. Xing, Fault-tolerant control strategies for T-type three-level inverters considering neutral-point voltage oscillations, IEEE Trans. Ind. Electron. 66 (4) (2019) 2837–2846, https://doi.org/10.1109/TIE.2018.2842731.

[53] S.E. Damian, et al., Review on the challenges of hybrid propulsion system in marine transport system, J. Energy Storage 56 (2022) 105983, https://doi.org/10.1016/j.est.2022.105983.

[54] O.B. Inal, J.-F. Charpentier, C. Deniz, Hybrid power and propulsion systems for ships: current status and future challenges, Renew. Sust. Energ. Rev. 156 (2022) 111965, https://doi.org/10.1016/j.rser.2021.111965.

[55] D. Tamunodukobipi, D. Tamunodukobipi, A. Sidum, Design analysis of a lightweight solar powered system for recreational marine craft, World J. Eng. Technol. 6 (2018) 441–456.

[56] S. Nitonye, S. Adumene, U.U. Howells, Numerical design and performance analysis of a tug boat propulsion system, J. Power Energy Eng. 05 (11) (2017) 80–98, https://doi.org/10.4236/jpee.2017.511007.
[57] M. Okubo, T. Kuwahara, New Technologies for Emission Control in Marine Diesel Engines, Butterworth-Heinemann/Elsevier, 2020.
[58] N. Shakeri, M. Zadeh, J. Bremnes Nielsen, Hydrogen fuel cells for ship electric propulsion: moving toward greener ships, IEEE Electrific. Mag. 8 (2) (2020) 27–43, https://doi.org/10.1109/MELE.2020.2985484.
[59] T. Tanuma, Advances in Steam Turbines for Modern Power Plants, Woodhead Publishing/Elsevier, 2017.
[60] S.A. Stout, Z. Wang, Standard Handbook Oil Spill Environmental Forensics, second ed., Elsevier, Rockland, MA, 2016.
[61] H.I.H. Saravanamuttoo, G.F.C. Rogers, H. Cohen, P. Straznicky, Gas Turbine Theory, seventh ed., Pearson Education, 2008.
[62] S. Danielsen, O.B. Fosso, M. Molinas, J.A. Suul, T. Toftevaag, Simplified models of a single-phase power electronic inverter for railway power system stability analysis—development and evaluation, Electr. Power Syst. Res. 80 (2) (2010) 204–214, https://doi.org/10.1016/j.epsr.2009.09.003.
[63] D. Serrano-Jiménez, L. Abrahamsson, S. Castaño-Solís, J. Sanz-Feito, Electrical railway power supply systems: current situation and future trends, Int. J. Electr. Power Energy Syst. 92 (2017) 181–192, https://doi.org/10.1016/j.ijepes.2017.05.008.
[64] H.H. Alnuman, D.T. Gladwin, M.P. Foster, E.M. Ahmed, Enhancing energy management of a stationary energy storage system in a DC electric railway using fuzzy logic control, Int. J. Electr. Power Energy Syst. 142 (2022) 108345, https://doi.org/10.1016/j.ijepes.2022.108345.
[65] K. Almaksour, et al., Comparison of dynamic models for a DC railway electrical network including an AC/DC bi-directional power station, Math. Comput. Simul. 184 (2021) 244–266, https://doi.org/10.1016/j.matcom.2020.05.027.
[66] C.H. Bae, A simulation study of installation locations and capacity of regenerative absorption inverters in DC 1500V electric railways system, Simul. Model. Pract. Theory 17 (5) (2009) 829–838, https://doi.org/10.1016/j.simpat.2009.02.003.
[67] N.M. Tabatabaei, A.J. Aghbolaghi, N. Bizon, F. Blaabjerg, Reactive Power Control in AC Power Systems, Springer Berlin Heidelberg, New York, NY, 2017.
[68] S. Paul, P.-W. Han, J. Chang, Y.-D. Chun, J.-G. Lee, State-of-the-art review of railway traction motors for distributed traction considering South Korean high-speed railway, Energy Rep. 8 (2022) 14623–14642, https://doi.org/10.1016/j.egyr.2022.10.411.
[69] Fuji Electric Corp. of America, Electrical Equipment for High Speed Rolling Stock, Fuji Electric, 2016.
[70] J.-P. Lee, B.-D. Min, T.-J. Kim, D.-W. Yoo, J. Yoo, Input-series-output-parallel connected DC/DC converter for a photovoltaic PCS with high efficiency under a wide load range, J. Power Electron. 10 (1) (2010) 9–13, https://doi.org/10.6113/JPE.2010.10.1.009.
[71] H. Hayashiya, K. Kondo, Recent trends in power electronics applications as solutions in electric railways, IEEJ Trans. Electr. Electron. Eng. 15 (5) (2020) 632–645, https://doi.org/10.1002/tee.23121.
[72] M. Zhang, S. Wang, Y. Sun, Z. Guo, J. Bao, A phase-shifted full-bridge converter used for DC charging pile, IOP Conf. Ser. Mater. Sci. Eng. 768 (6) (2020) 062010, https://doi.org/10.1088/1757-899X/768/6/062010.
[73] P. Wheeler, T.S. Sirimanna, S. Bozhko, K.S. Haran, Electric/hybrid-electric aircraft propulsion systems, Proc. IEEE 109 (6) (2021) 1115–1127, https://doi.org/10.1109/JPROC.2021.3073291.
[74] B. Rahrovi, M. Ehsani, A review of the more electric aircraft power electronics, in: 2019 IEEE Texas Power and Energy Conference (TPEC), 2019, pp. 1–6, https://doi.org/10.1109/TPEC.2019.8662158.
[75] L. Dorn-Gomba, J. Ramoul, J. Reimers, A. Emadi, Power electronic converters in electric aircraft: current status, challenges, and emerging technologies, IEEE Trans. Transp. Electrif. 6 (4) (2020) 1648–1664, https://doi.org/10.1109/TTE.2020.3006045.
[76] T. Yang, F. Gao, S. Bozhko, P. Wheeler, Power Electronic Systems for Aircraft, Elsevier, The Netherlands, 2018.
[77] Ayse Kocalmis BILHAN and Sedat SUNTER, Simulation and Design of Three-Level Cascaded Inverter Based on soft computing method, Teh. Vjesn. Tech. Gaz. 27 (2) (2020), https://doi.org/10.17559/TV-20180503115747.
[78] M. Doff-Sotta, M. Cannon, M. Bacic, Predictive energy management for hybrid electric aircraft propulsion systems, IEEE Trans. Control Syst. Technol. (2022) 602–614, https://doi.org/10.1109/TCST.2022.3193295.
[79] J. Benzaquen, J. He, B. Mirafzal, Toward more electric powertrains in aircraft: technical challenges and advancements, CES Trans. Electr. Mach. Syst. 5 (3) (2021) 177–193, https://doi.org/10.30941/CESTEMS.2021.00022.
[80] J. Hoelzen, et al., Conceptual Design of Operation Strategies for hybrid electric aircraft, Energies 11 (217) (2018).
[81] J.W. Kolar, T. Friedli, J. Rodriguez, P.W. Wheeler, Review of three-phase PWM AC–AC converter topologies, IEEE Trans. Ind. Electron. 58 (11) (2011) 4988–5006, https://doi.org/10.1109/TIE.2011.2159353.
[82] T. Friedli, J.W. Kolar, J. Rodriguez, P.W. Wheeler, Comparative evaluation of three-phase AC–AC matrix converter and voltage DC-Link Back-to-Back converter systems, IEEE Trans. Ind. Electron. 59 (12) (2012) 4487–4510, https://doi.org/10.1109/TIE.2011.2179278.

C H A P T E R

12

Automotive applications of power electronics

Anushree Ramanath
Schneider Electric, San Jose, CA, United States

12.1 Introduction

The automotive industry sets high standards for electronic components, necessitating the development of highly reliable, cost-effective, and efficient traction inverters. As the automotive landscape undergoes a transformative shift toward cleaner vehicle technologies, power electronic interfaces emerge as pivotal drivers of this evolution. Advancements in power electronics are therefore imperative for modern automotive powertrains, shaping the trajectory of clean energy adoption in the transportation sector. Against the backdrop of escalating climate change concerns and mounting environmental pressures, the imperative to transition toward cleaner vehicle technologies has never been more urgent for automobile manufacturers. In response to these challenges, governments worldwide are actively promoting the development and adoption of zero-emission vehicles (ZEVs) through legislative regulations and incentives, further catalyzing industry efforts toward sustainable mobility solutions. By championing innovation and collaboration across the automotive ecosystem, stakeholders can collectively pave the way for a greener, more sustainable future, driving toward a world where clean energy technologies power the vehicles of tomorrow.

Power electronics serve as the backbone for supporting a diverse array of loads critical for ensuring the safe operation of automobiles, enhancing passenger comfort, and enabling advanced technical features for drivers. In today's automotive landscape, electrical systems must seamlessly accommodate an assortment of loads, spanning mechanical, electrical, electromechanical, and electronic domains. The cornerstone for integrating newer varieties of loads, thereby facilitating feature additions and functional enhancements, lies in the advancement of power electronic systems. This necessitates continuous innovation not only in power electronic interfaces but also in associated control strategies and battery technologies. Essential performance metrics such as efficiency, reliability, size, cost, driving range, and charging time underscore the significance of optimizing power electronic systems for automotive applications. Addressing the inherent limitations and adversities pertaining to these parameters poses significant design challenges, driving ongoing research and development efforts in the field. By confronting these challenges head-on, stakeholders can unlock new avenues for innovation, propelling the automotive industry toward a future defined by enhanced performance, efficiency, and sustainability.

This chapter discusses the design, implementation, application, and control of power electronics in automotive applications. Section 12.2 provides a sneak peek into the history of automotive vehicles and state-of-the-art architectures of the automotive electric power system. Section 12.3 describes the significance of power electronic systems especially in automotive applications. Section 12.4 details the design considerations used for the design of automotive power electronics including voltage and current ranges, temperature values, compliance and regulatory requirements, environmental variables, and other performance parameters. Section 12.5 describes the different components that make up an automotive power electronic system including the air conditioning, lighting, heating, actuators, sensors, motors, and other loads. Section 12.6 elaborates on the process of power conversion and a glimpse of automotive power electronics in action. Section 12.7 lists the applications of power electronics in electric vehicles and hybrid electric vehicles. Section 12.8 delves into the current state of automotive power electronic systems by detailing the automobile manufacturers that focus on clean energy, their offerings, and key specifications. Section 12.9 provides a glimpse into the prospective future trends and some planned configurations of automotive power electronic systems. Section 12.10 provides the summary of this chapter revisiting the key points for automotive applications of power electronic systems.

Handbook of Power Electronics in Autonomous and Electric Vehicles
https://doi.org/10.1016/B978-0-323-99545-0.00015-4

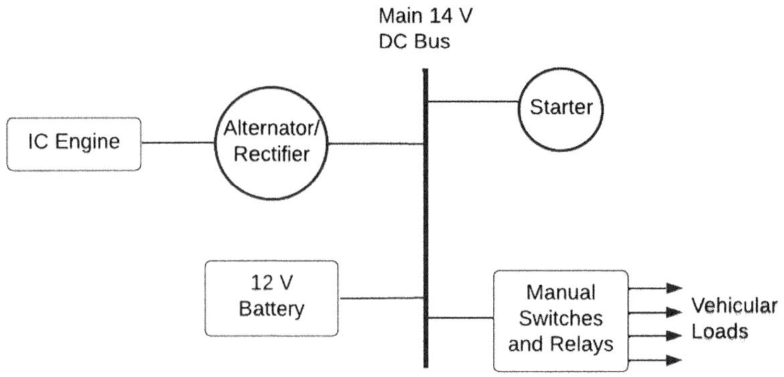

FIG. 1 Conventional 14 V DC power system architecture [3].

12.2 Historical evolution of the automotive industry

During the mid-1950s, the automotive industry decided to go for 12 V electrical power systems for vehicles, since the then-popular 6 V system was rapidly becoming obsolete by the increasing vehicular load demands. The battery became a six-cell module instead of three cells, at approximately the same energy rating. The electrical system demand had risen from 100 W in the early 1900s to typically about 1 kW by the late 1990s, as more and more electrically powered devices were installed [1].

The conventional electrical system in an automobile can essentially be divided into the architectural elements of energy storage, generation, starting, and distribution. The distribution system of a conventional 14 V power system satisfies vehicular loads such as interior/exterior lighting, electric motor-driven fans/pumps/compressors, and instrumentation subsystems [2]. The block representation of the conventional 14 V electric power system architecture is shown in Fig. 1.

The conventional power system has a single 14 V DC voltage level with the vehicular loads being controlled by manual switches and relays. The present average power demand in an automobile is approximately 1 kW. The voltage in a 14 V system actually varies between 9 and 16 V at the battery terminals, depending on the alternator output current, battery age, state of charge, and various other factors [4]. This results in overrating the loads at nominal system voltage. Also, the 14 V system cannot handle future electrical loads to be introduced in the more electric environment of future vehicles, as it would be expensive and inefficient to do so.

The development of EVs has been through a lot of flux. In 1990, General Motors announced its electric vehicle (EV) called Impact, which could accelerate up to 62 miles/h in just 8 s and had a top speed of 100 miles/h with a range of about 150 miles between charges. Electronic systems for lighting, ignition, engine management, entertainment, and other automobile applications have been prevalent for decades. However, in recent years, electronic systems have grown in terms of the functionalities offered, thus making them a critical component of the total automobile cost. In the 1950s, electronic systems constituted about 1% of the cost of the entire automobile; this value grew to about 30% in 2010 [5]. This number is anticipated to grow to around 50% by 2030.

The key drivers for multi-voltage electric systems in automobiles are the need for higher power to enable sophisticated functionalities and the wish to improve the environment through the use of cleaner technologies. The enabling technology for driving the cost of power electronic circuitry down are the advances in the semiconductor industry. The costs of solid-state power converters and switches have seen steady declines in the 1990s [6]. This has enabled the use of complex circuitry in automobiles to offer a wide variety of features and support technological advances with enhanced functional offerings.

12.3 Significance of power electronic systems in automotive applications

Automotive industry has a high demand for cost-effective, reliable, and compact power electronic systems. As the installation space within the engine compartment is less, modular, highly integrable, and small-size power electronics are preferred. Embedding the power switches, control chips, and other passive components onto printed circuit boards (PCBs) helps in making modules more compact. Also, this helps in better thermal management and power density offered. Embedding also helps in the reduction of wires and solder inefficiencies. This directly translates to better modularity with shorter paths, faster reaction times, isolation due to lamination, enhanced thermal performance, and efficient packaging.

Power electronic systems in automobiles play a crucial role in the basic operation of the engine, functioning as starters and supporting essential features such as engine control, spark ignition, and cruise control. These systems ensure the fundamental requirements for a functional automobile are met, including vital components like lighting, wipers, horns, and blower fans. Beyond these necessities, power electronic systems also enhance passenger comfort with amenities such as seat warmers, air conditioning, and entertainment systems. These various loads are efficiently managed by connecting them in parallel to the battery and alternator, and in series with switching controls and the fuse box, thereby enabling seamless operation and improved vehicle functionality.

12.4 Design considerations for power electronic systems

The design of automotive power electronic equipment is strongly impacted by the adverse environmental conditions that the automobile has to survive. The key aspects include static and transient voltage ranges, extreme temperatures, mechanical vibrations, electromagnetic interference and electromagnetic compatibility (EMI/EMC) requirements, and other environmental conditions.

12.4.1 Voltage ranges

The range of static voltages under normal operating conditions is expected to be around 11–16 V [7]. That is the range of voltage the bus is maintained at as well. To address potential issues such as incorrect battery installation with reversed polarity, faulty regulators, or other static over-voltage conditions, automotive electronics are designed to accommodate a wider range of voltages. This ensures robust and reliable performance under various adverse conditions, enhancing the overall durability and safety of the vehicle's electrical system. It might not be practical to always over design for worst case but there are always healthy margins accommodated to prevent any common human or system errors. The range of static voltages that can be expected in the automotive electric system is summarized in Table 1.

12.4.2 Temperature ranges

The EVs need to operate in extreme weather conditions and during adverse situations without breaking down (see Table 2). The most challenging environmental characteristic is the extreme range of temperatures that occur in the automobile. However, the performance of the system is temperature dependent. The higher the temperature is, the higher the losses will be; since this is positive feedback, it holds till breakdown. In the passive components present in power electronic devices, high temperatures can lead to damage to isolating materials, conductive materials, and magnetic components. This in turn affects the device hardware and the driving range of EVs. Typically, 125°C

TABLE 1 Static voltage range for automotive electric system [7].

Static voltage condition	Voltage (V)
Nominal voltage with engine on	14.2
Nominal voltage with engine off	12.6
Maximum normal operating voltage	16
Minimum normal operating voltage	9
Minimum voltage during starting	4.5
Jump-start voltage	24
Reverse battery voltage	−12
Maximum voltage with alternator-regulator failure followed by battery failure	130

TABLE 2 Automotive temperature extremes by location [7].

Vehicle location	Min temp. (°C)	Max temp. (°C)
Exterior	−40	85
Chassis		
Isolated	−40	85
Near heat source	−40	121
Drivetrain high-temperature location	−40	177
Interior		
Floor	−40	85
Rear deck	−40	104
Instrument panel	−40	85
Instrument panel top	−40	177
Trunk	−40	85
Under hood		
Near radiator support structure	−40	100
Intake manifold	−40	121
Near alternator	−40	131
Exhaust manifold	−40	649
Dash panel (normal)	−40	121
Dash panel (extreme)	−40	141

is the nominal upper limit for silicon-based devices that are used in most of the power electronic systems while ambient temperatures as low as −40°C may be evidenced during the operation of the EV.

12.4.3 Mechanical vibrations

Electric vehicles (EVs) typically experience fewer mechanical vibrations and shocks under adverse operating conditions compared to conventional vehicles. Despite this advantage, the design of automotive and associated power electronic systems must still account for these factors, ensuring that the mechanical design can withstand such adversities. Additionally, EVs generally feature fewer mechanical components susceptible to wear and tear in the drivetrain. To ensure compliance and reliability, it is crucial to conduct thorough vibration and dynamic analysis of both the EV and its power electronic systems. This analysis is directly related to minimizing external noise, reducing the need for frequent maintenance, and addressing potential durability issues.

12.4.4 EMI/EMC requirements

It is imperative that the power electronic circuits designed for automotive applications must exhibit electromagnetic compatibility and not suffer from electromagnetic interference. This means that the conducted and radiated emissions generated by the electronic circuit must not interfere with the other equipment on the vehicle. Also, the circuit must be designed such that it is inherently immune to radiated and conducted disturbances. There have been several technical reports in this regard and standard codes, which need to be followed. EVs that are built need to be compliant with the relevant standard set, which includes immunity to power lead transients and limitations around conducted emissions. Also, limits for the amount of EMI that an automotive electronic component can generate are fixed. The limits for both conducted and radiated emissions are specified in SAE standards [7].

12.4.5 Other environmental adversities

Designing automotive electronics requires careful consideration of numerous adverse operating conditions from the conceptualization phase through implementation. Key environmental factors influencing these designs include temperature, humidity, mechanical shock, immersion, exposure to sand, and various chemical substances. Environmental design goals are established to provide quantitative guidelines for these conditions. Given the significant impact of ambient temperature on power electronic systems, it is essential to collaborate closely with vehicle manufacturers to establish precise temperature specifications for specific applications. Besides temperature extremes, thermal cycling and mechanical shock are critical considerations due to their direct effect on component reliability. Additionally, high humidity, frost, particle intrusion (such as sand and gravel), and corrosion from external factors must all be addressed to ensure the robustness and durability of automotive power electronic equipment.

12.5 Components of automotive power electronic systems

The core of power electronic systems comprises electric machines (EMs), which operate independently or work in conjunction with an internal combustion engine (ICE). The most commonly employed electric machines are the AC machines, which are induction type or permanent magnet based. In the motoring mode, a power electronics-based traction inverter drives them. It enables the conversion of direct current to alternating current at a variable frequency while extracting the energy from the battery. In the generating mode, energy is returned to the battery as the electric machine functions as a generator and the inverter now operates as a rectifier.

Over the last few decades, power electronic systems have been instrumental in introducing several functionalities in EVs such as traction control, anti-lock braking system, advanced traction control, and electrification of some of the existing features in the automobile. There are several essential functions such as speed control, and power steering that the power electronics facilitate along with passenger comfort features such as heating, and air conditioning. This list will keep growing but it is essential to have an understanding of how power electronic systems play a crucial role in all of these aspects [8].

12.5.1 Lighting

Lighting is a crucial aspect of an automobile. From headlights and fog lights to interior lighting, there is a need for different kinds of lights in an automobile. Initially, halogen lamps and the like were used for such applications. However, high-intensity discharge (HID) lamps with over three times the luminous efficacy than that of halogen lamps are now being used [9]. It also has higher reliability, longer life (about three times more than that of halogen lamps), and greater styling flexibility compared to traditional halogen lamps. These lamps require power electronic ballasts and a high-voltage spike to ignite. Usually, a step-up DC converter like a boost or fly-back converter is employed to elevate the voltage from 12 V to the necessary steady state, typically ranging between 10 and 30 V. Subsequently, an H-bridge is utilized to generate the AC voltage that maintains lamp illumination at a steady state. Additionally, a starter circuit is integrated to initiate lamp activation.

Incandescent lamps are another alternative to lighting with about an order of magnitude lesser cost than that of HID lamps. Although incandescent lamps that can operate at higher voltages might not be feasible due to longer thinner filament implementation that might not be all that reliable, they can still be made to work at higher voltages using pulse-width modulation (PWM). This makes it a candidate for use in headlight applications. In PWM-based lighting system, the switches are modulated to apply a periodic pulsed voltage to the lamp filament thus facilitating lighting of the incandescent lamp at a higher voltage of around 42 V. The advantages of such lighting include control of lighting intensity irrespective of the bus voltage, ability to adapt low-voltage 12 V lights to high-voltage 42 V bus cost-effectively, implementation of multiple intensities due to PWM-based control, and enhanced reliability accompanied with robustness.

12.5.2 Air conditioning

In an electric air conditioner, an electric motor is used to drive the compressor. The motor is typically a three-phase brushless DC motor driven by a three-phase transistor bridge [10]. This has typically been an engine-driven function but it clearly benefits from electrification because the speed of the compressor is independent of the engine speed and electrification gives the flexibility to place it anywhere in the car. It also eliminates belts and pulleys and facilitates improved design accompanied by control while allowing the operation even when the engine is off. This makes a good case for replacing some of the engine-driven functions of the vehicle with its electrically driven counterparts.

12.5.3 Power steering

Another function that was engine-driven but benefits from being electrically driven is that of power steering. A hydraulic power steering system can be replaced with an electric power steering (EPS) system that utilizes a brushless DC motor to provide steering assistance [11]. Needless to say, this is more efficient as the motor operates only when needed and it alone can yield about 80% energy savings as compared to the conventional system. However, cost and reliability are the two other aspects that need to be looked into before making the transition for such a critical subsystem.

12.5.4 Speed control

There are motors in the vehicles that require variable speed control. They are typically permanent magnet DC motors with a squirrel cage fan, and the speed can be controlled by varying the resistance connected in series with the motor winding. This could lead to significant power loss as in the case of a blower motor that provides air flow to the passenger compartment. An alternative method to control the speed is by using semiconductor devices and power electronic systems that incur lower losses comparatively. The control is accomplished by pulse-width modulation that is switching of the transistor on and off with different duty ratios for different speed settings along with an input filter. This setup corresponds to that of a variable output DC-DC converter. Instead, a three-phase brushless DC motor can also be employed for controlling the speed of the motor by controlling the switches in the DC-AC converter that drives the motor.

12.5.5 Actuators

Ultrasonic motors are employed as actuators for window lifts, seat positioning, and head restraints in automobiles [12]. Among various ultrasonic motors, the traveling wave type stands out as one of the most favored due to its compact design. Operating on the principle of converting piezoelectrically induced ultrasonic vibrations within an elastic body into unidirectional motion, these motors boast exceptional torque density and dynamics, coupled with silent operation, rendering them highly desirable. To activate the ultrasonic motor, power electronics play a crucial role in the drive circuit. A typical power electronic drive setup involves two H-bridges, meticulously controlled to generate waveforms precisely 90 degrees out of phase with each other, ensuring efficient operation and optimal performance of the ultrasonic motor.

12.5.6 Engine valves

Engine valves are also actuators but are electromagnetics based and are more desirable for automotive applications. That is due to the fact that they can be more easily controlled by a microprocessor to provide more precise control. An application of electromagnetic actuators that is of particular interest is the replacement of the camshaft and tappet valve assembly by electromechanically driven engine valves [13]. The opening and closing of the intake and exhaust valves can be controlled to achieve optimum engine performance and improved fuel economy over a wide range of conditions determined by variables such as speed, load, altitude, and temperature. Power electronic circuits come in handy for driving the solenoids for calve actuation. This can be accomplished using two switches and two diodes for solenoids that require unidirectional currents through their coils and using four switches for solenoids that require bidirectional currents through their coils across the voltage source.

12.6 Power conversion in electric vehicles—An example

A power conversion system (PCS) is typically a device used for bidirectional conversion of electrical energy connected between the battery and the loads. There are several power electronic components involved in the process, and this section details the nuances of design, implementation, and control of such devices and how they fit into the drivetrain.

12.6.1 Inverter components

The inverter comprises power electronic components that facilitate switching and inversion along with a microcontroller or digital signal processor (DSP) to implement the motor control software. Control algorithms are deployed on the microcontroller and typically on a separate control board within the inverter dedicated to motor control and modulation. As the features offered have increased, the control algorithms have increased rapidly leading to higher complexity and switching speed requirements. In such cases, field programmable gate array (FPGA) hardware is employed for modulation and DSP is kept only for achieving motor control [14].

Semiconductor switches are one of the most important components of the inverter. Typically, insulated gate bipolar transistors (IGBTs) have been heavily employed for traction inverters due to their ease of availability, high power capability, and low-cost options. To achieve even higher power capabilities, power modules are built using these devices and are connected in parallel.

To switch the IGBTs per the requirement, gate drivers are utilized. They support the dynamic behavior of the switches and help increase the system's efficiency. They are typically integrated circuits that enable switching of the transistors by means of control signals and provide feedback through the freewheeling diodes for protection. The design of gate drivers needs to consider the aspects related to the switching performance of the power electronic component while providing isolation using passive circuits or closed-loop circuits [15].

Sensors play a pivotal role in the inverter control board, providing essential excitation signals and enabling precise conversion and feedback control. Position sensors are employed to ascertain the rotor angle of the electric machine, while current sensors accurately measure the inverter output current. In certain cases, Hall effect sensors may also be utilized for open-loop operation and isolation purposes, enhancing the versatility and functionality of the system. These sensors collectively ensure optimal performance and efficiency of the power electronic system.

DC link capacitor is another key component of the inverter and it serves as a protection device. It helps filter out the current and voltage spikes both on the input and output sides. The choice of the capacitor is based on the voltage and current ripple requirements that need to be adhered to. Film capacitors can be used in such cases as they are efficient, low in cost, and have high power capability. It is also found to be better than the use of electrolytic capacitors that are more sensitive to higher temperatures [16].

Thermal management systems such as heat sinks are crucial for any inverter. Due to the increased power demands, there have been concerns related to heat dissipation leading to overheating, failure, and less reliability [17]. However, the precise design of heat sinks is very complicated due to the requirements of loss and effective thermal models. Thus, cooling techniques are researched in parallel, and liquid cooling is also carried out to dissipate the heat generated along with housing necessary thermal management systems.

High-power inverters also employ busbars to handle current instead of printed circuit boards connected using wires. During design, switching events, voltage and currents generated and operating environment conditions are accounted for. This makes the design slightly complex even though it offers higher power density and better current handling.

12.6.2 Inverter design

The production of EVs in recent years typically uses three-phase voltage source inverters (VSIs) that are made of insulated gate bipolar transistors (IG-BTs). This is because they offer high efficiency and reliability at low cost while enabling easy manufacturability and simple control of switching devices. The battery pack is either connected directly to the inverter DC input or a boost converter can be used in between to step up the battery voltage [18]. A DC link capacitor is connected in parallel to smooth the ripple current by reducing high-frequency current harmonics and ensuring nearly constant voltage.

VSI is one of the most mature topologies that has been used extensively in the industry thus making it more reliable and robust. However, there are other alternative topologies suggested for the inverters that offer additional benefits in terms of the features or performance but have adverse effects on control design, ability to manufacture, and cost—mainly due to the increased number of components required. Some of the alternative inverter topologies include current source inverters (CSIs), Z-source inverters (ZSIs)—which is a combination of VSI and CSI, and three-level inverters. Also, soft-switching inverters are employed in some cases to reduce the losses incurred by hard-switching.

12.6.3 Converter topologies

The interface between a battery and an inverter in a low-voltage EV is typically a converter. This is a DC-DC interface typically added to boost the voltage or enhance efficiency while fixing instability issues. Bidirectional DC-DC converters in a lot of variants have been studied in the past for this application along with catering to additional critical constraints [19]. However, to achieve improved power density, higher efficiency, and compact size, it is recommended to use full-bridge bidirectional soft-switching DC-DC converters for this application. Also, current-fed topologies are preferred over voltage-fed topologies to attain lower ripple current values.

12.6.4 Control techniques

Closed-loop control of the inverter is achieved by phase current and rotor position feedback values. Pulse width modulation (PWM) techniques are used for switching the power semiconductor devices and enable control of the switching components as per desired pattern of operation. This technique is applied to the inverter AC output voltages that are fed to the electric machine to accomplish closed-loop operation.

The most commonly used PWM technique is called sinusoidal PWM (SPWM) as it is simple to implement. However, it generates high harmonic losses due to a higher value of total harmonic distortion (THD) and the use of its DC link voltage is limited. To overcome these concerns, space vector PWM (SVPWM) modulation is utilized despite the fact that it is computationally intensive. This technique demonstrates better performance on both the fronts and is preferred for EV applications due to its superlative performance. There is also a six-step modulation technique that is an extension of the SVPWM, which is known to improve the electric machine performance further. However, this is seldom used in the industry or is used in conjunction with SVPWM for a very specific purpose such as increasing the torque capability in flux-weakening regions. The control of torque is extremely challenging in this scenario due to lower-order current harmonics.

12.6.5 EV drivetrain

The use of power electronics is extended to the implementation of EV drivetrain as well. The power conditioning system in this case is restricted by the need for adapting to dynamic loads and supporting specific values of voltage, power capacity, and drive range. A variable DC bus system with high-voltage capability has a simple architecture with less number of components. The battery is connected to the input of the bidirectional DC-DC converter, which is then connected to the induction machine via a DC link capacitor and three-phase inverter. This has partial isolation and regulation, which makes control very challenging and the high-voltage operation is not suitable for compact vehicles. The typical propulsion system components of an EV powertrain are shown in Fig. 2.

Another alternative architecture involves a low-voltage variable DC bus system, featuring a unique setup with a DC high-frequency AC coupled with high-frequency AC and pulsating DC stage, replacing the conventional DC link

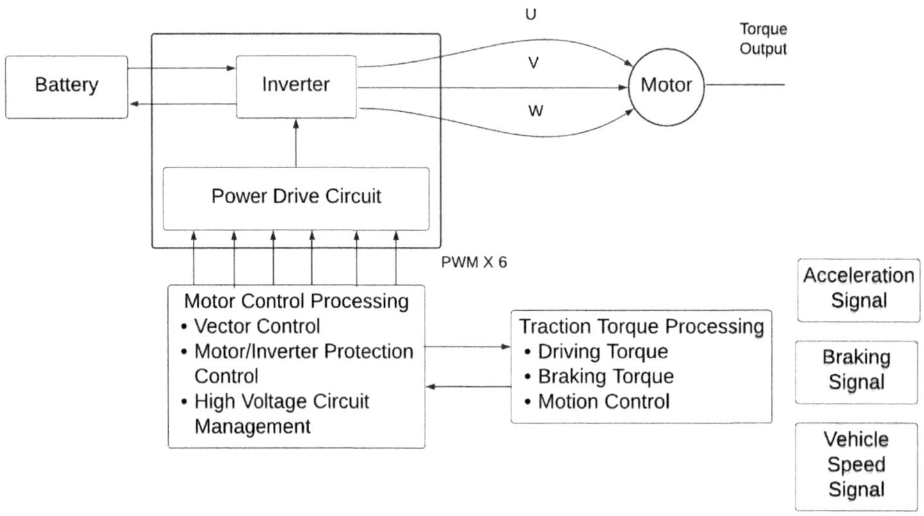

FIG. 2 Typical propulsion system components of an EV powertrain [20].

capacitor. Although more intricate, this configuration offers enhanced isolation and reduced switching losses, achieved by eliminating the DC link capacitor and implementing a six-pulse modulation technique. On the other hand, a low-voltage fixed DC bus system presents a different approach, incorporating an additional converter at the battery input side compared to the previous architecture. This design simplifies the overall system design and control, thanks to its regulated DC bus voltage. Each of these architectures presents distinct advantages and trade-offs, catering to specific application requirements and design considerations.

12.7 Electric and hybrid electric vehicles

Electrified vehicles (EVs) have been growing at a promising rate due to the improvement they offer to the fuel economy while remaining a reliable, efficient, and cleaner alternative to conventional automobiles. There are several types of EVs including hybrid electric vehicles (HEVs), plug-in hybrid electric vehicles (PHEVs), battery electric vehicles (BEVs), and fuel cell vehicles [18]. The pressure to produce clean energy-based vehicles is rapidly increasing due to the legislation requirements and increasing adversities due to the climate crisis.

The concept of EVs is not very new and the ones developed in the 19th century did achieve high speeds. In general, they were more expensive and most of them had a smaller drive range compared to the more recent EVs. The advances in battery technology have improved a lot of electronics related to EVs but facilities related to the recharging of the battery are still actively researched and being worked on. The general EV layout is shown in Fig. 3.

The concept of a combined power source vehicle, or hybrid vehicle, is elegantly straightforward. Conventional internal combustion engines, while ubiquitous, are notorious for their emissions and inefficiency, particularly at partial loads. Conversely, electric drives boast minimal emissions but are hampered by limited drive range. The true advantage lies in harmonizing the strengths of both systems while mitigating their weaknesses. This entails employing sophisticated controls that seamlessly integrate the motor and engine, optimizing their performance under specific conditions. By leveraging the best of both worlds and minimizing their drawbacks, hybrid vehicles emerge as a compelling solution for sustainable and efficient transportation.

12.7.1 Power specifications for traction inverters

Power specifications are critical parameters for power inverters and they determine the design space, especially for deciding the type of active and passive components required for implementation. It also decides the thermal management system, integration efforts, and the scope for optimization in the future generations of EVs. Based on the degree of hybridization, HEVs can be categorized into micro, mild, and full-hybrids [22]. In the case of micro hybrids, the power rating of the electric propulsion vehicle is very small as it is required only for applications such as the starting of the vehicle and the capturing of partial energy from regenerative braking. In the case of mild hybrids, they use an electric motor alongside a traditional internal combustion engine - more like a boost during accelerating. Here, the electric machine can be focused on providing more energy to improve performance during acceleration and can also capture more energy during regenerative braking. In the case of full hybrids, the electric machine provides enough power for it to cater to all the functionalities independently and it can work in pure electric mode.

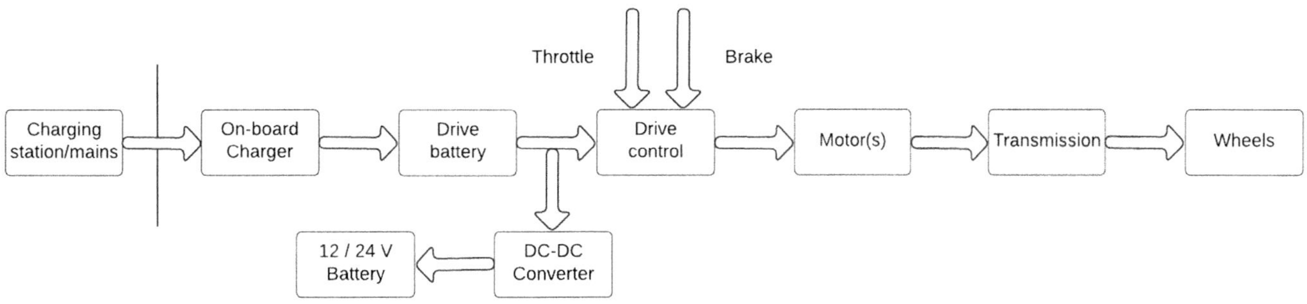

FIG. 3 General EV layout [21].

12.7.2 Battery chargers

A PHEV is a hybrid EV with a storage system that can be recharged by connecting to an external electric power source through an AC or DC charging system. AC charging system is typically an on-board charger mounted inside the vehicle and is connected to the grid. DC charging system is mostly an off-board charger mounted to provide regulated DC power directly to the batteries inside the vehicle [23].

The front-end AC-DC converter is a critical component of the charging system and typically a two-stage approach is ideal for use in PHEVs. A single-stage approach is more suited for low-power applications and due to the large low-frequency ripple generated, it demands higher safety precautions and isolation - which makes it a not-so-good candidate for use in PHEVs. In the case of a two-stage approach, power rating is high and are mostly accomplished by lithium-ion (Li-ion)-based storage systems—front end PFC circuit is followed by the DC-DC section.

The PFC stage helps in rectification of AC input voltage and transfer to regulated intermediate DC bus while achieving power factor correction. The DC-DC section then converts the DC bus voltage into regulated output DC voltage, which is used for charging the batteries and is compliant with the regulatory and transient requirements. Front-end AC-DC converter needs to possess high-power density, high-conversion efficiency and achieve power factor correction while meeting regulatory requirements. Thus, conventional boost topology is mostly popular for PFC applications.

A dedicated diode bridge performs rectification and this is followed by voltage boosting. Some of the commonly used topologies include interleaved boost PFC converter, which consists of two boost converters operating 180 degrees out of phase [24]: dual boost PFC converter that is a bridgeless topology [25] and bridgeless interleaved boost PFC converter with fast diodes [26]. The subsequent DC-DC section may comprise a variety of configurations, such as a zero voltage switching (ZVS) full-bridge converter with a capacitive output filter, an interleaved ZVS full-bridge with a voltage doubler, or a straightforward full-bridge LLC resonant converter. Each option offers unique advantages and trade-offs, catering to specific system requirements and performance criteria.

12.8 Current state of automotive power electronic systems

Modern automobiles are marvels of engineering, boasting an impressive array of electric loads that support numerous features encompassing drive efficiency, stability, passenger safety, travel comfort, and entertainment. It is astounding to note that these vehicles can now accommodate hundreds of individual electric loads, with power capabilities extending up to a kilowatt—a feat that seemed unimaginable just a decade ago. This remarkable progress owes much to the collaborative efforts of national laboratories, dedicated researchers, component suppliers, and original equipment manufacturers (OEMs). Through their collective expertise and innovation, they have continuously pushed the boundaries, enhancing the performance and reliability of power electronic systems tailored for automotive applications. As a result, today's automobiles deliver unprecedented levels of functionality, efficiency, and user experience, setting new standards for the future of transportation.

The most critical component of the power electronic system is the traction inverter and there have been several industry-grade high-power inverters designed and implemented over the past decade. As the industry is trying to progress toward fully electric power trains, it is very relevant to study the PHEVs and BVs implemented in recent years. Several OEMs offer electrified or fully electrified EV models, and the implementation details of the same are crucial to understand. A lot of this information is available through the technical white papers published by the manufacturers or through research findings based on reverse engineering the EV. A summary of some of the key EV models along with critical performance data is summarized in Table 3 [27]. That is followed by a sneak peek into nuances of some of the models released by key manufacturers listed alphabetically.

12.8.1 BMW

BMW's first mass-produced zero emissions EV featured a lithium-ion-based battery with a permanent magnet motor. It has a two-mode hybrid drivetrain. Two-mode hybrids typically contain several planetary gear sets and can provide more operational modes. The power ratings of the two electric machines in two-mode HEVs are almost the same as they could work either as motors or as generators. It could also work as a motor and a generator in which case, the operations switch frequently in accordance with the road conditions. One of the first models introduced by BMW was i3.

TABLE 3 Recent EVs and HEVs.

Automobile maker	Model	Powertrain	Battery
BMW	ActiveHybrid X6	Two-mode Hybrid	Ni-MH
Ford	Fusion 2011	Hybrid Synergy	Ni-MH 275 V
General Motors	Chevrolet Volt 2011	Series Hybrid	Li-ion 345.6 V
General Motors	Chevrolet Tahoe	Two-mode Hybrid	Ni-MH 300 V
Honda	Insight 2012	Parallel Hybrid	Ni-MH 100.8 V
Honda	Civic 2012	Parallel Hybrid	Li-ion 144 V
Honda	Accord Hybrid 06	Parallel Hybrid	Ni-MH 144 V
Hyundai	Sonata Hybrid 2012	Parallel Hybrid	Li-polymer 270 V
Toyota	Prius 2011	Hybrid Synergy	Ni-MH 201.6 V
Toyota	Prius PHEV 2012	Hybrid Synergy	Li-ion 345.6 V
Toyota	Camry 2012	Hybrid Synergy	Ni-MH 244.8 V
Toyota	Highlander 2012	Hybrid Synergy	Ni-MH 288 V

12.8.2 Ford

Ford Escape Hybrid is one of the most well-known hybrid EVs, which is fuel-efficient and one of the most well-designed SUVs. The system is a hybrid synergy drive. As a hybrid EV, it uses a combination of a four-cylinder gasoline engine and an electric motor. The Escape hybrid recovers a substantial portion of what would otherwise be lost energy by means of regenerative braking. The Escape hybrid is a full hybrid that is able to run on either its internal combustion engine and/or its electric motor—depending on which will deliver the most efficient fuel performance.

12.8.3 General motors

General Motors (GM) has arguably led the motor industry in EV development since the 1960s. It has a long history of electrification starting from 1996 and it is known for developing the world's first specifically designed production electric vehicle that went for sale in the United States, EV-1. The key to the success of the EV-1 is its electrical power-train, based on a 103 kW (137 HP) three-phase AC induction motor with an integral, single-speed, dual-reduction gear-set driving the front wheels. The unit requires no routine maintenance for over 100,000 miles. The battery pack uses 26, 12 V maintenance-free lead-acid batteries, giving a total voltage of 312 V and a range of 70 miles per charge in urban conditions and 90 miles on the open road. However, new nickel-metal-hydride (NiMH) batteries were phased into production in 1998, almost doubling the EV-1's range to 140 miles in the city and 160 miles on highways. Full recharging can be carried out safely in all weather conditions and takes 3–4 h using a 220 V standard charger or 15 h using the onboard 110 V convenience charger. Compared with normal fossil fuels, the lower cost of domestic electricity means operating costs are relatively low. EV-1 comes with traction control, cruise control, anti-lock brakes, dual airbags, power windows, door locks and outside mirrors, AM/FM CD/cassette, tire inflation monitor system, and numerous other features [21].

The first generation of its automobiles had a power density of 3 kW/L [28] whereas the second generation model from 2016 gives a power density of 17.3 kVA/L [29]. As an enhancement for the second generation of inverters, they were placed directly on the transmission side thus enabling better integration and reduced component count, design complexity, and cable cost. The 2016 model, Chevrolet MY2016 Volt housed dual voltage source inverters for the motor and generator along with another smaller one for the engine oil pump. An innovative regenerative braking system helps to extend that range still further by converting the energy used when braking back into electricity to recharge the battery pack partially. The next design was built upon the previous Volt power-train design but had the ability to cater to higher performance requirements. This had the same number and type of traction inverters as the previous design but had a slightly different implementation. As a large hybrid power train, the Cadillac MY2016 CT6 model enabled compact design by the use of higher efficiency and reduced thermal impedance of the IGBT modules [30].

12.8.4 Honda

Honda is known for its mild HEVs and they are some of the most successful manufacturers of the same. There are several models of Honda mild HEVs including Insight, Civic, and Accord. Typically, parallel architecture and permanent magnet (PM) machines are used in these mild hybrid EVs. The power rating of the motor is around 10–20 kW whereas the nominal DC bus voltage is about 100–200 V [31].

12.8.5 Nissan

Nissan had initial trials in the United States in 1998. The Altra was built focusing on the US market and was the first EV to be sold outside Japan. It comprises a water-cooled, permanent magnet, synchronous electric motor made of a special alloy called neodymium-iron-boron (Nd-Fe-B). The Altra has hydraulic power steering driven by an electric power, rather than a mechanical or hydraulic pump, thus operates only when power assistance is required. A standard 12 V lead-acid battery charged via a water-cooled DC-DC transformer from the main battery powers the auxiliary systems. Heating, ventilation, and air conditioning consume 50% of the energy of a conventional system in air conditioning mode and 66% when heating the cabin.

The Nissan LEAF inverter is a single VSI powering the vehicle's one electric machine. The inverter is situated directly on top of the motor thus eliminating external phase cables. As the vehicle has only a single EM and no ICE, the packaging size of the inverter did not appear to be a significant priority in the design, thus the power density is somewhat low. The most recent Nissan Leaf 2023 is shown in Fig. 4.

12.8.6 Tesla

Tesla introduced the Model S in 2012. The key components of the same are highlighted in Fig. 5. It is a rear-wheel drive full-size luxury Sedan with an 85 kWh battery pack. This is one of the first EVs from Tesla followed by an all-wheel drive version of the same in 2014. In the all-wheel drive version of Model S, there is an electric machine on both the front and rear axles.

12.8.7 Toyota

The hybrid synergy system was initially developed by Toyota and it was equipped successfully on several different hybrid models under its brand. In a hybrid synergy system, the transmission combines the ICE with two electric machines. It has achieved huge market success since it was first released in 1997 and it is currently the most widely used hybrid system. The battery bank is connected to the DC bus through a boost converter thus making the DC bus voltage independent of the battery voltage. Toyota's most promising model has been the Prius, which was introduced in 1997. They have continued to focus on hybrid models with some expansion into the PHEV market with the Prius Prime in 2012.

Toyota has integrated two VSIs, a DC-DC boost converter and a DC-DC buck converter for the auxiliary power module into one unit. The inverter housing is mounted to the transaxle to eliminate external phase cables and improve

FIG. 4 Nissan Leaf 2023. *Source: Nissan.*

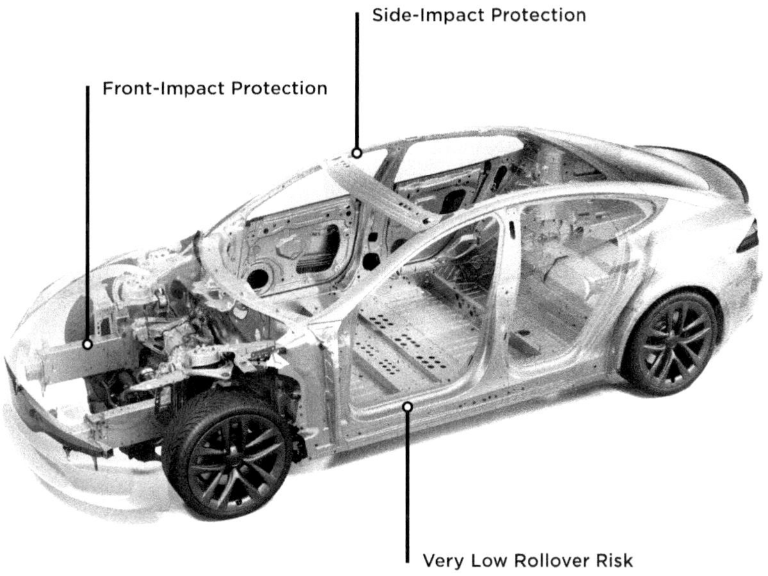

FIG. 5 Tesla Model S. *Source: Tesla.*

the collision robustness. Due to the mounting location, additional considerations for vibration resistance were required, including rubber mounting bushings, elastic washer PCB mounting, and vibration simulations during the design of the signal interconnect.

12.9 Future trends

There is a lot of potential and possibilities to achieve greater kW/L inverter, which is super helpful in introducing more functionalities in an EV. It can also help to move some of the critical functionalities from engine-operated to electrified independent operation. The inverter target set by the Department of Energy for 2025 is 100 kW/L and there have been active research studies ongoing to find the strategies and implementation ideas to get there successfully. Integration of inverter sub-components, advanced packaging, enhanced thermal design, better electrical isolation, newer materials, and many such strategies can help get significantly higher switching frequencies, power density, and reliability. There have been multiple industry and research trends focused on gate driver design, use of wide bandgap devices, component packaging, system integration, and manufacturing techniques to improve the performance of traction inverters [20].

12.9.1 Wide bandgap devices

The automotive industry is producing a wide range of electric vehicles that use clean energy and are thus environment friendly. The electrically powered vehicles include HEVs, PEVs, and fully electric ones. Most of the initial power electronics utilized silicon (Si)-based switching devices as the technology is mature and offers a wide range of choices in terms of component selection. Further, it has been proven that silicon carbide (SiC) is another promising material for semiconductor devices used in power electronics due to its ability to operate at high temperatures efficiently while yielding lower losses. This helps in reducing the size and weight of the power electronics thus improving the power density. However, implementation using SiC devices is quite challenging and there has been tremendous research related to device topologies, process technologies, failure analysis, crystal growth techniques, and application to on-board systems.

Toyota and a few other manufacturers have looked into SiC-based technology and this has led to more powerful power conditioning units [32]. The high-frequency operation ability accompanied by low losses makes it an attractive candidate in spite of challenges to its implementation. However, SiC-based technology is not highly scalable yet with limited supply, fabrication, and manufacturing platform constraints. The other candidate for replacing silicon-based transistors is gallium nitride (GaN)-based power devices. They have a projected 100× performance advantage over silicon, a 10× performance advantage over SiC, and are game changers for power electronics in general. They have excellent material properties that enable a reduction in conduction and switching losses mostly due to low on-state resistance and fast switching. General Motors and a few other OEMs have looked into GaN-based technology for the automotive industry.

The wide bandgap-based devices, mainly SiC and GaN power electronics, promise lower losses, higher power density, and performance advantages. However, there is a need for a paradigm shift in circuit design and implementation to realize the benefits of such fast-switching devices. Parasitics also need to be reduced to achieve high switching speeds thus introducing challenges for the technology on both performance and reliability fronts. There has been a lot of research in this direction and some implementation as well. However, higher acceptance is based on manufacturability in large volumes at competitive prices without compromising on performance and reliability aspects. The widespread adoption is still ongoing but the future looks promising for wide bandgap-based devices in the automotive sector.

12.9.2 Advanced packaging technologies

Multiple approaches can be taken to improve the packaging of switching devices, which can result in superior thermal and electrical performance while reducing the size and weight. In recent times, there are two cooling loops in the commercial HEVs—one for engine cooling and the other for power electronics cooling, both maintained at different temperatures per the requirement. There is a need and trend to integrate the two cooling systems in the automotive industry to be more cost-effective, which poses a design challenge for the thermal design aspect of power electronic systems. Some of the areas of research to this end include a focus on advanced packaging to reduce the thermal resistance between IGBT silicon dies and coolant, and the application of power devices with higher allowable junction temperature. In addition, reliability improvements can also be achieved by improved packaging and enhanced thermal design.

12.9.3 Power electronic enhancements

Traction inverters have predominantly adhered to conventional two-level voltage source inverter (VSI) topologies, with multiphase inverters being deployed in scenarios demanding high torque. However, with the expanding array of electrified functionalities, the demand for supporting higher DC-link voltages has surged, necessitating the adoption of multilevel inverters. While multilevel inverters offer promising prospects for enhanced performance, there remains ample scope for further research and development in this domain. Despite the potential benefits, challenges such as cost escalation due to additional switching devices and practical limitations in consumer automotive applications must be addressed. Nevertheless, ongoing exploration in this area holds the key to unlocking new frontiers in automotive power electronics, driving innovation and advancement in electric vehicle technology.

With the advancement in traction inverter topologies, there is a need for parallel development in the design of transistors such as the most commonly used IGBTs and the gate driver circuits that are used to switch them per requirements. The trend is to move toward smarter gate drivers that can adapt the turn-on and turn-off slopes of IGBT based on the operating conditions. This also demands better control techniques so that switching losses are kept in check, while surge voltage and EMI emissions are minimized. It also fosters better monitoring and protection options while enabling high-temperature operation, high current densities, and faster switching capabilities.

In addition to improvements in the device package and cooling aspects, there is a need for research on removing heat from devices. As the device power density increases, it gets more challenging to extract heat with a conventional liquid cold plate design thus demanding the need for improved thermal design. This is followed by the need for tighter integration of the traction inverter and the electric machine so that manufacturing is more efficient in the case of the production of EVs.

12.9.4 More electric vehicles (MEVs)

The concept of more electric vehicles (MEVs) is a trend toward expanding electrical loads and replacing mechanical, hydraulic, and other nonelectric loads with more electrical systems. These loads include lights, pumps, fans, and electric motors for various functions. In addition, they also include some advanced, electrically assisted vehicular loads, such as power steering, air conditioner/compressor, electromechanical valve control, active suspension/vehicle dynamics, and catalytic converter. Furthermore, additional advanced vehicular loads include anti-lock braking, throttle actuation, ride-height adjustment, and rear-wheel steering, which will be driven electrically in the future [3].

Most of the future advanced electric loads require power electronic systems and controls. In advanced EVs, power electronics is forecasted to perform three major tasks. The first task is simple on/off switching of loads, which is performed by mechanical switches and relays in conventional cars [4]. The second task is to act as a suitable controller for electric traction motors. The third is the requirement that power electronics intensive power systems will not only be used for the obvious task of changing system voltage levels, but also for converting electrical power from one form to another, using DC-DC, DC-AC, and AC-DC converters. Thus, MEVs need to be highly reliable and fault tolerant to deliver high-quality power from the source to the electrical loads. Also, the future MEVs are expected to be comprised of a single main voltage bus (high voltage) with a provision for hybrid (DC and AC), multi-voltage-level distribution, intelligent energy, and load management [33].

12.10 Summary

The imperative for cleaner technologies has witnessed an incessant rise, especially within the automotive sector, where the demand for cost-effective, reliable, and space-efficient power electronic systems has surged. Power electronics play a pivotal role in enabling various essential functions crucial for the safe operation of automobiles and ensuring passenger comfort. The performance of power electronic systems is gauged against a spectrum of key metrics, including efficiency, reliability, size, cost, driving range, and charging time. Addressing the inherent limitations associated with these parameters presents significant design challenges, serving as focal points for ongoing research and development endeavors. By tackling these challenges head-on, the automotive industry can unlock innovative solutions that propel the evolution of cleaner, more sustainable transportation technologies.

Section 12.2 delves into the rich history of automotive vehicles, tracing the evolution of their power electric systems from inception to their current state-of-the-art architectures. The journey of electric vehicles (EVs) has been marked by significant flux, reflecting the dynamic landscape of technological advancements and shifting consumer preferences. In dissecting the conventional electrical system of an automobile, it becomes evident that it comprises distinct architectural elements, including energy storage, generation, starting, and distribution. Within this framework, the distribution system of a conventional 14 V power system plays a pivotal role in catering to essential vehicle loads, encompassing lighting, motor-driven subsystems, and pertinent instrumentation. Understanding the nuances of these architectural components provides valuable insights into the intricate workings of automotive power electric systems

Section 12.3 sheds light on the paramount importance of power electronic systems, particularly within automotive applications. These systems serve as the backbone of automobile functionality, facilitating essential operations such as engine ignition, control, and propulsion. From igniting the spark for combustion to regulating cruise control, power electronic systems are indispensable for ensuring the smooth operation of vital vehicle functions. Beyond the core functionalities, power electronic systems address a spectrum of ancillary needs, ranging from basic essentials like lighting, wipers, horns, and blower fans to enhancing passenger comfort with amenities like seat warmers, air conditioning, and entertainment systems. Moreover, in the realm of electric vehicles (EVs), power electronic interfaces and systems assume even greater significance, underpinning the entire EV drivetrain and charging infrastructure. In essence, the seamless integration and optimization of power electronic technologies are pivotal for advancing automotive innovation and meeting the evolving demands of modern transportation.

Section 12.4 delves into the intricate design considerations crucial for automotive power electronics, encompassing various factors such as voltage and current ranges, temperature thresholds, regulatory compliance, and environmental variables. Additionally, it explores performance parameters essential for optimizing system functionality and reliability. In normal operating conditions, the static voltage range typically hovers between 11 and 16 V. However, to preempt potential issues like incorrect battery installation or static over-voltage scenarios, a broader voltage range is factored into the design process. Temperature emerges as a critical determinant of electric vehicle (EV) performance, with the typical upper limit set at 125°C and ambient temperatures extending as low as −40°C. Furthermore, meticulous attention is paid to mechanical considerations, as automotive power electronic systems must withstand and mitigate the impact of mechanical vibrations. Robust mechanical design strategies are thus essential to ensure system durability and longevity. Moreover, achieving electromagnetic compatibility (EMC) is paramount, as power electronic circuits in automotive applications must not succumb to electromagnetic interference (EMI). By adhering to stringent EMC standards, automotive power electronics can operate seamlessly and reliably amidst diverse environmental conditions, ensuring optimal performance and safety on the road.

Section 12.5 provides a comprehensive overview of the diverse components comprising an automotive power electronic system, encompassing crucial elements such as air conditioning, lighting systems, actuators, sensors, motors, and various other loads. At the heart of electrified power electronic systems lie electric machines (EMs), which play a pivotal role either independently or in tandem with an internal combustion engine (ICE). Among the most prevalent electric machines are AC machines, available in induction or permanent magnet configurations. During motoring mode, these electric machines are propelled by a power electronics-based traction inverter. This sophisticated system facilitates the seamless conversion of direct current into alternating current at variable frequencies, effectively harnessing energy from the battery. Conversely, in generating mode, surplus energy is redirected back to the battery, with the electric machine transitioning into a generator and the inverter functioning as a rectifier. These dynamic power electronic systems support a myriad of functionalities critical for vehicle operation, including lighting, air conditioning, power steering, speed control, actuation mechanisms, engine valves, and more. By orchestrating the harmonious interaction of these components, automotive power electronic systems enable optimal performance, efficiency, and user experience on the road.

Section 12.6 delves into the intricate process of power conversion, offering a compelling insight into the dynamic realm of automotive power electronics in action. At its core, a power conversion system (PCS) serves as the linchpin, enabling bidirectional energy conversion between the battery and various loads. This section meticulously explores the multifaceted landscape of power electronic components, delving into the intricacies of their design, implementation, and control within the drivetrain architecture. Central to this discussion is the inverter, a critical component comprising an array of power electronic elements tasked with switching and inversion functions. Complementing the inverter is a sophisticated microcontroller, orchestrating precise motor control software execution. From semiconductor switches and gate drivers to sensors, DC link capacitors, thermal management systems, and busbars, each component plays a pivotal role in ensuring optimal system performance and reliability. Embracing a three-phase voltage source inverter topology, the inverter design is further bolstered by robust closed-loop control mechanisms, enhancing system stability and efficiency. By unraveling the intricacies of power conversion within automotive power electronics, this section offers invaluable insights into the foundational technologies driving the next generation of electric vehicles and sustainable transportation solutions.

Section 12.7 provides a comprehensive overview of the myriad applications of power electronics in electric and hybrid electric vehicles (EVs and HEVs). The burgeoning adoption of electrified vehicles stems from their remarkable advancements in enhancing fuel economy while offering a reliable, efficient, and environmentally cleaner alternative to traditional automobiles. Within the realm of EVs, various configurations exist, including hybrid electric vehicles (HEVs), plug-in hybrid electric vehicles (PHEVs), battery electric vehicles (BEVs), and fuel cell vehicles. Crucially, power specifications serve as foundational parameters for power inverters, delineating the design space and informing critical decisions regarding active and passive component selection. These specifications also govern the thermal management system, integration efforts, and pave the way for optimization opportunities in future generations of EVs. Charging systems, underpinned by power electronics, constitute another indispensable component of the EV infrastructure. The AC charging system typically features an onboard charger housed within the vehicle, facilitating connection to the grid. In contrast, the DC charging system, often an offboard charger, delivers regulated DC power directly to the vehicle's batteries, ensuring efficient charging and seamless integration with the EV ecosystem. By illuminating the diverse applications of power electronics in electric and hybrid vehicles, this section underscores the pivotal role of this technology in driving the transition toward sustainable mobility solutions and shaping the future of transportation.

Section 12.8 offers an insightful exploration into the current landscape of automotive power electronic systems, shedding light on the pioneering efforts of automobile manufacturers dedicated to clean energy solutions. Today's automobiles stand as remarkable feats of engineering, boasting the capacity to support hundreds of individual electric loads while delivering a plethora of advanced features encompassing drive efficiency, stability, passenger safety, travel comfort, and entertainment. The remarkable evolution of automotive power electronics, capable of delivering power capabilities up to a kilowatt, represents a significant milestone that was once deemed unattainable just a decade ago. This transformation has been made possible through the collaborative endeavors of national laboratories, researchers, component suppliers, and OEMs, united in their mission to enhance the performance of power electronic systems for automotive applications. Highlighting recent electric and hybrid electric vehicles (EVs and HEVs) from the past decade, this section offers a glimpse into the technological advancements and milestones achieved by leading manufacturers. By delving into the details and histories of these vehicles and their manufacturers, it provides valuable insights into the evolutionary journey that has culminated in the present state of automotive power electronic systems, offering a compelling narrative of progress and innovation.

Section 12.9 offers a tantalizing glimpse into the exciting possibilities for enhancing the functionality of electric vehicles (EVs) and achieving greater power density in inverters. Through the integration of inverter sub-components, advanced packaging techniques, refined thermal design strategies, enhanced electrical isolation methods, and the exploration of novel materials, a plethora of avenues emerge to elevate switching frequencies, power density, and overall reliability. Industry and research communities have been at the forefront of exploring innovative trends and strategies aimed at optimizing gate driver designs, harnessing the potential of wide bandgap devices, refining component packaging methodologies, advancing system integration techniques, and innovating manufacturing processes. These concerted efforts are geared toward unlocking the full potential of traction inverters, ensuring they meet the evolving demands of the automotive landscape and pave the way for next-generation electric vehicles. By embracing these cutting-edge advancements and pushing the boundaries of innovation, the automotive industry stands poised to usher in a new era of electrified transportation, characterized by enhanced performance, efficiency, and sustainability.

References

[1] M. Ehsani, et al., Modern Electric, Hybrid Electric, and Fuel Cell Vehicles: Fundamentals, Theory, and Design, CRC Press, 2004.

[2] P.T. Krein, et al., Packaging and performance of an IGBT-based hybrid electric vehicle, in: Proceedings of 1994 IEEE Workshop on Power Electronics in Transportation, 1994, pp. 47–52, https://doi.org/10.1109/PET.1994.572355.

[3] A. Emadi, S.S. Williamson, A. Khaligh, Power electronics intensive solutions for advanced electric, hybrid electric, and fuel cell vehicular power systems, IEEE Trans. Power Electron. 21 (3) (2006) 567–577, https://doi.org/10.1109/TPEL.2006.872378.

[4] J.G. Kassakian, D.J. Perreault, The future of electronics in automobiles, in: Proceedings of the 13th International Symposium on Power Semiconductor Devices and ICs. IPSD '01 (IEEE Cat. No.01CH37216), 2001, pp. 15–19, https://doi.org/10.1109/ISPSD.2001.934550.

[5] Statista, Automotive electronics cost as a percentage of total car cost worldwide from 1970 to 2030, Martin Placek, Jan 2023. https://www.statista.com/statistics/277931/automotive-electronics-cost-as-a-share-of-total-car-cost-worldwide/.

[6] J.G. Kassakian, et al., Automotive electrical systems circa 2005, IEEE Spectr. 33 (8) (1996) 22–27, https://doi.org/10.1109/6.511737.

[7] SAE Electronic Systems Committee, Recommended Environmental Practices for Electronic Equipment Design, SAE Recommended Practice SAE J1211, 1978.

[8] I.A. Khan, D.J. Perreault, K.K. Afridi, Automotive Applications of Power Electronics, Elsevier, 2018.

[9] Y. Shun-Hsiang, et al., Smart automotive lighting for vehicle safety, IEEE Commun. Mag. 51 (12) (2013) 50–59, https://doi.org/10.1109/MCOM.2013.6685757.

[10] J.L. Oldenkamp, D.M. Erdman, Automotive electrically driven air conditioner system, in: Automotive Power Electronics, 1989, pp. 71–72, https://doi.org/10.1109/APE.1989.97158.

[11] R.J. Valentine, Electric steering power electronics, in: Power Electronics in Transportation, 1996, pp. 105–110, https://doi.org/10.1109/PET.1996.565917.

[12] J. Wallaschek, Piezoelectric ultrasonic motors, J. Intell. Mater. Syst. Struct. (1995).

[13] R. Henry, M.A. Theobald, B. Lequesne, Control of engine load via electromagnetic valve actuators, in: International Congress and Exposition, SAE Paper 920447, Detroit, MI, 1994.

[14] L. Diao, et al., An efficient DSP–FPGA-based implementation of hybrid PWM for electric rail traction induction motor control, IEEE Trans. Power Electron. 33 (4) (2018) 3276–3288, https://doi.org/10.1109/TPEL.2017.2707639.

[15] J. Groeger, et al., Optimized dv/dt, di/dt sensing for a digitally controlled slope shaping gate driver, in: 2017 IEEE Applied Power Electronics Conference and Exposition (APEC), 2017, pp. 3564–3569, https://doi.org/10.1109/APEC.2017.7931209.

[16] J. Guo, J. Ye, A. Emadi, DC-link current and voltage ripple analysis considering antiparallel diode reverse recovery in voltage source inverters, IEEE Trans. Power Electron. 33 (6) (2018) 5171–5180, https://doi.org/10.1109/TPEL.2017.2738327.

[17] V. Smet, et al., Ageing and failure modes of IGBT modules in high-temperature power cycling, IEEE Trans. Ind. Electron. 58 (10) (2011) 4931–4941, https://doi.org/10.1109/TIE.2011.2114313.

[18] J. Reimers, et al., Automotive traction inverters: current status and future trends, IEEE Trans. Veh. Technol. 68 (4) (2019) 3337–3350, https://doi.org/10.1109/TVT.2019.2897899.

[19] M. Marchesoni, C. Vacca, New DC–DC converter for energy storage system interfacing in fuel cell hybrid electric vehicles, IEEE Trans. Power Electron. 22 (1) (2007) 301–308, https://doi.org/10.1109/TPEL.2006.886650.

[20] K. Hamada, Present status and future prospects for electronics in EVs/HEVs and expectations for wide bandgap semiconductor devices, Mater. Sci. Forum 600 (2009) 889–893.

[21] T. Denton, Automobile Electrical and Electronic Systems, fifth ed., Institute of the Motor Industry, n.d.

[22] S.M. Lukic, A. Emadi, Effects of drivetrain hybridization on fuel economy and dynamic performance of parallel hybrid electric vehicles, IEEE Trans. Veh. Technol. 53 (2) (2004) 385–389, https://doi.org/10.1109/TVT.2004.823525.

[23] S.S. Williamson, A.K. Rathore, F. Musavi, Industrial electronics for electric transportation: current state-of-the-art and future challenges, IEEE Trans. Ind. Electron. 62 (5) (2015) 3021–3032, https://doi.org/10.1109/TIE.2015.2409052.

[24] Y. Jang, M.M. Jovanovic, Interleaved boost converter with intrinsic voltage-Doubler characteristic for universal-line PFC front end, IEEE Trans. Power Electron. 22 (4) (2007) 1394–1401, https://doi.org/10.1109/TPEL.2007.900502.

[25] Y. Jang, M.M. Jovanovic, A bridgeless PFC boost rectifier with optimized magnetic utilization, IEEE Trans. Power Electron. 24 (1) (2009) 85–93, https://doi.org/10.1109/TPEL.2008.2006054.

[26] F. Musavi, W. Eberle, W.G. Dunford, A phase-shifted gating technique with simplified current sensing for the semi-bridgeless AC–DC converter, IEEE Trans. Veh. Technol. 62 (4) (2013) 1568–1576, https://doi.org/10.1109/TVT.2012.2231709.

[27] H. Ye, Y. Yang, A. Emadi, Traction inverters in hybrid electric vehicles, in: 2012 IEEE Transportation Electrification Conference and Expo (ITEC), 2012, pp. 1–6, https://doi.org/10.1109/ITEC.2012.6243442.

[28] K. Rajashekara, Present status and future trends in electric vehicle propulsion technologies, IEEE J. Emerging Sel. Top. Power Electron. 1.1 (2013) 3–10, https://doi.org/10.1109/JESTPE.2013.2259614.

[29] M. Anwar, et al., Development of a power dense and environmentally robust traction power inverter for the second-generatio chevrolet VOLT extended-range EV, in: 2015 IEEE Energy Conversion Congress and Exposition (ECCE), 2015, pp. 6006–6013, https://doi.org/10.1109/ECCE.2015.7310502.

[30] M. Anwar, et al., Compact and high power inverter for the Cadillac CT6 rear wheel drive PHEV, in: 2016 IEEE Energy Conversion Congress and Exposition (ECCE), 2016, pp. 1–7, https://doi.org/10.1109/ECCE.2016.7854931.

[31] J. Wang, Practical design considerations of power electronics in hybrid and fuel cell vehicles, in: 2008 IEEE Vehicle Power and Propulsion Conference, 2008, pp. 1–6, https://doi.org/10.1109/VPPC.2008.4677673.

[32] Toyota, Toyota Develops New Silicon Carbide Power Semiconductor with Higher Efficiency, 2014. https://global.toyota/en/detail/2656842/.

[33] S.C. Smithson, S.S. Williamson, Constant power loads in more electric vehicles - an overview, in: IECON 2012 - 38th Annual Conference on IEEE Industrial Electronics Society, 2012, pp. 2914–2922, https://doi.org/10.1109/IECON.2012.6389432.

CHAPTER

13

Autonomous solar electric vehicles

Mohammad Reza Maghami and Arthur Guseni Oliver Mutambara
Institute for the Future of Knowledge (IFK), The University of Johannesburg, Johannesburg, South Africa

Nomenclature

Tc	temperature of the PV cell, °C
$Tc_{,\ S.T.C.}$	temperature of the standard test condition, °C
$T_{.C.,NOCT}$	nominal operating cell temperature, °C
T_α	ambient temperature, °C
D_{PV}	derating factor, %
$G_{.T.}$	solar radiation, W/m^2
$G_{.T.,\ S.T.C.}$	incident irradiation at the S.T.C., W/m^2
SOC	state of battery charge, %
I_{bat}	current rate of battery, A
R_{pv}	PV capacity rate, kW
DOD	depth of discharge,%
ω_{EV}	angular velocity
M_{EV}	torque at the wheels
$P_{battery}$	battery power
P_{ch}	charging rate
P_{dis}	discharging rate
η	converter efficiency

Abbreviations

CC	charge cycle
ESS	energy storage system
LF	load following
MPP	maximum power point
PV	photovoltaic
REs	renewable energies
SEV	solar powered electric; vehicle

Greek letters

ρ_0	self-discharge rate, %/month
σ	temperature coefficient, %
α_P	solar module efficiency, %
η_C	solar transmittance, %
$\tau\alpha$	solar transmittance, %

13.1 Introduction

In the face of growing concerns about climate change and the depletion of fossil fuel reserves, there has been a global push for sustainable transportation alternatives. Among these alternatives, solar powered electric vehicles have emerged as a highly promising technology that could potentially revolutionize the automotive industry [1]. By harnessing the power of the sun, solar powered electric vehicles (SPEVs) offer a clean and renewable source of energy for transportation, significantly reducing greenhouse gas emissions and lessening our reliance on traditional energy

Handbook of Power Electronics in Autonomous and Electric Vehicles
https://doi.org/10.1016/B978-0-323-99545-0.00013-0

sources. Over the past few years, there have been remarkable advancements in the field of SEVs [2]. These advancements have been driven by breakthroughs in solar cell efficiency, energy storage systems, and electric vehicle components. By integrating lightweight and flexible solar panels into the design of vehicles, coupled with advancements in battery technology, the driving range of SPEVs has substantially increased, while enhancing their overall performance. Additionally, the development of smart charging infrastructure and innovative power management systems has addressed some of the challenges associated with solar charging. The potential applications of solar electric vehicles extend far beyond personal transportation [3]. They can play a vital role in various sectors, including public transportation, logistics, and delivery services. In urban environments, where short commuting distances and frequent stops are common, SPEVs are particularly well-suited due to their ability to absorb solar energy during these stops. Moreover, the integration of solar electric vehicles (SEVs) into renewable energy microgrids strengthens the stability and resilience of the power grid while simultaneously reducing the overall carbon footprint. This chapter serves as a comprehensive introduction to the importance of SEVs, recent developments in the field, and their potential applications. It lays the foundation for the subsequent chapters, which will delve into specific aspects of SEVs, including solar cell technologies, energy storage systems, vehicle design considerations, charging infrastructure, and policy frameworks. By exploring these topics in detail, this book aims to provide readers with a thorough understanding of SEVs and their role in shaping the future of sustainable transportation [3].

13.2 Overview of solar electric vehicle

The power of the sun is harnessed through a photovoltaic array. The photovoltaic (PV) array comprises PV cells that convert solar energy into usable electricity. This electricity is subsequently employed to drive electric motors, providing the necessary propulsion for the vehicle. However, to account for periods when sunlight is not available or insufficient, solar electric vehicles commonly integrate energy storage solutions in the form of batteries. These batteries serve a dual purpose: they store excess electricity generated during peak sun exposure for later use, and they act as a reliable power supply during times of low solar input. By incorporating batteries into the solar electric vehicle (SEV) system, a consistent supply of electricity is ensured, enabling uninterrupted operation regardless of solar availability. This integration of solar panels and batteries in SEVs represents a synergistic approach that optimizes energy capture, storage [3], and utilization, ultimately enhancing the overall efficiency and viability of solar-powered transportation [4]. In 1955, William G. Cobb, an executive at General Motors Company, created the first solar powered vehicle called the "Sunmobile." This miniature vehicle utilized 12 selenium-based photoelectric cells [5]. Although small in scale, it marked an early exploration of solar energy for transportation and paved the way for advancements in solar powered vehicles. Today, solar electric vehicles have become more efficient and practical, offering sustainable transportation options. Solar races have played a significant role in the advancement of roadworthy SEVs. Solar car races have served as a driving force for solar design teams globally, spurring them to achieve greater efficiencies, enhance aerodynamics, and reduce energy consumption in their vehicles. Solar car competitions such as the World Solar Challenge and the American Solar Challenge have proven to be influential in motivating solar design teams globally. The World Solar Challenge presents a formidable test, spanning a grueling 3000-km route across the Australian continent [6]. Similarly, the American Solar Challenge covers the distance from Omaha, Nebraska to Bend, Oregon. These prestigious races have served as catalysts for innovation, prompting teams to strive for higher levels of efficiency, improved aerodynamics, and reduced energy consumption in their solar car designs [6].

Over the last decade, numerous automobile manufacturers and startups specializing in electric vehicle technology have introduced their unique iterations of SEVs. One notable example is Aptera Motors, who has developed solar powered vehicles that harness the energy of the sun, eliminating the need for regular charging during typical daily usage and enabling a daily range of up to 40 miles. The pricing of Aptera vehicles ranges from $25,900 to $46,000 or more, depending on the specific range and optional features selected, which is high for many people. Venturi Automobiles introduced solar charging with a range of 7 km/day at a top speed of 50 km/h [7]. Eindhoven University developed the "Stella Lux," a four-seater vehicle designed as the successor to Stella for the 2015 World Solar Challenge [8]. With a remarkable top speed of 125 km/h and an impressive European range of 1000 km, the Stella Lux represents a significant accomplishment compared to conventional electric vehicles. The Lightyear is a luxurious sedan with a sleek design, capable of traveling up to 725 km using solar and battery power [9,10]. The Clean Motion is recognized as one of the remarkable solar cars available. It takes the form of a compact, modular delivery van and is equipped with approximately 600 watts of solar panels. On a bright sunny day, this solar car has the potential to extend its range by an impressive 130 km per day [11]. The Squad, a micro car, is capable of achieving a range of approximately 20 km solely powered by solar energy. Considering the average usage of microcars in Europe, which typically covers around 12 km, the Squad has the potential to be completely sustainable in terms of energy consumption [11]. Additionally, a case study in India exemplified the successful implementation of solar electric vehicles, including solar cars, e-rickshaws, tricycles, buses, and vans, demonstrating the potential for widespread adoption of SEVs. These research efforts underscore the growing interest in SEVs and the potential for solar energy to transform the transportation sector. For instance, a study investigated the efficiency and cost of different electrical architectures for SEVs, highlighting the importance of maximizing solar energy capture.

In their study, H. Koten et al. examined key designs for EVs, emphasizing the importance of factors such as minimizing drag resistance, achieving lightweight construction, and reducing rolling resistance. The authors presented an intriguing alternative by investigating the use of solar powered Stirling engine technology, rather than traditional PV technology [12]. In another research, the vehicle incorporated 480 PV cells, by considering maximum power point trackers for optimized solar generation. The car was equipped with a highly efficient brushless DC motor, boasting 90% efficiency. Despite the absence of sunlight, the vehicle, supported by a 5 kWh battery pack, demonstrated an impressive range of over 75 miles [13]. In another study, they presented innovative design solutions for solar powered electric mobility applications. They addressed the design challenges involved in creating an SEV [14]. The challenges highlighted in the research revolved around three main aspects: limited space for integrating PV cells, the need for effective visual communication of the concept's intended function, and technical limitations associated with certain proposed charging methods. In the study conducted by S. Ahmed et al., an innovative design for a prototype solar powered lightweight vehicle was introduced, specifically engineered to accommodate two passengers. The research not only focused on the vehicle's design aspects but also included a comprehensive economics analysis. This analysis aimed to evaluate the long-term cost implications of owning and operating a solar car compared to a traditional gas-powered vehicle. The findings of the economic analysis were particularly insightful. Over a span of 20 years, it was determined that the cost of ownership for the solar car amounted to only 1/9th of the total cost of ownership for the gas-powered car. This remarkable cost disparity showcases the potential economic benefits of solar powered vehicles, as they offer significantly reduced expenses over their operational lifespan. Overall, the advancements in SPEV technology and the growing interest in solar powered mobility highlight the potential for a sustainable and efficient future of transportation. Continued research and development in this field will contribute to further improvements in SEV, making them increasingly practical and widespread.

13.3 Solar electric vehicle model

13.3.1 Modeling electric vehicle movements

Modeling the movements of electric vehicles involves developing mathematical or computational models that simulate various aspects of the behavior of electric vehicle (EV), such as acceleration, deceleration, energy consumption, and range estimation. These models take into account factors such as vehicle dynamics, powertrain characteristics, driving conditions, and energy management systems. There are different approaches to modeling EV movements, ranging from simple physics-based models to more complex data-driven or machine learning-based models. Physics-based models utilize fundamental principles of mechanics, such as Newton's laws of motion, to simulate the vehicle's behavior under different driving conditions. These models typically require input parameters such as vehicle mass, aerodynamic properties, tire characteristics, and powertrain efficiency. The total resistance force on a solar electric vehicle includes aerodynamic drag, rolling resistance, and gravitational force. These forces affect the vehicle's speed, energy consumption, and overall performance. Optimizing these factors is important for improving the efficiency and range of solar electric vehicles (SEVs).

$$F_{\text{total}} = F_f + F_w + F_a$$

Coulomb's law states that the force between two charged objects is proportional to the product of their charges and inversely proportional to the square of the distance between them. In the context of an electric vehicle, this law does not directly describe the forces associated with the vehicle's movement or efficiency. However, just as Coulomb's law considers the interaction between charged objects, modeling an electric vehicle involves understanding and simulating the interactions between different components and forces acting on the vehicle. This includes factors such as the vehicle's mass, aerodynamics, road conditions, and powertrain characteristics. Coulomb's law is used to describe the rolling resistance force and it can be calculated as follows:

$$F_f = m \times g.\cos\alpha \times f, N,$$

In the given equation, the variables m, g, α, and f represent the weight of the electric vehicle in kilograms, the acceleration due to gravity on Earth's surface (9.81 m/s^2), the angle of inclination of the support surface (road) in radians, and the rolling resistance coefficient of the tire, respectively. In cases where specific terrain data are unavailable, the angle α is assumed to be 0, indicating a flat road surface. The aerodynamic drag force, acting as a resistance, opposes the motion of the EV.

$$F_w = \frac{C_x S_m \rho_{\text{air}} V^2}{2}$$

In the given equation, C_x, S_m, and ρ_{air} represent the aerodynamic resistance coefficient of the EV, the midsection area in square meters, and the air density (typically 1.2 kg/m^3 under normal conditions), respectively. $\dot{V}$ represents the speed of the vehicle's linear motion in meters per second.

The force of inertia of EV refers to the resistance it experiences to changes in motion due to its mass and velocity. It plays a role in the vehicle's acceleration, deceleration, and handling. By accounting for inertia, EV models can accurately simulate the vehicle's behavior and optimize its performance. It can be calculated as follows:

$$F_a = m \times a$$

where a represents an acceleration of the SEV,

$$a = \frac{dV}{dt}, \text{m/s}^2$$

13.3.2 Modeling of wheel

The rolling resistance coefficient of a tire is dependent on the speed at which the tire is moving in a straight line. As the translational speed increases, the rolling resistance coefficient may vary, affecting the overall resistance experienced by the tire. The correlation between the rolling resistance coefficient and the translational speed of the tire can have a direct impact on the vehicle's energy consumption as a whole.

$$f = f_o + k_f \times V^2$$

where f_o, k_f represent the rolling resistance coefficient of a wheel at low speed, the increase of rolling resistance coefficient of a wheel as it increases with speed. Additionally, the equation considers the torque applied to the wheel axis and integrates the wheel's moment of inertia into the calculations.

$$M_w = F_{\text{total}^r} + a \times \frac{I_w n_w}{r}$$

represent the average radius of the EV wheels, the moment of inertia of the EV wheel (measured in kg·m^2), and the number of EV wheels, respectively.

13.3.3 Transmission model

In transmission modeling, the focus is on representing the mechanical system responsible for transmitting power from the engine or motor to the wheels of a vehicle. The transmission model typically includes components such as gears, shafts, clutches, and differentials. To model the transmission, various parameters and characteristics need to be considered, such as gear ratios, torque capacity, efficiency, and shifting dynamics. These parameters determine how power is transferred and distributed throughout the system, affecting the vehicle's speed, acceleration, and overall performance. The modeling approach can vary depending on the level of detail required. Simplified models may use basic mathematical relationships and assumptions to estimate performance, while more advanced models may incorporate dynamic simulations and control algorithms. By accurately modeling the transmission system, engineers and researchers can analyze and optimize various aspects, such as gear selection, power distribution, energy efficiency, and overall drivetrain performance in vehicles. By considering this simplified model, the angular velocity of the EM can be determined.

$$\omega_{EV} = \frac{V_{utr}}{r}\ \text{rad/s},$$

where *utr* refers to the transmission ratio of the gear.

The torque transmitted to the shaft of an EV from the wheel through the transmission system can be determined using the following equation. This equation considers factors such as the wheel's radius, wheel's moment of inertia, the number of wheels on the EV, and other relevant parameters. By considering the torque transfer mechanism within the transmission system, the calculation allows for the determination of the torque exerted on the shaft. This torque value plays a critical role in understanding the power distribution and overall performance of the EV. It can be calculated as follows:

$$M_{EV} = \frac{M_w}{u_{tr}} \eta_{tr}^{sgn\,(Mw)},\ \text{N} \cdot \text{m}$$

where M_w, u_{tr}, are torque at the wheels, N·m, and transmission efficiency. *Sgn* (*Mw*) is wheel torque sign function: mechanical efficiency is determined by the value of friction, and forces, so when accelerating it increases the load moment on the EM (sign −1), and when recuperating reduces it (sign +1).

The moment of inertia of the rotor of the EM, given to the torque on the axis of the EM is as follows:

$$M_I = I_{EM} \frac{d\omega_{EV}}{dt} = I_{EV} \frac{dV}{dt} \frac{utr}{r} = I_{EV} a \frac{utr}{r}$$

where I_{EM} is represents the moment of inertia of the rotor of the EM, kg·m^2.

13.3.4 Modeling of electric motor

Modeling an electrical motor involves considering the torque on the motor shaft, which is the sum of M_{EM} and M_I. The motor's efficiency can be represented by a table-valued function based on the angular velocity and load moment:

$$\eta_{EV} = f(\omega_{EV}, M_{EM} + M_I,)$$

The power consumed by an electrical motor (EM) can be calculated by considering its input power and efficiency:

$$P_{EV} = (M_{EV} + M_I) \times \omega_{EV} \times \eta_{EV}^{sgn\,(M_{EV} + M_I)}$$

where η_{EV} is the efficiency of EV.

13.3.5 Solar photovoltaic model

$$P_{pv}(t) = R_{pv} D_{pv} \left(\frac{G_T(t)}{G_{T,STC}} \right) [1 + \alpha_p (T_{cell}(t) - T_{cell,STC})$$

where $P_{pv}(t)$, R_{pv}, and D_{pv} denote, respectively, the output power of the PV generator during hour *t* of the day, rated capacity (kW), and the PV derating factor (%). G_T and $G_{T,STC}$ denote the incident solar radiation (kW/m^2), incident radiation in STC (standard test conditions). α_p refers to the power temperature coefficient (%/°C) and T_{cell} and $T_{cell,STC}$ are the cell temperature (°C) under operating and STC conditions, respectively.

By connecting the solar cells in series and parallel, the PV module is obtained. However, mostly the solar cells are connected in series to make a PV module. If N_S solar cells are connected in series, the output current is equal to the current of each individual cell and the output voltage is equal to N_S time of voltage of each individual cell. Moreover, if N_P solar cells are connected in parallel, the output voltage is equal to the voltage of each individual cell and the output current is equal to N_P times the output current of each individual cell [15]. By considering the as discussed earlier notes, the voltage V_M and current I_M of a PV module formed by $N_S \times N_P$ solar cells are defined as follows:

$$I_M = N_P \times I$$

$$I_{SCM} = N_P \times I_{SC}$$

$$V_M = N_S \times V$$
$$V_{OCM} = N_S \times V_{OC}$$

In the given context, *NS* represents the number of cells connected in series, while *NP* represents the number of columns of cells connected in parallel within a module. The subscript "*M*" indicates the module, while subscripts without "*M*" pertain to individual solar cells. The resistance for the PV module is expressed as follows:

$$R_{SM} = \frac{N_S}{N_P} R_S$$

By substituting the as discussed earlier equations, the output current of a module is written as follows:

$$I_M = I_{SCM} - N_P I_S \left[\exp\left(\frac{q(\mathrm{V}_M + \mathrm{R}_{SM}\mathrm{I}_M)}{A.K_B.T.N_S}\right) - 1\right]$$

where it has been assumed that $I_{LM} \approx I_{SCM}$. Moreover, for an open circuit condition, I_M is equal to zero and saturation current can be written as follows:

$$I_O = \frac{I_{SCM}}{N_P\left[\exp\left(\frac{qV_M}{A.K_B.T \cdot N_S}\right) - 1\right]}$$

By substituting the equation, the output current of PV module can be presented as the following equation:

$$I_M = I_{SCM}\left[1 - \exp\left(\frac{q(\mathrm{V}_M + \mathrm{R}_{SM}\mathrm{I}_M - \mathrm{V}_{OCM})}{A.K_B.T.N_S}\right)\right]$$

In accordance with this study, we have established a fixed car model with a roof, hood, and trunk dimension of 1.8 m^2. The potential power that can be obtained is depicted in Fig. 2, which illustrates the relationship between voltage factor and irradiation value. Through simulations, we were able to generate the $I = f(\mathrm{v})$ and $P = f(\mathrm{v})$ curves for the PV field under varying solar radiation levels and temperatures.

To achieve the maximum power point, the maximum power point tracking is implemented as a controller. There are two types of perturb and observe algorithms; the first one is by using the voltage reference to produce the output and the second one is by using the duty cycle. Fig. 1 shows the flow chart of perturb and observe (P&O) method based on duty cycle. The direct duty cycle approach has the benefit of eliminating the requirement for a proportional-integral controller to handle overshoot and steady-state errors. In the P&O technique [16–19], the PV voltage is perturbed, and the resulting power is measured. If both voltage and power increase, the duty cycle is adjusted to raise the voltage. Conversely, if the voltage increases but the power decreases, the voltage is reduced. [16–19]. Likewise, when both the voltage and power decrease, the duty cycle needs to be adjusted to increase the voltage. Furthermore, when the voltage decreases but the power increases, reducing the voltage is necessary to shift the operating point toward the MPP. [20,21].

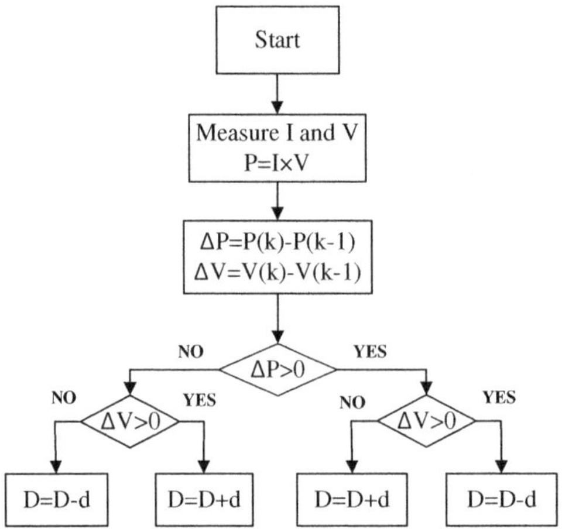

FIG. 1 Flowchart of perturb and observe algorithm.

13.3.6 Battery energy storage model

The battery model in an EV represents the behavior of the vehicle's energy storage system. It includes parameters such as capacity, voltage, internal resistance, and efficiency. Different types of models, ranging from simple empirical to complex physics-based models, are used to simulate battery performance and characteristics. Battery models are essential for estimating the state of charge, predicting range, and optimizing energy management in EVs.

$$P_{\text{battery}} = P_{ch} \times \eta - P_{dis} - P_{los}$$

where P_{ch}, η *and* P_{dis}, respectively, refer to electric power consumed by the converter, efficiency of the converter and the electric power consumed by the traction inverter (W). P_{los} denotes the total loss of electrical power.

13.4 Solar electric vehicle simulation model

To gain a deeper understanding of the aforementioned theories, let us examine an example of a solar electric vehicle model that has been simulated using MATLAB software. The model consists of five main sections: solar generation, battery storage, DC motor, engine, drive wheel, and controller. Fig. 2 provides a visual representation of the structure of this solar electric vehicle (SEV) model from an electrical point of view. The solar generation component is responsible for capturing solar energy through solar panels and converting it into usable electricity to power the vehicle. The battery storage section stores the generated energy for later use, ensuring a continuous power supply for the SEV. The DC motor serves as the primary propulsion system of the vehicle, converting electrical energy from the battery into mechanical energy to drive the wheels. The engine section, although not the primary power source in this case, may be included to provide additional power support if necessary. The drive wheel component transfers the mechanical energy from the motor to the road, enabling the vehicle's movement. Finally, the controller section plays a crucial role in monitoring and managing the power flow between the various components, optimizing energy usage, and ensuring efficient operation of the SEV. This example model provides a comprehensive framework for studying and analyzing the performance and behavior of an SEV. Through simulations conducted in MATLAB, researchers and engineers can evaluate the system's efficiency, and performance under different conditions, and identify areas for improvement. By utilizing such models, researchers can explore various design and optimization strategies, evaluate the impact of different variables, and ultimately contribute to the development of more efficient and sustainable SEV technologies.

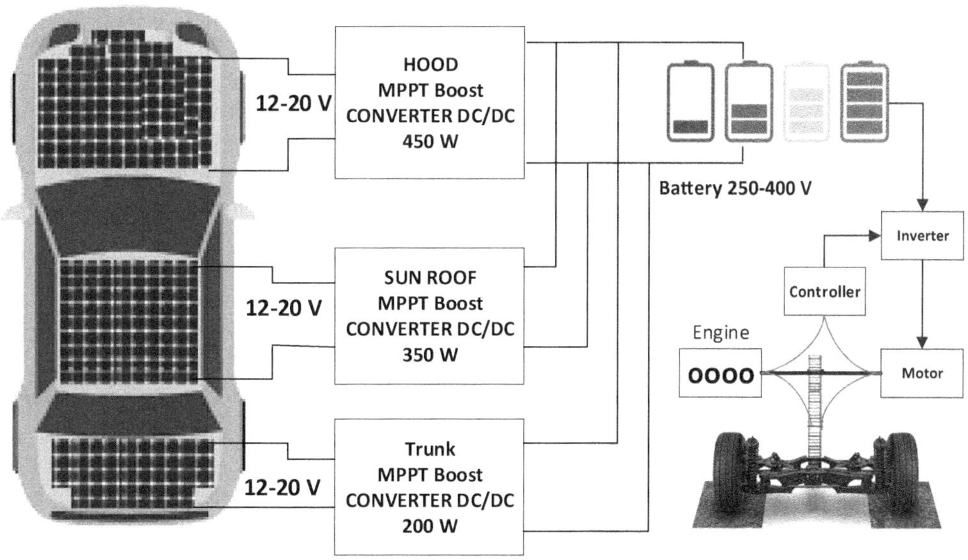

FIG. 2 SEV electrical subsystem structure.

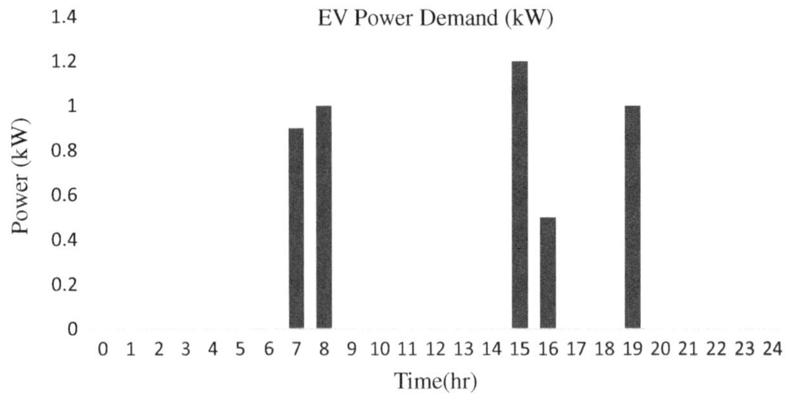

FIG. 3 Power demand of a solar electric vehicle over 2 days of drive.

13.4.1 Daily electric vehicle power demand

For this particular simulation, we consider a small car that is capable of generating 1 kWh of energy. The electric vehicle (EV) is assumed to be driven from 7 to 8 am at an average speed of 40 km per hour. The daily EV drive profile is displayed in Fig. 3. Upon examining the figure, it becomes evident that the EV is in motion three times throughout the day. The total power consumed by the EV amounts to 5 kWh. It is important to note that there is a 40% variation in the power demand from day to day. The power demand is influenced by various factors such as driving duration, speed, and vehicle characteristics. Understanding and accurately estimating the daily power demand is crucial for designing appropriate charging infrastructures and optimizing the energy management of EVs. By considering the variation in power demand from day to day, it becomes apparent that flexibility in charging strategies and energy storage solutions may be necessary to accommodate the fluctuating power requirements. Additionally, incorporating renewable energy sources, such as solar power, can contribute to reducing the reliance on the grid and further enhance the sustainability of EV operations.

13.4.2 EV specification

SEV motors are electric motors specifically designed to be powered by solar energy in EVs. They leverage solar panels to capture sunlight and convert it into electricity that is then used to drive the motor and propel the vehicle. Key specifications to consider for a solar EV motor are as follows: the rated power of the motor refers to the maximum power output it can deliver is 2.5 kW. It determines the overall performance and acceleration capabilities of the solar EV. The rated torque of the motor represents the maximum twisting force it can generate. It is crucial to provide sufficient torque to move the vehicle efficiently, especially during acceleration and climbing inclines. SEV motors operate at specific voltage levels, typically determined by the battery system and power electronics in the vehicle, which in this case is determined to be 24 V. The voltage rating of the motor should align with the overall electrical system of the EV. It is important to note that the specifications of a SEV motor can vary based on the specific design, size, and intended application of the motor. Manufacturers typically provide detailed specifications for their SEV motors, taking into account factors such as weight, size constraints, and power requirements to meet the unique needs of DEVs. The following table shows the current SEV specification (Table 1).

13.4.3 Energy management strategy

In this study, the focus is on utilizing the MATLAB software tool to achieve optimization through various dispatch strategies. However, particular emphasis is placed on employing a load-following dispatch strategy due to its effectiveness in reducing excess energy and resulting in lower values of battery. Fig. 4 illustrates the control strategy that governs the operation of the EV. The power system configuration design was determined using a mathematical optimization method. The optimization method utilized in this study considers three crucial factors: the building's load

TABLE 1 Motor specification.

Components	Variables	Specification
Motor	Power size	8kW
	Motor torque	140Nm
	Max. speed	80km
Solar capacity	Power	1kW
	Lifetime	25ye
	OSV	36.4 Volt
Battery	Rate	21kWh
	Max.	90%
	Min.	10%

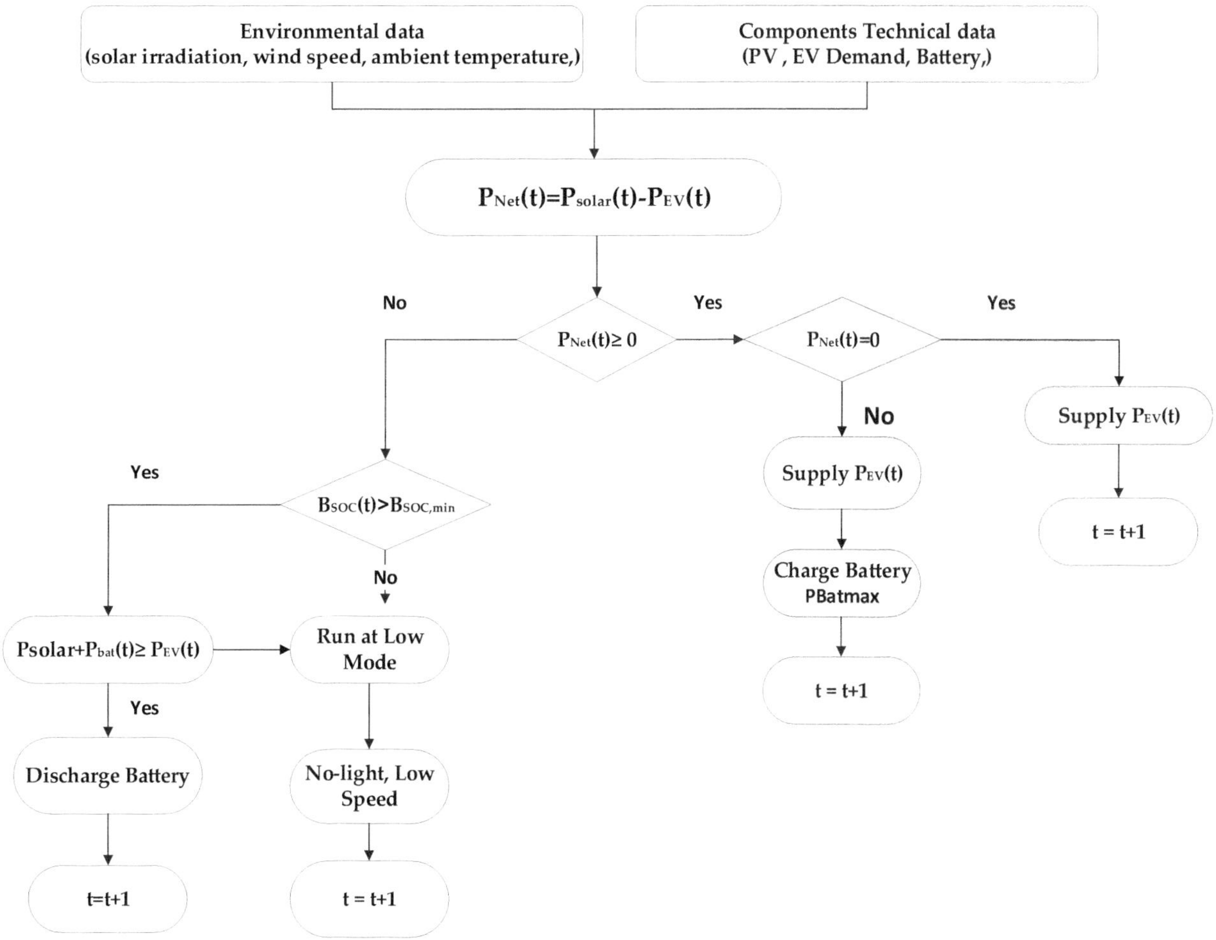

FIG. 4 Electric vehicles' battery charge control strategy.

profile, the capacity of solar energy generation, and the system design constraints. These factors play a pivotal role in determining the potential solar energy generation. The search space in this context is determined by the sizing of various components in the power system such as EV power demand, battery lifetime, and weight of SEV. Simulations were conducted on various power system architectures. The power system was optimized according to the desired requirements (constraints). Each simulation involved 1276 data points per year and a time step of 60min. The control algorithm begins by incorporating input parameters that encompass electric and thermal loads, as well as meteorological data such as temperature, irradiance, and wind speed. These parameters are utilized to evaluate and determine the optimal energy generation and storing system behavior:

- when E_{NET} is larger than the EV demand. It indicates that the energy generation exceeds the demand, there is an excess of energy available after fulfilling the electric demand. This excess generation is then stored in the batteries, up to its E_{Bmax}.
- when E_{NET} is equal to EV demand. It indicates that energy generation matches the EV demand, the solar generation fulfills the EV demand.
- when E_{NET} is less than the EV demand, it indicates a deficit in EV demand, resulting in no electric generation. To address this deficit, the controller discharges the battery system to supply EV demand. In situations where the battery shortage persists, the EV low mode is utilized.

The following energy management flowchart, calculates constraints and objective functions at each time step. The load-following strategy ensures that the generator meets the load demand without charging the battery, while the surplus power from renewable sources is used to charge the battery.

13.5 Result and discussion

In this section, we will discuss our results and examine the variables that impact the energy system of electric vehicles. From an electrical standpoint, the generation of solar power and the storage of that power are critical considerations for every electric vehicle (EV) system. Fig. 5 shows the performance under different solar irradiation levels, the changes in power-voltage *P-V* output characteristics of the power-voltage (PV) module as the irradiation levels vary, while keeping the temperature constant; similarly, in the current-voltage output characteristics of the PV module under varying irradiation levels at a constant temperature. An observation from the graphs is that as the irradiation increases, both the current output and voltage output of the PV module also increase. Consequently, this leads to a net increase in the output power of the PV module when the irradiation levels increase, while maintaining a constant temperature.

The monthly generation of solar electric vehicle is dependent on various factors, such as the amount of solar irradiation received and additional parameters that influence solar energy conversion efficiency. These factors can include the orientation and tilt angle of the solar panels, shading effects, panel efficiency, and environmental conditions like temperature. By considering these factors, it is possible to estimate the amount of solar energy that can be generated by an electric vehicle over a specific time period, such as a month. In Fig. 6, it is clear that the global solar irradiation that reaches the surface of a solar electric vehicle (SEV) shown by the orange color, is significantly reduced when converting to power output. Many factors impact solar generation in SEVs such as the angle of solar irradiation, wind speed, location, etc.

Fig. 7 illustrates the EV's solar generation and the state of charge of the battery over a two-day period. Upon analyzing the figure, we observe that the state of charge (SOC) of the battery experiences a decrease in the morning from 7 to 9 am due to the movement of the EV. However, thereafter, the SOC of the battery rises as a result of solar power generation. The graph further reveals two additional reductions in SOC due to EV movement, occurring from 2 to 3 pm

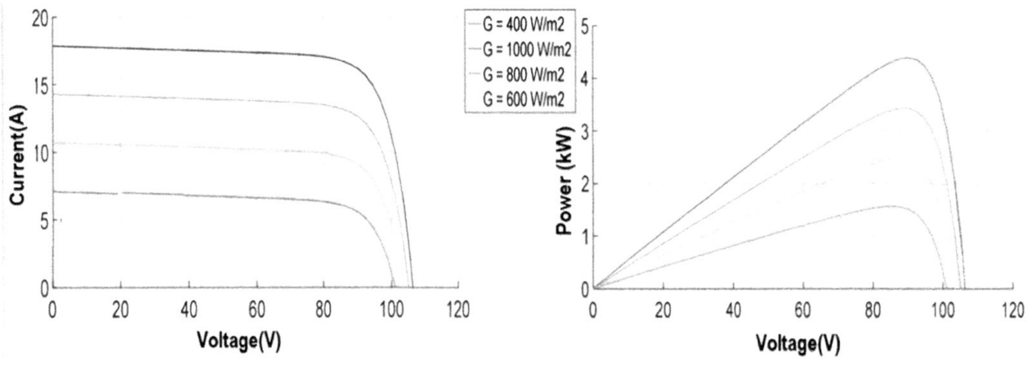

FIG. 5 Solar electric vehicle performance under different solar irradiation levels on power, voltage, and current.

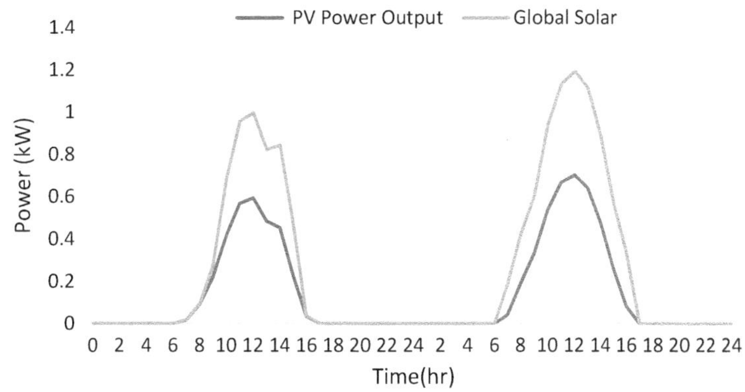

FIG. 6 Comparing the amount of solar irradiation reaching a solar electric vehicle and power generation.

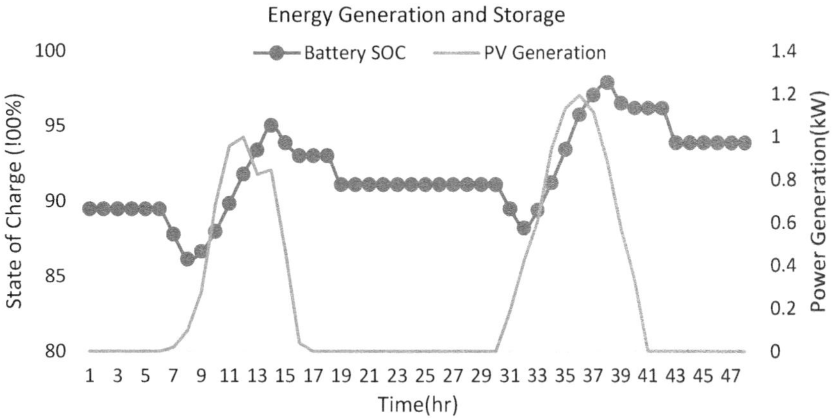

FIG. 7 Solar and state of charge of battery performance.

and 7 to 8 pm. On the subsequent day, assuming the EV exhibits a similar behavior with a 15% variation throughout the year, we notice a comparable pattern in the SOC of the battery and solar generation. These findings highlight the dynamic nature of the EV energy system and its dependence on variables such as solar generation and EV movement. The SOC of the battery fluctuates in response to the EV's activity, while solar power generation plays a crucial role in replenishing the battery's energy. Understanding and effectively managing these variables are essential for optimizing the performance and efficiency of EV energy systems. By analyzing and adapting to the behavior and variations observed in the SOC and solar generation, it is possible to design effective energy management strategies that maximize the utilization of solar power and ensure optimal performance of EVs throughout the year.

In SEV, the speed of the vehicle and the road angle are critical variables that significantly influence the power demand. The power demand of an SEV varies depending on these two factors. Fig. 8 illustrates the impact of SOC and SEV's speed on the power demand of the SEV. The figure clearly demonstrates that a 20% reduction in SEV's speed results in a decrease in EV electrical demand from 4 to 3 kW. This decrease in speed also leads to an increase in the SOC of the battery from 40% to 63%. Conversely, when the SEV's speed is increased by 20%, the power demand rises to 5 kW while the SOC decreases to 35%. These findings highlight the direct relationship between SEV's speed, power demand, and battery's SOC. Slowing down the SEV results in lower power demand and higher battery's SOC, enhancing the vehicle's energy efficiency. Conversely, increasing the SEV's speed leads to higher power demand and lower battery's SOC, which may impact the vehicle's overall performance. Understanding the impact of speed and road angle on power demand is crucial for efficient energy management in SEVs. By considering these variables and optimizing the vehicle's speed and driving conditions, it is possible to enhance the SEV's energy efficiency, maximize battery life, and improve overall performance.

Energy management is a crucial aspect to consider for the electrical side of SEVs. Its primary goal is to minimize the mismatch between power generation and demand, reduce power losses, and enhance the reliability of SEVs, while optimizing the number of batteries required. Fig. 8 demonstrates the power demand and the electrical aspects of the SEV, including power generation, battery charging, and discharging over a two-day period. Notably, during

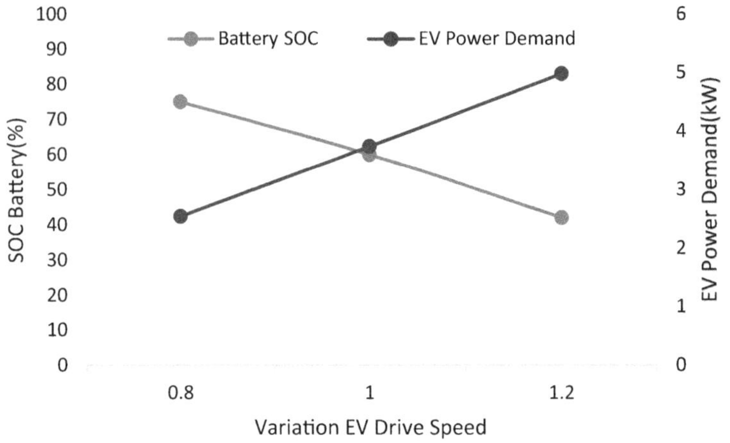

FIG. 8 Impact of speed on state of charge of battery and power demand.

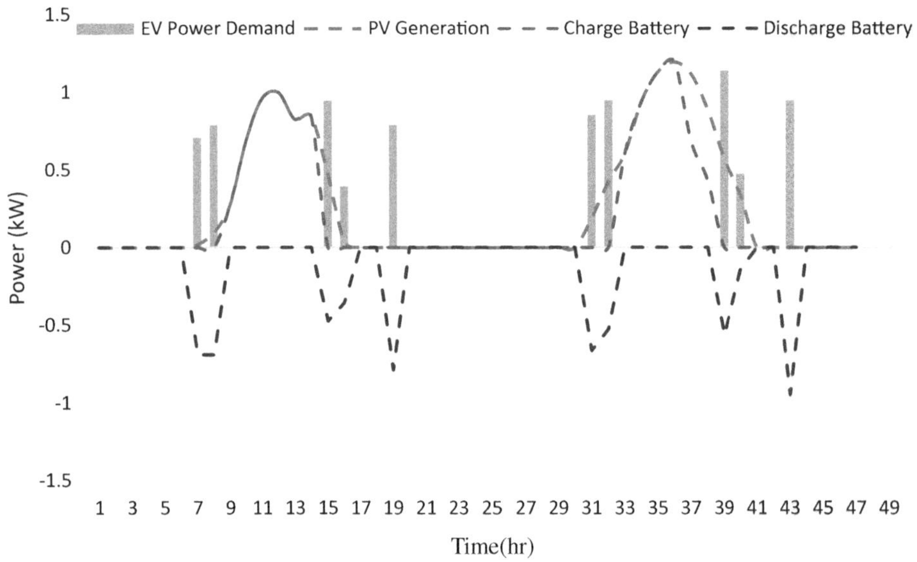

FIG. 9 Electrical side performance of solar electric vehicle.

periods of low solar generation, such as from 4 to 6 pm and at night, the battery discharges to meet the SEV's power demands. Therefore, the sizing of the battery, either oversizing or undersizing, significantly impacts the performance of the SEV. Appropriately addressing energy management considerations is vital to ensuring optimal SEV performance and efficiency (see Fig. 9).

This study looked at how the weight of SEVs affects how well they perform. It found that SEVs with lower weight can drive more and need fewer energy storage units and converters. Lighter SEVs also lose less power compared to heavier ones. The study noticed that making the SEV lighter makes it go faster. So, reducing the weight of SEVs makes them perform better overall. From Table 2 it is clear that an SEV with the weight of 800 kg has the potential to drive up to 87 km with a minimum number of energy storage and converters with 14 and 3, respectively. So, reducing the weight of SEV significantly impacts SEV performance but also reduces the cost of the SEV.

As depicted in Fig. 10, SEVs characterized by lower weight not only attain higher speeds but also demonstrate an extended range in terms of driving kilometers. This advantageous attribute enables these vehicles to traverse more significant distances, contributing to their overall efficiency and practicality in various transportation scenarios. By maintaining higher speeds and covering more ground, SEVs with lower weight configurations offer enhanced mobility and expanded travel capabilities, potentially catering to diverse user needs across varying travel distances. It is observed that the SEV with a weight 800 kg was able to drive 12 km in a speed of 70 km whereas others two 8 km and 3 km for SEV with weights 950 and 1100 kg, respectively.

TABLE 2 Comparison impact of SEV weight on variables.

Variables	Weight 800 kg	Weight 950 kg	Weight 1100
Kilometers	87	78	49
No. of batteries	14	17	24
No. of DC/DC motors	3	4	5
Power losses, %	8	12	18
Max. speed	70	65	60
PV generation kW/yr	1750	1850	2010

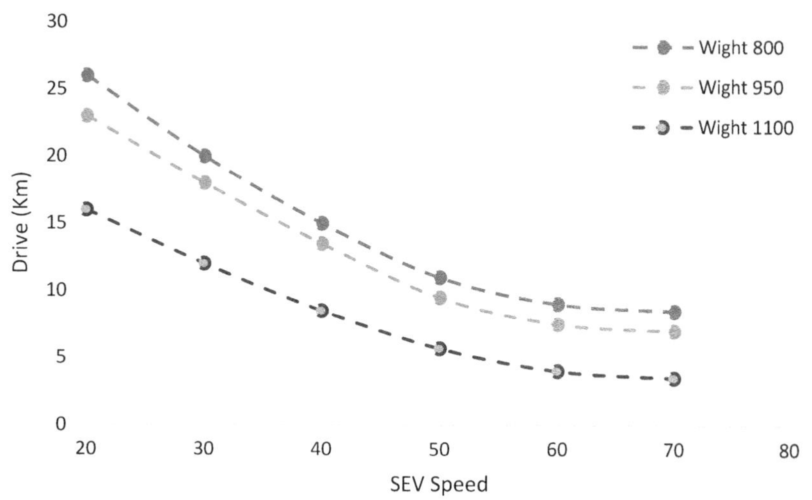

FIG. 10 Comparison of solar electric vehicles' weights on speed and driving distance.

13.6 Conclusions

In conclusion, solar electric vehicles hold great promise for significant growth in the next decade. The overview of solar electric vehicles (SEVs) highlights their utilization of solar energy for propulsion, with solar panels capturing sunlight and converting it into electricity. In this chapter by considering factors such as power demand, speed, distance, weight and road angle, SEVs demonstrate their efficiency and range capabilities. However, for continued development and expansion, improvements and extensions in solar panel efficiency, lightweight design, battery technology advancements, vehicle-to-grid integration, and intelligent power management systems are essential. By pursuing these avenues, the performance of SEVs can be further enhanced, making them a compelling solution for sustainable transportation and reducing carbon emissions in the years to come. SEVs have the potential for significant growth in the next decade. Here are several factors that contribute to this potential:

Advancements in solar technology: The efficiency and cost-effectiveness of solar panels have been improving over the years. Continued advancements in solar technology are likely to make solar panels more efficient, lightweight, and affordable, making them more practical for integration into vehicles.

Growing environmental concerns: The increasing awareness of climate change and the need to reduce greenhouse gas emissions have led to a greater emphasis on clean energy solutions. SEVs offer a sustainable transportation option by utilizing renewable energy for propulsion, thereby reducing carbon emissions.

Government Support and Incentives: Many governments worldwide are implementing policies and incentives to promote the adoption of EVs. This includes financial incentives, tax credits, and infrastructure development. As SEVs fall under the EV category, these policies and incentives will also benefit SEVs.

Advancements in battery technology: The development of advanced battery technology, such as higher energy density and faster charging capabilities, is crucial for the growth of SEVs. Improved batteries will enable SEVs to store solar energy efficiently and provide longer driving ranges, addressing one of the current limitations of solar powered vehicles.

Infrastructure development: The expansion of charging infrastructures is essential for the widespread adoption of electric vehicles, including SEVs. As the demand for electric vehicle charging stations increases, it is expected that more solar powered charging stations will be deployed, allowing SEV owners to conveniently recharge their vehicles using solar energy.

Cost reduction: Over time, the cost of solar panels and associated technology has been decreasing. As economies of scale and technological advancements continue, it is anticipated that the cost of SEV components will further decline, making SEVs more affordable and accessible to a larger consumer base.

Integration with smart grids: The integration of SEVs with smart grids allows bidirectional energy flow, enabling vehicles to not only draw power from the grid but also feed excess energy back into it. This integration facilitates better energy management and grid stability, while maximizing the utilization of solar energy for both transportation and grid balancing purposes.

Considering these factors, it is expected that SEVs will experience significant growth in the next decade as technology improves, costs decrease, and support from governments and infrastructure development increases. However, the exact pace of growth will depend on various factors, including policy decisions, market demand, and technological advancements.

References

[1] S. Manivannan, E. Kaleeswaran, Solar powered electric vehicle, in: 2016 First International Conference on Sustainable Green Buildings and Communities (SGBC), IEEE, 2016.

[2] F. Mohammadi, Design, analysis, and electrification of a solar-powered electric vehicle, J. Sol. Energy Res. 3 (4) (2018) 293–299.

[3] M. Nivas, et al., Modeling and analysis of solar-powered electric vehicles, Int. J. Power Electron. Drive Syst. 13 (1) (2022) 480.

[4] C. Duan, et al., A solar power-assisted battery balancing system for electric vehicles, IEEE Trans. Transp. Electrif. 4 (2) (2018) 432–443.

[5] M.-L. Kamberg, Working as a Mechanic in Your Community, The Rosen Publishing Group, Inc, 2015.

[6] M.A. Green, Event report. Brain sport: the 1996 world solar challenge solar Car race across Australia, Prog. Photovolt. Res. Appl. 5 (1) (1997) 69–76.

[7] P. Babalola, O. Atiba, Solar powered cars-a review, in: IOP Conference Series: Materials Science and Engineering, IOP Publishing, 2021.

[8] B.B. de Wit, Vehicle Stability Analysis of a Solar Car, Stella Lux, n.d. https://scholar.google.com/scholar?hl=en&as_sdt=0%2C5&q=Vehicle+Stability+Analysis+of+a+Solar+Car%3A+Stella+Lux&btnG=.

[9] A. Husnain, Design and Analysis of a Micro Solar Electric Vehicle for Application in Pakistan, Memorial University of Newfoundland, 2022.

[10] M. Arieffadillah, I. Kumara, Y. Divayana, The Development of Solar PV Car to Reduce Carbon Emissions from the Transport Sector. https://scholar.google.com/scholar?hl=en&as_sdt=0%2C5&q=The+Development+of+Solar+PV+Car+to+Reduce+Carbon+Emissions+from+the+Transport+Sector&btnG=.

[11] A. Husnain, M.T. Iqbal, System design and PV sizing of a micro solar electric vehicle for Pakistan, in: 2022 IEEE Conference on Technologies for Sustainability (SusTech), IEEE, 2022.

[12] H. Koten, M. Yilmaz, M.Z. Gul, Design Consideration of Solar Powered Cars, 2011. https://scholar.google.com/scholar?hl=en&as_sdt=0%2C5&q=H.+Koten%2C+M.+Yilmaz%2C+M.Z.+Gul%2C+Design+Consideration+of+Solar+Powered+Cars%2C+2011&btnG=.

[13] R. Mangu, et al., Design, development and optimization of highly efficient solar cars: Gato del Sol I-IV, in: 2010 IEEE Green Technologies Conference, IEEE, 2010.

[14] A. Sierra, A. Reinders, Designing innovative solutions for solar-powered electric mobility applications, Prog. Photovolt. Res. Appl. 29 (7) (2021) 802–818.

[15] Y.-H. Ji, et al., C-language based PV array simulation technique considering effects of partial shading, in: IEEE International Conference on Industrial Technology, 2009, ICIT 2009, IEEE, 2009.

[16] N. Femia, et al., A technique for improving P&O MPPT performances of double-stage grid-connected photovoltaic systems, IEEE Trans. Ind. Electron. 56 (11) (2009) 4473–4482.

[17] I.V. Banu, R. Beniuga, M. Istrate, Comparative analysis of the perturb-and-observe and incremental conductance MPPT methods, in: 2013 8th International Symposium on Advanced Topics in Electrical Engineering (ATEE), IEEE, 2013.

[18] R. Faranda, S. Leva, V. Maugeri, MPPT techniques for PV systems: energetic and cost comparison, in: 2008 IEEE Power and Energy Society General Meeting-Conversion and Delivery of Electrical Energy in the 21st Century, IEEE, 2008.

[19] A.K. Abdelsalam, et al., High-performance adaptive perturb and observe MPPT technique for photovoltaic-based microgrids, IEEE Trans. Power Electron. 26 (4) (2011) 1010–1021.

[20] O. Wasynezuk, Dynamic behavior of a class of photovoltaic power systems, IEEE Trans. Power Syst. 9 (1983) 3031–3037.

[21] A. Al-Amoudi, L. Zhang, Optimal control of a grid-connected PV system for maximum power point tracking and unity power factor, in: Seventh International Conference on Power Electronics and Variable Speed Drives, 1998 (Conf. Publ. No. 456), IET, 1998.

CHAPTER

14

Advanced electrical energy storage in power electronics systems

Anushree Ramanath
Schneider Electric, San Jose, CA, United States

14.1 Introduction

A wide variety of electric vehicles (EVs) have made it to the market over the past decades. Traction inverter is certainly one of the key components of the EV, but the other equally critical component is that of energy storage. Batteries are deemed critical as they are also one of the weakest points of EVs. As much as different designs and implementations of EVs have been introduced in the market, all of them aim to solve the carbon emission concerns, thus leading to cleaner environment. Even though technological advances in power electronic design, cost reduction in semiconductors, and enhanced reliability of electronic systems in the automobiles have made EVs more promising, detection of accurate battery health and its state of charge (SoC) have remained as issues of concern. The lawmakers are pushing toward cleaner alternatives, and researchers are focusing toward vehicle autonomy, which could help drive EVs to the consumer market more effectively.

The choice of battery or energy storage system is thus a crucial aspect of the EV due to increased focus on electrical performance, reliability, and range of operation [1]. They should ideally be a trade-off between drive performance and reliability without compromising on cost. Drive performance accounts for output power and the vehicle autonomy offered. Reliability aspect includes increased lifetime, high degree of safety, less or maintenance required, and energy regeneration capabilities [2]. Thus, the ideal characteristics for a battery used in EV applications are high-energy density, high charge-discharge efficiency, long life, no maintenance, ability to sustain wide range of temperatures, good load characteristics, fast charging, minimal self-discharge, low internal resistance, good recycle ability, low cost, high reliability, and safety [3].

This chapter discusses the nuances of design and implementation of electrical energy storage in power electronics-based systems. Section 14.2 provides an overview of the state-of-the-art techniques used for electrical energy storage while touching upon the history behind its development over the years and the classification of energy storage devices. Section 14.3 describes the significance of power electronics-based systems in energy storage by elaborating on the advantages they offer and the applications in which they are essential. Section 14.4 delves into the details of the electrical energy storage by explaining the design and creation of battery models for EVs, other storage devices that can be used along with batteries like ultracapacitors and fuel cells, and necessary control strategies to have them all working together. Section 14.5 describes the batteries used in different kinds of EVs by getting into the details of the battery chemistry and discusses the battery management system that is used to control the necessary performance parameters. Section 14.6 discusses the charging solutions offered at the power electronic interfaces in an EV and describes different ways of implementing conductive charging and inductive charging. The future trends are highlighted in Section 14.7. Section 14.8 provides a comprehensive summary detailing the different aspects pertaining to the design, implementation, and applications of electrical energy storage in power electronics-based systems as relevant to automotive applications.

Handbook of Power Electronics in Autonomous and Electric Vehicles
https://doi.org/10.1016/B978-0-323-99545-0.00004-X

14.2 State-of-the-art

The electric power and transportation industries form a major portion of the primary energy consumption with consumption of few trillion kilowatthours of energy and emission of a thousand million metric tons of carbon dioxide in US alone. This is in a span of a few months, which translates to the challenges we are facing for reducing the carbon footprint while pushing for cleaner energy technologies, especially in automotive industry. The transition to cleaner energy sources is only possible with significant advances in the design and implementation of semiconductors, power electronic components, electric machines, and battery technologies [4].

One of the major bottlenecks for EV development is the battery pack that consumes most of the space in the EV and also increases the weight of the vehicle. It also demands higher electricity from the transportation sector and the related infrastructure that is required for procurement, storage, and distribution of these vehicles. All these concerns demand the need for more efficient storage systems to be used in EVs and also to carefully consider the entire life-cycle of these vehicles for optimization.

14.2.1 History of energy storage-based devices

Since the beginning of the 19th century, a number of energy storage devices have been used for several applications. The use of power electronic components in conjunction with traditional energy storage devices is relatively new, but dates back to 1960s. The key point to observe is that since the power electronics-based energy storage devices have been introduced, they have gained significant momentum in replacing the conventional energy storage devices based on mechanical or electrochemical technologies.

The main disadvantages observed in the conventional solutions were high cost, low reliability, low conversion efficiency, slow response time, bad power quality, bulky size or low-power density, and less tolerance to extreme environmental conditions. Power electronics-based storage solutions helped overcome some of these while reducing the issues on other fronts, thus becoming the choice of system for both low- and high-voltage applications in automotive sector. They also offer flexibility in terms of operation as they can be connected in series or parallel, and can be controlled to serve as voltage or current generators based on the application in hand.

14.2.2 Classification of energy storage devices

The modern power electronics-based energy storage devices have several advantages compared to their traditional counterparts. They can also be controlled to act as current or voltage generators having an energy storage media that is capable of providing active power for a stipulated amount of time as required. Also, it can be connected either in series or in parallel with the power system based on the application. Power electronics-based energy storage systems are broadly classified into electrical energy storage systems, electrochemical energy storage systems, mechanical energy storage systems, and hybrid energy storage systems as shown in Fig. 1.

Energy storage technologies have been constantly evolving. They also have diverse classifications based on their storage media, based on the functions or services they provide, and so on. Based on the storage media, they can be

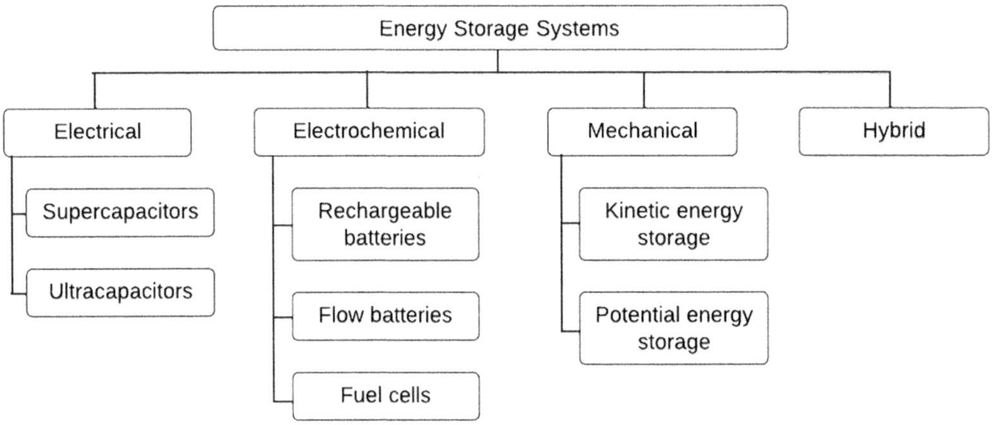

FIG. 1 Classification of energy storage devices.

classified into electrical—including capacitors, supercapacitors, and magnetic energy storage; electrochemical—including rechargeable batteries, flow batteries, and fuel cells; and mechanical—including potential energy storage and kinetic energy storage. Based on the functionalities they support, they can be categorized based on their energy density, their power density, their cycle time, their efficiency, or their response time.

The design of power electronic systems has been focused toward specific functionalities. Under the grid code compliance, they have been used for power quality improvement and to cater to ancillary services. This translates to the ability of providing protection against power supply interruptions, compensating voltage-based distortions like voltage harmonics, voltage fluctuation, and voltage unbalances, and current-based distortions like current harmonics and current imbalances. In terms of the ancillary services provided, it includes voltage support, frequency support by means of containment, reactive power control, black start capability, system restoration after islanding, fault ride through capability, and others. The other major aspect is the ability to support development of clean energy by means of backup power supply, peak shaving, and load shifting.

14.3 Significance of power electronics-based systems in energy storage

Power electronics-based energy storage devices are one of the fastest growing technologies that help in providing ancillary services and solving power problems while enabling access to clean energy-based technologies for multiple segments. They have proven to be advantageous for use in a wide variety of segments spanning from renewable energy sources, electric vehicles to heavy machinery, and a range of dynamic loads. They are flexible, compact, modular, and cost-effective and offer superior performance compared to its counterparts. They can be used for critical applications as they can respond instantaneously; have good stability; are robust, reliable; and have less losses comparatively. They also comply with most of the power quality standards and grid codes, thus making it very sustainable for use.

The introduction of energy storage devices with power electronics-based switches revolutionized storage as they started serving as replacement technology for conventional solutions like mechanically switched energy storage media and stand by generators that have traditionally faced issues with power quality and adherence to codes and standards. Not to forget the fact that the conventional solutions have other disadvantages like limited power quality, less efficient, slow response time, and extremely bulky. Also with the cost factor coming into play, the cost of power electronics-based switches started becoming more effective thus facilitating the transition to these devices in low- and high-voltage applications.

14.4 Electrical energy storage (EES)

Internal combustion engine (ICE) has been the power source of choice for vehicle propulsion due to the high-energy density offered and the low cost of petroleum. An acceptable range and performance are achieved even with peak tank to wheel efficiency of less than 30%. However, as we are moving toward the adoption of cleaner technologies, drivetrain electrification is being considered to reduce oil consumption. Also, due to the recent technological advancements and increasing comfort needs, the number and electric power drawn to cater to convenience type of loads is increasing. This is directly affecting the energy consumption and is resulting in larger energy storage system requirements.

The applications of electrical energy storage are increasing tremendously as the number of EVs and its variants are increasing. This necessitates more thought into the choice of energy storage system. There are different types of energy storage systems that can be used in EV applications. Batteries are the primary energy storage devices especially for ground vehicles, but in case of hybrid vehicles, ultracapacitors (UC) and/or fuel cells (FC) can also be employed. In fact, a combination of batteries, UCs and FCs, could be most appropriate for advanced hybrid EV energy storage systems. Power electronics plays a key role in all of these options and facilitates energy storage. Batteries are discussed in much detail in the upcoming subsections, but an overview of UCs and FCs is provided right after.

14.4.1 Supplementary storage devices for HEVs

The use of EVs is limited mostly because of restricted range, increased refueling or recharging time, and high cost. All the three of these concepts are closely tied to the energy storage system. As we have been discussing, the main considerations for any energy storage device for EV applications are power density, energy density, weight,

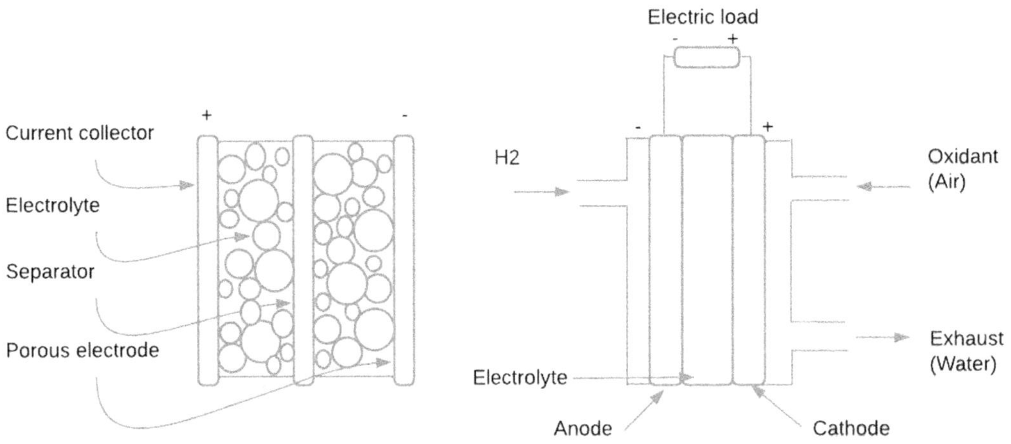

FIG. 2 Cross-section of an individual UC and a hydrogen FC [6].

compactness, cycle life, and cost. The other considerations include safety, recycling capability of materials used, maintenance requirements, and operating temperature. The main energy storage device is battery for most of the EVs. The other supplementary storage devices used in conjunction with batteries are UCs and FCs.

The EV performance parameters are very closely tied with the electrical requirements. For example, power density determines the acceleration ability of the EV, energy density gives an indication of the potential drive range of the EV, and cycle life measures how often the battery can be recharged to its full capacity while determining the reliability aspect by the battery lifetime, while weight and volume directly affect the efficiency of the entire system. Cost on the other hand is determined by the availability of resources, technology, and manufacturability [5]. The following Fig. 2 shows the cross-section of an individual UC and a hydrogen FC.

14.4.1.1 Ultracapacitors (UCs)

Ultracapacitors (UCs) store energy by physically separating positive and negative charges. The setup is of two parallel plates that store charges and is separated by an insulator. Positive plate functions as a current collector, and the plates are porous in nature. The insulator serves as a separator and is surrounded by electrolyte. The applied potential on the positive electrode attracts the negative ions in the electrolyte, whereas the potential built up on the negative electrode attracts positive ions. UCs typically have a long life-cycle due to the absence of chemical variations on the electrodes but suffers from low-energy density. The power density of the UC is considerably higher than that of the battery as the charges are physically stored on the electrodes.

The terminal voltage of the UC is directly proportional to the SoC. This makes it a good backup energy storage device especially for HEVs. In urban driving conditions, there are several instances of stop and go driving, and in such cases, the required total power is relatively low. UCs are the best at such scenarios as they capture electricity from regenerative braking almost instantaneously and responds by delivering power for acceleration immediately. This is feasible due to its very fast charge and discharge rates. Also, the development of power electronics interface allows the UC to operate well throughout its variable voltage range, making it a viable candidate for HEVs.

There are several UC technologies that are being actively researched and developed including carbon or metal based fiber composites, foamed carbon, a carbon particulate with a binder, doped conducting polymer films on a carbon cloth, and mixed metal oxide coatings on a metal foil [7]. Higher-energy density can be achieved by use of carbon composite electrode by using an organic electrolyte rather than using carbon or metal fiber composite electrode with an aqueous electrolyte. All these technologies still point toward UCs having higher power densities than batteries, which have higher-energy densities. Also, the long lifetime and low maintenance requirement of UCs make it cost-effective and an ideal candidate for combined use with batteries in HEVs to maximize the benefits of both the batteries and UCs.

14.4.1.2 Fuel cells (FCs)

The fuel cell (FC) generates electricity form the fuel on the anode and the oxidant on the cathode while reacting in the electrolyte. During electricity generation process, the reactants flow into the fuel cell, whereas the products of the reaction flow out. The advantages of using the FC include long life, reliability, high efficiency, less emission, quiet operation, and ability to recover waste heat. One of the most popular types of FC is a hydrogen FC that has hydrogen as the fuel. It is advantageous as hydrogen is nonpolluting and has highest energy density compared to other fuels, and the end product of cell reaction is water. Other combinations of fuels and oxidants are also possible for FCs like hydrocarbons and alcohols as fuels with chlorine and its oxide form as oxidants.

The efficiency of a FC is directly dependent on the amount of power drawn from it which means higher the power drawn, lower the efficiency of the FC. Also, the reactants must be refilled before they are emptied making it necessary to carry a fuel tank on the EV, which is not very convenient. The response time of FCs is relatively longer compared to that of batteries and UCs while it costs a fortune more—about 5 times the cost of conventional systems, making it a difficult choice for use in HEVs even though it is technologically feasible.

14.4.2 Battery models for EVs

Due to the ever-growing applications of battery storage in EV market, there is a constant need for innovating and experimenting with different battery chemistries. Simulation is one of the most commonly used techniques to evaluate the design cost effectively and reduce the development cycle time [8]. This enables efficient parallel operation as the same model can be deployed on multiple setups to evaluate different use-cases without the need for expensive test and measuring equipment. There are several types of modeling work that has been done in the past, including the mathematical ones, electrochemistry-based models, physics-based models that mostly use equations, and electrical equivalent models that use circuits for analysis. The goal is to build models that are simple, accurate, and as close to reality as possible.

An example for the use of battery models can be that of a battery management system (BMS) implementation in an EV. BMS typically has limited amount of memory and computational ability thus demanding a simple yet accurate model representation of the same [9]. This model could come in handy for designing algorithms that predict the state of the battery in terms of state of charge (SoC) and also helps with the implementation of controls. Thus, there is possibility of building different types of models that cater to different end uses based on the fidelity level. However, it is vital to focus on the trade-offs between complexity, performance, and parameterization efforts [10].

The most commonly used electrical battery models are as shown in the Fig. 3 [11]. The models include the open circuit voltage, resistance of the contacts, electrodes and the electrolyte, and the resistances and capacitances that

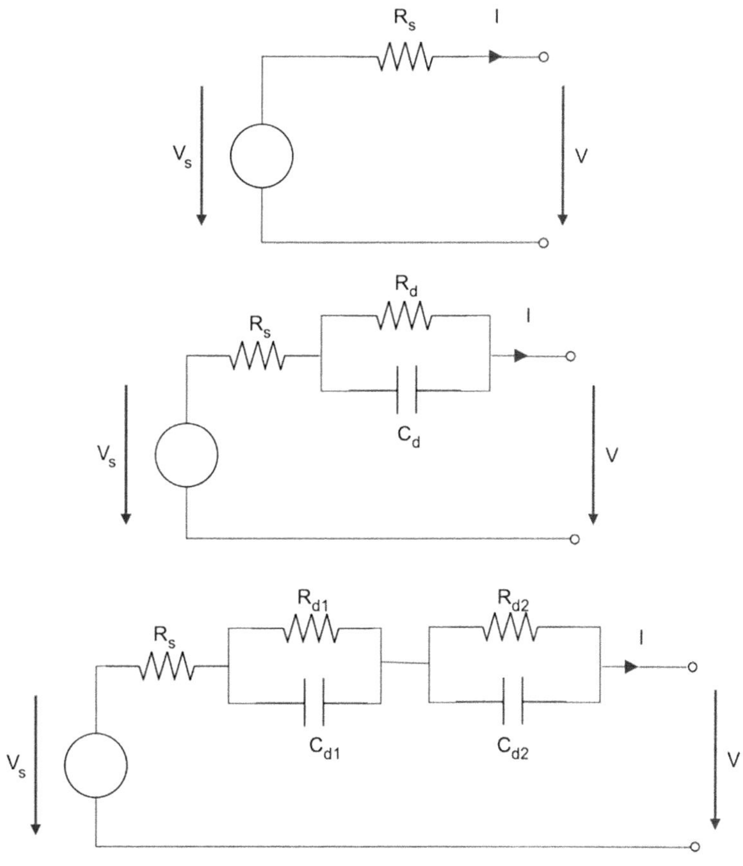

FIG. 3 Battery models [11].

characterize the transient response of the battery cell electrodes. The parameterization efforts for these models can be carried out either in a linear manner or by use of look-up tables. These result in differences in terms of accuracy and the complexity, which is driven by the use-case in hand. The linear models typically require significantly less parameterization time but do suffer from a small loss of accuracy in the output. The battery model presented with one resistor and capacitor combination, and uses linear parameterization of the internal impedance seems to offer the best trade-off between the parameterization efforts and accuracy of the results.

14.4.3 Control strategies

The battery-based energy storage systems typically require bidirectional AC-DC converters and DC-AC converters for its operation. This demands highly efficient converters in place for the EES along with sophisticated controls to cater to efficient operation of the system. However, the electric power sources for EVs are now including batteries, UCs, FCs, and other energy storage systems along with it. With the multiple sources of energy, the design of intelligent control strategies is crucial, inevitable, and also a critical issue. The control strategies employed depend on the type of EV, application, range or operation, and several other factors. However, most of the strategies of the past have focused on improving battery life and overall vehicle efficiency. It is important to also consider power demand, range of velocity, and SoC of all the power sources employed along with batteries. This makes the design of optimal controls a very important part of the design and implementation of any EV.

To implement optimal controls, it is important to understand the configuration of the electric power train. The hybrid drivetrains are typically configured in a series, parallel, or series-parallel fashion. Parallel and series-parallel configurations work best for smaller vehicles due to the space constraints, whereas larger EVs can go for series configuration. A typical hybrid energy storage system comprises of batteries and UC along with it [12]. The configuration can further be differentiated as active or passive in nature.

The actively connected drive train regulates the transfer of energy to and from each component by means of power electronic components through a specially designed controllable power electronics buffer that is placed between batteries and the UC. However, control overcharging and discharging of components can only be achieved in parallel configuration by maintaining the voltage on the bus constant.

14.5 Batteries for EVs, HEVs, and PHEVs

It is evident that the on-board energy storage system of the vehicle is the heart of the EV. Batteries are the primary electrical energy storage devices in most of the ground EVs with ultracapacitors and fuel cells supporting the vehicular fuel economy. The most important characteristics of vehicular EVs are power density, cost-effectiveness, reliability, maintenance-friendly, and energy density. The overall performance of the EV depends not only on the ability to store large amounts of energy but also to be able to release it quickly and effectively based on the load demands [13].

Lithium-based technologies and lithium-ion batteries are leading the way to meet the requirements of EV/HEVs. The most prevalent chemistry is the carbon/graphite anode and lithium metal oxide cathode, with the metal being either cobalt, manganese, nickel, or a mixture of these, with lithium salt dissolved in an organic solvent. Another prevalent chemistry is a lithium titanate anode with a lithium manganese cathode [14].

14.5.1 Battery chemistry

Batteries typically have higher-energy densities and help store majority of the on-board electric energy, whereas ultracapacitors can be used along as a hybrid source due to its high-power density, reliability, and faster response time. Table 1 highlights the characteristics of most commonly used batteries in EVs. The most commonly used battery types are discussed in the following subsections.

TABLE 1 Characteristics of commonly used batteries in EVs [14].

Battery type	Lead-acid	NiMH	Lithium-ion
Energy density (Wh/kg)	30–40	50–80	100–150
Power density (W/kg)	120–200	250–1000	1000–1500
Cycle life	200–300	300–500	500–1000

14.5.1.1 Lead-acid batteries

Lead-acid batteries are available commercially and are quite mature in terms of its technology, thus making it one of the simplest choices. Lead oxide serves as the positive active material while spongy lead works as the negative active material with diluted sulfuric acid as the electrolyte. However, it is found that at deep rate of state of charge, the battery's life-cycle reduces drastically making it less fit for higher discharges of its rated capacity [15].

14.5.1.2 Nickel-metal hydride (NiMH) batteries

The NiMH battery is known to have double the energy density as compared to the lead-acid battery [16]. It is also safe to operate at high voltage, has wide operating temperature range, and long cycle life while facilitating safer operation as the batteries can be recycled without harming the environment. It is composed of nickel hydroxide on the positive electrode and engineered alloy of vanadium, titanium, nickel, and other metals as the negative electrode with alkaline solution as the electrolyte [6]. It has been the most commonly used energy storage system for EV applications since the beginning due to its fuel efficiency.

14.5.1.3 Other nickel-based batteries

Nickel-zinc (Ni-Zn) batteries have high-energy and power densities, deep cycling capability, recyclable, and ability to operate under severe temperatures. However, they suffer from poor life cycles making it unsuitable for EV applications.

Nickel-cadmium (Ni-Cd) batteries have good range of operation, are recyclable, and have a long lifetime but is expensive to be used in EV applications.

14.5.1.4 Lithium-ion batteries

Li-ion batteries have high-energy density, are recyclable, have long battery life, and ability to sustain high external temperatures. It has been proven to have great performance in other critical fields like medical devices, safety equipment, and consumer electronics. Given the promising features and potential, it might replace NiMH batteries in the EV sector all together. The positive electrode is made of an oxidized cobalt material, and the negative electrode is made of a carbon material with lithium salt in an organic solvent used as an electrolyte. This has huge potential and serves as the best option in terms of safety and compliance with regulatory requirements (Table 2).

14.5.2 Battery management system (BMS)

The goal of battery management system (BMS) is to achieve maximum system performance while minimizing the power consumption to extend the battery life. To achieve these objectives, it is crucial for BMS to accurately communicate the state of the battery to the system controller. The key functions performed by BMS include undervoltage protection, overvoltage protection, short-circuit protection or maximum current limit, and thermal protection [2]. The additional functions performed by BMS may include monitoring of state of charge (SoC), state of health (SoH) and state of function (SoF), and cell balancing or equalization on the cell or module level [7].

The block diagram of a generic BMS is as shown in Fig. 4 [7]. The voltage information gathered by the sensing subsystem is sent to the balancing circuit, which uses the information to ensure that all cell voltages are within the allowed limit. The balancing is typically done on the module level, rather than on the cell-to-cell level. The sensing block also records temperature and current values. Voltage, current, and temperature information along with the battery history are then used to determine the state of charge of the battery.

TABLE 2 Characteristics of commercial batteries for HEV applications [17].

Battery type	Capacity (Ah)	Voltage (V)	Resistance (mΩ)	Usable SoC (%)
NiMH	6.5	7.2	11.4	40
Li-ion	12	4	7	20
Lead-acid	25	12	7.8	28

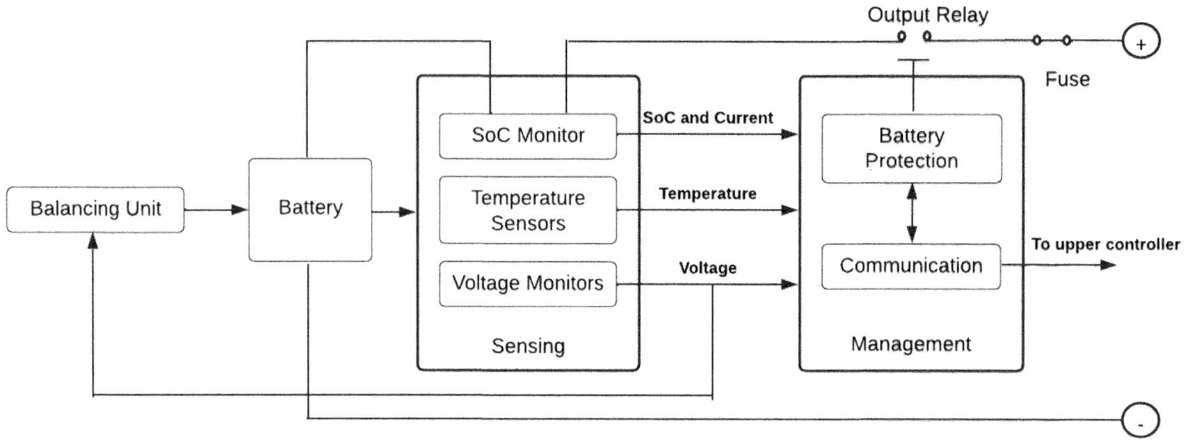

FIG. 4 Block diagram of a generic BMS [7].

14.6 Charging solutions

The need for developing drivetrains that are either fully or at least partially electrified is rising due to the increased demand for cleaner and fuel-efficient vehicles. These are referred to as electric vehicles (EVs) and plug-in hybrid electric vehicles (HEVs). The plug-in EVs (PEVs) typically have larger on-board storage accompanied with power electronics that is used for charging and discharging of the battery as compared to HEVs. The adoption of PEVs is highly dependent on the type of charging solution thus making it a crucial design aspect.

A PHEV is a hybrid EV with a storage system that can be recharged by connecting to an external electric power source through an AC or DC charging system. AC charging system is typically an on-board charger mounted inside the vehicle and is connected to the grid. DC charging system is mostly an off-board charger mounted to provide regulated DC power directly to the batteries inside the vehicle [18]. PEV chargers are typically classified based on the nature of charging—conductive or inductive, stages of power conversion—integrated single stage or two/multiple stages, power levels and type of semiconductor devices used—silicon (Si), silicon carbide (SiC), or gallium nitride (GaN) [19].

The front-end AC-DC converter is a critical component of the charging system, and typically, two-stage approach is ideal for use in PHEVs. Single-stage approach is more suited for low-power applications, and due to large low-frequency ripple generated, it demands higher safety precautions and isolation—which makes it a not so good candidate for use in PHEVs. In case of two-stage approach, power rating is high and is mostly accomplished by lithium-ion (Li-ion)-based storage systems—front-end PFC circuit is followed by DC-DC section. There are some single converter integrated charger topologies studied in the past. The critical parameters of the same are summarized in Table 3.

The two distinct functionalities within a PEV are that of propelling and charging, which is controlled by suitable power electronic interface (PEI). Propelling functionality is achieved by motor inverter as the PEI, whereas charging and discharging can be accomplished by bidirectional DC-DC converter [6]. Also, there are two charging approaches—conductive charging which has hard-wired connection between power supply and PEI, and inductive charging that has transmitter and receiver coils for wireless power transmission.

14.6.1 Conductive charging

Conductive charging has physical wiring between supply and the power electronics while typically housing power factor correction (PFC) circuit with AC-DC rectifier followed by DC-DC converter for boosting. They can also be integrated together to enhance the performance while reducing volume, cost, and weight. This also means that a single-stage converter can potentially be used for charging, motoring, and regenerative braking applications.

There are different ways of classifying conductive chargers—based on supported power levels, and based on location of the charger on the PEV. On-board chargers, as the name suggests, reside inside the PEV and comprises of rectification and current regulation power stages. Off-board chargers on the other hand are installed on external charging infrastructure and are high-power charging solutions with fast response time.

The specifics in terms of the topology of on-board or off-board conductive chargers are determined by the structure of the rectifier circuit and the type of the DC-DC converter used. The role of front-end AC-DC converter is to convert

TABLE 3 Comparison of single converter integrated charger topologies [19].

Parameter	Positive buck/boost rectifier	Three-level AC/DC	Bridgeless direct AC/DC	Buck/boost diode rectifier
No. of inductors	1	2	1	1
No. of switches	6	11	9	4
No. of diodes	9	14	3	7
Diode bridge	Yes	No	No	Yes
Bidirectional operation	No	Yes	Yes	No
Control complexity	Moderate	High	Moderate	Low
THD suppression	Low	High	Moderate	Moderate

TABLE 4 Comparison of various boost PFC topologies [20].

Parameter	Conv. boost	Sym. boost	Inter-leaved boost	Three-level boost	Half bridge
No. of boost inductors	1	1	2	1	1
Size of boost inductor	L	L	L/2	L	L
Output voltage	V_0	V_0	V_0	$2V_0$	$2V_0$
Max. no. of semiconductors in current path	3	2	3	2	1
Existence of diode bridge	Yes	No	Yes	No	No
No. of switches	1	2	2	2	2

the AC voltage on the supply side to an intermediate DC link voltage while enabling power factor correction, harmonic reduction, and input current ripple reduction. Typically, boost topologies are employed as PFC AC-DC rectifiers and the choice of the variant is governed by the efficiency, cost, robustness, passive component count, total harmonic distortion (THD), control complexity, and other performance factors [20]. Conventional boost rectifier operating in continuous conduction mode (CCM) is the easiest to implement but suffers from high conduction losses and output diode recovery losses. However, higher power ratings demand interleaving of converters, which in turn improves the power density and reduces the inductor size. Table 4 summarizes the comparison of various boost PFC topologies, including the half-bridge and full-bridge variants.

DC-DC converters help in increasing or decreasing the voltage levels based on the application in hand. Isolated DC-DC converter topologies alter the voltage level along with providing isolation between the input and output terminals. Half-bridge and full-bridge configurations can be designed to be current-fed or voltage-fed or even combinations of both. Resonant converters and dual-active bridges are some other variants of DC-DC converters along with those listed previously. Zero voltage switching (ZVS) can be accomplished on the current-fed side, whereas zero current switching (ZCS) or ZVS can be achieved on the voltage-fed side [21]. Table 5 summarizes the comparison of commonly used isolated DC-DC converter topologies.

The on-board chargers utilize the previously discussed rectifiers and DC-DC converters as part of the two-stage charging. The current shaping stage for minimizing THD and achieving unity power factor (UPF) is made possible by the rectifier stage, whereas the regulation of the battery charging current is achieved by the isolated DC-DC converter. Furthermore, the integrated chargers combine the charging stage with the DC-DC bidirectional converter that is used in PEV powertrain between the storage system source and inverter DC link. This combination is preferred as it offers advantages like use of single converter for all modes of operation, lesser number of components, smaller size, less weight, and cost. There are several other topologies studied in great detail mostly with additional passive components, additional DC-DC stages, or other added power electronics to cater to very specific applications.

However, for higher power charging of greater than 20 kW, placing of conventional type of high-power chargers on-board could not be feasible due to increased cost, volume, and weight. In such scenarios, off-board chargers that are

TABLE 5 Comparison of commonly used isolated DC-DC converters [21].

Parameter	VCFFB	DAFB	Resonant
Switching frequency (kHz)	50	50	50
Circulating current	No	Yes	No
Number of switches	9	8	8
Number of diodes	9	8	12
Short-circuit current limit	No	Yes	Yes
Soft switching	ZCS/ZVS	ZVS	ZVS
Control complexity	Moderate	Simple	Complex

located on external charging stations can be employed. These could be linked either with a common AC link or a common DC link. Possible converter topologies that can be utilized for high-power charging include three phase AC-DC power converters and its variants.

14.6.2 Inductive charging

Inductive charging does not use the wired connection between the supply and the power electronic circuitry thus also referred to as contactless charging. In early days, contactless stationary transformer was used. Recently, resonant converters are employed to transfer power using the principle of magnetic induction. There is also scope for use of advanced wide band-gap devices in charging circuits. However, it has to be understood that the extent of PEV penetration depends on the type of charging solution employed.

The inductive chargers are classified based on the locations of power transfer, the magnetic structure or shape of the primary and secondary coils, and the arrangement of passive elements like resistors and capacitors within the resonant circuit. Based on the power transfer, inductive chargers are classified as lumped and distributed. Based on the shape of the coil, there are E-, U-, or I-shaped cores.

Inductive chargers work on the principle of inductive power transfer (IPT) where a varying magnetic field is used to transfer power across an air gap to a connected load without any physical contact unlike the conductive charging that requires metal to metal contact or trailing of cables. Energy transfer in this case is achieved magnetically similar to the principle used in transformers and also provide galvanic isolation. The other advantages include durability, robustness, and less wear and tear.

On the input side, AC in from the main supply is converted into high-frequency AC. This is then transferred to the vehicle inlet where it is rectified to DC in order to charge the batteries. The SoC of the batteries is monitored by charge controllers. Initially, contactless transformers were used to facilitate IPT, which was then moved to be based on resonant converters. Also, IPT systems are typically classified based on the configuration of the magnetic circuit. If the power transfer happens only at discrete locations, then it is called lumped or concentrated. This is used for offline EV charging system or stationary systems. On the other hand, if the primary coil is spread out such that the power transfer happens at multiple locations, then it is called as online EV charging or distributed IPT system.

14.7 Future trends

The future of EV battery could be based on lithium-air technology. These batteries have an anode made of lithium and an air cathode made of a porous material that draws in oxygen from the surrounding air. When the lithium combines with the oxygen, it forms lithium oxide and releases energy. As the oxygen does not have to be stored in the battery, the cathode is much lighter than that of a lithium-ion battery, which gives lithium-air batteries their high-energy density. These batteries seem to have the capability to significantly increase the range of EVs due to their high-energy density.

14.7.1 Energy storage system challenges

One of the major bottlenecks for increasing the development of EVs is the energy storage system. The most common issues related to rechargeable lithium cells is that of cell equalization where the capacity and internal resistance of all the cells stacked up are not equal leading to numerous balancing and performance issues. Power electronics-based battery cell voltage equalizers help in battery cycle life improvement, support increase in power, reliability, and safety. It is a controller that helps equalize the voltage in each cell by continuously measuring the actual capacity, internal resistance, and state of charge (SoC). This helps ensure that the SoC of each cell is same during charging and discharging, thus ensuring that they will degrade equally over time. The equalizers can be resistive, inductive, or capacitive in nature.

Safety is of utmost importance and is one of the critical issues in electrical energy storage system. It is inevitable to adhere to standards and comply with regulatory requirements while ensuring that the materials used are of the highest quality and safety is carefully monitored at every stage of development and installation. All the critical battery parameters like cell level voltages, currents, SoC, overall temperature, and others need to be monitored along with taking necessary corrective measures when the values are out of range. Also, it has been found that the cell equalizers not only help with maintaining SoC and cycle life of the cells, but also improves the overall calendar life while ensuring system safety.

Since the development of EVs is driven by environmental concerns, the recyclability of the materials used in the energy storage system is also an important consideration. Also, in EV applications, there is need to operate at higher voltages, and thus, several battery modules need to be connected in series or parallel or series/parallel combinations. This leads to more challenging problems around charge equalizing, safety, and reliability. It is important to holistically address these concerns while designing, developing, and implementing the next generation of energy storage systems.

14.7.2 Technological advancements

Although fuel cell technology seems to be promising, vehicles completely relying on fuel cells are very limited or even restricted to research and/or demo purposes. There are significant issues related to cost, manufacturing capabilities, reliability, robustness of the technology, hydrogen production, hydrogen infrastructure, and so on. As there is advancement in polymer electrolyte membrane and solid oxide fuel cell technologies, FCs could be used as range extenders instead of ICE driven generators in series hybrid vehicles. This is huge as the dependency on batteries alone will be alleviated to a great extent. Thus, plug-in fuel cell vehicles (PFCVs) consisting on a smaller FC and a larger battery may be the future direction for automobiles for certain markets [14].

Several battery companies and automobile manufacturers are investing heavily to be able to build cheaper, denser, and compact batteries. The innovation in terms of possible battery chemistries is still ongoing. Research includes building up on the existing battery chemistries, looking at newer materials, or modifying the form factor or battery assembly aspects to realize performance gains or cost advantages, even if incremental. There is also scope for taking a deeper look at other energy storage technologies—batteries, electrochemical storage solutions, flywheels, advanced chemistry batteries, flow batteries, and different grades of capacitors.

14.8 Summary

Electrical energy storage device is a critical component in the EV and is one of the weakest points as well. Even though technological advances in power electronic design, cost reduction in semiconductors and enhanced reliability of electronic systems in the automobiles have made EVs more promising, detection of accurate battery health, and its state of charge (SoC) have remained as issues of concern. The choice of battery or energy storage system is thus a crucial aspect of the EV due to increased focus on electrical performance, reliability and range of operation. The ideal characteristics for a battery used in EV applications are high-energy density, high charge-discharge efficiency, long life, no maintenance, ability to sustain wide range of temperatures, good load characteristics, fast charging, minimal self-discharge, low internal resistance, good recycle ability, low cost, high reliability, and safety.

Section 14.2 provides an overview of the state-of-the-art techniques used for electrical energy storage while touching upon the history behind its development over the years and the classification of energy storage devices. One of the major bottlenecks for EV development is the battery pack that consumes most of the space in the EV and also increases the weight of the vehicle. It also demands higher electricity from the transportation sector and the related infrastructure that is required for procurement, storage, and distribution of these vehicles. All these concerns demand the need for more efficient storage systems to be used in EVs and also to carefully consider the entire life-cycle of these vehicles for optimization. Power electronics-based storage solutions helped overcome some of these while reducing the issues on other fronts, thus becoming the choice of system for both low- and high-voltage applications in automotive sector. They also offer flexibility in terms of operation as they can be connected in series or parallel, and can be controlled to serve as voltage or current generators based on the application in hand.

Section 14.3 describes the significance of power electronics-based systems in energy storage by elaborating on the advantages they offer and the applications in which they are essential. Energy storage systems in general are electrical, mechanical, electromechanical, or hybrid in nature. The introduction of energy storage devices with power electronics-based switches revolutionized storage as they started serving as replacement technology for conventional solutions like mechanically switched energy storage media and stand by generators that have traditionally faced issues with power quality and adherence to codes and standards.

Section 14.4 delves into the details of the electrical energy storage by explaining the design and creation of battery models for EVs, other storage devices that can be used along with batteries like ultracapacitors and fuel cells, and necessary control strategies to have them all working together. The applications of electrical energy storage are increasing tremendously as the number of EVs and its variants are increasing. This necessitates more thought into the choice of energy storage system. There are different types of energy storage systems that can be used in EV applications. Batteries are the primary energy storage devices especially for ground vehicles, but in case of hybrid vehicles, ultracapacitors (UC) and/or fuel cells (FC) can also be employed. In fact, a combination of batteries, UCs and FCs, could be most appropriate for advanced hybrid EV energy storage systems.

Section 14.5 describes the batteries used in different kinds of EVs by getting into the details of the battery chemistry and discusses the battery management system that is used to control the necessary performance parameters. It is evident that the on-board energy storage system of the vehicle is the heart of the EV. Batteries are the primary electrical energy storage devices in most of the ground EVs with ultracapacitors and fuel cells supporting the vehicular fuel economy. The most important characteristics of vehicular EVs are power density, cost-effectiveness, reliability, maintenance-friendly, and energy density. The overall performance of the EV depends not only on the ability to store large amounts of energy but also to be able to release it quickly and effectively based on the load demands. The goal of battery management system (BMS) is to achieve maximum system performance while minimizing the power consumption to extend the battery life.

Section 14.6 discusses the charging solutions offered at the power electronic interfaces in an EV and describes different ways of implementing conductive charging and inductive charging. The two distinct functionalities within a plug-in hybrid EV are that of propelling and charging, which is controlled by suitable power electronic interface (PEI). Propelling functionality is achieved by motor inverter as the PEI, whereas charging and discharging can be accomplished by bidirectional DC-DC converter. Also, there are two charging approaches—conductive charging which has hard-wired connection between power supply and PEI, and inductive charging that has transmitter and receiver coils for wireless power transmission.

The future trends are highlighted in Section 14.7. Current challenges for energy storage system development are discussed, including safety concerns, monitoring of critical battery parameters, cell balancing, and possible solutions for these concerns. Looking forward, use of fuel cells in hybrid powertrain, considering other battery chemistries for EVs and other research aspects are reviewed.

References

[1] T.J. Mller, Lithium ion battery automotive applications and requirements, in: Seventeenth Annual Battery Conference on Applications and Advances. Proceedings of Conference (Cat. No.02TH8576), 2002, pp. 113–118, https://doi.org/10.1109/BCAA.2002.986381.

[2] A. Affanni, et al., Battery choice and management for new-generation electric vehicles, IEEE Trans. Ind. Electron. 52 (5) (2005) 1343–1349, https://doi.org/10.1109/TIE.2005.855664.

[3] Y. Nishi, et al., The development of lithium-ion secondary battery systems for EV and HEV, in: Thirteenth Annual Battery Conference on Applications and Advances. Proceedings of the Conference, 1998, pp. 31–36, https://doi.org/10.1109/BCAA.1998.653834.

[4] M. Ferdowsi, Plug-in hybrid vehicles: a vision for the future, in: 2007 IEEE Vehicle Power and Propulsion Conference, 2007, pp. 457–462, https://doi.org/10.1109/VPPC.2007.4544169.

[5] C.C. Chan, An overview of electric vehicle technology, Proc. IEEE 81 (9) (1993) 1202–1213, https://doi.org/10.1109/5.237530.

[6] A. Khaligh, Z. Li, Battery, ultracapacitor, fuel cell, and hybrid energy storage systems for electric, hybrid electric, fuel cell, and plug-in hybrid electric vehicles: state of the art, IEEE Trans. Veh. Technol. 59 (6) (2010) 2806–2814, https://doi.org/10.1109/TVT.2010.2047877.

[7] S.M. Lukic, et al., Energy storage systems for automotive applications, IEEE Trans. Ind. Electron. 55 (6) (2008) 2258–2267, https://doi.org/10.1109/TIE.2008.918390.

[8] H.L. Chan, A new battery model for use with battery energy storage systems and electric vehicles power systems, in: 2000 IEEE Power Engineering Society Winter Meeting. Conference Proceedings (Cat. No.00CH37077), Vol. 1, 2000, pp. 470–475, https://doi.org/10.1109/PESW.2000.850009.

[9] M. Einhorn, et al., Comparison, selection, and parameterization of electrical battery models for automotive applications, IEEE Trans. Power Electron. 28 (3) (2013) 1429–1437, https://doi.org/10.1109/TPEL.2012.2210564.

[10] U.S. Department of Energy, FreedomCAR Battery Test Manual for Power-Assist Hybrid Electric Vehicles, 2003.

[11] O. Tremblay, L.A. Dessaint, A.-I. Dekkiche, A generic battery model for the dynamic simulation of hybrid electric vehicles, in: 2007 IEEE Vehicle Power and Propulsion Conference, 2007, pp. 284–289, https://doi.org/10.1109/VPPC.2007.4544139.

[12] K.P. Yalamanchili, M. Ferdowsi, Review of multiple input DC-DC converters for electric and hybrid vehicles, in: 2005 IEEE Vehicle Power and Propulsion Conference, 2005, pp. 160–163, https://doi.org/10.1109/VPPC.2005.1554613.

[13] H. Li, C. Liao, L. Wang, Research on state-of-charge estimation of battery pack used on hybrid electric vehicle, in: 2009 Asia-Pacific Power and Energy Engineering Conference, 2009, pp. 1–4, https://doi.org/10.1109/APPEEC.2009.4918709.

[14] K. Rajashekara, Present status and future trends in electric vehicle propulsion technologies, IEEE J. Emerging Sel. Top. Power Electron. 1.1 (2013) 3–10, https://doi.org/10.1109/JESTPE.2013.2259614.

[15] A. Cooper, P. Moseley, Progress in the development of lead-acid batteries for hybrid electric vehicles, in: 2006 IEEE Vehicle Power and Propulsion Conference, 2006, pp. 1–6, https://doi.org/10.1109/VPPC.2006.364282.

[16] M.C. Wehrey, What's new with hybrid electric vehicles, IEEE Power Energy Mag. 2 (6) (2004) 34–39, https://doi.org/10.1109/MPAE.2004.1359016.

[17] A.F. Burke, Batteries and ultracapacitors for electric, hybrid, and fuel cell vehicles, Proc. IEEE 95 (4) (2007) 806–820, https://doi.org/10.1109/JPROC.2007.892490.

[18] S.S. Williamson, A.K. Rathore, F. Musavi, Industrial electronics for electric transportation: current state-of-the-art and future challenges, IEEE Trans. Ind. Electron. 62 (5) (2015) 3021–3032, https://doi.org/10.1109/TIE.2015.2409052.

[19] A. Khaligh, S. Dusmez, Comprehensive topological analysis of conductive and inductive charging solutions for plug-in electric vehicles, IEEE Trans. Veh. Technol. 61 (8) (2012) 3475–3489, https://doi.org/10.1109/TVT.2012.2213104.

[20] J.P.M. Figueiredo, F.L. Tofoli, B.L.A. Silva, A review of single-phase PFC topologies based on the boost converter, in: 2010 9th IEEE/IAS International Conference on Industry Applications - INDUSCON 2010, 2010, pp. 1–6, https://doi.org/10.1109/INDUSCON.2010.5740015.

[21] D. Yu, et al., Review of high power isolated bi-directional DC-DC converters for PHEV/EV DC charging infrastructure, in: 2011 IEEE Energy Conversion Congress and Exposition, 2011, pp. 553–560, https://doi.org/10.1109/ECCE.2011.6063818.

CHAPTER

15

Battery charging technologies for EV

Muhammad H. Rashid[a], Sreedhar Madichetty[b], B. Mohan Krishna[b], and N.B. Shanthi Kumar[b]

[a]Florida Polytechnic University, Lakeland, FL, United States
[b]Mahindra University, Hyderabad, India

15.1 Introduction

Electrical vehicles (EVs) are gaining traction in today's world. Being energy conscious and emitting lesser greenhouse gases make EVs more popular. This makes EVs eco-friendly. The energy storage devices (batteries) are used to store the energy from the grid (or some other energy sources) and drive the EVs using the energy stored in the batteries. Therefore, battery chargers and battery charging technologies play a crucial role in electric vehicle technologies. The primary considerations for charging the batteries are the type of batteries used, chargers used, charging speed, power density, and proper maintenance of grid (single phase/three phase), [1] etc. In 1900, EVs had nonrechargeable batteries. The invention of rechargeable batteries made the EV technologies simpler and facilitated the use of batteries without any replacement. Before 2000, most areas did not have uninterrupted access to electricity, which made it impossible to charge the batteries frequently. Come the early 2000s, greater power generation, availability of power at a lower cost, improvement in battery charger topologies, and improvement in efficiency made EVs more popular [2]. Fig. 1 shows the classifications of EV battery chargers.

15.2 Charging modes

There are four charging modes based on voltage, current and power levels, and charging time for battery charging application [3].

(a) Mode I charging:

In Mode I charging, the single-phase AC system is used to charge the batteries with current and power ratings of 13–20 A and 2–3.7 KW, respectively. This type of charging scheme is not too popular because it takes about 5–8 h to charge the batteries and is therefore used at homes, garages, etc.

(b) Mode II charging:

In Mode II charging, the single-phase and three-phase AC systems are used to charge the batteries with current and power ratings of 16–32 A and 7 KW (for single phase) to 22 KW (for three phase), respectively. This type of charging scheme is used at public parking/charging areas, etc.

(c) Mode III charging:

In Mode III charging, the single-phase AC systems are used to charge the batteries with current and power ratings of 32–63 A and 22–43 KW, respectively. This type of charging scheme is used at public parking/charging areas, etc.

Handbook of Power Electronics in Autonomous and Electric Vehicles
https://doi.org/10.1016/B978-0-323-99545-0.00010-5

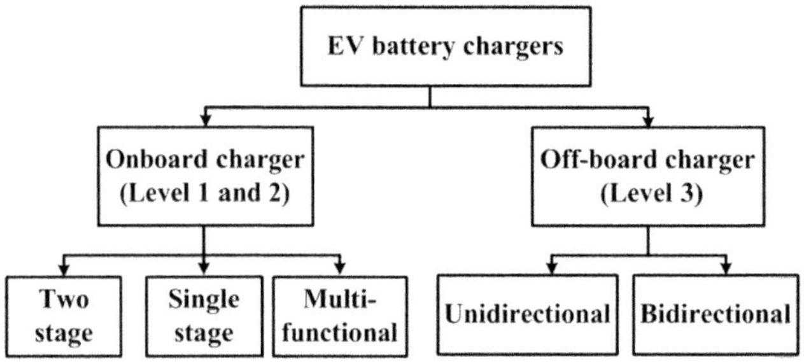

FIG. 1 Classifications of EV battery chargers.

(d) Mode IV charging:

In Mode IV charging, the 200–500 V DC system is used to charge the batteries with current and power ratings of 200–400 A and 50–240 KW, respectively. This type of charging scheme is in demand as it takes less than 30 min to charge the batteries, and it requires large charging stations with a protection system.

15.3 Charging methods

Electric vehicles (EVs) are gaining traction in India. The main component of an EV is its batteries. Nowadays, most of the batteries used in EVs are Li-Ion known as lithium-ion batteries. These batteries are widely used due to the fact that they have long life cycles, low self-discharge, and no memory effect and also have a high energy to weight ratio. The EV charging companies are mostly using Li-Ion batteries in their vehicles. Battery charging mainly consists of batteries and battery management system (BMS). The BMS's main function is to charge the batteries by following different types of algorithms. The various types of algorithms used to charge the batteries are CC-CV charging algorithm and pulse charging algorithm, which are discussed in the subsequent sections.

15.3.1 CC-CV charging method

CC-CV charging method is the frequently used algorithm in order to charge the batteries, because this method has less complexity and requires lesser components. In this method, two modes are performed one after the other. First is the constant current (CC) mode and following which is the constant voltage mode (CV) [4]. The charging algorithm of this method is shown in Fig. 2.

Initially, the battery voltage V_b is checked in an open circuit. If this voltage is in the specified voltage value (V_{min}), then the battery is kept charging in CC mode. If V_b is less than the specified voltage (V_{min}) level, then the battery is charged with 10% of total current capacity of the battery. This type of charging is known as trickle charging. With this type of charging, the V_b reaches to specified voltage value (V_{min}). After trickle charging, the battery is kept charging in CC mode.

In this mode, we charge the battery using 60% of its total current capacity. The battery is charged in CC mode until the V_b reaches predetermined value (V_{set}). When V_b reaches predetermined voltage (V_{set}), the CC mode changes to CV mode, and the current starts decreasing slowly from the specified limit. In CV mode, V_b gradually reaches the maximum voltage (V_{max}) of the battery. When the battery attains maximum voltage, the current flowing through the battery reaches 1% of its total current capacity of the battery, and then, the charging of the battery is terminated.

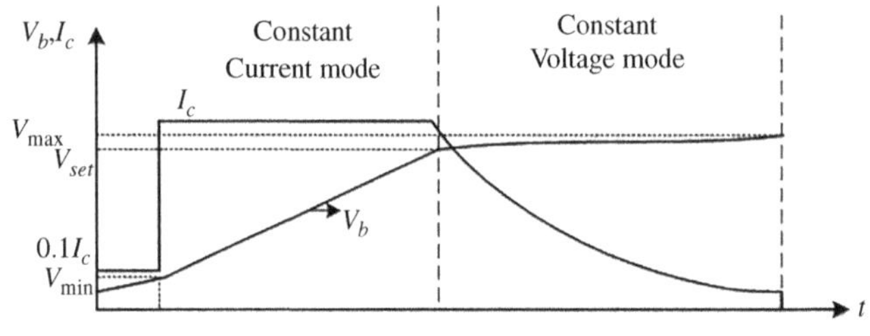

FIG. 2 CC-CV charging method.

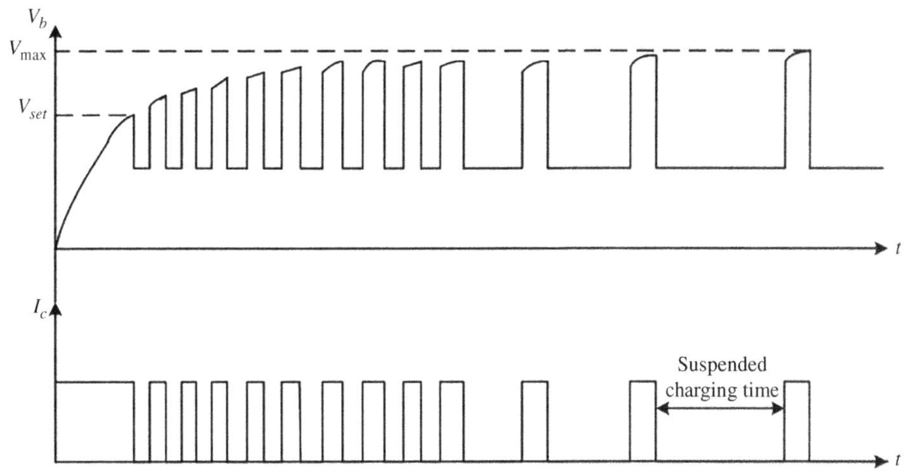

FIG. 3 Pulse-current charging method.

15.3.2 Pulse charging method

Pulse charging method (PCM) is used to fast charge the batteries. This PCM method charges the battery with 14% more speed as compared to the CC-CV method. In PCM, the battery is charged using pulsed current [5], which is shown in Fig. 3. This PCM method charges the battery with alternating periods of charging and suspension of charging. Suspension of charging means the current flowing through the battery is zero. During this period, the V_b is measured, which is open circuit (OC) (V_{OC}) voltage of the battery. This OC voltage is compared to the specified limit of voltage, after which the battery is once again charged with pulse current (I_C) if this voltage is not equal to the maximum voltage (V_{max}) of the battery. The time between the instant (I_C) is zero, and the instant the I_C starts charging is known as suspended charging time. When the time taken by the battery reaching its open circuit voltage to fall below the V_{max} of battery is greater than the specified suspended charging time, then the battery is fully charged.

Similarly, the time taken by the battery for its open circuit voltage to attain V_{min} is greater than the specified suspended charging time. Then, we can assume that the battery is removed.

15.4 Types of charging connectors

15.4.1 AC charging connectors

There are three types of AC charging connectors. Fig. 4 shows the three types of AC charging connectors [3,6].

1. Type 1 connector (A) is mainly used in the United States and Japan. This type of connector is used for charging the EV with a single-phase AC system with maximum voltage and current ratings of 120/240 V and 80 A, respectively.
2. Type 2 connector (B) is mainly used in Europe. This type of connector is used for charging the EV with both a single-phase and a three-phase AC system with maximum voltage and current ratings of 240 V and 80 A for single phase and 400 V and 63 A for three phase, respectively.
3. Tesla US connector (C) is used in EVs, which are manufactured by Tesla. It has a unique feature that the same connector is used for both AC and DC systems. It has a maximum AC power of 17.2 KW when connected to 240 V AC.

15.4.2 DC charging connectors

There are four types of DC charging connectors. Fig. 5 shows the four types of DC charging connectors [3,6].

1. Combined charging system (CCS) also called Combo 1 and Combo 2 connectors (Fig. 5A and B) are mainly used in the United States and Europe, respectively. Power line communication (PLC) is used for communication between the EV and charger. The maximum voltage, current, and power ratings are 1000 V, 350 A, and 350 KW, respectively.

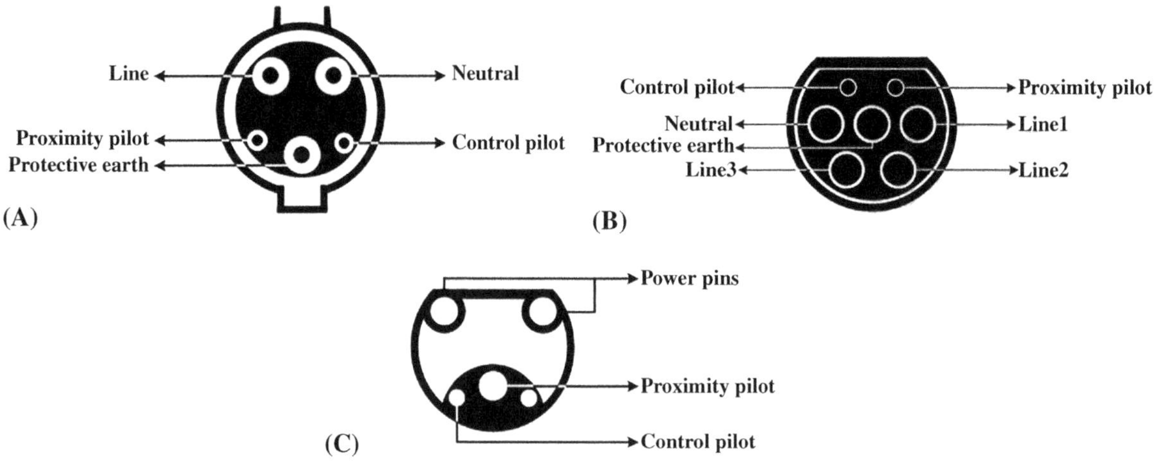

FIG. 4 AC EV battery charging connectors. (A) Type 1 connector, (B) Type 2 connector, and (C) Tesla US connector.

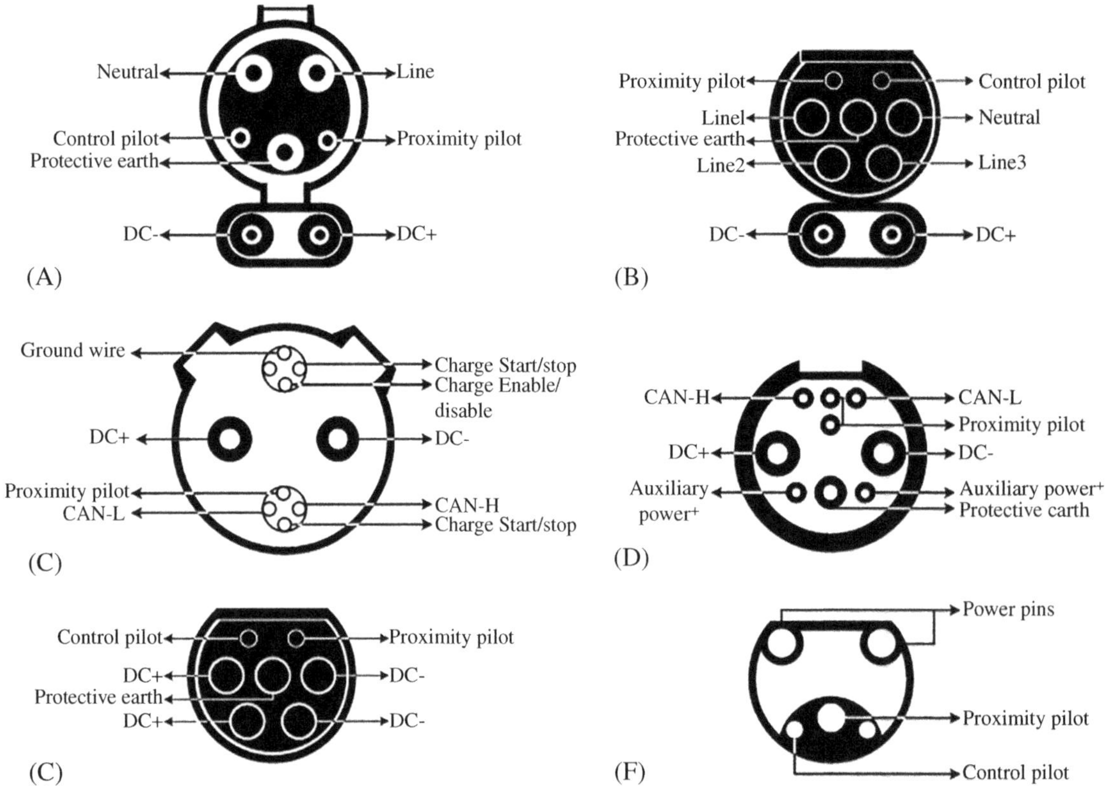

FIG. 5 DC EV battery charging connectors. (A) CCS Combo 1 connector, (B) CCS Combo 2 connector, (C) CHAdeMO connector, (D) GB/T standard connector, (E) Tesla Europe connector, and (F) Tesla US connector.

2. CHAdeMO connector (Fig. 5C) is used in EVs, which are manufactured in Japan. Control Area Network (CAN) protocol is used for communication between the EV and the charger. The maximum voltage, current, and power ratings are 1000 V, 400 A, and 400 KW, respectively.
3. Tesla connector (Fig. 5E and F) is used in EVs, which are manufactured by Tesla. It has a unique feature that the same connector is used for both AC and DC systems. The maximum DC power rating of the connector is 120 KW.
4. GB/T connector (Fig. 5D) is mainly used in China and India. Control Area Network (CAN) protocol is used for communication between the EV and the charger. The maximum voltage, current, and power ratings are 1000 V, 250 A, and 250 KW, respectively.

15.5 On-board charger

On-board charger is a device used to convert AC supply from any source to DC to charge the battery inside the vehicle. So, this type of charger is used to charge the vehicle with our home outlets. This type of charger has limitation of size, weight, and volume. This type of charger comes under Level 1 and Level 2 categories, which doesn't have high power capability.

On-board charger is basically of two types:

(a) Two-stage
(b) Multifunctional

15.5.1 Two-stage

On-board charger with two stage comprises of two converters with front end and back end. Front-end converter basically consists of AC-DC conversion, whereas back-end converter consists of DC-DC conversion.

Front-end converter is essentially required to convert the AC voltage of the supply to the DC bus voltage, which is approximately equal to 390 V. Then, this DC bus voltage regulates to 48 V for charging the batteries using DC-DC converter.

Front-end converters are built with power factor correction (PFC) in order to shape the input current waveforms to synchronize with input voltage of the converter. A perfect PFC circuit has the input current synchronizing with input voltage as a resistor. To achieve PFC, boost topologies are used. PFC is used to achieve a high power factor and low harmonic distortion. There are three types of topologies used to achieve good PFC.

15.5.1.1 Interleaved boost PFC

In this type of topology, the converter consists of full bridge to convert the AC supply to rectified DC, and this rectified DC voltage is stepped up by using the interleaved boost converter [7] shown in Fig. 6. Interleaved boost PFC consists of two or more boost converters connected in parallel so that the output voltage ripple reduces while converting the current into a sinusoidal shape. The boost converters are 180° out of phase. This converter has high losses as input stage consists of bridge configuration, which reduces the efficiency of the converter.

15.5.1.2 Bridgeless totem-pole PFC

Bridgeless totem-pole PFC has a higher efficiency topology compared to interleaved boost PFC as it consists of switches instead of bridge configuration [8] at the input side of the converter. This topology consists of four switches S_1, S_2, S_3, and S_4. The switches S_1 and S_2 operate with high frequency (nearly 100 KHz), and S_3 and S_4 switches operate with line frequency, i.e., 50 Hz. The schematic diagram of bridgeless totem-pole PFC topology is shown in Fig. 7. As the

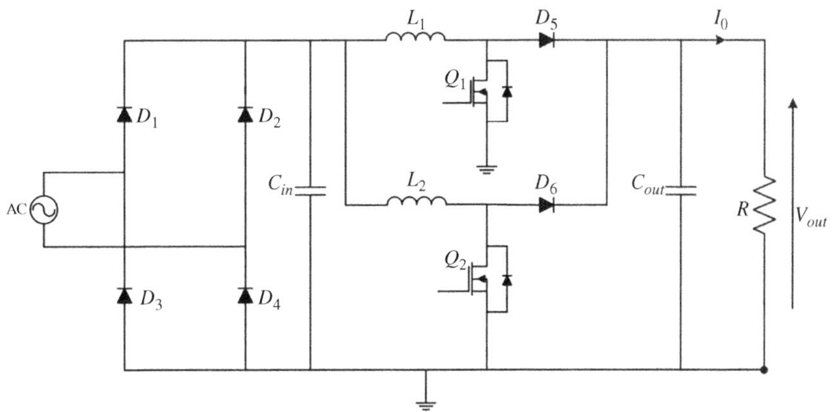

FIG. 6 Schematic circuit of interleaved boost PFC.

circuit consists of switches with less R_{DS}, it reduces the losses and helps to improve the efficiency of the circuit. The circuit has fewer parts, i.e., four switches and an inductor. The size of the converter is also small. The circuit has all switches used for bidirectional power flow as well. This topology has fewer limitations as this circuit operates in either discontinuous conduction mode or critical conduction mode as it has poor reverse recovery performance of Si MOSFET body diodes in totem-pole configuration, which makes Q_{rr} losses greater during turn on in continuous conduction mode. This converter fits only for low power applications.

15.5.1.3 Interleaved totem-pole PFC

Interleaved totem-pole PFC is a bidirectional bridgeless converter, is designed using silicon carbide (SiC) switches, and has low recovery charge and low R_{DS} in continuous conduction mode, which helps to increase efficiency, as losses are less using this converter for on-board charger [9]. As this circuit is bidirectional, we can transfer the power from G2V or from V2G using this type of converter. This converter has expertise in terms of low current ripple, high efficiency, low reverse recovery current, simple gate drive, and simple circuit structure. The schematic diagram of interleaved totem-pole PFC is shown in Fig. 8.

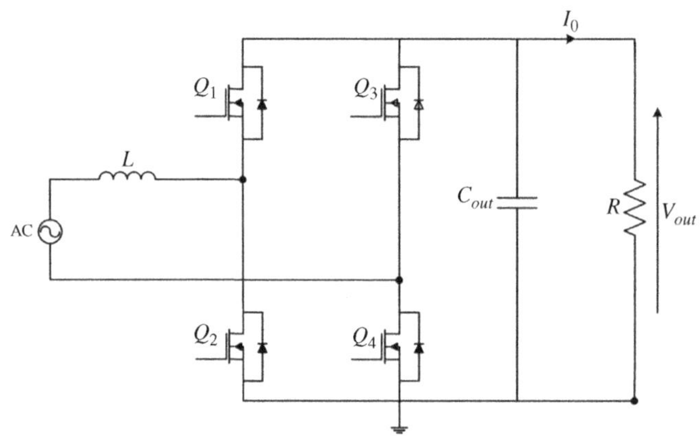

FIG. 7 Schematic circuit of bridgeless totem-pole PFC.

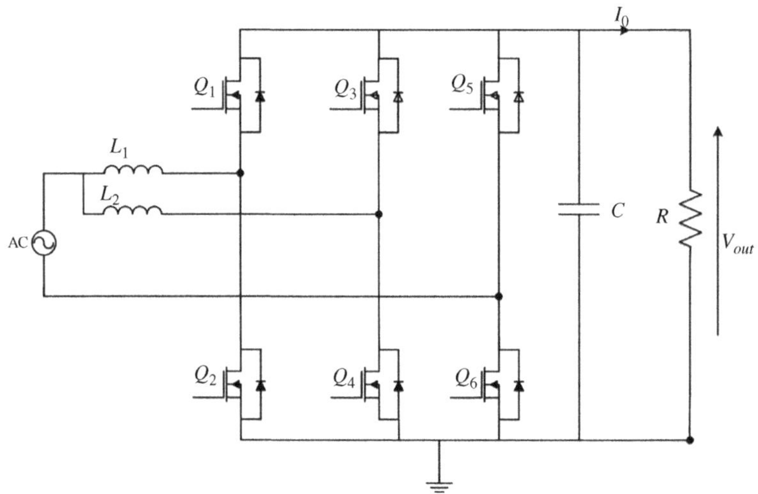

FIG. 8 Schematic circuit of interleaved bridgeless totem-pole PFC.

This converter consists of six SiC MOSFETs Q_1, Q_2, Q_3, Q_4, Q_5, and Q_6, where Q_1,Q_2, and L_1 form one totem-pole PFC phase and Q_3, Q_4, and L_2 form another totem-pole PFC. In each phase, two switches are present at high side and low side, and these switches operate in a complementary manner. In the circuit, switches Q_1 and Q_2 operate complementary to each other in one phase, and switches Q_3 and Q_4 operate complementary to each other in another phase and operate with high switching frequency. These two phases are 180° out of phase with each other. The switches Q_5 and Q_6 are located in one leg at high side and low side, respectively, and operate with line frequency, which is equal to 50 Hz.

The operating principle of one leg of interleaved totem-pole PFC is illustrated using different modes:

Mode I:

In this mode of operation, switch Q_6 turned on, which connects the low potential of AC line to the ground. Initially, switch Q_2 is made to turn it in on position. The simplified circuit is shown in Fig. 9A. Then, the AC line charges the inductor L_1 through switches Q_2 and Q_6. The load is supplied by the output capacitor. The inductor current starts to increase linearly with respect to time. The inductor current is expressed as follows:

$$\frac{di_{L1}}{dt} = \frac{v_{ac}}{L_1}$$

$$i_{L1} = \int \frac{v_{ac}}{L_1} \cdot dt$$

$$i_{L1} = \frac{v_{ac}}{L_1} t$$

When Q_2 is turned off, dead time is introduced in between Q_2 turn off and Q_1 turn on to prevent short circuit of the leg. The simplified schematic diagram is shown in Fig. 9A.

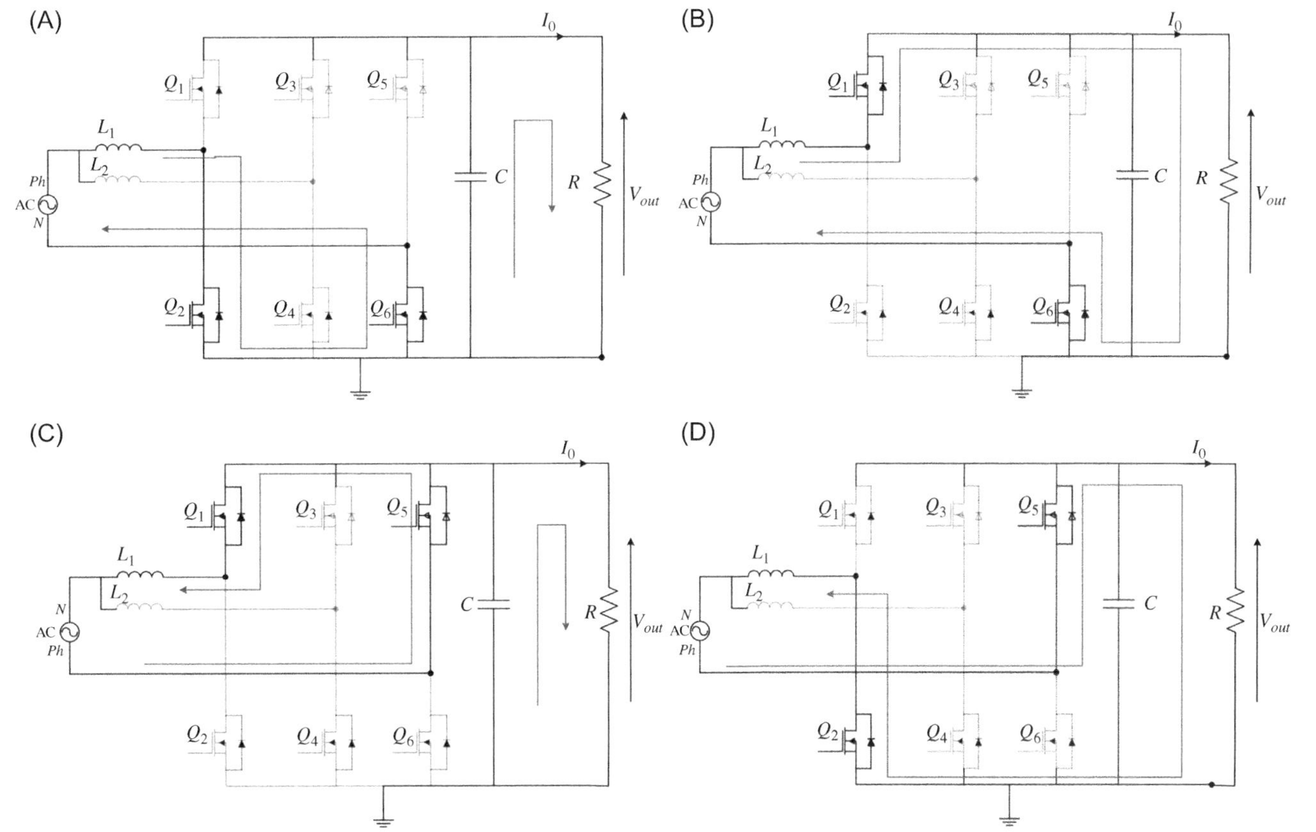

FIG. 9 Interleaved bridgeless totem-pole PFC: (A) Mode I, (B) Mode II, (C) Mode III, and (D) Mode IV.

Mode II:

During this mode, Q_1 and Q_6 switches are turned on, stored charge in the inductor flows through the output capacitor (C). The inductor current i_{L1} starts decreasing with respect to time. The simplified schematic diagram is shown in Fig. 9B. The i_{L1} expressed as follows:

$$\frac{di_{L1}}{dt} = \frac{v_{ac} - V_0}{L_1}$$

$$i_{L1} = \int \frac{v_{ac} - V_0}{L_1} \cdot dt$$

$$i_{L1} = \frac{v_{ac} - V_{DC}}{L_1} t$$

The switching time of Q_2 switch is determined by using control algorithm duty cycle D, and Q_1 switch is complementary to Q_2, i.e., (1-D).

In these two modes, switching cycle is approximately half of the AC line frequency.

Mode III:

In this mode of operation, the AC voltage changes its polarity, so Q_6 is turned off, and Q_5 is turned on, which connects to positive potential of AC line to DC link. As Q_1 is turned on, the inductor L_1 charges through Q_5 and Q_1. The simplified schematic diagram is shown in Fig. 9C.

The i_{L1} expressed as follows:

$$i_{L1} = \frac{v_{ac}}{L_1} t$$

In between Mode III and Mode IV, dead time is created, which is used to prevent short circuit currents in the converter.

Mode IV:

During this mode, Q_5 and Q_2 switches are turned on. Stored energy in the inductor L_1 will supply the load. The simplified circuit diagram is shown in Fig. 9D. The i_{L1} starts decreasing linearly with respect to time. The i_{L1} is expressed as follows:

$$i_{L1} = \frac{v_{ac} - V_{DC}}{L_1} t$$

The switches Q_3 and Q_4 act in another interleaved phase. Switching of these two switches is similar to the switching of Q_1 and Q_2. These switches are operating 180° out phase with the switches Q_1 and Q_2.

This operation is explaining the G2V charging. This converter converts the 85VAC-265Vac to the 300–600 V DC voltage.

15.5.1.4 Full-bridge LLC-type DC-DC converter

The back-end converter is a DC-DC converter, which regulates the 300–600 V DC bus voltage to 48 V for charging the batteries in the vehicle. Resonant-based converters are used for DC-DC converter, which has high efficiency and low output voltage ripple. By this converter, it is easy to achieve ZVS of switches, which helps to improve efficiency as losses are less. Because of zero voltage switching operation, the converter needs to operate at a higher frequency, which helps to reduce magnetic components in the circuit and improve the power density ratio of the converter.

In series resonant converters (SRC), the resonant tank is connected to the load. With light or no loads in order to regulate the output voltage, a high switching frequency is required (approximately tending toward infinity). In parallel resonant converters (PRC), the resonant tank is connected in parallel with the load. In addition, it is easy to regulate the output voltage. However, it has high energizing losses. To overcome these issues, a combination of SRC and PRC is used. The most common converter is LLC converter [10], which has higher efficiency (as switches operate with ZVS), low EMI losses, and higher power density ratio.

The LLC-type DC-DC converter is shown in Fig. 10A. It consists of a full bridge converter at the input side to convert DC voltage from bus voltage to square waveform. The LLC resonant tank circuit is connected in series with a full bridge rectifier, which contains resonant inductance L_r, resonant capacitance C_r in series, and magnetizing inductance L_m in parallel with the circuit. The series capacitor C_r has two functions. First, it blocks the DC component that is coming from the square waveform and also forms a resonant circuit with L_r. The high-frequency transformer is placed after the resonant tank in order to provide isolation of the circuit. Transformer secondary side consists of full bridge (FB) rectifier used to convert AC to rectified DC, and a filter capacitor is used to provide a ripple-free output voltage to the load.

The resonant tank is provided to eliminate higher-order harmonics from square waveform and provide pure sinusoidal current to the transformer. The series capacitor C_r with series inductor L_r provides a switching frequency (f_s), and the C_r and L_r parallel with inductor L_m provide another frequency, namely, f_m.

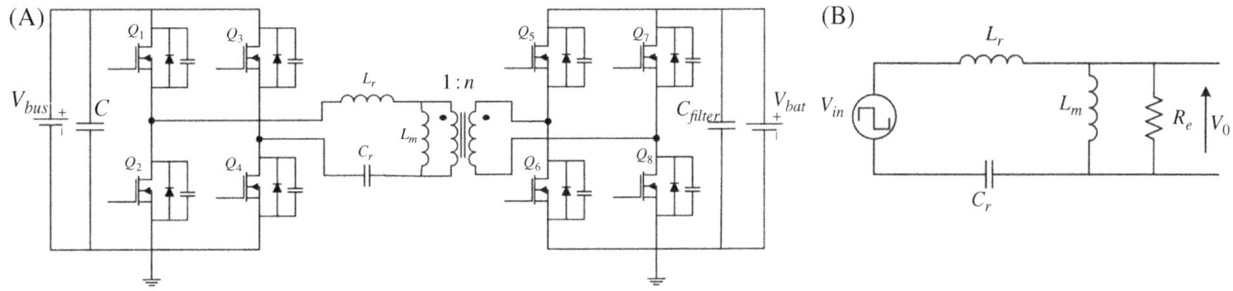

FIG. 10 (A) Schematic circuit full bridge LLC DC-DC converter and (B) AC equivalent circuit of LLC converter.

$$f_s = \frac{1}{2\pi\sqrt{L_r \cdot C_r}}$$

$$f_m = \frac{1}{2\pi\sqrt{(L_r + L_m) \cdot C_r}}$$

This converter operates between these two frequencies in order to achieve constant output voltage for various inputs.

To maintain the constant voltage at output and higher efficiency, we need to design proper LLC parameters for maximum and minimum voltages. For that, we need gain of the converter.

To find the gain of the system, we use first harmonic approximation circuit, which is shown in Fig. 10B. In this circuit, the transformer and rectifier stage are indicated by equivalent resistance R_e

$$R_e = \frac{8n^2}{\pi^2} R_{Load}$$

where R_{Load} is the load resistance that represents the battery.

The gain of the system is calculated as:

$$M = \frac{nV_0}{V_{in}} = \frac{\frac{R_e \cdot \omega L_m}{R_e + \omega L_m}}{\omega L_r + \frac{1}{\omega C_r} + \frac{R_e \cdot \omega L_m}{R_e + \omega L_m}}$$

By simplifying the earlier equation, we get:

$$M = \frac{L_n \cdot f_n^2}{\left[(L_n + 1)f_n^2 - 1\right] + j\left[\left(f_n^2 - 1\right) \cdot f_n \cdot Q_e \cdot L_n\right]}$$

where L_n is inductance ratio, Q_e is quality factor, and f_n is normalized frequency.

$$L_n = \frac{L_m}{L_r}, f_n = \frac{f_s}{f_r} \text{ and } Q_e = \frac{\sqrt{L_r \cdot C_r}}{R_e}$$

Gain value is dependent on different parameters like L_n, Q_e, and f_n.

Low quality factor can achieve higher booster gain, while for minimum gain, the frequency modulation is more, which in turn increases the losses. High quality factor reaches the minimum gain with a small increase in switching frequency, but it is unable to reach the maximum gain. So, we need to choose a moderate Q_e value in order to satisfy both conditions. But we don't rely on switching the maximum Q_e in order to reach maximum gain. We rely on L_n in order to find maximum gain.

Lower values of L_n reach higher booster gain and also have less switching frequency modulation. However, lower L_n has lower magnetizing inductance, which improves the magnetizing current ripple and causes higher conduction losses. For this, we can use the higher m values to achieve higher gain and also low switching frequency modulation.

For greater L_n values, we can achieve higher magnetizing inductance and lower magnetizing current ripple, which will improve the efficiency of the converter.

After selecting L_n and Q_e, the gain can be varied by varying the f_s of the converter. By properly choosing f_s, we can calculate the LLC parameters in the converter as follows:

$$C_r = \frac{1}{2\pi f_s Q_e R_e}$$

$$L_r = \frac{1}{(2\pi f_s)^2 \cdot C_r}$$

$$L_m = L_n \cdot L_r$$

In LLC converter, the switches at inverter side operate with ZVS, and the switches at rectifier side operate with synchronous rectification, i.e., switches operate in complementarily. The converter operation is illustrated in different modes and is discussed as follows:

Mode I: $t_0 < t < t_1$

In this mode of operation, at $t = t_0$, the switches Q_1, Q_4 at inverter side are turned on with ZVS. Q_5, Q_8 at rectifier side are also in on position. The simplified circuit diagram is shown in Fig. 11A. The voltage at resonant inductor is:

$$V_{Lr} = V_{in} - V_{Cr}$$

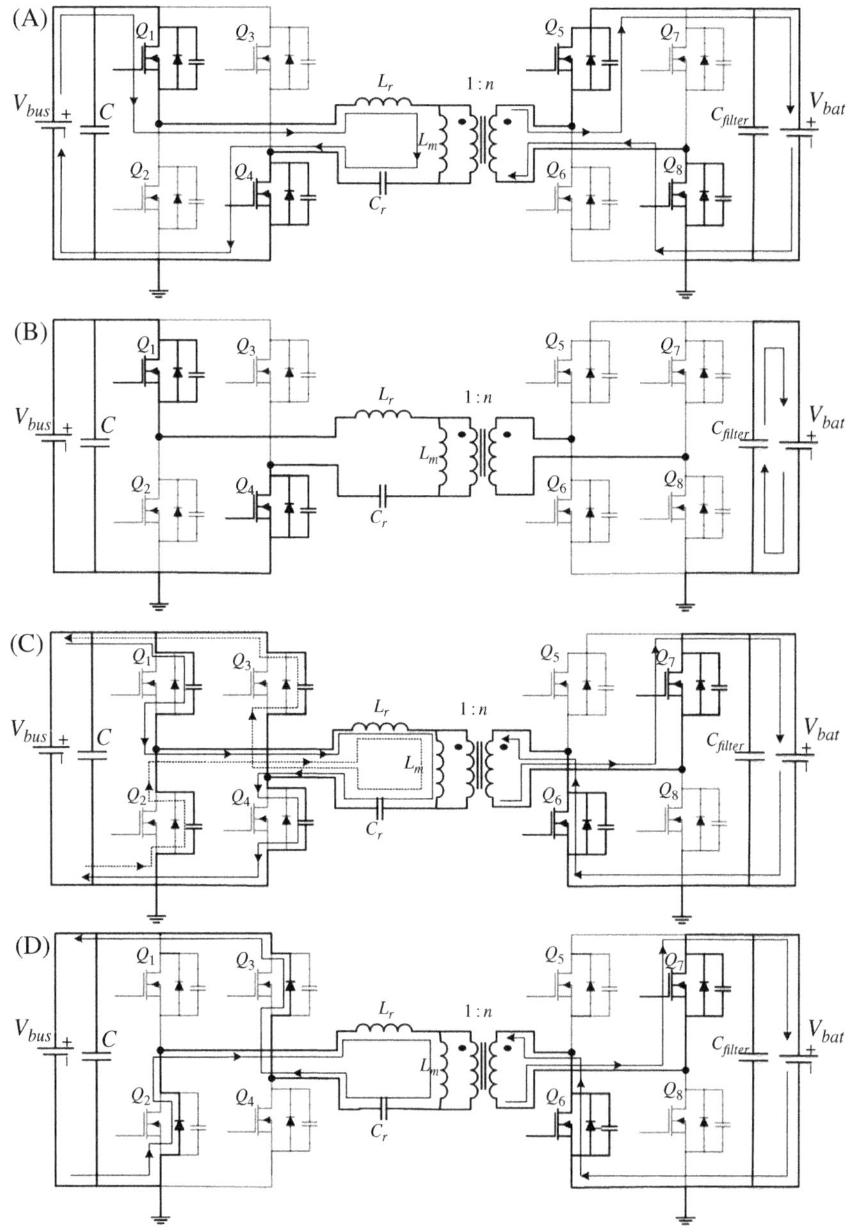

FIG. 11 Full bridge LLC DC-DC converter: (A) Mode I, (B) Mode II, (C) Mode III, and (D) Mode IV.

The primary voltage of transformer is clamped to nV_0. The inductor current i_{Lr} rises sinusoidally due to resonance between resonant capacitor C_r and resonant inductor L_r, and it is greater than the i_{Lm}. As Q_5, Q_8 are switched on at rectifier side, energy is transmitted from input to the load.

Mode II: $t_1 < t < t_2$

In this mode of operation, at $t = t_1$, L_m resonates with C_r and L_r at that time the current i_{Lm} reaches to i_{Lr}. The load current is supplied by output filter capacitor as no energy is transmitted from primary to secondary as switches Q_5, Q_8 are turned off. The simplified schematic diagram is shown in Fig. 11B.

Mode III: $t_2 < t < t_3$

In this mode of operation, at $t = t_2$, the switches Q_1, Q_4 are turned in off position at the inverter side. The capacitances of switches Q_1, Q_4 and switches Q_3, Q_2 are charged and discharged, respectively, through resonant inductor current i_{Lr}. Energy is transferred from primary to secondary as Q_6, Q_7 switches are in on position. The simplified schematic circuit is shown in Fig. 11C.

Mode IV: $t_3 < t < t_4$

During this mode of operation, at $t = t_3$, the voltages across switches Q_3, Q_2 are zero. Resonance occurs between C_r and L_r, which decreases the primary current, and the primary voltage of the transformer is $-nV_0$. The energy is transformed from primary to secondary. The simplified circuit diagram is shown in Fig. 11D.

The next cycle follows from Mode I to Mode IV in the opposite direction.

The complete circuit diagram of a two-stage OBC with interleaved totem-pole PFC at front end and full bridge LLC-type DC-DC converter at back end is shown in Fig. 12.

15.5.2 Multifunctional OBC

OBC provides bidirectional charging either from G2V or V2G. Multifunctional OBC has three different types of functions. In the first function, the battery is charged using grid supply. The second function is that the excess power supply from the battery is fed to the grid. The third function is charging a low voltage (LV) battery from the high voltage (HV) battery. This LV battery is used to supply power to the electric fields in a vehicle such as head lights, air conditioners, and music systems.

This multifunctional OBC has a combination of OBC and LVDC. OBC works when EV is in idle position, whereas LVDC works when EV is in running motion. Initially, low-voltage batteries are charged using an alternator. This low-voltage battery has two functions: the first one is to supply power to electric fields and the second one is to give high supply voltage to high-voltage battery for starting.

The schematic circuit of multifunctional OBC [11] is shown in Fig. 13. This circuit consists of three different stages. The first one is a full-bridge circuit, the second one is a dual-active bridge circuit, and the third one is a dual-function circuit. This dual-function circuit acts for both LV charging and active power decoupling (APD).

As we know at the end of inverter side, electrolytic-type capacitors are used to smoothen the battery current ripple. To smoothen the battery current ripple to a specified limit, large electrolytic capacitors are used. The alternate solution for this is to use APD circuit [12]. The capacitances required in APD are smaller than the capacitors used in a conventional circuit, which reduces size and cost and improves power density ratio of the converter. The APD circuit is used in charging the battery using the grid. This circuit consists of two switches parallel with DC bus of inverter and is coupled with the inductor.

In multifunctional OBC, two switches are present: one at the grid side named as SW1, which is a DPDT switch; and another, SW2, which is a DPST switch at APD circuit. When the SW1 is in the closed position, normal operation is done

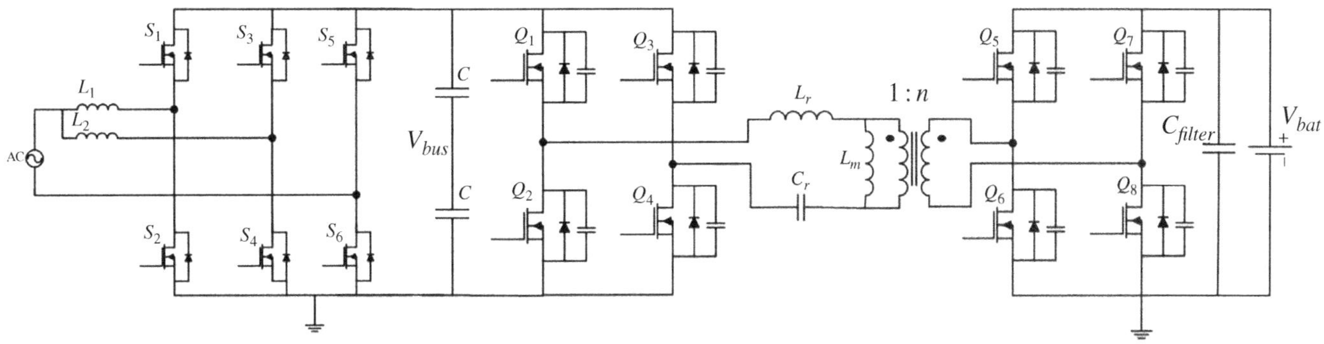

FIG. 12 Two-stage onboard charger of interleaved bridgeless totem-pole PFC with full bridge LLC converter.

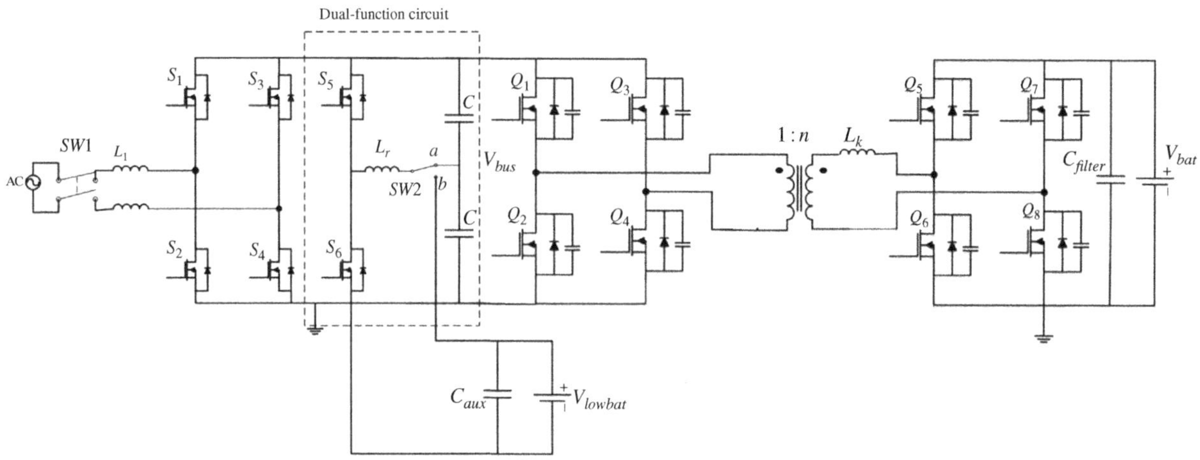

FIG. 13 Circuit of multifunctional on-board charger.

as battery is charged using the grid. In this operation, the dual-functioning circuit behaves like an APD circuit as the SW2 is connected to a point. The excess energy from the battery flows to the grid.

When SW1 is open and SW2 is connected to b position, the HV battery charges the LV battery. In this operation, the dual-function circuit behaves like a buck converter. The DAB circuit provides galvanic isolation between high-voltage battery and low-voltage battery.

15.6 Off-board EV charger

Level 3 chargers are usually fast chargers because the power ratings are very high. The level 3 chargers (known as off-board chargers) are located at the exterior of the EV. The charging speed of a level 3 off-board charger is very high (180–1200 mph) compared with a level 1 charger (3–5 mph) and a level 2 charger (12–80 mph). The level 3 off-board charger comprises of two stages: (a) front-end AC to DC conversion stage and (b) back-end DC to DC conversion stage.

(a) Front-end stage (AC to DC conversion):

Front-end stage (AC to DC conversion) consists of an AC to DC converter. The input of the converter is connected to the grid (single phase/three phase), and the output is connected to the input of the DC to DC converter (back-end stage).

(b) Back-end stage (DC to DC conversion):

Back-end stage (DC to DC conversion) consists of DC to DC converter. The input of the converter is connected to output of an AC to DC converter (front-end stage), and the output is connected to the EV for battery charging application.

15.6.1 Unidirectional AC to DC converter (front-end stage)

Vienna rectifier is most commonly used as an AC to DC converter [13] because it has more advantages in terms of harmonics, power factor, and efficiency compared with the conventional rectifier. Vienna rectifier draws the sinusoidal current from the grid, which leads to less harmonic content at the grid side. It improves the power factor and efficiency of the EV battery charger. It controls the input current and output voltage with reduced ripple. It reduces the switching losses and provides low voltage stress on switches. It transfers the power from the grid to the EV for battery charging application. The schematic circuit of the one phase out of three phases Vienna rectifier is shown in Fig. 14.

Modes of operation:

There are four modes of operation in a one phase out of three phases Vienna rectifier.

Mode I:

The simplified circuit of Mode I is shown in Fig. 15A. When the transistor T_1 is turned on, during the positive half cycle of the AC grid supply, the inductor starts charging, and the current flows in phase through inductor L_{in}, diode D_1, transistor T_1, and diode D_4 to the neutral. In this mode, the output capacitors C_{O1} and C_{O2} discharge to the resistive load R. The output voltage is constant because the output capacitors C_{O1} and C_{O2} are assumed to be large.

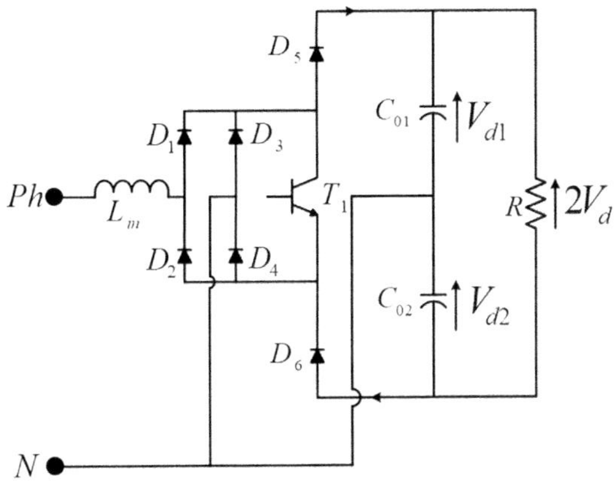

FIG. 14 Schematic circuit of the one phase out of three phases Vienna rectifier.

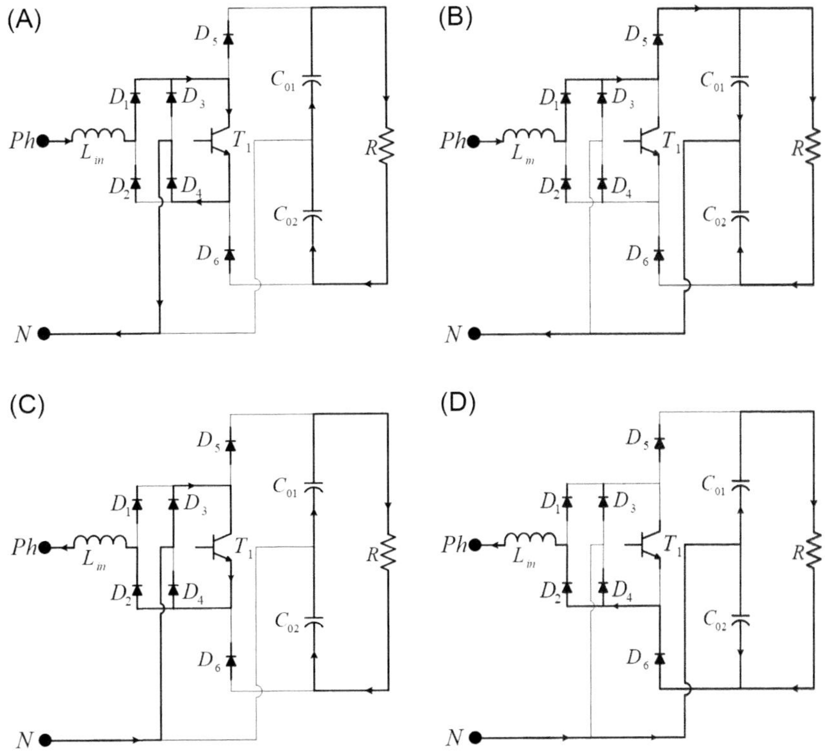

FIG. 15 One phase out of three phases Vienna rectifier: (A) Mode I, (B) Mode II, (C) Mode III, and (D) Mode IV.

Mode II:

The simplified circuit of Mode II is shown in Fig. 15B. When the transistor T_1 is turned off, during the same positive half cycle of the AC grid supply, the inductor starts discharging, and the current flows in phase through diode D_1, diode D_5, and capacitor C_{O1} to the neutral, and also some amount of current flows in phase through diode D_1, diode D_5, resistive load R, and the capacitor C_{O2} to neutral. The output capacitor C_{O2} slightly discharges.

Mode III:

The simplified circuit of Mode III is shown in Fig. 15C. When the transistor T_1 is turned on, during the negative half cycle of the AC grid supply, the inductor starts charging in the reverse direction, and the current flows from neutral through diode D_3, transistor T_1, diode D_2, and inductor L_{in} in phase. The two capacitors C_{O1} and C_{O2} discharge to the resistive load R.

Mode IV:

The simplified circuit of Mode III is shown in Fig. 15D. When the switch T_1 is turned off, during the negative half cycle of the AC grid supply, the inductor starts discharging, and the current flows from neutral through capacitor C_{O2}, diode D_6, and diode D_2 to the neutral, and also some amount of current flows through capacitor C_{O1}, resistive load R, diode D_6, and diode D_2 to neutral. The capacitor C_{O1} slightly discharges.

Similarly, the mode analysis of the next two phases are same with a phase shift of 120°.

15.6.2 Unidirectional DC to DC converter (back-end stage)

Phase-shifted full bridge converter is the most commonly used DC to DC converter because it has more advantages in terms of power density, efficiency, and magnetic interference compared with the conventional full bridge PWM DC to DC converter. It has high power density, high efficiency, and low magnetic interference. It reduces the switching losses and provides low voltage stress on switches. It transfers the power from the grid to the EV for battery charging application. In a conventional full bridge converter, the two diagonal switch pairs are driven alternatively. The output transformer is fed with an AC rectangular voltage. By applying a phase-shifting approach, a deliberate delay can be introduced between the gate signals and the switches [14] in order to achieve ZVS in full bridge converter. The schematic of phase-shifted full bridge converter is shown in Fig. 16. The converter consists of four switches at the input side, namely, Q_1, Q_2, Q_3, and Q_4. The switches with leg pair Q_1 and Q_2 turned on complementarily with less than 50% duty cycle, as dead time is present between them. Similarly, Q_3 and Q_4 turned on complementarily with less than 50% duty cycle. This will create a phase shift in between pulses, which is used to provide ZVS to switches for turn on. The operating gate control signals and steady-state performance of transformer primary and secondary waveforms are shown in Fig. 17. The phase-shifted full bridge converter operates in various modes that are illustrated as follows:

Mode I: $(t_0 < t < t_1)$

In this mode of operation, the switches Q_1 and Q_4 are turned on. Q_1 is turned on with ZVS. As the two switches are conducting, the energy transfers from primary to secondary. Now, the secondary voltage equals nV_{in}. As the secondary voltage is positive, the diode D_1 conducts, and the primary current rises and reaches to nI_{L0}. As we know, the two switches are conducting; hence, the voltages across these two switches are zero. The simplified circuit diagram of this mode is shown in Fig. 18A.

Mode II: $(t_1 < t < t_2)$

In this mode of operation, the switch Q_4 is turned off, which in turn charges the capacitance of switch Q_4 and discharges capacitance of switch Q_3 from 0 to V_{in} and V_{in} to 0, respectively. As soon as the capacitance of switch Q_3 discharges, the body diode of switch Q_3 conducts in order to achieve ZVS turn on of switch Q_3. In this mode, the secondary voltage becomes zero. The simplified circuit diagram is shown in Fig. 18B.

Mode III: $(t_2 < t < t_3)$

In this mode of operation, the switch Q_3 is turned on with ZVS. The primary current of transformer free wheels through switches Q_1 and Q_3. As the secondary voltage is zero, the voltage across the inductor at the output discharges by the load. The simplified schematic is shown in Fig. 18C.

Mode IV: $(t_3 < t < t_4)$

In this mode of operation, the switch Q_1 is turned off, which in turn charges the capacitance of switch Q_1 and discharges capacitance of switch Q_2 from 0 to V_{in} and V_{in} to 0, respectively. As soon as capacitance of switch Q_2 discharges, the body diode of switch Q_2 conducts in order to achieve ZVS turn on of switch Q_2. In this mode, the secondary voltage becomes zero. The simplified circuit diagram is shown in Fig. 18D.

Mode V: $(t_4 < t < t_5)$

In this mode of operation, the switches Q_2 and Q_3 are turned on. Q_2 is turned on with ZVS. As two switches are conducting, the energy transfers from primary to secondary, and the negative voltage is applied at the primary of the

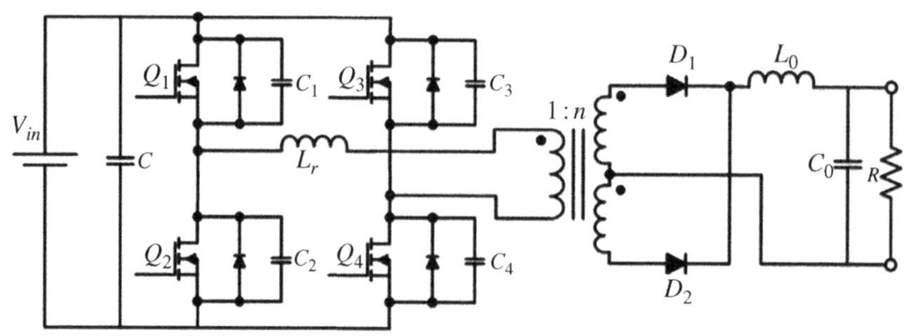

FIG. 16 Schematic circuit of phase-shifted full bridge converter.

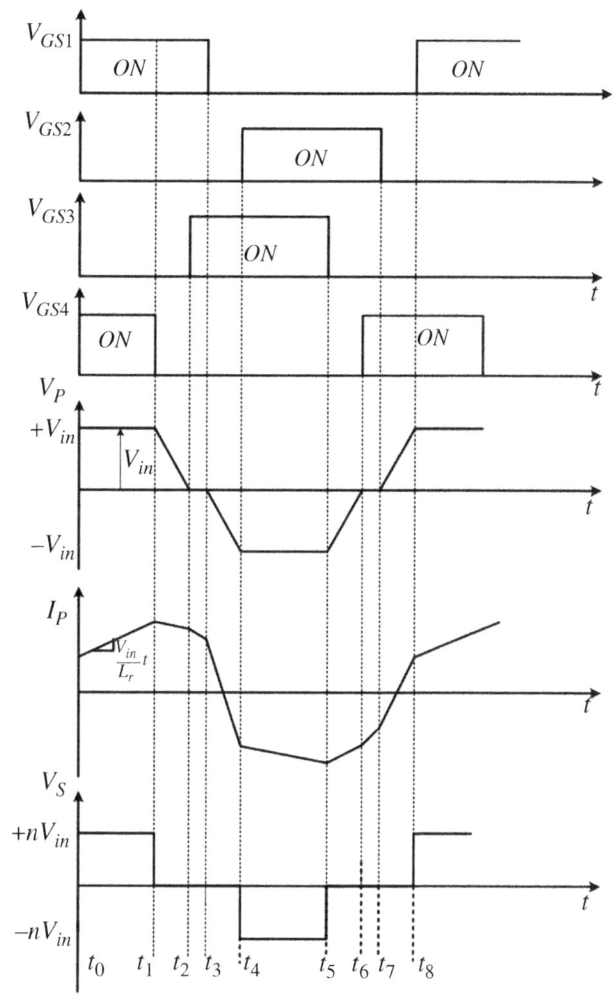

FIG. 17 The steady state performance of transformer primary and secondary waveforms.

transformer. Now, the secondary voltage is equal to $-nV_{in}$. As the secondary voltage is negative, the diode D_2 conducts, and primary current starts to rise in negative direction. As we know, the two switches are conducting; hence, the voltages across these two switches are zero. The simplified circuit diagram of this mode is shown in Fig. 18E.

Mode VI: $(t_5 < t < t_6)$

In this mode of operation, the switch Q_3 is turned off, which in turn charges the capacitance of switch Q_3 and discharges capacitance of switch Q_4 from 0 to V_{in} and V_{in} to 0, respectively. As soon as the capacitance of switch Q_4 discharges, the body diode of switch Q_4 conducts in order to achieve ZVS turn on of switch Q_4. In this mode, the secondary voltage becomes zero. The simplified circuit diagram is shown in Fig. 18F.

Mode VII: $(t_6 < t < t_7)$

In this mode of operation, the switch Q_3 is turned on with ZVS. The primary current of the transformer free wheels through switches Q_1 andQ_3. As the secondary voltage is zero, the voltage across the inductor at the output is discharged by the load. The simplified circuit shown is in Fig. 18G.

Mode VIII: $(t_7 < t < t_8)$

In this mode of operation, the switch Q_2 is turned off, which in turn charges the capacitance of switch Q_2 and discharges the capacitance of switch Q_1 from 0 to V_{in} and V_{in} to 0, respectively. As soon as the capacitance of switch Q_1 discharges, the body diode of switch Q_1 conducts in order to achieve ZVS turn on of switch Q_1. In this mode, the secondary voltage becomes zero. The simplified circuit diagram is shown in Fig. 18H.

The operation repeats from Mode I to Mode VIII for different cycles.

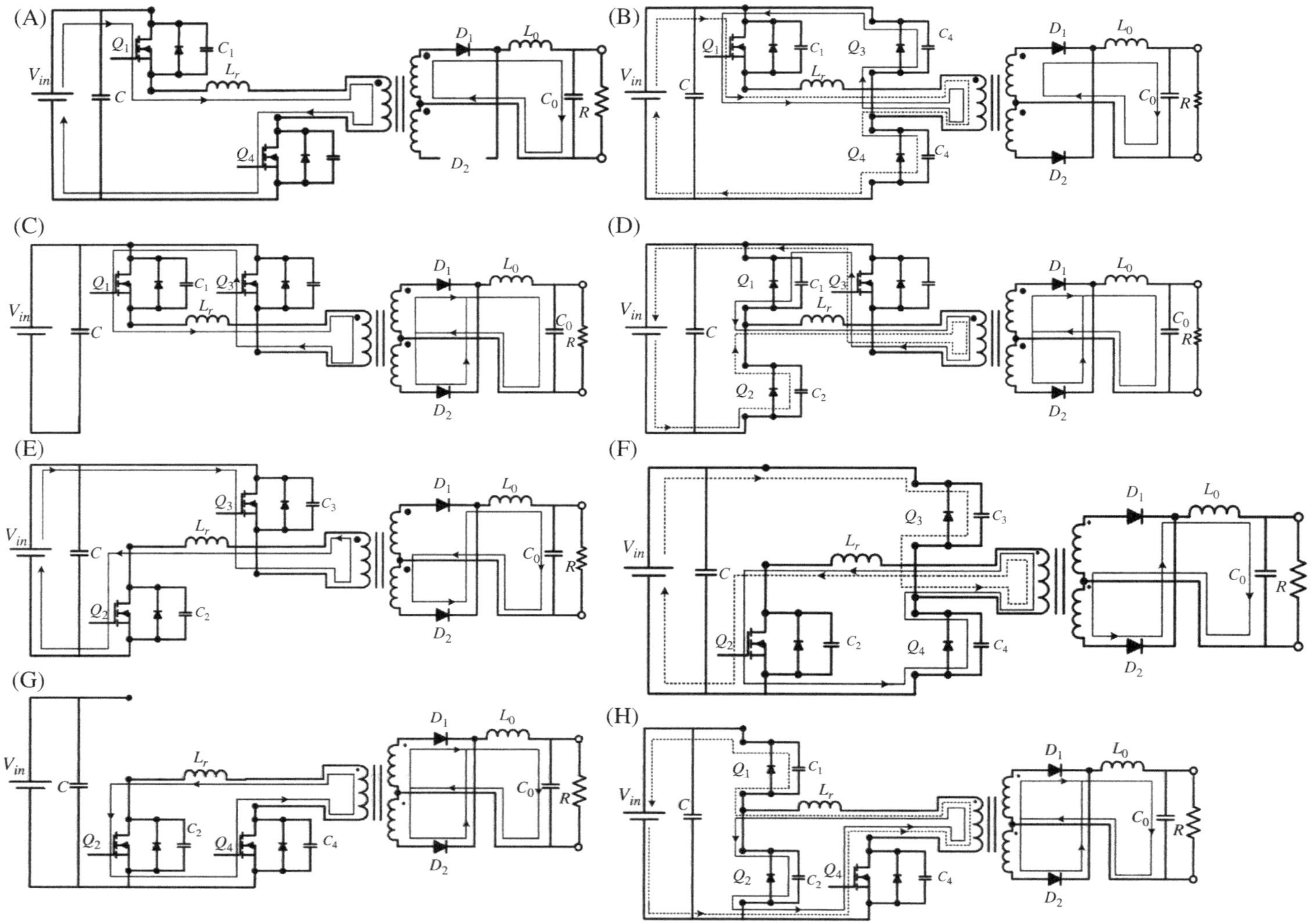

FIG. 18 Phase-shifted full bridge converter: (A) Mode I, (B) Mode II, (C) Mode III, (D) Mode IV, (E) Mode V, (F) Mode VI, (G) Mode VII, and (H) Mode VII.

References

[1] M. Brenna, F. Foiadelli, C. Leone, M. Longo, Electric vehicles charging technology review and optimal size estimation, J. Electr. Eng. Technol. 15 (2020) 2539–2552.

[2] S.M. Arif, T.T. Lie, B.C. Seet, S. Ayyadi, K. Jensen, Review of electric vehicle technologies, charging methods, standards and optimization techniques, Electronics 10 (16) (2021) 1910, https://doi.org/10.3390/electronics10161910.

[3] R. Raff, V. Golub, D. Pelin, D. Topić, Overview of charging modes and connectors for the electric vehicles, in: 2019 7th International Youth Conference on Energy (IYCE), 2019, pp. 1–6, https://doi.org/10.1109/IYCE45807.2019.8991586.

[4] S.J. Thomson, P. Thomas, R. Anjali, E. Rajan, Design and prototype modelling of a CC/CV electric vehicle battery charging circuit, in: 2018 International Conference on Circuits and Systems in Digital Enterprise Technology (ICCSDET), IEEE, 2018, pp. 1–5.

[5] T. Hagino, Pulse Charging Method for Rechargeable Batteries, 1998. U.S. Patent 5,808,447.

[6] Q.-S. Jia, T. Long, A review on charging behavior of electric vehicles: data, model, and control, Control Theory Technol. 18 (2020), https://doi.org/10.1007/s11768-020-0048-8.

[7] J.P.M. Figueiredo, F.L. Tofoli, B.L.A. Silva, A review of single-phase PFC topologies based on the boost converter, in: 2010 9th IEEE/IAS International Conference on Industry Applications-INDUSCON 2010, IEEE, 2010, pp. 1–6.

[8] W. He, Y. Wei, P. Zhu, W. Peng, Research on Totem-pole bridgeless PFC based on synchronous reference frame, in: 2018 Joint International Advanced Engineering and Technology Research Conference (JIAET 2018), Atlantis Press, 2018, pp. 482–487.

[9] E. Firmansyah, S. Tomioka, S. Abe, M. Shoyama, T. Ninomiya, An interleaved totem-pole power factor correction converter, Res. Rep. Inf. Sci. Electr. Eng. Kyushu Univ. 15 (2010) 19–24.

[10] M.H. Lokesha, S.G. Srivani, LLC resonant converter design and development, in: 2014 Annual IEEE India Conference (INDICON), IEEE, 2014, pp. 1–5.

[11] H.V. Nguyen, D.-C. Lee, Single-phase multifunctional onboard battery chargers with active power decoupling capability, in: 2018 IEEE Applied Power Electronics Conference and Exposition (APEC), IEEE, 2018, pp. 3434–3439.

[12] I. Serban, C. Marinescu, Active power decoupling circuit for a single-phase battery energy storage system dedicated to autonomous microgrids, in: 2010 IEEE International Symposium on Industrial Electronics, IEEE, 2010, pp. 2717–2722.

[13] P. Shanmugam, R. Karpagam, T. Thangavelu, Modelling and control of VIENNA rectifier a single phase approach, IET Power Electron. 8 (2015), https://doi.org/10.1049/iet-pel.2014.0503.

[14] General Electric Company, Full-Bridge Lossless Switching Converters, U.S. Patent 4,864,479, 1989.

C H A P T E R

16

Autonomous and electric vehicles charging schemes without an operator

S. Kodeeswaran[a,b], M. Nandhini Gayathri[b], A. Kannabhiran[b], and P. Sanjeevikumar[c]

[a]Department of Electric Power and Energy Conversion, Saveetha School of Engineering, SIMATS Deemed to be University, Chennai, Tamil Nadu, India

[b]School of Electrical and Electronics Engineering, SASTRA Deemed to be University, Thanjavur, Tamil Nadu, India

[c]Department of Electrical Engineering, IT and Cybernetic, University of South-Eastern Norway, Porsgrunn, Norway

16.1 Introduction

Electric vehicles (EVs) and autonomous vehicles (AVs) are becoming popular in this modern world and smart city technology development. These EVs and AVs are using very efficient electric motors, high-potential batteries, and electric power trains [1–3]. The development of EVs and AVs is mostly depending on charging time, driving range, and charging point availability. Most EVs are charged by the plug-in method. If the vehicle increases, then it creates power requirements and charging problems, which may lead to waiting at the charging stations. To avoid such problems, charging without an operator is a good idea for the EV and AV emergence. The wireless power transfer (WPT) method is following the charging process without an operator [4]. The advantage of this WPT method is that easy to adopt, safe, and easy to charge when the vehicles are near the charging pad. Fig. 1 shows the EVs' and AVs' charging-based WPT approach.

WPT methods permit the automatic charging for EVs and AVs, which will be the future of IoT for the upcoming AVs. The modes of WPT methods are (1) static, (2) quasi-dynamic, and (3) dynamic charging [4]. The static method eliminates the electric shock problems and is easy to locate in residential and office parking areas. The quasi-dynamic method may help to charge the vehicle while waiting for the traffic signal. The dynamic method is used to charge the vehicle while moving, and it may be installed on highways. A separate lane is proposed for vehicle charging and to avoid disturbances for the other vehicles that may not require charging. The dynamic charging method helps to increase the driving range and can reduce the battery size.

The WPT method is further classified as inductive power transfer (IPT), capacitive power transfer (CPT), and resonant inductive power transfer (RIPT) methods. These methods already exist and apply to EVs. Most of the EVs are charged by the IPT wireless charging method, which uses magnetic coils to transfer the power [5]. The power is transferred based on the mutual inductance principle. The transmitter coil is placed on the ground, and the receiver coil is placed on the vehicle side to charge the EVs. The ground side circuit is called primary, and the vehicle side may be called secondary. Both sides use the compensation circuits to reduce the reactive power present in the source power and transmitted power. The CPT method uses aluminum plates on the transmitter and receiver sides, which uses an electric field to transfer the power [6]. The CPT system is similar to the 000IPT system, but the coils are replaced by aluminum plates.

This chapter explains the WPT methods and their operation very elaborately. First, the basic requirements, different modes, and methods are explained for a better understanding. Then, the IPT charging pad was designed in ANSYS Maxwell simulation. The rounded coil is taken for the analysis, and based on the simulated results, the values of the components are calculated. The detailed circuit analysis based on SS topology is derived for the calculation of power and efficiency. The remaining topologies for IPT are discussed and tabulated for comparison. Based on the calculations, the IPT method is simulated in MATLAB, and the graphs are drawn.

Handbook of Power Electronics in Autonomous and Electric Vehicles
https://doi.org/10.1016/B978-0-323-99545-0.00016-6

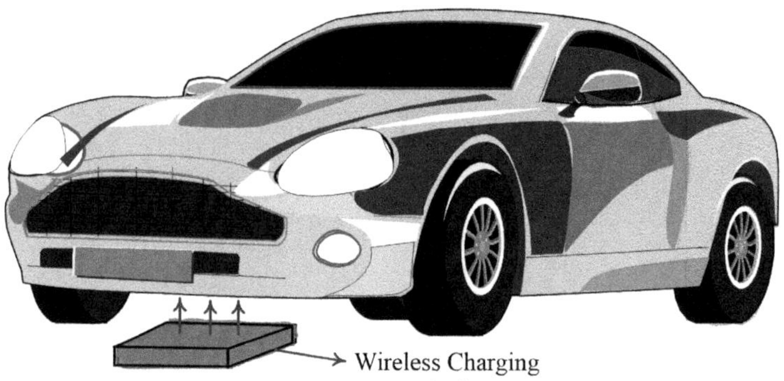

FIG. 1 EV and AV charging system without an operator.

Similarly, the CPT method is discussed, and the plates are designed for the specific dimension to calculate the mutual and self-capacitances. The circuit analysis for CPT based on LC topology is discussed, and the output power is calculated. In addition, the different combination topologies were discussed and tabulated for the different circuits. The overall circuit diagram is simulated in MATLAB, and the output power and efficiency are calculated. The comparison of IPT and CPT methods is discussed in Section 16.7. This chapter helps to identify the different methods for EV and AV charging for researchers, engineers, and workers.

16.2 Wireless power transfer (WPT)

The WPT technology is even now an active method that is utilized in low-power devices, like mobiles, robots, and electrical machines, and recently applied in drones. In recent days, the WPT method is also appropriate for autonomous and electric vehicle charging. The WPT method has been the subject of recent research for charging electric vehicles, and this will be the future IoT for autonomous vehicles. The main advantage of this method is no need for bulky wires or cables and no need to plug the wires. When a vehicle is in parking condition near the charging pad, the vehicle batteries start the charging process without an operator [3]. The AC source is given to the AC-DC converter to produce DC and then given to a high-frequency inverter to produce a high-power AC source. The high-frequency AC converter for this WPT is a major requirement and works under resonance conditions. The compensation circuits are used on both the transmitter and receiver sides to compensate for the power by reducing the reactive power present in the transmitted power. The compensation circuits have different combinations based on the requirement it is designed and used in WPT. The coupling devices are designed based on power requirements, which are positioned on the ground side and vehicle side at some distances. The distance may vary based on the vehicle used like cars, buses, and trucks. The rectifier is used on the receiver side to connect the battery with the DC source. Figs. 2A and 2B shows the complete block diagram of WPT based on AC and DC sources [1].

16.2.1 Essential requirements and important factors consideration of WPT

Before entering to design the WPT system, it is important to study the factors affecting the overall system. The purpose of this consideration is to give more ideas and motivate the researchers.

(1) *Efficiency:* The efficiency of the WPT system is classified as power efficiency and system efficiency. Power efficiency is further described as energy efficiency, economic efficiency, and managing efficiency. The WPT efficiency is a very significant factor in the nontechnical analysis. The efficiency of the system is usually defined as the fraction of the output power to the input power, and it is not fundamentally too large.

(2) *Cost and size:* The price of the system is the next crucial factor, which also covers the extensive manufacturing of WPT chargers for electric and autonomous vehicles. The materials used in WPT components, which may be light and small, tend to cause EMI issues. Therefore, substituting with the component arrangement, temperature effect, reliability, and electromagnetic interference (EMI) should be done rather than just reducing the cost and volume.

(3) *Reliability:* When evaluating the system's reliability, the volume of power transfer and operation are crucial to consider. When the system's temperature rises and mechanical stress is created, the WPT's electronic components'

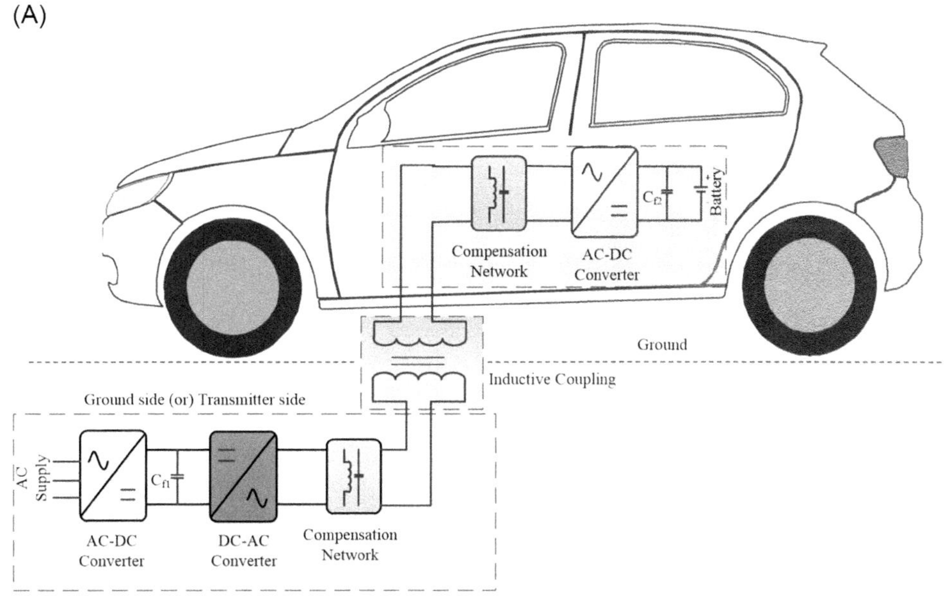

FIG. 2A WPT block diagram based on AC source.

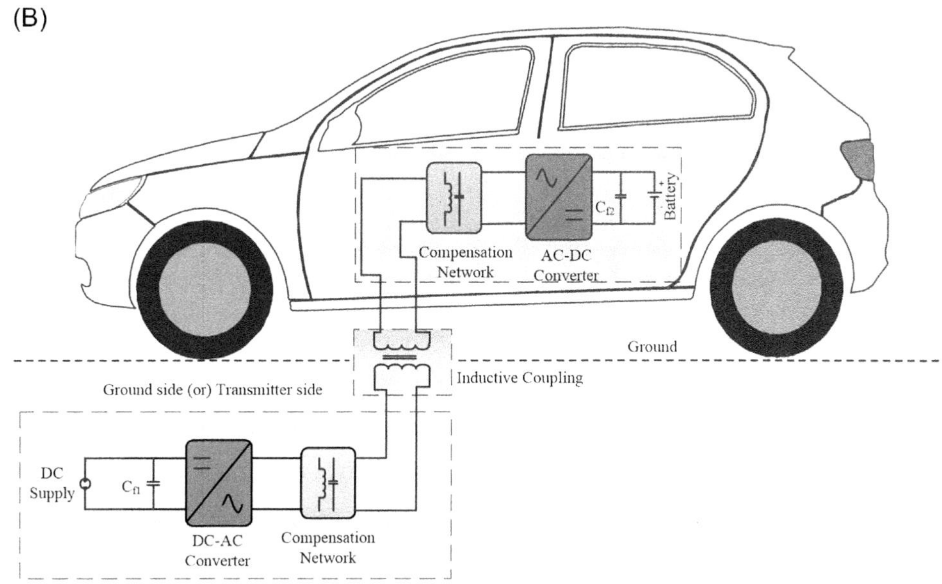

FIG. 2B WPT block diagram based on DC source.

performance may be impacted. For the WPT design, the marginal value needs to be fixed to increase reliability. The amount of power transferred from WPT was decreased due to the variations in coupling values formed by the transmitter and receiver, as well as the orientation of the transmitter and receiver, transfer distance, and displacement. As a result, maintaining the reliability of the WPT system is made possible by controlling the operating frequency of the high-frequency inverter and rectifier.

(4) *Climate changes:* Environmental changes shouldn't have an impact on WPT commercialization. The WPT system will be impacted by the local climate, humidity, and temperature. Therefore, the system's hardware should be able to withstand environmental changes. The ICNIRP standard confirms that the emission of an electric field, magnetic field, and radio magnetic waves should be kept to a safe level. According to the ICNIRP 2010 guidelines, the magnetic field should be kept to a maximum of 27 μT for frequencies between 3 kHz and 10 MHz.

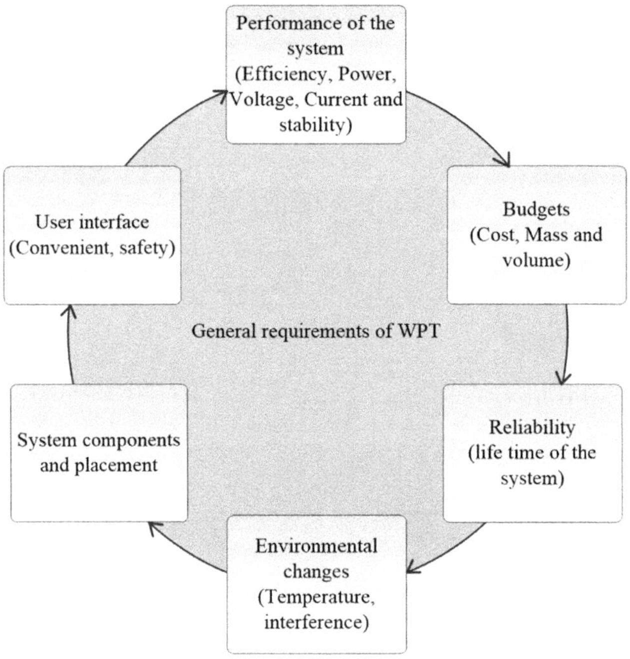

FIG. 3 Requirements of WPT.

(5) *System performance:* Performance measures for a system include voltage and current variations, harmonic distortion, output and input power levels, transient responses, and frequency changes under various operating conditions. Further complicating the design of the WPT system were the variations or movements in the transmitter and receiver, magnetic material properties, switching converters, and controllers. The earlier-mentioned points are very important for the design, and still research work is going on based on these points. So, the development of WPT is a never-ending process.

(6) *System components and placement:* For preventing EMI issues, temperature changes, and thermal design, the system components' rating is a crucial concept. Additionally, the placement of the components calls for precise mechanical and electrical interfaces before connecting to the electric vehicles. In terms of price and potential development use, the WPT infrastructure is a significant concern. The converter and circuit components are necessary for the transmitter's in-ground placement, which can vary depending on the application and charging method. So, the construction of the WPT system is a very important consideration.

(7) *User interface:* It is important to reduce the emission by mechanical and electrical component interfaces with safe limits to implement the WPT system and allow users to experience its benefits.

In Fig. 3, the mentioned points are very important and deal with WPT system implementation and commercialization. The users should feel the usage of the WPT by increasing the power level of the system.

16.3 WPT operating modes and types

16.3.1 Based on modes of operation

The WPT technology can be applied in EV, and its mode is classified into three types depending on the operation: static, quasi-dynamic, and dynamic modes.

(1) *Static mode:* The battery typically starts charging in static mode when the car is parked, and the engine is off. The transmitter sends power to the vehicle when it is parked near the charging pad. Usually, supermarkets, office parking, and residential areas can use this kind of charging system. To prevent misalignment issues, this method uses various control techniques. This type of controller is created by Bombardier in Berlin and runs at 200 kW when the bus is in static mode [4]. The power electronic circuits are located in the corner of the road or charging areas, and the charging pad is placed on the ground side (Fig. 4A).

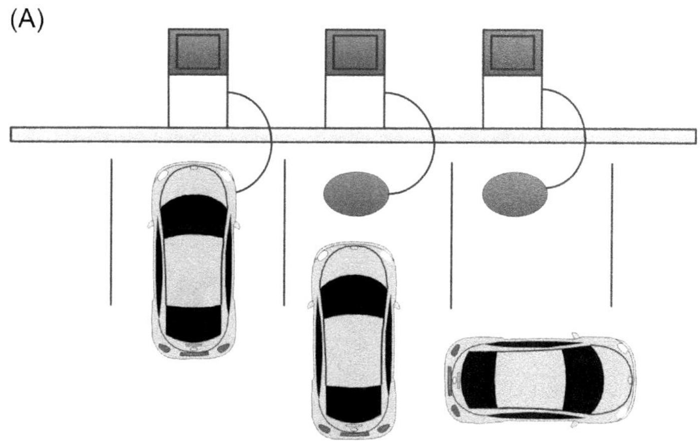

FIG. 4A Static charging mode.

(2) *Quasi-dynamic or stationary mode:* When the EV is not moving, but the engine is still running, it operates in the quasi-dynamic or stationary mode. Although the battery can be charged using this method for a while, it cannot be fully charged. As a result, this kind of charging can be installed in the immediate area of a traffic signal, and while a vehicle is waiting for a signal, it can also be used to charge batteries. Additionally, while passengers are being dropped off and picked up at the bus stop, public transportation may use this type of charging mode [4] (Fig. 4B).

(3) *Dynamic mode:* Dynamic charging is applied when the EV is in running on the road [4]. The charging circuits are installed in the corner or inside the road, and the transmitter may be fixed on the roadside. When the vehicle is running in the charging lane, the battery starts the charging process (Fig. 4C).

16.3.2 Based on types

The WPT system uses an inverter on the transmitter side and a rectifier on the receiver side to charge the battery. However, the system creates the fields depending on the materials used to transmit power. Based on the creation of the field, it is classified into:

(1) *Inductive charging method:* The inductive charging method uses a core coil on the transmitter and receiver sides as shown in Fig. 5A. The IPT method works on the principle of mutual induction, the same as that of the transformer.

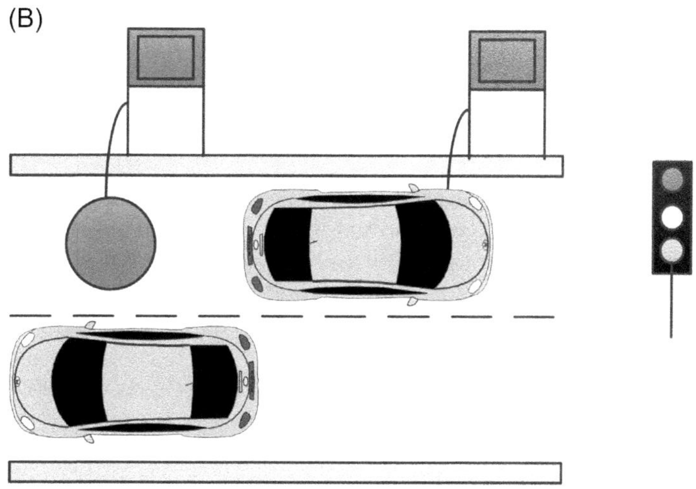

FIG. 4B Quasi-dynamic charging.

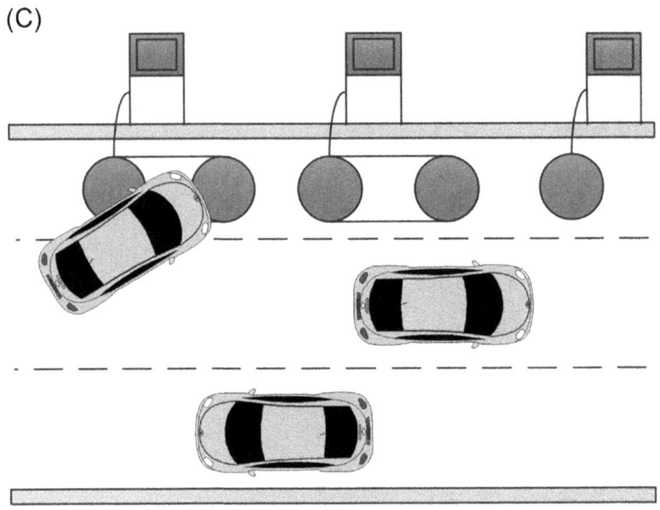

FIG. 4C Dynamic charging.

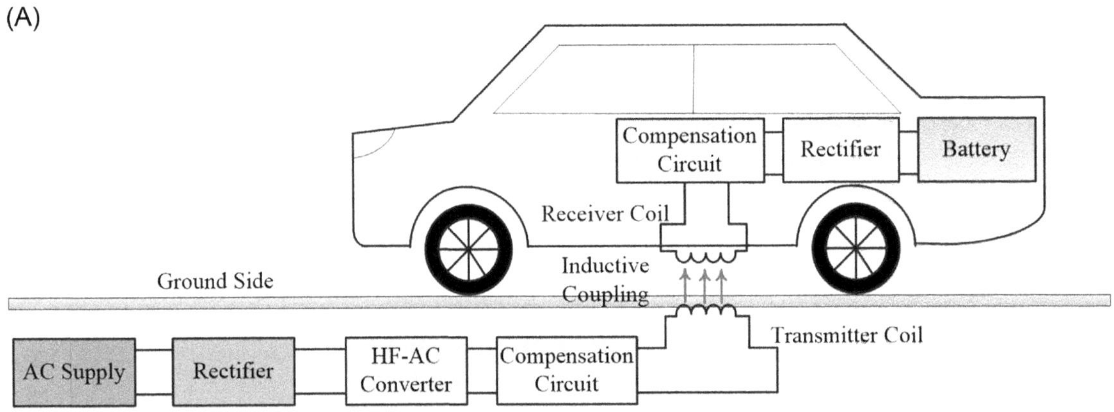

FIG. 5A Inductive charging method.

When current is flowing through the coil the magnetic field is produced, and it transfers the power from the ground side to the vehicle side. The dimensions, distance, and the number of turns used in the coils play a major role to determine the rating of circuit power [6,7]. Because these factors help to determine the mutual inductance and self-capacitance of the system. The transmitter and receiver coils have low weak magnetic coupling due to the distance and oriental position of the coil. Due to the presence of a core that magnetic coupling may increase at the same time losses in the coil may be the drawback of this charging method. However, the operating frequency does not affect the magnetic coupling, but this will produce parasitic resistance and capacitance.

(2) *Capacitive charging method:* The capacitive charging method uses an aluminum plate on the transmitter and receiver sides as shown in Fig. 5B. When the voltage is formed in the plates, the electric field produces in between the plates and transfers the power from the transmitter to the receiver side. Here also, the dimension, distance, and voltage rating help to determine the mutual capacitance and self-capacitance of the circuit [6]. The power rating of the whole circuit determines these important factors. The voltage stress between the plates may reduce system performance due to the high electric field.

(3) *Resonant charging method:* The inductive and capacitive charging methods use resonant circuits to amplify the power and energy by the Quality factor (Q-factor). The LC network is used as a resonant circuit, which is commonly denoted as a compensation circuit. The compensation circuits are used on both the transmitter and receiver sides. If the required power is more, the resonant circuit is mandatory, but for the small amount of power transfer, it is not needed, which means the voltage drop across the inductance is small due to the low current rating.

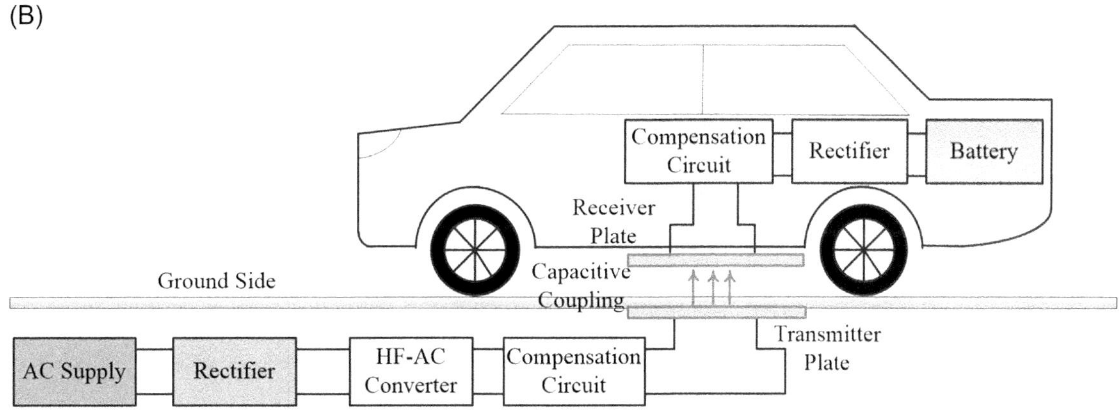

FIG. 5B Capacitive charging method.

Depending on the inductance and capacitance connection, it is classified into Series-Series (SS), Series-Parallel (SP), Parallel-Series (PS), and Parallel-Parallel (PP) in the inductive power charging, and LCC topologies were very familiar methods [8]. The capacitive charging method uses LC, LCL, LCLC, and CLLC topologies [9].

16.4 Renewable energy integration with WPT

Renewable energy sources (RES) are a fantastic and emerging technology for electric vehicle charging. Grids that connect to the charging stations as shown in Fig. 6A were interfaced with using wind and solar energy sources. The interface with the grids and charging stations relies heavily on power electronic components. The block diagram of renewable energy sources connected to EV charging stations can be seen in Fig. 6A. When wind and solar power systems are not used to produce energy, charging stations are connected to the AC power grid [7]. In China and the United States, these technologies play a major role in charging EVs.

Residential and commercial buildings use 32% of the energy produced overall. The development of the microgrid depends heavily on the efficient use of these renewable energy sources, which are used to power EV charging. As seen in Fig. 6B, an electric vehicle's battery is connected to a microgrid that is powered by solar and wind when it is parked. These infrastructures will make it possible to charge more vehicles simultaneously while they are parked in residential, commercial, and public spaces.

16.5 Inductive charging pad and circuit design

16.5.1 Calculation of circuit components

Fig. 7 shows the circuit diagram of a resonant inductive power transfer circuit with SS topology [10]. For simplification, the following assumptions were made:

(1) The system should function at 100% efficiency overall. Until the coil's resistance becomes negligible and infinite, this statement is true.
(2) Other higher-order harmonics are negligible, and only the fundamental harmonics of the input and output voltage are considered.
(3) All the converters provide an efficiency of 100%.
(4) The input voltage V_1 is the equal voltage magnitude of the grid.
(5) The primary and secondary side resonance is assumed to be ideal.

The desired output power is P_{out}. The output voltage is assumed to be the battery voltage as mentioned by the manufacturer because it is a charging circuit [11]. Now, the load resistance is:

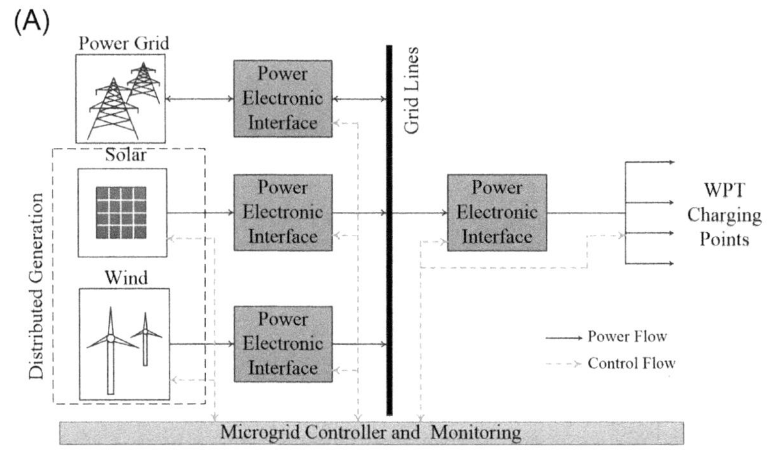

FIG. 6A Block diagram of renewable energy interface.

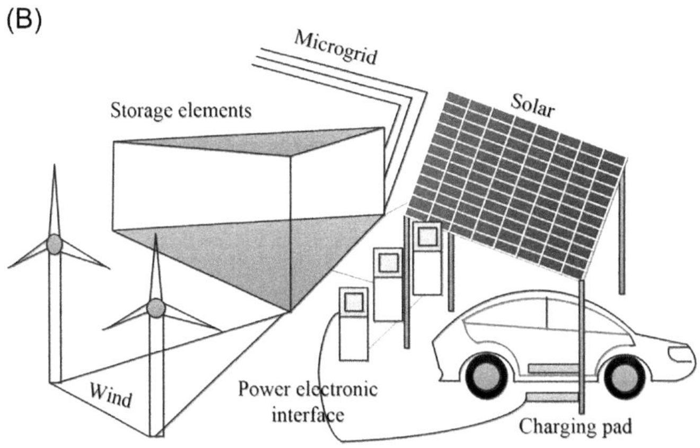

FIG. 6B Proposed WPT charging point with RES interface in smart cities.

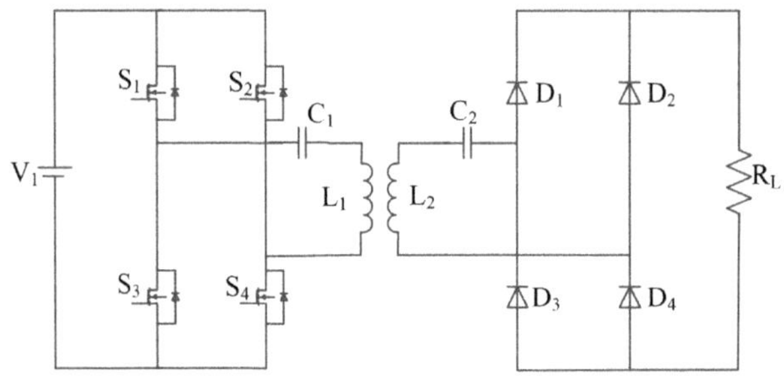

FIG. 7 Circuit diagram of IPT SS topology.

$$R_L = \frac{{V_2}^2}{P_{out}} \tag{1}$$

The RMS value of voltage and current due to the capacitive filter is:

$$V_{2rms} = 2\sqrt{2}\frac{V_2}{\pi} \tag{2}$$

$$I_{2rms} = \frac{V_{2rms}}{R_L} \tag{3}$$

The output power is equal to the input power ($P_{out} = P_{in}$). Since V_2 has been assumed to be equal to the grid voltage, the RMS value of the primary current is:

$$I_1 = \frac{P_{in}}{V_{1rms}} \tag{4}$$

When the rate of the primary and secondary current is obtained, then the value of mutual inductance can be calculated for the desired output power by applying KVL on the secondary.

The resonant frequency, $\omega = \frac{1}{\sqrt{L_1C_1}} = \frac{1}{\sqrt{L_2C_2}}$.

The value of coupling coefficient k is calculated by:

$$k < \frac{1}{Q_2}\sqrt{1 - \frac{1}{4Q_2{}^2}} \tag{5}$$

Normally, the value of the Q-factor is assumed to be 2–10, when the value of k is calculated, the primary inductance,

$$L_1 = \frac{M^2}{L_2k^2} \tag{6}$$

16.5.2 Circular inductive coil design

Fig. 8A shows the 3D view design of the inductive coil. One coil placed on the ground acts as a primary coil, and another coil placed in the electric vehicle acts as a secondary coil. n is the number of turns, D_{in} is the inner diameter of the coil, D_{out} is the exterior diameter of the coil, and d is the distance between the primary and secondary coil as shown in the figure. Table 1 shows the design values for this setup. In an electric vehicle charging application, the air gap distance d is usually 150–170 mm to analyze the performance. To calculate the mutual inductance of the coil, ANSYS Maxwell simulation was used. Mutual inductance is calculated based on the following formula:

$$M = k\sqrt{L_1L_2} \tag{7}$$

It can be noticed in Figs. 8A–8D that the self-inductances L_1 and L_2 increase when air gap distance increases and mutual capacitance decreases. Whenever the air gap distance increases, the flux of the coils decreases. The flux is inversely proportional to inductance, so that self-inductance of the coil increases. Inductance L_1 value is high compared with L_2 because of different dimensions.

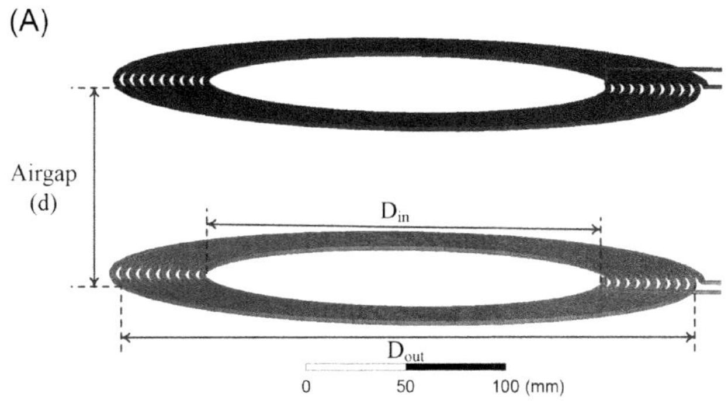

FIG. 8A 3D view of inductive coil.

TABLE 1 Design parameters.

Winding	No of turns	Inner diameter D_{in} (mm)	Outer diameter D_{out} (mm)	Thickness (mm)	Air gap (d) (mm)
Primary	38	450	600	3.2	150
Secondary	20	300	400	3.2	

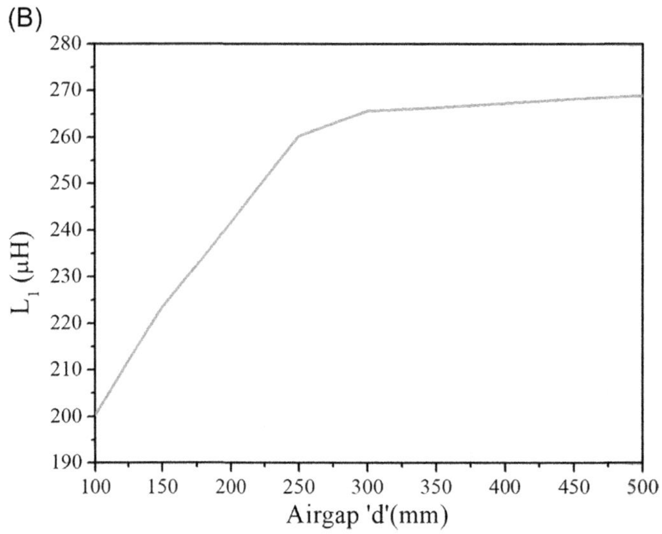

FIG. 8B Maxwell simulated result for self-inductance *L*1 for different air gap distance.

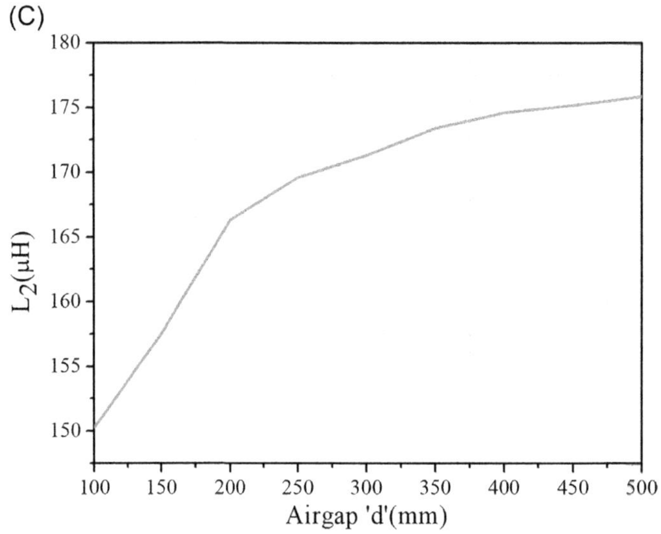

FIG. 8C Maxwell simulating result for self-inductance *L*2 at different air gap distance.

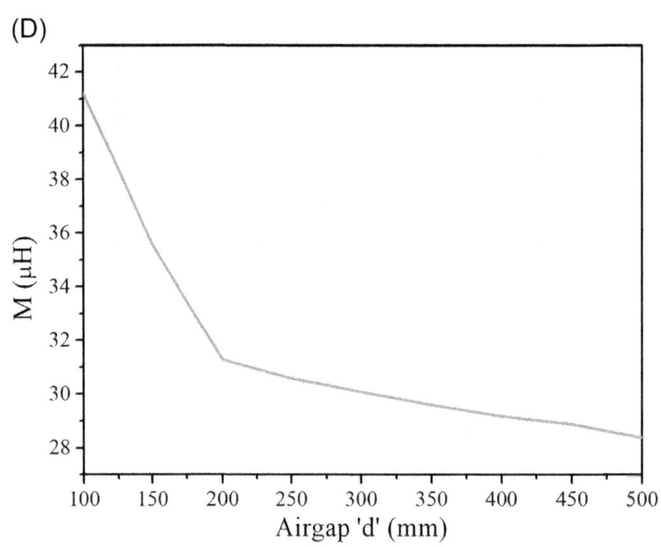

FIG. 8D Maxwell simulating result for mutual inductance *M*.

16.5.3 Misalignment

The position of the primary and secondary coil changes (misalignment) as shown in Fig. 9A, then the value of self-inductance and mutual inductance of the system may vary. Due to this variation, the value of output power is reduced. So, it is significant to analyze the misalignment problem [12] in the IPT method. The value of misalignment changes from 10 to 100 mm, and its inductance variation is shown in Fig. 9B, which indicates that the value of inductance is reduced if the misalignment is increased.

16.5.4 Compensation networks

The compensation circuits are placed in between the inverter and transmitter coil on the primary side. In addition, the compensation circuits are present in between the receiver coil and rectifier on the secondary side. The high leakage inductance and mutual inductance require a high voltage-ampere (VA) rating from the input supply to transfer the power, and this VA rating increases losses [13]. To eliminate this high reactive current, the compensation capacitors are coupled either in parallel or series to cancel the leakage inductance. So, the leakage inductance is canceled by compensation capacitance and vice versa in the CPT method.

16.5.4.1 Analysis of IPT series-series (SS) topology

Fig. 10A illustrates the simplified circuit of the IPT circuit with SS topology. The inductors L_1 and L_2 are connected in series with the capacitance C_1 and C_2, respectively. The voltage drop across the load R_L is taken as V_2. By applying KVL,

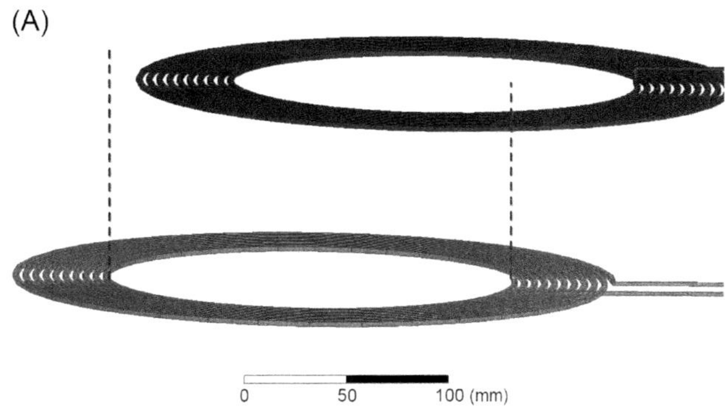

FIG. 9A Misalignment between the primary and secondary coil.

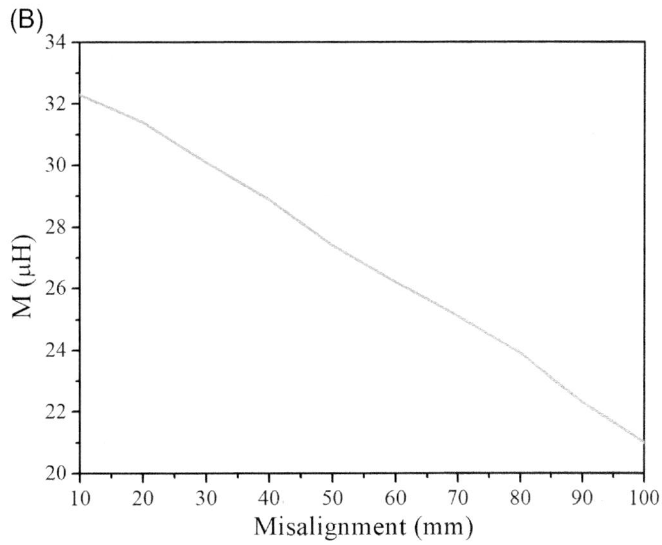

FIG. 9B Misalignment vs mutual inductance graphical structure.

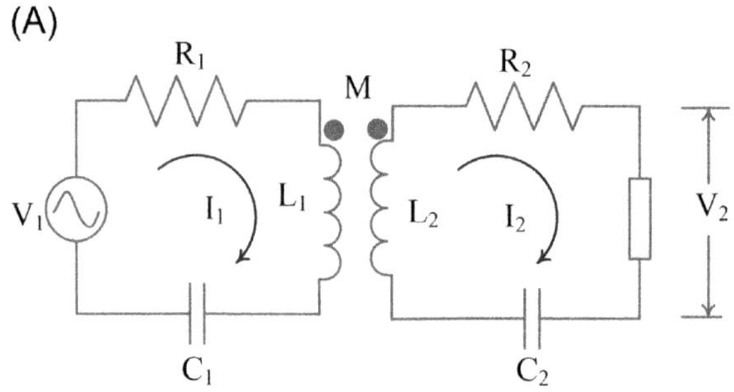

FIG. 10A Simplified circuit diagram for SS topology.

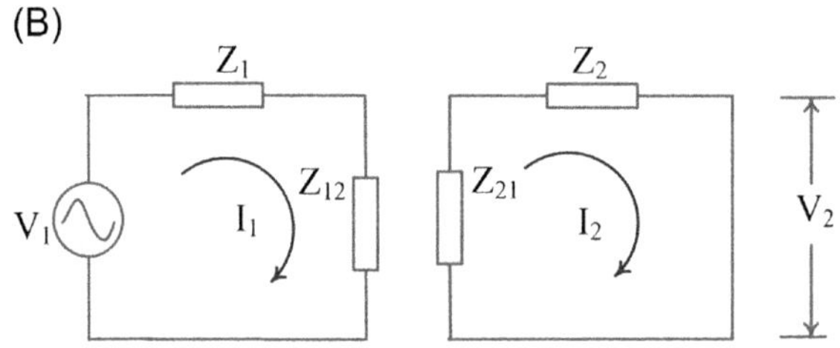

FIG. 10B Equivalent circuit.

$$V_1 = I_1\left(R_1 + j\omega L_1 + \frac{1}{j\omega C_1}\right) - I_2 j\omega M \tag{8}$$

$$V_2 = I_2\left(R_2 + j\omega L_2 + \frac{1}{j\omega C_2}\right) - I_1 j\omega M = 0 \tag{9}$$

$$I_2 = \frac{j\omega M I_1}{R_2 + R_L + j\omega L_2 + \frac{1}{j\omega C_2}}$$

Now Eq. (8) becomes,

$$V_1 = I_1\left(R_1 + j\omega L_1 - \frac{j}{\omega C_1}\right) - \frac{(j\omega M)^2 I_1}{\left(R_2 + R_L + j\omega L_2 - \frac{j}{\omega C_2}\right)}$$

$$\frac{V_1}{I_1} = R_1 + j\left(\omega L_1 - \frac{1}{\omega C_1}\right) + \frac{\omega^2 M^2}{R_2 + R_L + j\left(\omega L_2 - \frac{1}{\omega C_2}\right)}$$

For simplification, the equivalent circuit is considered as shown in Fig. 10B. By taking impedance,

$$Z_1 = R_1 + jX_1 \tag{10}$$

$$Z_2 = R_2 + jX_2 + R_L \tag{11}$$

$$Z_{12} = \frac{\omega^2 M^2}{Z_2} \tag{12}$$

$$Z_{21} = \frac{\omega^2 M^2}{Z_1} \tag{13}$$

$$V_2 = \frac{j\omega M V_1}{Z_1}$$

Take,

$$Z_{total} = \frac{V_1}{Z_1} = R_1 + jX_1 + \frac{\omega^2 M^2}{R_2 + R_L + jX_2} \tag{14}$$

where,

$$X_1 = \omega L_1 - \frac{1}{\omega C_1} \tag{15}$$

$$X_2 = \omega L_2 - \frac{1}{\omega C_2} \tag{16}$$

At resonance condition, $\omega L = \frac{1}{\omega C}$, that is, $X_1 = 0$ and $X_2 = 0$

Now, $Z_1 = R_1;\ Z_2 = R_2 + R_L;\ Z_{12} = \frac{\omega^2 M^2}{R_2 + R_L};\ Z_{21} = \frac{\omega^2 M^2}{R_1}$

$$I_1 = \frac{V_1}{Z_1 + Z_{12}} = \frac{V_1}{R_1 + \frac{\omega^2 M^2}{R_2 + R_L}}$$

$$I_1 = \frac{V_1(R_2 + R_L)}{R_1(R_2 + R_L) + \omega^2 M^2} \tag{17}$$

$$I_2 = \frac{V_2}{Z_2 + Z_{21}} = \frac{V_2}{R_2 + R_L + \frac{\omega^2 M^2}{R_1}}$$

$$= \frac{V_2(R_1)}{R_1(R_2 + R_L) + \omega^2 M^2}$$

$$= \frac{\left(\frac{j\omega M V_1}{Z_1}\right) R_1}{R_1(R_2 + R_L) + \omega^2 M^2}$$

$$= \frac{\left(\frac{j\omega M V_1}{R_1}\right) R_1}{R_1(R_2 + R_L) + \omega^2 M^2} \qquad [\text{Since, } Z_1 = R_1]$$

$$I_2 = \frac{j\omega M V_1}{R_1(R_2 + R_L) + \omega^2 M^2} \tag{18}$$

$$P_{in} = V_1 I_1 = \frac{{V_1}^2(R_2 + R_L)}{R_1(R_2 + R_L) + \omega^2 M^2} \tag{19}$$

$$P_{out} = {I_2}^2 R_L = \frac{\omega^2 M^2 {V_1}^2 R_L}{\left[R_1(R_2 + R_L) + \omega^2 M^2\right]^2} \tag{20}$$

Now overall efficiency,

$$\eta = \frac{P_{out}}{P_{in}} = \frac{\omega^2 M^2 R_L}{(R_2 + R_L) \times \left[R_1(R_2 + R_L) + \omega^2 M^2\right]} \tag{21}$$

16.5.4.2 Other compensation topology analysis

Table 2 illustrates the different compensation topologies used in the IPT system and the efficiency equation.

16.5.5 IPT overall circuit simulation and result

The simulation circuit is drawn, and the performance of the whole setup is tested in MATLAB. The distance between the primary and secondary coil is taken as 150 mm, and the values of L_1 and L_2 are calculated from ANSYS, and the remaining values are calculated by circuit simplification. The fundamental frequency of the whole circuit is taken as 85 kHz, and at this frequency, the value of capacitance C_1 and C_2 is calculated. To transfer the power and minimize the losses, the primary and secondary resistances R_1 and R_2 are assumed to be zero. The output power is calculated as 3180 W, and the efficiency of the setup is achieved at 96%. The calculated values are tabulated in Table 3, and the simulation results are shown in Fig. 11.

TABLE 2 Compensation topologies.

S. no	Topology	Circuit diagram	Efficiency
1	SS [14]	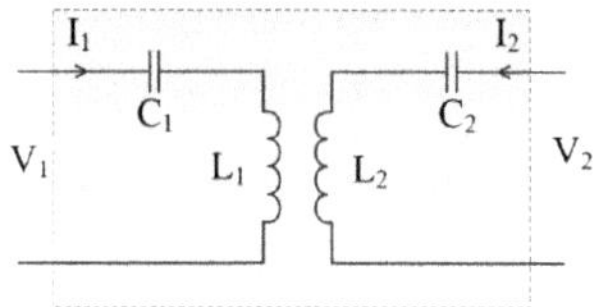	$\eta = \frac{\omega^2 M^2 R_L}{(R_2 + R_L)\left[R_1(R_2 + R_L) + \omega^2 M^2\right]}$
2	SP [15]	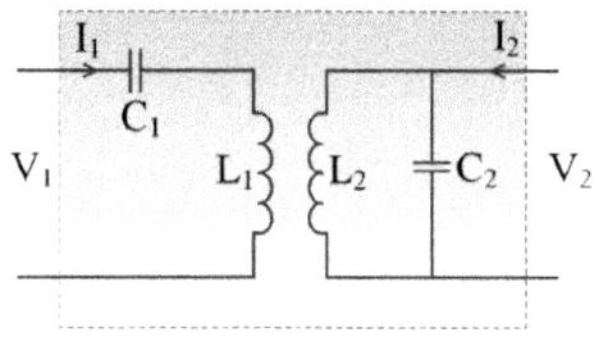	$\eta = \frac{j\omega^2 M^2 R_L X_2}{(R_2 + R_L + jX_2)\left[R_1(R_2 + R_L + jX_2) + \omega^2 M^2\right]}$
3	PS [11]	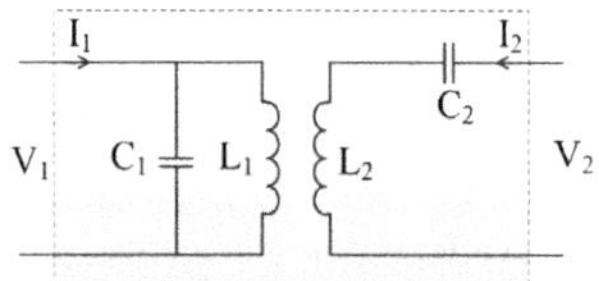	$\eta = \frac{\omega^2 M^2 R_L}{(R_2 + R_L)\left[(R_1 + jX_1)(R_2 + R_L) + \omega^2 M^2\right]}$
4	PP [16]	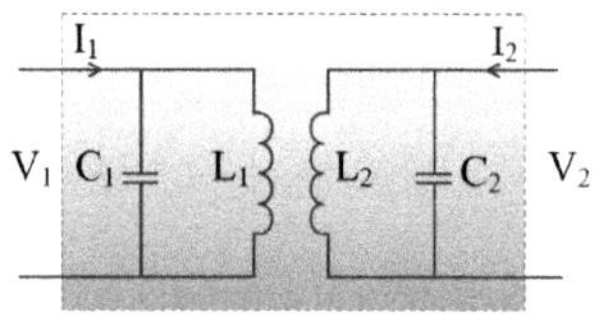	$\eta = \frac{j\omega^2 M^2 R_L X_2}{(R_2 + R_L + jX_2)\left[(R_1 + jX_1)(R_2 + R_L + jX_2) + \omega^2 M^2\right]}$

TABLE 3 Simulation circuit design values.

Side	Parameters	Values
Primary	Input voltage (V_1)	300 V
	Capacitance (C_1)	6.6 nF
	Inductance (L_1)	220.3 μH
Secondary	Capacitance (C_2)	7.3 nF
	Inductance (L_2)	158.2 μH
	Load resistance (R_L)	30 Ω

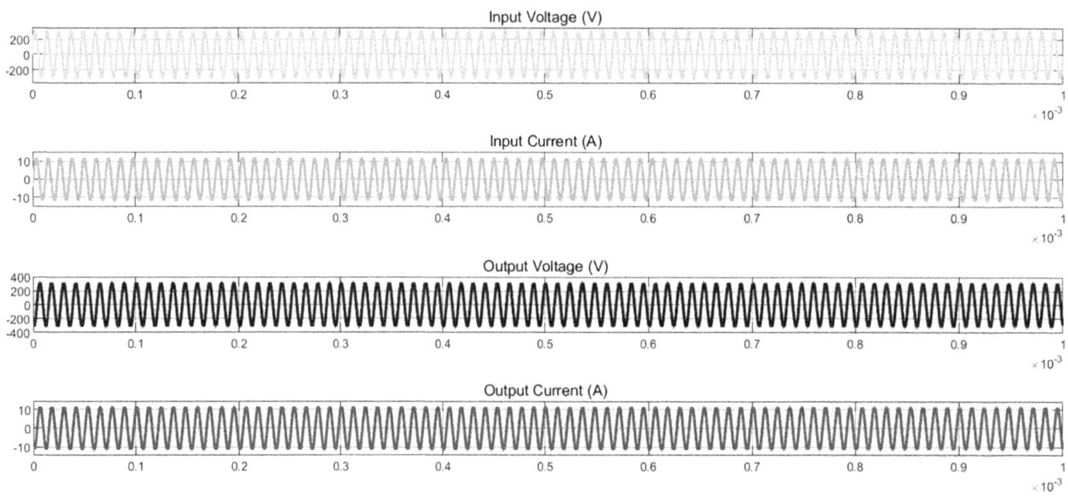

FIG. 11 MATLAB simulation result for IPT.

16.6 Capacitive charging pad and circuit design

The overall circuit diagram for the CPT charging system is shown in Fig. 12. The full-bridge high-frequency inverter is used to convert the input DC source to a high-frequency AC source. The circuit is designed for a 1-MHz operating frequency. The LC-based compensation circuit is used on both sides to reduce the reactive power presented from the transmitter circuit. The receiver compensation circuit is connected to the rectifier to convert a DC, and the battery starts the charging process. Here, the resistive load is considered as a battery for the simulation.

16.6.1 Capacitive plate design

The aluminum plates are taken to design a capacitive coupler design as shown in Fig. 13A. The plates P_1 and P_3 are connected to the ground side, and the plates P_2 and P_4 are connected to the vehicle side [9]. The electric field is induced in the plate by applying the voltage. Therefore, the power is transferred from the primary to the secondary circuit at an operating frequency of 1 MHz. The plate length is taken as L, the spacing between pair of plates is d_s, and the air gap distance is d (Fig. 13B). All the parameter values are taken in millimeters, and the design values are tabulated in Table 4. The ANSYS Maxwell simulation is used to find the mutual capacitance between the plates, and it is calculated as:

$$CM = \frac{C_{MA} \cdot C_{MB}}{C_{MA} + C_{MB}} \tag{22}$$

The value of C_{MA} and C_{MB} is equal due to their similar structure. The designed plate is simulated in ANSYS Maxwell to find the required self and mutual capacitances. The primary side self-capacitance C_P is found for different air gap distances, and it is shown in Fig. 13C. The value of capacitance is decreased concerning the distance increases. Similarly, the secondary side self-capacitance C_S is calculated and shown in Fig. 13D. Due to the symmetrical view on the primary and secondary sides and their dimensions, the value of capacitances on both sides does not change. The calculated mutual capacitance for the different air gap distances is represented in Fig. 13E.

16.6.2 Misalignment

The position of the primary and secondary plate changes may produce misalignment problems. Even though the CPT method has fewer misalignment problems compared to IPT technology, which is the big advantage of the CPT system, the cost is less. Fig. 14 shows the analysis of misalignment for different positions and the mutual capacitance values. The graph indicates that the mutual capacitance reduces concerning the position changes from 10 to 100 mm from the left side.

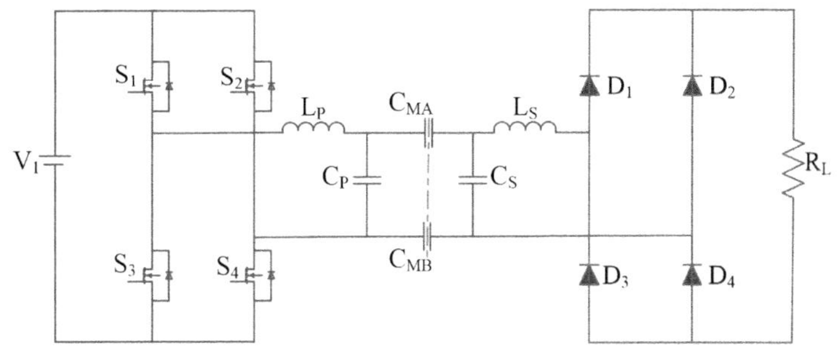

FIG. 12 Overall circuit diagram for the CPT system.

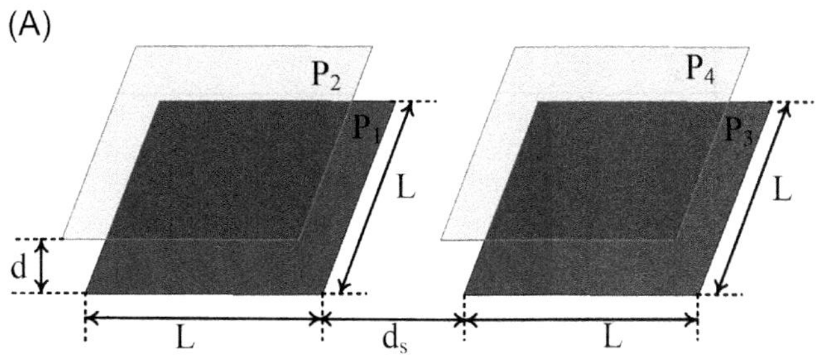

FIG. 13A Capacitive plate design in 3D view.

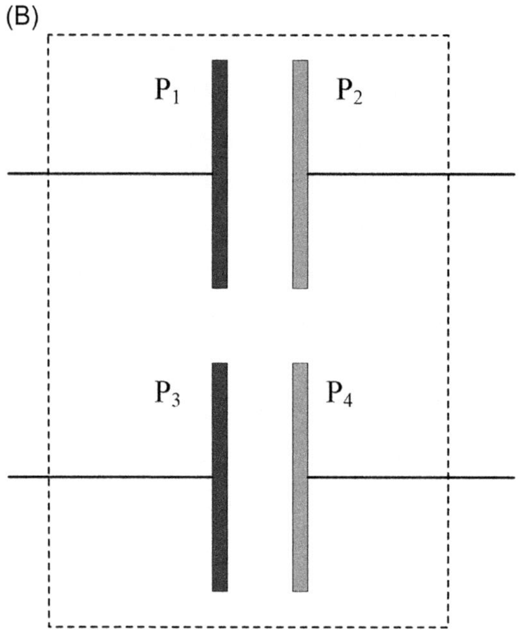

FIG. 13B Capacitance formation between transmitter and receiver.

TABLE 4 Plate design values for CPT.

Plates	Plate length (L) (mm)	Spacing between plates (d_s)	Thickness (t) (mm)	Air gap distance (d) (mm)
Primary	500	300	2	150
Secondary	500	300	2	

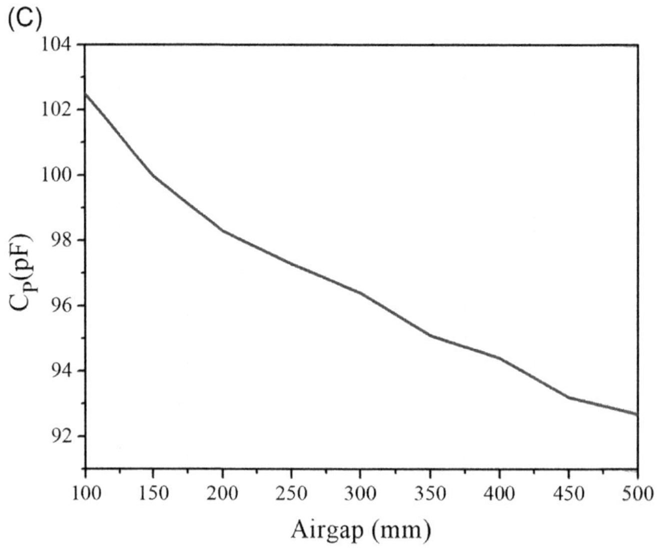

FIG. 13C Maxwell simulated capacitance C_P in pF.

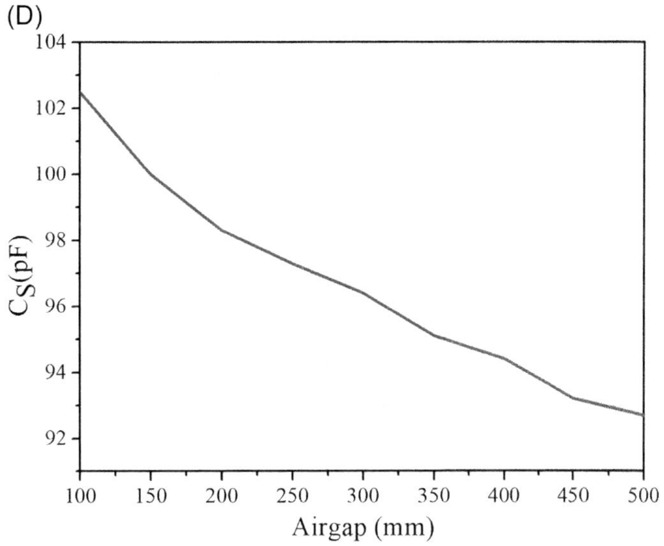

FIG. 13D Maxwell simulated capacitance C_S in pF.

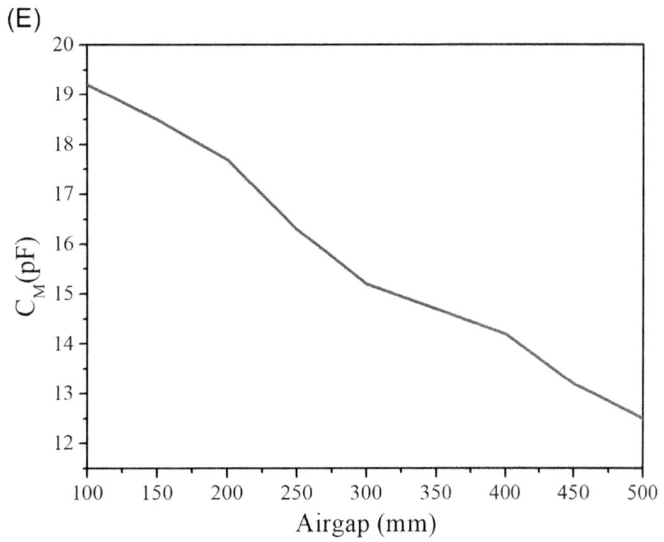

FIG. 13E Mutual capacitance C_M in pF.

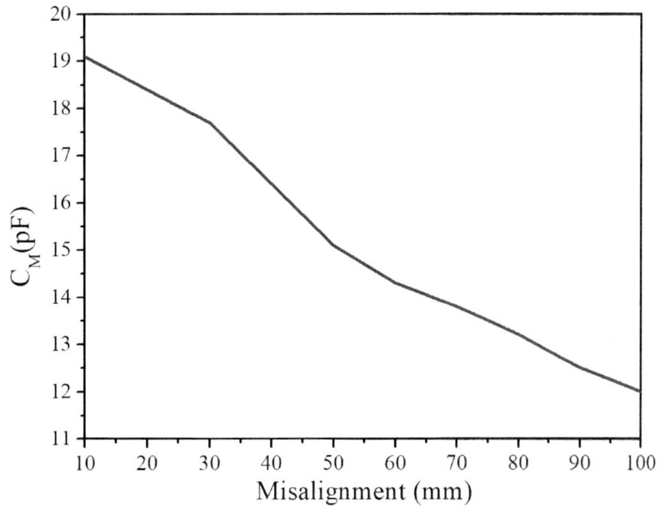

FIG. 14 Misalignment vs mutual capacitance C_M.

16.6.3 Analysis of CPT-LC circuit

Like IPT technology, the compensation circuit is used in CPT technology. This section explains the CPT system with LC circuit topology [9], and it is shown in Fig. 15.

The inductance L_P and L_S are calculated as,

$$L_P = \frac{1}{\omega_S^2 C_P} \tag{23}$$

$$L_S = \frac{1}{\omega_S^2 C_S} \tag{24}$$

where ω_S is the switching frequency of the CPT system. For simplification, capacitance-voltage V_{CS} is calculated as,

$$V_{CS} = -V_P.\left(C_P + \frac{C_M C_S}{C_M + C_S}\right) \tag{25}$$

Now,

$$\begin{aligned} -I_2 &= V_{CS}.\frac{1}{j\omega_S L_P} \\ &= -V_P.\left(C_P + \frac{C_M C_S}{C_M + C_S}\right).\frac{1}{j\omega_S L_P} \\ &= -V_P.\left(C_P + \frac{C_M C_S}{C_M + C_S}\right).\frac{1}{\frac{j}{\omega_S C_P}} \end{aligned} \tag{26}$$

$$-I_s = j\omega_s.C_S.\left(C_P + \frac{C_M C_S}{C_M + C_S}\right).V_P \tag{27}$$

$$V_{CP} = V_S.\left(C_S + \frac{C_M C_P}{C_M + C_P}\right) \tag{28}$$

$$-I_P = j\omega_s.C_P.\left(C_S + \frac{C_M C_P}{C_M + C_P}\right).V_S \tag{29}$$

$$P_{in} = P_{out} = \omega_S.C_S.\left(C_P + \frac{C_M C_S}{C_M + C_S}\right).|V_P|.|V_S| \tag{30}$$

16.6.4 Other compensation topologies

Table 5 depicts the types of topologies used in the CPT system and its output power equation.

16.6.5 MATLAB simulation result for CPT

The overall circuit diagram is simulated in MATLAB. The values of C_P and C_S are calculated from ANSYS, and in this calculation, the value of resonant inductance is calculated at an operating frequency of 1 MHz. Fig. 16 shows that the input voltage and current are in phase with each other. Similarly, the output voltage and current are equal phases.

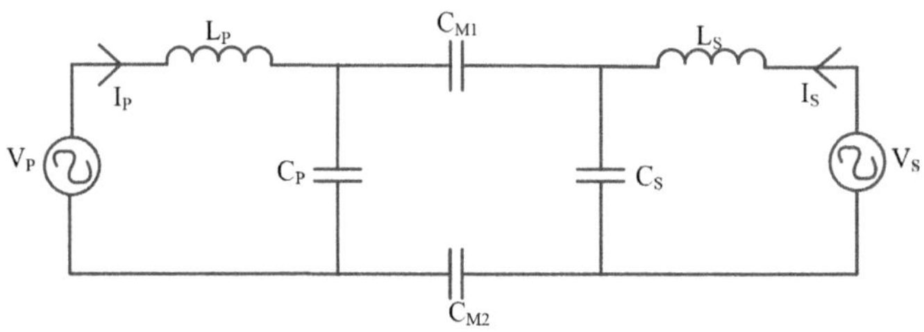

FIG. 15 LC compensation circuit for CPT system.

TABLE 5 Various topologies for CPT.

S. no	Types	Output power
1	Series L circuit [17]	$P_{out} = V_P{}^2/R$
2	LC circuit [18]	$P_{out} = \omega_S.C_S.\left(C_P + \frac{C_M C_S}{C_M + C_S}\right).\lvert V_P\rvert.\lvert V_S\rvert$
3	LCL circuit [19]	$P_{out} = \omega_S C_S\left(\frac{C_M.C_P}{C_M + C_P}\right)\lvert V_P\rvert.\lvert V_S\rvert$
4	LCLC circuit [20]	$P_{out} = j\omega C_P C_S \frac{C_M}{C_P C_S + C_P C_M + C_S C_M}.\lvert V_1\rvert.\lvert V_2\rvert$

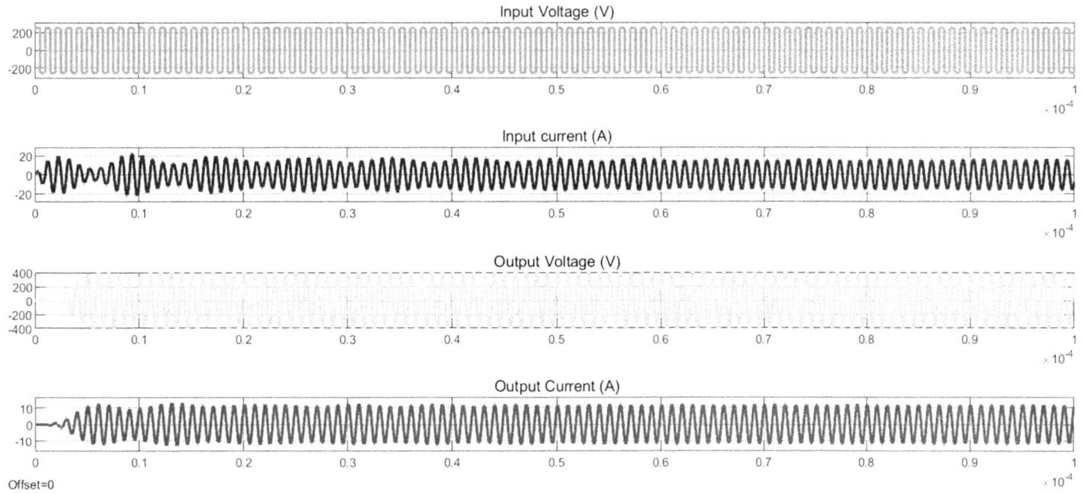

FIG. 16 MATLAB simulation result for CPT.

TABLE 6 Circuit components values for CPT system.

Side	Parameters	Values
Primary	Input voltage (V_P)	300 V
	Inductance (L_P)	11.6 μH
	Capacitance (C_P)	100 pF
Secondary	Inductance (L_P)	11.6 μH
	Capacitance (C_P)	100 pF
	Load resistance (R_L)	30 Ω
Mutual capacitance (C_M)		38 pF

But the input voltage and output voltage are 90° lagging. The overall circuit diagram achieves the output power of 2873 W with an efficiency of 91%. The design values are tabulated as shown in Table 6.

16.7 Comparison between IPT and CPT technology

Both IPT and CPT technology have been applied in low-power and high-power applications [21]. But it is very important to study the comparison between these two technologies when applied in electric vehicles and autonomous vehicles [22]. Each method has some limitations, and these are listed in Table 7.

TABLE 7 Comparison between IPT and CPT technologies.

S. no	Parameters	IPT	CPT
1	Switching frequency	85 kHz	1 MHz
2	Coupling field	Magnetic	Electric
3	Material	Litz wires, ferrites	Copper/aluminum plates
4	Cost	High	Low
5	Safety	Good	Excellent
6	Size	Small	Large
7	Misalignment	Poor	Good
8	Efficiency	Excellent	Excellent
9	Voltage stress	Medium	High
10	Power level	High	Medium
11	Stationary or dynamic	Better for stationary	Both
12	Foreign objects (metal)	Will generate heat	Will not generate heat

16.8 Research opportunities and challenges

To implement EV charging, this section clarifies few aspects of WPT development. WPT is already developed and used in mobile, laptops, and other devices, but to provide a low-cost charger for electric vehicles, it needs to conduct some economic analysis [23]. The following are some of the crucial points that should be emphasized to draw in customers:

(1) High power transfer with small coil sizes will lower the system's cost, as well as losses and field emission.
(2) To maximize system efficiency, automatic alignment of the transmitter with the location of the receiver or vice versa must be implemented.
(3) According to the standards, it is necessary to talk about the impact of field emission on both the charging side and the vehicle side.
(4) It is necessary to develop a foreign object detection system for the safety aspects.
(5) It is reducing the charging system's size and weight.
(6) The last and important one is reducing the charging time.

These are just a few of the crucial issues that require attention. The wireless charging system is more dependable, safe, and environment-friendly from a sustainable perspective. Additionally, the wireless charging system offers more advantages in terms of safety than the wired charging system. However, it is important to concentrate on (a) the grid management system, (b) the policies should entice the customer with more benefits, (c) load balancing between transmission and distribution infrastructure for wireless charging stations could be improved, and (d) standards and emissions could be lowered for human safety.

The environmental restrictions, costs, reliability, and efficiency must all be carefully considered when implementing dynamic charging, and performance must also be carefully assessed. To assist and control the grid, a better method of integrating renewable energy with WPT systems, V2G, and G2V technologies needs to be developed. The WPT will excel in sustainable electric mobility and transportation electrification over the coming 10 years.

16.9 Conclusion

The overview of wireless power transfer adopted for EV and AV charging is discussed in this chapter. The WPT modes and methods were discussed with a clear diagram for the EVs' application. The static, quasi-dynamic, and dynamic methods are very suitable for charging the EVs and AVs from understanding. The IPT charging pad is designed in ANSYS Maxwell simulation, and based on the simulated result, the overall circuit diagram is designed. The circuit model and analysis are discussed for the IPT charging system, and compensation topologies are discussed and tabulated. The circuit parameters are calculated, and the overall circuit diagram is simulated in MATLAB. The simulated results show that the output power is achieved at 3.18 kW at an efficiency of 96%. Similarly, the CPT circuit is designed, and the aluminum plate parameters are simulated in ANSYS Maxwell to find the self and mutual capacitances. After obtaining the results, the circuit parameters for CPT are calculated and simulated in MATLAB. The CPT overall circuit simulation shows that the output power achieves 2.87 kW at an efficiency of 91%. To conclude, both IPT and CPT systems are more suitable for EV and AV charging depending upon the rating used.

References

[1] A. Ahmad, M.S. Alam, R. Chabaan, A comprehensive review of wireless charging technologies for electric vehicles, IEEE Trans. Transp. Electrif. 4 (1) (2017) 38–63.

[2] D. Patil, M.K. McDonough, J.M. Miller, B. Fahimi, P.T. Balsara, Wireless power transfer for vehicular applications: overview and challenges, IEEE Trans. Transp. Electrif. 4 (1) (2017) 3–37.

[3] C. Qiu, K.T. Chau, C. Liu, C.C. Chan, Overview of wireless power transfer for electric vehicle charging, in: 2013 World Electric Vehicle Symposium and Exhibition, EVS 2014, 2014.

[4] C. Panchal, S. Stegen, J. Lu, Review of static and dynamic wireless electric vehicle charging system, Eng. Sci. Technol. Int. J. 21 (5) (2018) 922–937. Elsevier B.V.

[5] S. Li, C.C. Mi, Wireless power transfer for electric vehicle applications, IEEE J. Emerg. Sel. Top. Power Electron. 3 (1) (2015) 4–17.

[6] J. Dai, D.C. Ludois, Capacitive power transfer through a conformal bumper for electric vehicle charging, IEEE J. Emerg. Sel. Top. Power Electron. 4 (3) (2016) 1015–1025.

[7] R. Bosshard, J.W. Kolar, Inductive power transfer for electric vehicle charging: technical challenges and tradeoffs, IEEE Power Electron. Mag. 3 (3) (2016) 22–30.

[8] S. Zhou, C. Chris Mi, Multi-paralleled LCC reactive power compensation networks and their tuning method for electric vehicle dynamic wireless charging, IEEE Trans. Ind. Electron. 63 (10) (2016) 6546–6556.

[9] F. Lu, H. Zhang, H. Hofmann, C.C. Mi, A double-sided LC-compensation circuit for loosely coupled capacitive power transfer, IEEE Trans. Power Electron. 33 (2) (2018) 1633–1643.

[10] W. Li, H. Zhao, J. Deng, S. Li, C.C. Mi, Comparison study on SS and double-sided LCC compensation topologies for EV/PHEV wireless chargers, IEEE Trans. Veh. Technol. 65 (6) (2016) 4429–4439.

[11] H. Hao, G.A. Covic, J.T. Boys, A parallel topology for inductive power transfer power supplies, IEEE Trans. Power Electron. 29 (3) (2014) 1140–1151.

[12] K. Song, et al., Design of DD coil with high misalignment tolerance and low EMF emissions for wireless electric vehicle charging systems, IEEE Trans. Power Electron. 35 (9) (2020) 9034–9045.

[13] W. Zhang, C.C. Mi, Compensation topologies of high-power wireless power transfer systems, IEEE Trans. Veh. Technol. 65 (6) (2016) 4768–4778.

[14] Y.C. Hsieh, Z.R. Lin, M.C. Chen, H.C. Hsieh, Y.C. Liu, H.J. Chiu, High-efficiency wireless power transfer system for electric vehicle applications, IEEE Trans. Circuits Syst. II Express Briefs 64 (8) (2017) 942–946.

[15] K. Aditya, S.S. Williamson, Comparative study of series-series and series-parallel topology for long track EV charging application, in: 2014 IEEE Transp. Electrif. Conf. Expo Components, Syst. Power Electron. - From Technol. to Bus. Public Policy, ITEC 2014, 2014.

[16] Z. Pantic, S.M. Lukic, Framework and topology for active tuning of parallel compensated receivers in power transfer systems, IEEE Trans. Power Electron. 27 (11) (2012) 4503–4513.

[17] M. Kline, I. Izyumin, B. Boser, S. Sanders, Capacitive power transfer for contactless charging, in: Conference Proceedings - IEEE Applied Power Electronics Conference and Exposition - APEC, 2011, pp. 1398–1404.

[18] S. Kodeeswaran, M. Nandhini Gayathri, Performance investigation of capacitive wireless charging topologies for electric vehicles, in: 2021 Int. Conf. Innov. Trends Inf. Technol. ICITIIT 2021, 2021.

[19] H. Zhang, F. Lu, H. Hofmann, W. Liu, C.C. Mi, A four-plate compact capacitive coupler design and LCL-compensated topology for capacitive power transfer in electric vehicle charging application, IEEE Trans. Power Electron. 31 (12) (2016) 8541–8551.

[20] F. Lu, H. Zhang, H. Hofmann, C. Mi, A double-sided LCLC-compensated capacitive power transfer system for electric vehicle charging, IEEE Trans. Power Electron. 30 (11) (2015) 6011–6014.

[21] D. Vincent, P.S. Huynh, N.A. Azeez, L. Patnaik, S.S. Williamson, Evolution of hybrid inductive and capacitive AC links for wireless EV charging - a comparative overview, IEEE Trans. Transp. Electrif. 5 (4) (2019) 1060–1077.

[22] S.Y. Choi, B.W. Gu, S.Y. Jeong, C.T. Rim, Advances in wireless power transfer systems for roadway-powered electric vehicles, IEEE J. Emerg. Sel. Top. Power Electron. 3 (1) (2015) 18–36.

[23] J. Høj, L. Juhl, S. Lindegaard, V2G—an economic Gamechanger in E-mobility? World Electr. Veh. J. 9 (3) (2018) 35.

C H A P T E R

17

Fuel cell systems for transportation electrification

Mehmet Uzunoglu[a], *Mohammad S. Alam*[b], *and Bulent Vural*[c]
[a]Tyzen Trade LLC, Fulshear, TX, United States
[b]Texas A&M University—Kingsville, Kingsville, TX, United States
[c]Medikont Electronics, Istanbul, Turkey

17.1 Introduction

Recently, energy-related concerns have been increasing in terms of many subtopics such as environmental sustainability, population growth, energy independence, energy security, and political imbalance between developed/developing countries. Among these hot topics, environmental sustainability is a major issue as the conventional fossil fuel consumption-based energy production results in significant increase in the global greenhouse gas (GHG) emissions, which in turn may cause global warming [1,2]. Many developed/developing countries have set future targets for reduction in GHG emissions for meeting the targets provided by the Kyoto Protocol, the Paris Agreement, and the 2030 Agenda for Sustainable Development. Thus, the existing drawbacks of conventional energy sources utilization have led to the development of new, cleaner energy-based technologies in different areas, such as transportation and power generation. Among these application areas, special emphasis has been given on the development of a new generation of environmentally friendly vehicular systems because of their potential to offer reduced emissions in the transportation sector as well as to reduce the dependence of many countries on imported fuel [3,4]. The automotive industry is one of the largest global industries, and there are around 1 billion vehicles in use. This sector generates a value of more than $3 trillion per year while enabling the employment of nearly 10 million people [5]. In this regard, the transportation sector plays a crucial and growing role in the world energy use, and it is responsible for approximately 15%–20% of overall GHG emissions [1,5]. The abovementioned factors generated renewed interest in the development of alternative vehicle technologies.

In particular, the efforts for technological improvements and innovation in transport and various electrically driven vehicles (EDVs) have been developed by important vehicular system manufacturers around the world supported by public authorities, nonprofit organizations, researchers, and many private companies promoting sustainable development [1,4,6]. EDVs have significant potential for widespread use in the future which may include battery-based electric vehicles (BEVs), fuel cell (FC)-powered electric vehicles (FCEVs), hydrogen FC plug-in hybrid vehicles, or plug-in hybrid electric vehicles (PHEVs) [1].

Many studies in the literature are dedicated to examining the economic and environmental impacts of different EDV technologies, which are based on case studies for different places. Brady and O'Mahony [7] evaluated the possibility of vehicular system-based emission reduction for the city of Dublin under different EDV market penetration scenarios. The results proved that EDVs could be considered as long-term alternatives to conventional vehicles, with the potential for considerable emissions reduction. The electrification of transportation has the potential to lower the primary energy requirement and CO_2 emissions by 50% per km. Browne et al. [8] focused on the role that can be played by different fuel delivery structures and new EDVs in reducing freight traffic and its environmental impacts in towns and cities. Perujo and Cuiffo [9] found that the vehicle-to-grid (V2G) interaction of EDVs could contribute to the efficiency of the energy distribution system, which lead to better environmental results. Görbe et al. [10] described the idea of utilizing EDV batteries as storage devices in a low-voltage grid composed of distributed small-scale

Handbook of Power Electronics in Autonomous and Electric Vehicles
https://doi.org/10.1016/B978-0-323-99545-0.00011-7

renewable power plants to contribute to both grid operation and environmental impact reduction. Miotti et al. [11] explored the efforts promoting the use of EDVs to replace transportation-based fossil fuel utilization and reduce GHG emissions, with a focus on reducing energy and emissions, and reported the mid-term and long-term positive impacts of EDV penetration. If hydrogen is created from renewable electricity, FCVs can reduce the life cycle GHG emissions by 50% compared to traditional (fossil) fuel vehicles, according to the results from Miotti et al. [11].

Transportation-specific policies and research and development (R&D) studies are strongly dedicated to find competitive alternative vehicular options to meet the GHG reduction targets in the long term and to enable the political effects of energy independence. Köhler et al. [12] assessed the innovation structures for R&D in low-carbon vehicles in the European Union (EU) automobile industry, which are primarily based on conventional vehicle technologies. This study also compared BEVs to both FCEVs and hydrogen FC PHEVs. The ever-increasing energy consumption causing the depletion of fossil fuels and rising public awareness for environmental protection resulted in much of the research work to focus on alternative/renewable energy sources. Thus, it becomes a requirement to replace conventional fuels with new energy sources [13]. As a consequence, novel renewable and clean energy power sources must be considered, including vehicular systems that are primarily driven by conventional fuel types. Among the possible alternatives, as a part of the change toward the hydrogen-based energy system, hydrogen FC vehicles are given specific importance and will play a major role to realize sustainable transportation (see Fig. 1), improve the quality of life, reduce health care costs, and promote technological reform in automotive industry [14–18].

An FC is an energy conversion device that converts the chemical energy of a reaction directly into electricity with by-products of water and heat. There are several types of commercially available FCs, such as polymer electrolyte or proton exchange membrane (PEM) FCs, solid oxide FCs, alkaline FCs, molten carbonate FCs, phosphoric acid FCs, and direct methanol FCs. Among the various types of FCs, the polymer electrolyte fuel cell (PEMFC) technology has found widespread use, especially in vehicular applications [14].

The PEMFC technology is still under active R&D investigation for powering systems in many applications. However, the daily operation of systems such as vehicular loads and residential units with transient load changes may not be suitable for using a sole PEMFC system. This phenomenon is mainly caused by the limitation of FCs to track fast load variations due to their slow dynamics. These fast energy demand periods frequently occurring in the operation of daily utilized systems lead to high voltage drops in a short time, which is defined as starvation phenomenon [16]. In the case of fuel starvation, the FC performance degrades, and cell voltage drops. This issue may cause a significantly insecure operation for the FC stack. Thus, to utilize an FC in dynamic applications, its current or power slope must be limited to prevent the fuel-starvation problem. Therefore, hybridizing FC system with an energy storage system (ESS) decreases system cost, while sizing improves dynamic performance of the overall vehicle system, promotes FC lifetime, provides fuel economy owing to regenerative braking energy capturing and load sharing. Because of different characteristics of multiple power sources or ESS in a PEMFC-powered hybrid system, the efficiency and the fuel economy of these hybrid units mainly depend on a proper energy management strategy (EMS) and well-designed

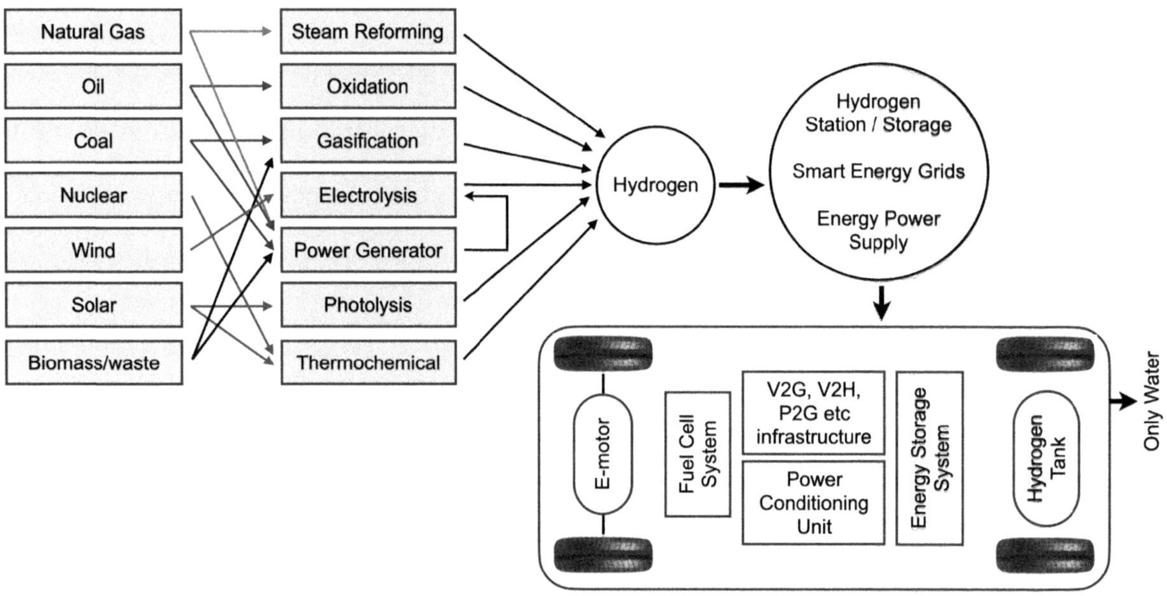

FIG. 1 Hydrogen energy chain and FCEV as a part of smart grids for sustainable transportation.

power electronics architecture [15]. However, the hybridization with ESS will cause the complexity of the vehicle and the control systems, weight increase, and extra storage cost [5].

Since FCEV is a completely new and complex scientific area with significant capital costs and administrative challenges, it needs systematic approach involving all stakeholders (central and local governments, transportation departments, energy companies, organizations, auto-makers, auto parts providers, universities, R&D institutes, etc.) [17,19]. Thus, coordination is a requirement about data collection and analysis, policy planning, infrastructure construction, safety, promotion, operation, management, maintenance, finance, etc. In addition, governments should support ambitious plans and incentivize regulations to introduce hydrogen-powered vehicles to the market. Hydrogen, which is one of the key elements of hydrogen-powered vehicles, can be produced from a variety of sources, including fossil fuels such as natural gas, oil, and coal, as well as renewable sources such as solar and wind power, biomass, waste, and nuclear energy.

Hydrogen is a secondary energy carrier (energy transporter and transmitter) that can be produced from a variety of primary energy sources, including fossil, nuclear, and renewable sources. By the evaluation of environmental and economic assessments of the hydrogen supply chain, renewable sources and conversion units such as photovoltaics (PVs), wind turbines (WTs), and FCs have been found to be promising energy sources toward building a sustainable energy economy in the near future [18,20]. These renewable energy sources, however, suffer from some deficiencies when used as stand-alone energy sources and grid-parallel sources. The power generated by WT and PV systems is highly dependent on weather conditions. In addition, it is difficult to store the power generated by a PV or WT system for future use. To alleviate these problems, WT and PV sources can be integrated with other alternative systems. Therefore, alternative energy conversion systems such as PV cells and wind turbines can be cooperated with FCs and FCEVs. Felgenhauer et al. [19] concluded that FCEVs can be used as a part of hydrogen infrastructure, and it may provide hydrogen to the natural gas supply (Power2Gas, P2G). Also, the V2G or vehicle to home (V2H) interaction of EDVs can contribute to the efficiency of the smart energy systems (as grid storage and P2G) or stand-alone applications such as emergency power sources [21]. Thus, higher efficiency can be obtained by making the best use of their features while overcoming their limitations [20]. In this case, when power is plentiful, FCEVs can be used both a storage and a flexible load resource [19]. The hydrogen produced by an electrolyzer using renewable power sources is used in the FC system and acts as an energy buffer. Thus, the effects of reduction and even the absence of available power from the renewable energy system can be easily tackled. By using the electrolyzer, hydrogen can be generated and stored for later use in FCV refueling [19].

This chapter focuses on the hydrogen-fueled FC technology, which is used in vehicle applications as an alternative energy converter (hydrogen to electric) for environmentally friendly sustainable transportation. This chapter is organized as follows: First, hydrogen infrastructure, hydrogen production methods, and hydrogen storage systems are described. Second, the main types of FCs are described in detail. Third, FC system components and their properties are explained. FC-powered automotive applications, their design and energy management architectures, current status and future trends, and safety issues are discussed in the final section.

17.2 Hydrogen infrastructure and vehicle hydrogen storage system

Hydrogen, which is the simplest and most abundant element in the universe, was discovered in the 1700s. The element is colorless, odorless, tasteless, and 14.4× lighter than air. Hydrogen does not occur free; it is found in compounds in nature. The most well-known compound is water. Hydrogen is not found naturally; it comes from fossil fuels, renewable sources, and nuclear energy by using a reformer, electrolyzer, or fuel processor.

Hydrogen is a chemical energy carrier capable of generating up to 33.3 kWh/kg of usable energy. When comparing to other energy sources, hydrogen has the highest energy content in terms of weight. For instance, 1 kg of hydrogen approximately has the energy of 2 kg natural gas or 3 kg oil. However, hydrogen has very low volumetric energy content which requires costly production, storage, and delivery of hydrogen. The capital cost of hydrogen production needs to be solved economically and environmentally friendly for a successful hydrogen economy [5]. Recent advancements in hydrogen-powered applications make hydrogen as an indispensable source of energy for the hydrogen economy.

In future energy systems, renewable energy sources will be used to generate hydrogen, and power demand might be satisfied using renewable sources and FCs since the hydrogen economy is a vision of the future. Therefore, alternate energy conversion systems such as PV cells and WTs can be cooperated with FCs. Here, to sustain the power demand and solve the energy storage problem when too much renewable electricity is being produced compared to the demand, which exists at that moment, electrical energy can be stored in the form of hydrogen. By using an electrolyzer, hydrogen can be generated and stored for future use. Since the process is providing hydrogen to the natural gas supply (Power2Gas, P2G) and combined heat power (CHP) approach, in the near future, they can be a competitive and convenient solution. It has been reported that with the growth in demand, the FC vehicle infrastructure would substantially reduce the cost [19].

17.2.1 Hydrogen production methods

Hydrogen can be produced from both renewable (hydro, wind, solar, biomass, geothermal) and nonrenewable (coal, natural gas, nuclear) resources [22]. Hydrogen is mainly produced through fossil fuel processing. In the entire world, 48% of hydrogen is produced from natural gas, 30% from oil, 18% from coal, and only 4% from renewable sources [23].

The main hydrogen production methods are steam reforming, gasification, partial oxidation, thermochemical cycling, photolysis, and electrolysis. Steam reforming may be used with natural gas, gasification with coal and waste, partial oxidation with oil, thermochemical cycling with nuclear and solar, photolysis with solar, electrolysis with wind and solar, etc. [23,24]. Producing hydrogen with a low cost means reforming it from fossil fuel such as natural gas, which may give rise to pollution by forming carbon dioxide. Manufacturing hydrogen is relatively more expensive [18,25].

Hydrogen production through electrolysis of water is more promising as it avoids carbon emission. Water can be decomposed into its elementary components by passing electric current between two electrodes separated by an electrolyte. According to Faraday's law, hydrogen production rate of an electrolyzer cell is directly proportional to the electrical current in the equivalent electrolyzer circuit.

Electrolysis process has been proposed especially for renewable systems. The power generated by these systems highly depends on weather conditions. For example, during cloudy periods and at night, a PV system would not generate any power. In addition, it is difficult to store the power generated by a PV system for future use. To overcome this problem, a PV system can be integrated with other alternate power sources and/or storage systems, such as electrolyzer, hydrogen infrastructure, and FC systems [26]. When assessing recent developments in hydrogen applications, renewable energy sources have been used to generate hydrogen. In addition, energy demand/supply management can be satisfied using renewable sources and hydrogen infrastructure by including FCEVs, smart grids, and hydrogen fueling systems. The renewable hydrogen cycle coupled with smart grids is shown in Fig. 2.

17.2.2 Hydrogen storage

The driving range of the FCEVs depends on its onboard hydrogen storage capability and other energy management systems. According to the improvements in the hydrogen storage tanks, FCEVs can meet to match driver expectations [5,23]. The use of hydrogen as a vehicle fuel requires an onboard storage that offers inherent safety and both volumetric

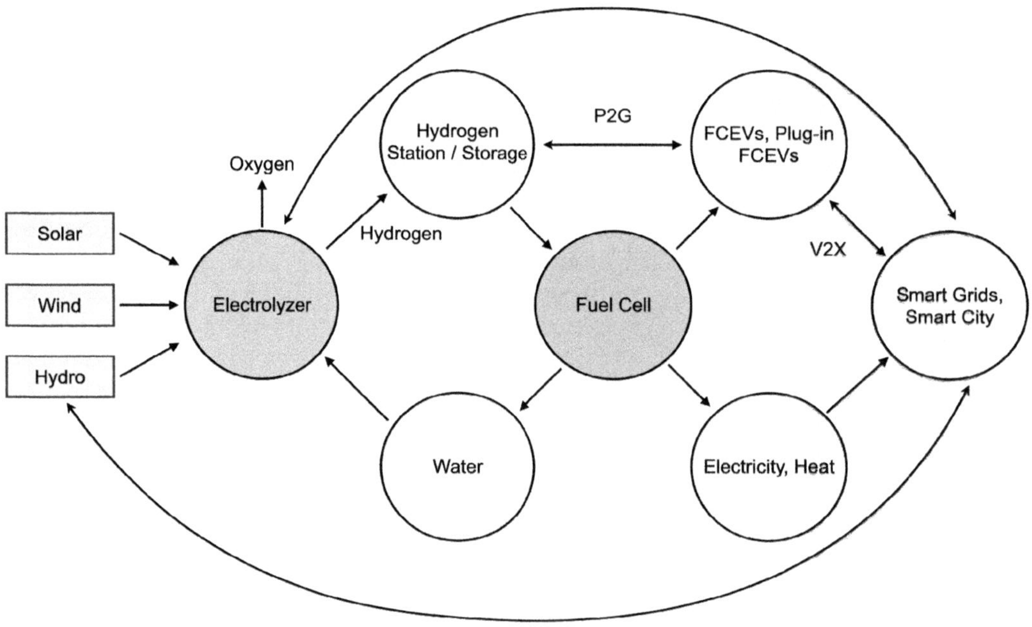

FIG. 2 Renewable hydrogen cycle coupled with smart grids.

and gravimetric efficiency. Onboard storage in vehicular systems is one of the major barriers to the acceptance of hydrogen-powered vehicles [27]. Since onboard storage systems increase safety concerns especially for drivers and passengers during dangerous situations such as car accidents, the storage system gives a hazardous impression. The problems associated with hydrogen storage change from one storage technology to another. Thus, hydrogen storage technologies are evaluated as the first step of the current deliverable presentation. There are four technologies available today to store hydrogen aboard vehicles [16,17,28]:

- Compressed hydrogen gas in high pressure tanks
- Liquified hydrogen
- Metal hydrides
- New methods currently under development

The most used technique for hydrogen storage is compression of hydrogen into different kinds of high-pressure tanks [16]. According to the Department of Energy (DoE), compressed hydrogen technique will continue to be the most popular type of hydrogen storage for vehicular systems until 2020. Most vehicles currently in the on-road testing phase and on the road commercially are equipped with such a composite high-pressure tank. This is because of its simple structure and charge-discharge easiness. In general, vehicular pressurized hydrogen storage tanks are made of very light and robust composite materials [29]. Storage density (gravimetric ratio) of above 0.05 kg of hydrogen per 1 kg of total weight, which means a stored hydrogen ratio of 5%, is easily achievable [16]. This ratio can reach up to 11.3% in this approach. There are two working pressures commonly used in commercialized vehicles such as 350 bar (35 MPa–5000 psi) and 700 bar (70 MPa–10,000 psi) [17]. Leading FCEV automakers have mounted the high-pressured storage tanks on their models with their own preferences [5,21].

In another approach, the hydrogen gas is liquified and stored under lower temperature and pressure compared to the high-pressure approach in liquefaction technique [16]. In this method, the ratio of stored hydrogen per tank weight is around 26%. However, the usage of liquid hydrogen storage system is limited because of its properties, cost of the manufacturing material, and cost of the tank. In addition, the whole process is relatively inefficient because a lot of energy is already consumed in the first stage of hydrogen liquefaction [30]. In this technique, nearly 26% of the stored hydrogen energy is consumed during the hydrogen liquefaction process [28]. In addition, liquid hydrogen tanks suffer from hydrogen leaks due to the unavoidable thermal losses that lead to pressure increase in the tank. This hydrogen self-discharge of the tank may reach to 3% daily, which means a 100% self-discharge in 1 month [31]. Hydrogen liquefaction and the use of liquid hydrogen are usually practiced only where achieving high storage density is absolutely essential, such as in aerospace applications [16,28]. Some prototype hydrogen-powered automobiles also use specially developed liquefied hydrogen tanks [16].

Metal hydrides offer alternatives to the storage of hydrogen in gaseous and liquid form. Storing hydrogen as hydride in metals provides some advantages, especially for safety in many applications. However, the tank weight provides a significant problem in this approach, especially for applications such as vehicular systems. The commercially available metal hydride hydrogen tanks offer a stored hydrogen ratio of 4.5% [27]. In addition, the repetition of hydrogen storage (absorption)/release (desorption) further decreases this ratio. Moreover, heat is released during the absorption process, which must be removed to achieve the continuity of the reaction. Similarly, during the hydrogen desorption process, heat must be supplied to the storage tank [16]. To overcome these limitations, some new techniques in metal hydride hydrogen storage approach are under investigation, which is expected to reach a hydrogen storage ratio of 10.5% in the near future [31].

New hydrogen storage methods have potential for providing higher energy densities than the conventional methods. These methods contain hydrogen storage in carbon nanotubes, storage in glass microspheres, boron-based storage, or storage in polyhydride chemical complexes [29]. Among them, the nanotube technique seems to be promising according to the recent developments in nanotechnology, which is anticipated to reach a hydrogen storage ratio of 68% [17]. As another novel hydrogen storage method, boron-based chemical storage materials (like borohydrides and ammonia borane) are promising compounds with their high hydrogen capacities. However, these hydrogen storage materials create ongoing problems from the kinetic and thermodynamic limitations in practice [32]. In addition, the current costs of these methods are extremely high that prevents the commercial utilization of these approaches [33,34].

17.2.3 Hydrogen delivery and fueling station

It is expected that the FCEVs will have a significant market share in upcoming years. To that end, an effective and efficient utilization of the currently existing hydrogen infrastructure should be reconsidered for widespread penetration of FCEVs. Considering that proof-of-concept experiment with R&D studies for hydrogen generation has been already shown, construction of hydrogen generation and distribution infrastructure is a requirement for commercialization of FCEVs to achieve a sustainable process [19,21].

Hydrogen has poor volumetric energy density, and hence, FCEVs have a limited storage capacity, which makes hydrogen generation and distribution infrastructure a must-have. In addition, to ensure acceptance by public, hydrogen-refueling stations should be constructed and mounted.

Despite the expensiveness and complexity of hydrogen storage, it has been reported that delivering hydrogen by pipeline or tanker trucks to the refueling stations is possible with some adaptive alterations to the current infrastructure [17]. It is significantly important for both developed and developing countries to realize these studies in a timely manner to prevent/reduce the possible future technological dependence on other countries about this issue. Several such studies are realized worldwide, especially in countries such as Japan, United States, Germany, China, South Korea, and Italy, where renewable sources are used with a high ratio with a vision of future hydrogen economy and hydrogen generation and distribution infrastructure has been already established or planned [17,19].

If hydrogen-fueled vehicles are used instead of natural gas delivery systems, using central hydrogen delivery system would be more convenient. Furthermore, decentralized and distributed hydrogen stations combined with electrolyzers and renewables will be built to meet the variable demands for hydrogen [5].

In this regard, the fueling and P2G operations of FCEVs should be controllable under the smart grids umbrella. In addition, hydrogen storage of FCEVs in parking condition can be utilized as a back-up gas unit when the gas demand increases or FCs of FCEVs with suitable power conditioning unit can supply power to electrical loads and appliances. Thus, FCEVs can be considered as multicarrier energy sources. Likewise, BEVs and PHEVs are also considered as both storage and a flexible load resource with a connection to smart grids (including renewables) under a suitable smart charging and discharging approach. Furthermore, smart grids have the potential of exploiting renewable sources as well as providing the required infrastructure for BEVs and FCEVs [19]. Additionally, CHP of the electrolyzer can be considered as a supplementary.

In this context, Honda proposed and developed novel and efficient complementary applications of hydrogen value chain such as FCEV, compact/smart hydrogen station (SHS), high-pressure storage tank, FC stack, electrolyzer, power conditioning units, etc. After accomplishing their proof-of-concept experiments, commercialization by Honda has already started and currently being deployed in some regions [21]. To this end, some governments and automakers are cooperating, aiming to support hydrogen infrastructure development by methods by partially subsidizing operation costs of infrastructure firms [21,35]. However, the aforementioned issues cannot be evaluated independent from smart energy technologies.

17.3 FCs and types

The fundamental principles of FC concept were described in 1838 by a Swedish scientist Christian Friedrich Schönbein. Sir William Grov, called the father of the FC, generated electricity by reversing the electrolysis of water in 1839, thus inventing the first FC system. In 1950, Francis Bacon from Cambridge University demonstrated the first 5-kW alkaline FC system. After the successful commercialization of alkaline FC technology, National Aeronautics and Space Administration (NASA) also started working on a weightless power system. In the 1970s, NASA developed a 12-kW alkaline fuel cell (AFC) system for use with the space shuttle. Furthermore, researchers started R&D studies to use different types of FCs on applications such as transportation systems. Currently, it is the most important development in the history of decentralized energy supply [5,24,36].

FC is an electrochemical device that generates electricity through a chemical reaction. It is an energy conversion device operating at high temperatures and using chemical energy stored in fuels, such as hydrogen, hydrocarbons, and alcohols without any direct combustion processes. Since the FC device is producing electricity and heat while altering hydrogen and oxygen into water, it emerged as one of the most promising power sources offering higher energy densities and energy efficiencies without any pollution [5,37,38].

FC is a converter, generating electric and heat energy with the electrochemical combination of hydrogen fuel and oxidant. This reaction emits only water. A FC system and chemical battery have a structural similarity. Unlike battery systems, the operation of FCs is restricted only by the existence of fuel and there is no charging limitation [13]. Since FCs convert the chemical energy directly into the electrical energy, FC efficiencies are not limited by theoretical Carnot efficiencies. Thus, FC has a higher conversion efficiency compared to mechanical and combustion processes [38]. If the waste heat/exhaust can be cogenerated with the purpose of heating and cooling (can be used in residential, commercial, and industrial applications), typical values of efficiency would range up to 85% [37].

The following are the main advantages of the FCs [5,37,39]:

- Potential to provide high operational efficiency and low operational maintenance
- Different fueling options
- Highly modular structure
- Offer zero or low environmental emissions and quiet operation
- No moving parts, except ancillaries (fans, pumps, etc.)
- Potential for high reliability

The following are the barriers to more widespread adoption of the FC systems [39]:

- Cost of FCs and the hydrogen infrastructure
- Lack of manufacturing and practical operating experience

FCs are generally classified depending on the type of electrolyte they use. Six main types of FCs are available in the market as listed below:

- Alkaline fuel cells (AFC)
- Molten carbonate fuel cell (MCFC)
- Polymer electrolyte fuel cell (PEMFC)
- Direct methanol fuel cell (DMFC)
- Phosphoric acid fuel cell (PAFC)
- Solid oxide fuel cell (SOFC)

The classification with some technical characteristics is shown in Table 1.

17.3.1 Alkaline fuel cell

AFC system is one of the first developed FC technology with the simplest structure. It is the first type widely used on space applications [37,41]. AFCs use a liquid solution of potassium hydroxide (KOH) as an electrolyte that conducts ions between two electrodes: anode and cathode. Because the electrolyte is alkaline, the ion conduction mechanism has differences from PEMFCs [24]. At the basic structure of AFCs as given in Fig. 3, at the anode side, hydrogen undergoes oxidation reaction that turns hydrogen into a positively charged ion and negatively charged electron. The negatively charged electrons travel through an external circuit that generates electricity and unifies with oxygen atoms at the cathode side [42]. The alkaline electrolyte only lets the composed hydroxyl (OH^-) go through the anode side. By this reaction, hydrogen ions and hydroxyl reunite, producing water. Eventually, water, heat, and the electricity will be obtained.

Some key advantages of AFC systems can be specified as follows [43]:

- Low operation temperatures (50–200°C)
- Quick start-up range (<0.1 h)

TABLE 1 Main types of fuel cells with some technical characteristics [5,25,37,40,41].

Types of FC	Operating temperature [°C]	Fuel	Charge carrier	Electrolyte	Oxidant	Start-up time (h)	Efficiency (%)
AFC	50–200	Pure H_2	OH^-	Liquid solution of KOH ($LiAlO_2$)	O_2 in air	<0.1	50–70
MCFC	650	H_2, CO, CH_4, other hydrocarbons	CO_3^{-2}	Lithium and potassium carbonate	O_2 in air	5–10	40–85 with cogeneration
PEMFC	50–100	Pure H_2	H^+	Solid polymer membrane (Nafion)	O_2 in air	<0.1	40 (reformed fuel)–60 (direct H2)
DMFC	60–200	CH_3OH	H^+	Solid polymer membrane	O_2 in air	<0.1	35–50
PAFC	160–210	Pure H_2	H^+	Phosphoric acid (H_3PO_4)	O_2 in air	1–4	35–85 with cogeneration
SOFC	800–1000	H_2, CO, CH_4, other hydrocarbons	O^-	Stabilized solid oxide electrolyte ($Y2O_3$, ZrO_2)	O_2 in air	1–5	45–60 with cogeneration

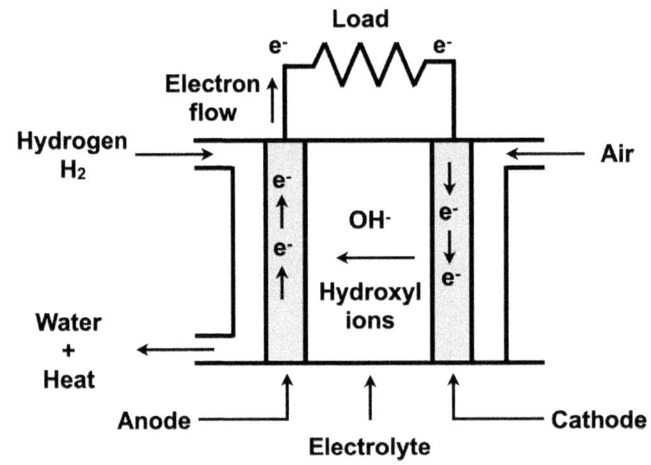

FIG. 3 Basic structure of an AFC [41].

- High efficiency (50%–70%)
- Low catalyst requirement reducing content costs
- Simple structure
- Low weight and volume

Some key disadvantages of AFC systems can be specified as follows [13,43–45]:

- Short operating life since they use a liquid solution of KOH as an electrolyte that is highly corrosive, thus erodes its parts
- Required oxygen must be provided after purification since they are very intolerant to carbon dioxide
- High cost for commercialized applications and current R&D studies are focusing on reducing costs

17.3.2 Molten carbonate fuel cell

The main principles of MCFCs are based on SOFCs and the other types of FCs. In the mid-1960s, Texas Instruments developed several MCFCs for US Army that led to a widespread use of MCFCs in the field [43].

MCFCs use lithium potassium carbonate salts as an electrolyte as shown in Fig. 4. They operate at high temperatures (650°C); therefore, catalysts can be made of low-priced metals to reduce cost and the heat/exhaust can be utilized [37]. If the heat/exhaust can be utilized (with CHP), the efficiencies of MCFCs can be increased from 60% to 85%. Additionally, higher temperature makes the FC system more tolerant to poisoning (carbon monoxide) than lower temperatures. These systems can use either carbon dioxide or oxygen as a fuel. Currently, there are many molten carbonate systems available with capacities ranging from 100kW up to 2MW.

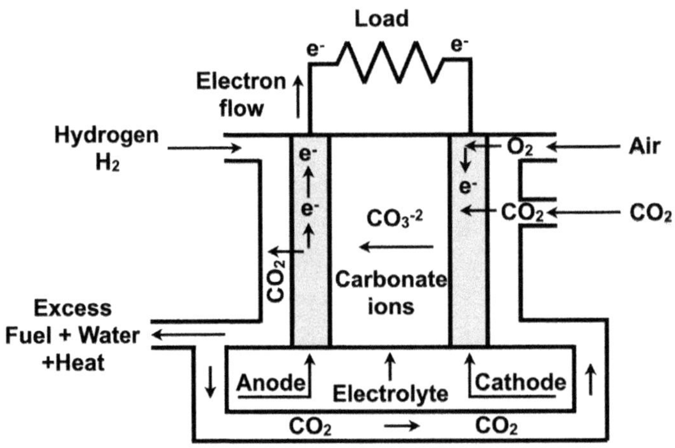

FIG. 4 Basic structure of an MCFC [41].

MCFCs do not demand an external reformer; thus, the primary fuel used may not be hydrogen gas. The basic components of natural gas, methane, and steam are processed at a reformer in the MCFC system and electricity will be produced. The internal reformer reduces the costs compared to the other FC technologies with external reformers.

Some key advantages of MCFC systems are specified below [43,46]:

- Using the fuel converted in the system
- Producing so much heat/exhaust that can be cogenerated
- High efficiencies
- Relatively fast response to load changes in steady-states conditions
- Require low-cost catalysts
- Lower sensitivity to poisoning

The following are some key disadvantages of MCFCs [46]:

- Slow start-up range (5–10 h)
- Demand redesigning with more resistant components to corrosion
- Problems in contenting the liquid electrolyte in the structure
- Requirement of preheating before starting work
- Short operating life especially when using for high power production systems

17.3.3 Polymer electrolyte fuel cell

Among the various next-generation power supplies, the polymer electrolyte or PEMFCs have been found to be one of the most promising energy sources due to their high efficiencies and environment friendly operations [47].

The basic structure of PEMFC, shown in Fig. 5, consists of three segments: the anode (negative characteristic electrode), the cathode (positive characteristic electrode), and the electrolyte membrane sandwiched between the two which allows only positively charged ions/protons to move between the two electrodes. The anode and the cathode have a catalyst, which speeds up the reactions at the electrode. Hydrogen goes to the anode side and oxygen from the air flow goes to the cathode side. At the anode side, hydrogen undergoes oxidation reaction that turns hydrogen into a positively charged ion and negatively charged electron. After the reaction, protons are being transferred from anode to cathode through the electrolyte membrane while the electrons travel through an external circuit. When the ions/protons reach the second electrode cathode, they will unify again with electrons and react with oxygen to form water. Eventually, electricity, water, and heat will be generated [39].

The following are some key advantages of PEMFCs [16]:

- Low operation temperatures (50–100°C)
- Quick start-up range (<0.1 h)
- High efficiencies (40%–60%)
- The structure of their electrolytes is more reliable, secure, and stable compared to other FCs
- High power density is the highest among all the available types of FCs
- High tolerance capability to differential pressures of the reactants
- Robust and simple structural design

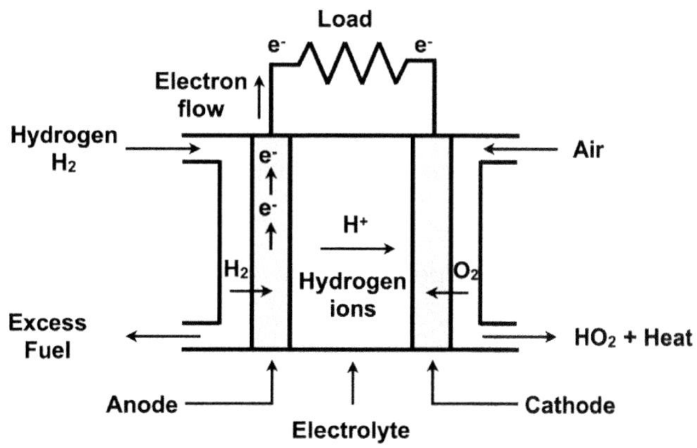

FIG. 5 Basic structure of a PEMFC system [41].

The following are some key disadvantages of PEMFCs are listed below:

- Possess a structure highly depending on the purity of hydrogen
- Very poor tolerance to carbon monoxide and sulfur particles
- Gases require to be moisturized before reacting
- Comprised of an expensive platinum catalyst and a solid polymer membrane

PEMFCs are good candidates to substitute conventional technologies with their higher efficiencies and lower environmental impacts. The cost of PEMFCs is expected to down to consumer-affordable levels in the near future [13].

Many advantages of PEMFCs, including fast response time, low operating temperature, high power density, and life cycling, have made the PEMFCs attractive and promising energy devices from the transportation and mobile applications to stationary applications as a distributed generation [5,16]. Therefore, lots of big companies such as Ballard, UTC, Nuvera, Plug Power, General Electric, Toshiba, Sanyo, etc., produce PEMFCs commercially. Lead automaker companies such as Honda, Toyota, Daimler-Chrysler, General Motors, BMW, Nissan, Ford, Hyundai, Renault, etc. and worldwide brandmark companies such as Samsung, IBM, etc. use this shining technology for their own commercial applications [48].

17.3.4 Direct methanol fuel cell

FCs either use hydrogen directly or reform hydrocarbon fuels such as methanol, which is the simplest organic fuel that can be most economically and efficiently produced from relatively plentiful fossil fuel [24]. Thus, the operation of the DMFC is based on oxidation of water mixed with methanol. After the oxidation, methanol goes to the FC anode side [37]. DMFCs use a liquid fuel methanol, which is easier to transport for public using with current infrastructure [41]. DMFCs are often used to supply power for portable electronics and emergency power supply.

Recently, concerns about the poisonous effects of methanol have emerged; thus, companies prefer using ethanol instead of methanol [37]. DMFCs have immature technology, low power density, slow power response, and low efficiency compared with direct hydrogen FCs [24].

17.3.5 Phosphoric acid fuel cell

PAFC is the first developed FC technology and first to be used commercially. They are considered as first generation of modern FC. PAFCs use liquid phosphoric acid as an electrolyte. The PAFC systems have about 40% energy production efficiency and high operating temperatures (150–220°C) [42]. With cogeneration of heat/exhaust, the efficiency can be increased up to 85% [43]. Currently, there are PAFCs with capacities ranging from 200 kW to 11 MW. Necessary improvements for widespread use of PAFCs will require reducing the operating cost while increasing the operating life.

These FC systems have a drawback of weight; therefore, these FCs are not preferable on transport applications. Some key advantages of PAFC systems can be specified as follows [43,46]:

- Tolerant till 30% carbon dioxide content, hence direct usage of atmosphere air
- High efficiency (85%) with cogeneration of heat/exhaust (CHP)
- Operate trouble-free at temperatures (150–220°C)

The following are some key disadvantages of PAFCs [42,43]:

- Slow start-up range (1–4 h)
- Low tolerant to carbon monoxide component
- Exhausting water content can damage acidic electrolyte (corrosion)
- Massive and heavy structure
- Requirement of external reformer
- Requirement of preheating before starting to operate

17.3.6 Solid oxide fuel cell

SOFCs operate at high temperatures (700–1000°C), and every single solid oxide cell produces 0.8–1 V voltages [43]. The heat/exhaust (around 500–850°C) generated by SOFCs is attractive for cogeneration. SOFC systems are currently

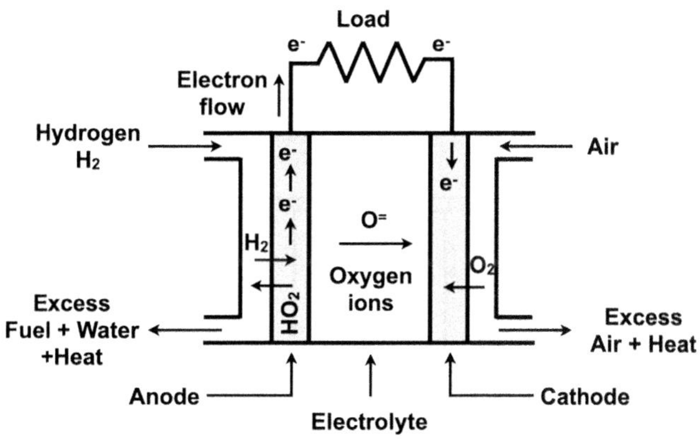

FIG. 6 Basic structure of an SOFC system [41].

used in stationary applications such as back-up power units. Furthermore, recent R&D studies are focusing on feasibility of SOFCs at transportation appliances [49,50].

SOFCs are promising with their inherent advantages for industrial applications such as its independence from pure fuel. Therefore, hydrogen as well as CO, CH_4, and other hydrocarbons can be used as fuel [37] in an SOFC. Since SOFCs operate at high temperatures, these systems can use low-cost catalyst components to enhance operational performance. SOFCs use a solid electrolyte, generally a ceramic material called yttria-stabilized zirconia (YSZ). Because the electrolyte does not involve a liquid phased reaction, a problem such as fuel flooding does not occur. Also, SOFCs can be manufactured for many configurations, although other FC types with liquid electrolyte cannot [46].

In SOFCs, fuel is supplied to the anode side and oxygen from the air flow goes to the cathode side as shown in Fig. 6. At the anode side, hydrogen undergoes oxidation reaction that turns hydrogen into a positively charged ion and negatively charged electron. The separated electrons travel through an external circuit from anode to cathode to produce electricity and oxygen atoms absorb the electrons. The solid electrolyte membrane only permits the transition of negative charged oxygen ions from cathode to anode. When oxygen ions reach anode, they react with hydrogen ions to form water.

SOFCs have about 45% efficiency which can reach 85% by capturing and utilizing the heat/exhaust. Additionally, the SOFC systems have low gas emissions [51].

Some key advantages of SOFC systems can be specified as follows [42,45,46]:

- Producing massive amount of heat/exhaust that can be utilized
- Fast chemical reactions
- High efficiency
- The solid electrolyte is more reliable than a liquid electrolyte
- Nonrequirement of external reformer
- No corrosion problems

Some disadvantages of SOFC systems include the following:

- Slow start-up range (1–5 h)
- Requirement of R&D studies on decreasing volumetric and gravimetric parameters
- Low tolerant to sulfur element
- Requirement of long-term proof-of-concept studies

17.4 FC system components

Among the various types of FCs, PEMFC has drawn the most attention due to its simplicity, viability, quick start-up range, higher power density, long cycle life, and low operational temperature (80–100°C). Since PEMFC is the best candidate for vehicular and commercial applications, in this section, the PEMFC system components are discussed in detail.

In practice, FC systems need auxiliaries (blowers, pumps, fans, ejectors, motors, turbines, compressors, valves, and regulators) to support their operation [24,52]. The main components of the FC system include the FC stack, fuel

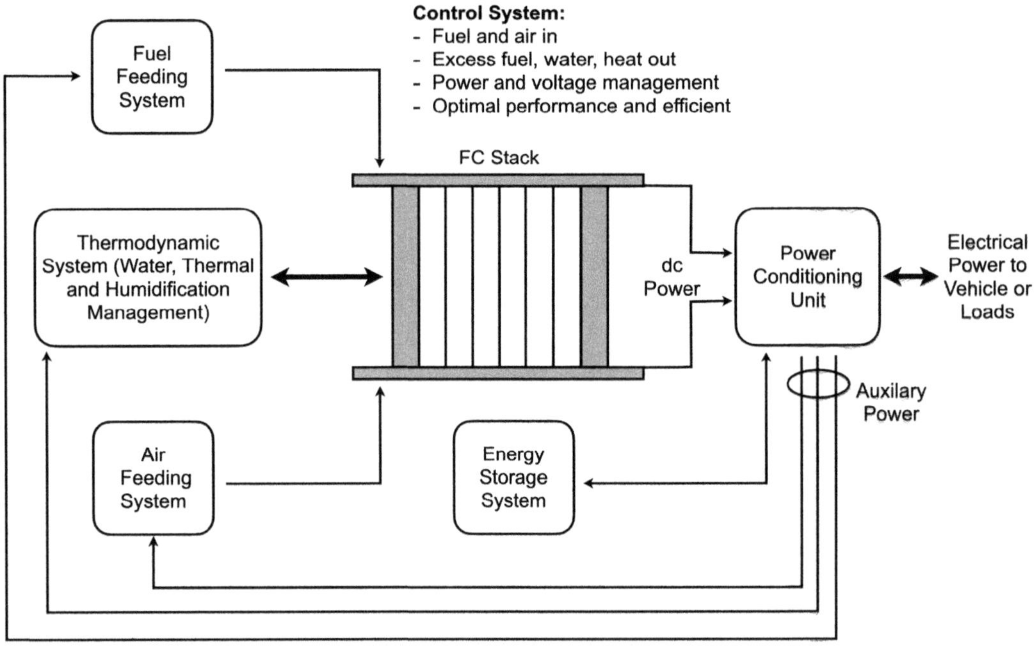

FIG. 7 FC system components.

feeding, air compressors, humidifiers, power conditioning unit with control systems and energy storage system as shown in Fig. 7 [25]. All of these components work together to provide power to an electrical load while regulating the proper amounts of fuel and ensuring system safety [52].

17.4.1 FC stack

Although the FC stack is the main component of the FC system, for the required electrical energy, all FC components must work in harmony. To that end, hydrogen fuel and air must be loaded into the stack as pressured gas. Thereafter, the excess hydrogen will be purged, and excess heat will be removed by circulating cooling water through the stacks during processing [23]. During operation, an FC produces DC electricity as well as water and heat through electrochemical process.

A FC stack is made by assembling cells to deliver the required amount of voltage and power. Since a single cell produces less than 1 V, this voltage ratio is insufficient for most of the applications [21,41]. A single FC consists of three metal plates called separators, and two membrane electrode assembly (MEA) plates, which generates electricity. Innovative MEA design can provide effective and efficient fuel flow [21]. The cell generates electricity in the form of direct current (DC) by reacting hydrogen and oxygen in the air at the MEA.

The ideal potential of a hydrogen/oxygen FC under standard state conditions (25°C and 1 at.) is 1.229 V with liquid water product. However, the actual FC potential is lower than the ideal potential value due to the irreversible voltage drops that occurs in the FC system. There are three types of irreversible voltage drops, namely, activation overvoltage, ohmic overvoltage, and concentration overvoltage. At low current densities, the activation overvoltage is responsible for the voltage drop of the FC. In addition, concentration overvoltage becomes more significant at high current densities. The effects of these voltage drops and the corresponding variance of FC voltage can be seen from the FC polarization curve shown in Fig. 8. Operating FC in this linear region is quite important for ensuring overall system efficiency.

The FC stack from combined assembling cells is the most important part of a FC power system, like an engine in the conventional vehicle. The FC stack provides the current required by governing of a fuel flow rate through the anode channels, i.e., the fuel flow rate varies according to the system power requirements [5]. The auxiliaries of FC system also consume 10%–15% of the total power output of the FC stack [24]. The amount of power produced by FC and its performance depends upon several factors, such as cell size, operating temperature, reactant gas stoichiometric flow rates, pressure of gases supplied to the cell, and anode and cathode humidification parameters [41,52].

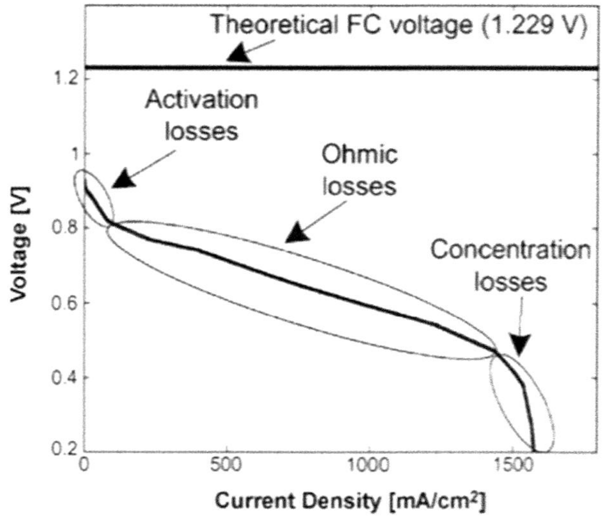

FIG. 8 Typical FC polarization curve [53].

17.4.2 Fuel and air feeding systems

Automotive applications with onboard hydrogen storage tank have control systems consisting of pressure regulators, relative humidity, fuel purity grade, and fluid-dynamic conditions [23]. As mentioned previously, FCs have some limitations to track fast load variations, such as acceleration and deceleration for vehicular application, due to their relative slow dynamics, which is defined as FC starvation and flooding phenomenon. Thus, to utilize an FC in dynamic applications, both the fuel feeding control system must be well managed and supported by a proper FC energy storage system. To improve FC performance and efficiency, design and control of the air compressor are very important to accumulate and efficiently remove water produced from the reaction [21,23,41]. However, some automakers like Hyundai use different design as near ambient air pressure to provide oxygen to the FC stack [54].

17.4.3 Thermodynamic system

In an FC system, water management, thermal management, and humidification can be increased to optimize the overall system performance and improve power generation [23]. These three components must be efficient and reliable for optimum functioning of overall system, which require proper design, control, and water balance. If these conditions cannot be achieved, problems such as membrane drying out, electrode flooding and conductivity inadequate, and excess water accumulation in the anode side would occur [23,25].

17.4.4 Power conditioning unit

Another part of the FC system is the power-conditioning unit (PCU). Because an FC produces DC electrical energy, it must be converted from unregulated low voltage level (voltage varies in a wide range depending on the load current) to regulated high voltage level using a DC/DC converter and/or inverted to alternating current (AC) using DC/AC inverter for AC electrical loads/drive-trains. The power-conditioning unit in such a vehicular system is a power electronic converter and is mainly composed of voltage regulator topologies and motor drive converter topologies. The voltage regulator converters are basically the DC/DC power converters. Nonisolated and isolated DC/DC converters are the two main categories. Between the input and the output, a nonisolated converter shares a common ground. The input and output terminals are electrically isolated in the isolated converter. Nonisolated and isolated converters are divided into two subcategories based on the direction of power flow through the converter: unidirectional and bidirectional converters. The DC/DC converter uses a transformer and coupled inductors to provide isolation. Which raises the conversion ratio of the converter, but also increases the cost, size, and losses. As a result, for minimizing the size and also cost of the transformer and linked inductor, high frequency is the better option. In unidirectional converters, power only flowsin one direction: from input to output.

The DC/DC converters used in FCEV topologies are the essential power-train components as they regulate FC voltage and allow load sharing between main energy source and energy storage systems (ESSs). For example, the FC power stage is assigned for a power tracking process to supply the determined power value from the FC system. In addition, ESS power stage can be devoted to DC link voltage control. Both power tracking and DC link voltage control processes can be implemented using a controller. Thus, this hybrid system topology provides the capability of delivering the desired power value from the FC system while keeping the bus voltage at desired intervals [55]. The driving motor can either be AC or DC motors such as induction motors, permanent magnet brushless motors, and switched reluctance motors [56]. The proper response of the FC system to the overall load dynamics can be achieved by generating appropriate power switching signals to the converters [47]. While these converters design and operate, they must be preferred with high voltage ratio, high efficiency, low electromagnetic interference (EMI), and high compactness the while keeping low cost [37].

17.4.5 Energy storage systems

A stand-alone FC system integrated into an automotive powertrain is not always sufficient to satisfy the load demands of a vehicle. Although FC systems exhibit good power capability during steady-state operation, the response of FCs during transient and instantaneous peak power demands is relatively poor. Therefore, when utilizing a PEMFC, it is recommended to associate it with an auxiliary power source to improve the dynamic performance of the whole system. There are many types of auxiliary units for compensating the slow response characteristics of the PEMFC unit and increasing the efficiency of the overall system. Energy storage units such as batteries, ultracapacitors (UCs), and flywheels can be classified in the auxiliary power units' group for use in hybrid systems including PEMFC. Specifically, energy storage units not only compensate the slow response phenomenon of PEMFC system, it also provides the advantage of recovering the reusable kinetic energy that occurs in vehicular applications during braking periods. Thus, the consumption of hydrogen for FC units can be significantly decreased.

In many applications, batteries are utilized as ESS. Among the various existing rechargeable batteries, lithium-ion batteries appear to occupy a prime position in various aspects. A battery is needed in an FC system to supply power during the time after a change in the flow of reactants occurs. An auxiliary power source such as a battery bank can be used to improve the FC dynamics. In addition, the use of batteries reduces the size and cost of the FC, increases the lifetime, and avoids overloading. The battery bank supplies auxiliary power during cold start of the FC and supports peak demand periods or transient events. They also compensate the tracking delays of FC system. The state of charge (SoC) is an indicator showing the emptiness or fullness of the battery. The battery current can either be negative or positive, with respect to the mode of operation, e.g., charging or discharging [57].

When the load power exceeds the maximum output of FC, battery starts to discharge through a converter located in the PCU and keeps discharging until the overloading ends. Also, if a step or ramp increase in the load demand occurs (even there is no overloading), battery bank helps the FC system until the FC power increases and matches the demanded power. The FC system has a relatively slow response in both step-up and step-down load conditions. When a step-down loading occurs, the output power of the FC slowly matches the reference power. In these conditions, the battery bank captures the differing power to successfully meet the load demands. These recaptured energies sometimes might not be sufficient to keep the battery SoC above a predefined level. Thus, another control loop is used to directly charge the battery bank from FC power plant with respect to the final SoC. The charge and discharge conditions of the battery are controlled with the aid of a DC/DC converter [57]. In the battery discharge current control, the key point is how good the FC power matches the active power requirement of the load. Thus, the positive difference between the FC reference power and the actual FC output power is processed to force the DC/DC converter to draw the reference current from the battery. The discharge current is limited using a saturation system to avoid battery overloading.

The battery can be charged when the FC reference power is above the power demand due to its slow response. While this can be evaluated as a naturally charging condition, another charging control loop is also employed. If all these conditions conform the respective criteria, battery can be charged. In the battery charge current control, the key point is instantaneous SoC of the battery when compared to its nominal SoC. The battery charge current should be inversely proportional to the depth of discharge [57]. Also, the battery has some effect on meeting the reactive power demand. We can express the reactive power as the oscillating power between source and load buses. In systems including DC/AC conversion such as FCs, we observe that the reactive power oscillates between the reversible device (battery or UC) of the DC bus and the AC load bus. The reactive power passes through the antiparallel diodes of the inverter circuit to reach the rechargeable device, and then passes to the load side through the next alternant. However, due to the reactive power losses on the contactors and the equivalent system impedance, there is always a DC power to be converted to the reactive power, which results in hydrogen consumption, stack DC voltage drop, as well as the load bus voltage drop [57].

UCs are also ESS devices that offer significantly better energy densities than conventional capacitors, better power densities than conventional batteries, and can be constructed in modular and/or stackable format. The charge and discharge time of a UC varies from fractions of a second to several minutes, while providing maintenance-free operation. UCs provide the lowest cost per Farad, extremely high cycling capability, and prove to be environmentally safe [47]. The capacitance of UCs may vary from a few Farads to several thousand Farads per cell. Because of the aforementioned unique characteristics, UCs are utilized for a wide range of applications. Therefore, a UC bank can effectively serve as a cost-effective alternative to batteries, especially during short peak demand periods.

The natural structure of UCs is appropriate to meet transient and instantaneous peak power demands. The UC bank is designed to provide the difference between the load demand and the output power generated by the FC system. Without the UC bank, the FC system must supply all power demand thus increasing the size and cost of the FC system.

The amount of energy drawn from the UC bank is directly proportional to the capacitance and the change in terminal voltage. The effective specific energy for a prescribed load can be supplied by various UC bank configurations. In practical applications, the required amount of terminal voltage and energy or the capacitance of UC storage system can be built using multiple UCs in series and parallel. The terminal voltage determines the number of capacitors, which must be connected in series to form a bank, and the total capacitance determines the number of capacitors, which must be connected in parallel in the bank. The natural structure of UC is appropriate to meet transient and instantaneous peak power demands.

As a result, while choosing an ESS, the following characteristics should be considered [5,57]: specific energy and energy density, electrical power, i.e., the electrical load required, volume and mass, storage capacity, reliability, durability, safety, cost, recyclability, environmental impact, response time, efficiency, self-discharge rate/charging cycles and excess discharge, sensitivity to temperature, charge-discharge rate life time, environmental effects, investing/operating cost, maintenance, availability, design complexity/voltage range, life cycle, and recharge rate of storage element.

17.5 FCs in automotive applications

FCs offer many advantages over internal combustion engines (ICEs) for vehicular applications because they are energy efficient, clean, and fuel flexible. Today, FCs have become smaller, more powerful, and more durable [54]. Since FCEVs provide much better air quality, health problems, such as incidence of asthma and other chest complaints will decrease [25]. Hydrogen FC systems have the potential to reach 60% peak efficiency [15]. Onboard the vehicle, conversion of hydrogen produces electricity for power train of vehicle and water. Hydrogen can be obtained from various sources with different methods [48]. Leading automobile industries are making every effort to make hydrogen-powered vehicles affordable, environmentally friendly, and safe [25,41].

Automobiles, buses, scooters, golf cars, utility vehicles (such as forklifts and airport vehicles), locomotives, tramways, boats, airplanes, and underwater vehicles can be considered as some of the applications of PEMFC systems in transportation area [15]. In automobile systems, almost all major automakers have demonstrated prototypes and commercially available FC vehicles and announced future plans for production and commercialization in the near future. On the other hand, buses seem to be the most likely type of road vehicles for an early market introduction of the FC technology. Also, scooters may be a significant market for FC technologies, especially in developing countries. To conclude, the transportation area seems to be the most promising sector of PEMFC technology.

The need to reduce emissions through the improvement of vehicle efficiency and the depleting fossil fuels has motivated researchers to investigate alternatives to conventional ICE-based vehicular systems. Among the alternatives, FC-powered electric vehicles using hydrogen as fuel is considered as the ultimate route to achieving sustainable long-term alternative propulsion systems [16]. Hydrogen is an energy carrier with many important unique features. It is considered as the lightest, most efficient, and cleanest fuel. One of its unique features is that it can be converted to electricity in FCs through electrochemical processes with higher efficiencies than conversion of fossil fuels to mechanical energy in ICEs. Thus, this particular characteristic allows higher efficiency and practically null emission of polluting agents for FC-powered vehicles compared to ICE-based vehicular systems. Particularly, these efficiency and environmental-friendliness advantages become more significant when renewable energy sources such as wind, solar, etc. are used for hydrogen production. In particular, passenger vehicles are mostly operated at part loads significantly below their rated power, thus vehicle running in urban or suburban areas demands a small fraction of the rated FC power most of the time. Thus, the efficiency of the overall system becomes higher as the FC is operated most efficiently. This issue is exhibited by Villatico and Zuccari [58] by comparing the performance of FC and ICE in real driving cycles. Moreover, FC-based vehicles have medium- and long-term potential to be the main-stream vehicle in the future because they have almost zero emission and compatible driving range to ICE-based vehicles [1,5,55]. For real utilization of EVs, many features should be considered, such as performance, cold-start, reliability, durability, cost, fueling/charging times, availability, and infrastructure [13,18]. Both FC and battery EVs have similar components, such as driving motors and power conditioners; also, they will coexist in the future. The FCEV is a promising solution because of faster refueling, cold-starting, faster warming up to operating temperature, and longer driving range features compared to BEV [24,25,59]. Generally, the FCEVs are proper for mid-size and long-range vehicles, while the BEV is suitable for small-size and short-range vehicles [5].

Among the various available FC types, PEMFC is the most likely candidate for automotive applications, thanks to its several attractive features, such as low operating temperatures, relatively low cost and quick start-up, simplicity, viability, packaging under the hood of vehicle, and high efficiency [55]. However, the unsteady operation of a vehicle may not be appropriate for the usage of a sole FC system. Moreover, a FC system (in the power train of an automotive) needs to be equipped with the combination of ESS, driving motor, converters, and various auxiliary devices [23]. The power demand of a vehicle motor undergoes significant variations due to acceleration and changes in road surface and traffic conditions. Thus, a sole FC system may not totally satisfy the power demand of a vehicle due to the limitation of FCs to track fast load variations because of their slow response dynamics. In addition, load demand fluctuations in vehicle operation may cause fuel starvation, flooding, membrane drying, and pressure imbalance across the FC membrane, which may damage the FC stack and decreases its lifetime [60]. Moreover, an advanced EMS is necessary to split the power demand of a vehicle in a suitable way for the onboard power sources to maximize the performance while promoting the fuel economy and endurance of hybrid system components.

17.5.1 Design architectures for PEMFC-powered hybridizations

In general, FCEVs are configured as series vehicles [59]. However, the architecture of a FCEV can be configured in series, series-parallel, and parallel. The selection of the hybrid power-train topology is very important for an efficient operation and better market penetration of FCEV systems. Effective and efficient power conversion technologies are strategic for the progress of FCEVs [61]. Interfacing the load/power-train unit requirements with the different operational modes of onboard generation and storage systems call for suitable power electronic converter configuration and control. The selection of power electronic converter for hybrid systems including PEMFC is based on some significant factors such as lower cost, space requirement, compact structure, higher efficiency, as well as ripple free and reliable operation [15,61].

Despite the fact that hybridization concept brings remarkable benefits, the main challenge is to find the appropriate combination of energy sources. Each energy source has unique characteristics and energy conversion efficiency. Thus, each combination is expected to show different performance, efficiency, and fuel/energy consumption. Moreover, in a hybrid topology, these energy sources can be connected to DC bus either directly or via a power converter. In case of utilizing a power converter, which allows higher flexibility in load sharing, an improved hydrogen economy may be expected. However, system efficiency may decrease due to additional conversion stage, which might worsen the equivalent fuel economy.

Considering the aforementioned factors as well as the load/drive-train requirements, various topologies may be used for the operation of PEMFC hybrid systems. One of the simplest architectures for the parallel operation of PEMFC hybrid systems includes the direct integration of the PEMFC system with the auxiliary power unit as shown in Fig. 9. During low power demand, the FC system generates up to its load limit, and the excess energy between the load demand and the FC output power is used to charge the auxiliary unit. In this period, the S1 switch is closed and S2 switch is open. During high power demand periods, both the FC system and the auxiliary units supply the load demand. During this period, S1 switch is kept open and S2 switch is closed. The direct integration topology is attractive, because it does not require a high power DC/DC converter, thus, the complexity, cost, weight, and volume of the system are significantly reduced [15]. For example, the UC system integration can be achieved with or without power electronic converter via series and parallel connection. Because of the high specific power and power density of UC bank, it may be possible to eliminate the DC/DC converter for voltage regulation so that it can deliver higher output. This topology is proposed by Honda [62] for a combined PEMFC/UC powered vehicular system. Also, the direct integration of the FC with the UC bank is possible for relatively low voltage applications with a DC bus of less than 50 V [63]. The UC bank and the load terminal voltages depend on the FC system terminal voltage, which prevents the power capability of the UC bank from being fully utilized. Therefore, the UC bank size is restricted by the terminal voltage of the FC system, as in the battery/UC hybrid system of Reference [64].

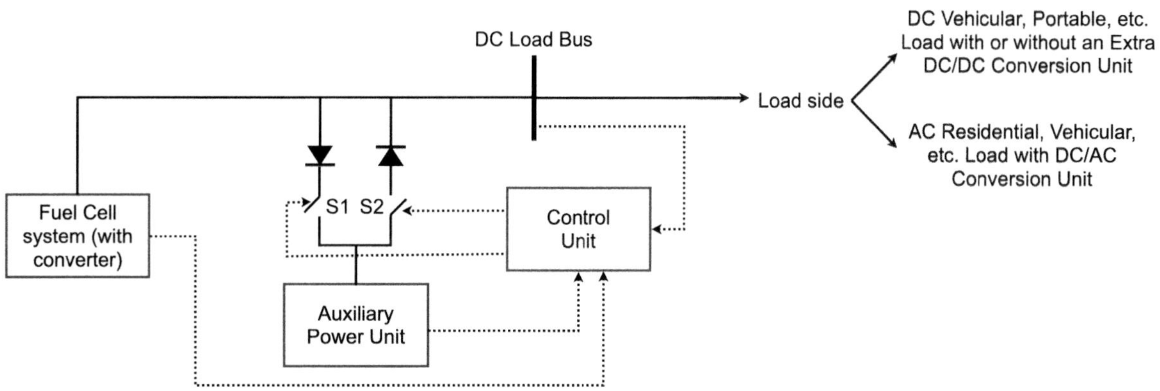

FIG. 9 Direct integration topology [15].

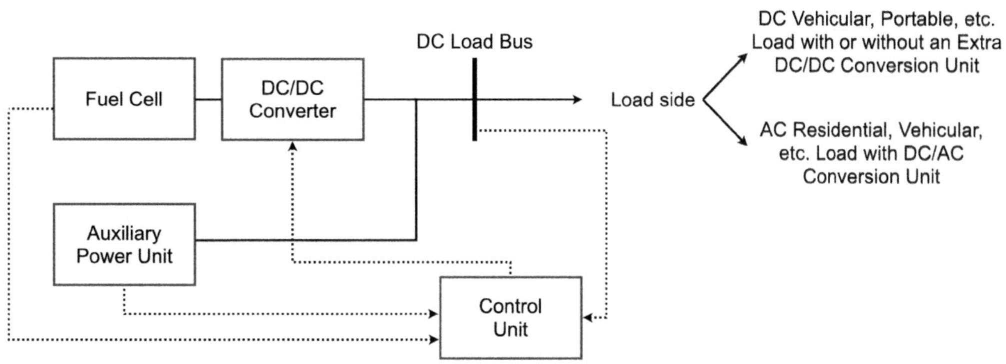

FIG. 10 Single converter-based topology [15].

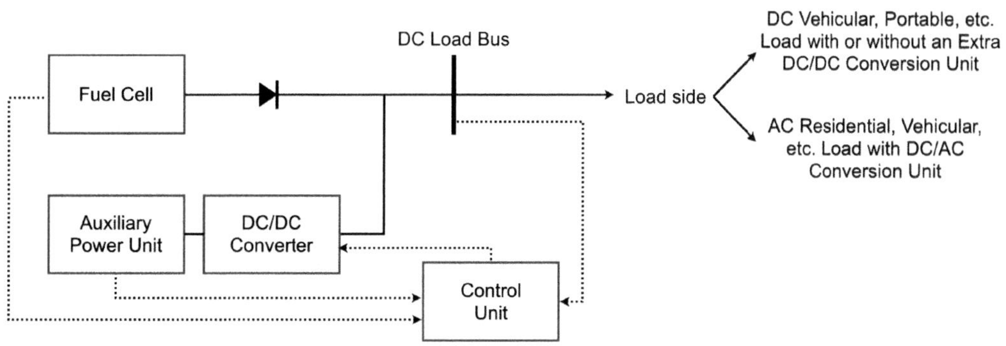

FIG. 11 Single converter-based power train topology [15].

Different designs of a topology with only one DC/DC converter unit are shown in Figs. 10 and 11. The topology in Fig. 10 regulates the output power of the FC system while the rest of the energy between the load demand and the FC output power is naturally supplied by the auxiliary unit. In addition, the topology in Fig. 11 is based on the regulation of the auxiliary system output power or the regulation of the DC bus voltage. There are many studies in the literature using these two topologies. Some previous studies on these topologies are reviewed and summarized in Erdinc and Uzunoglu [15].

The most common topology for hybrid systems is composed of multiple DC/DC converters as shown in Fig. 12. In this kind of topology, the DC/DC converter of one of the available power sources is employed for the DC bus voltage regulation, which is called voltage-oriented-control, and the rest of the converters are controlled for power tracking by power-oriented-control methodology. Thus, the proposed power-conditioning unit provides the capability of delivering the desired power value from the hybrid power sources while keeping the bus voltage at a desired level. Additional architectures for the usage of multiple converter topology for transportation area are available in [53,65–69].

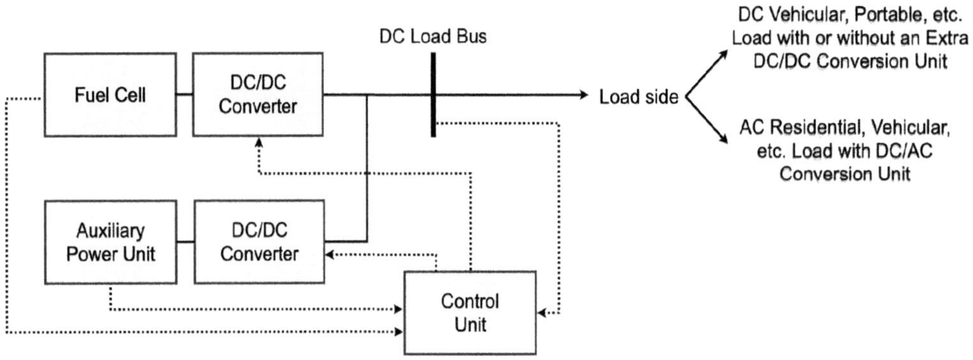

FIG. 12 Multiple converter-based power train topology [15].

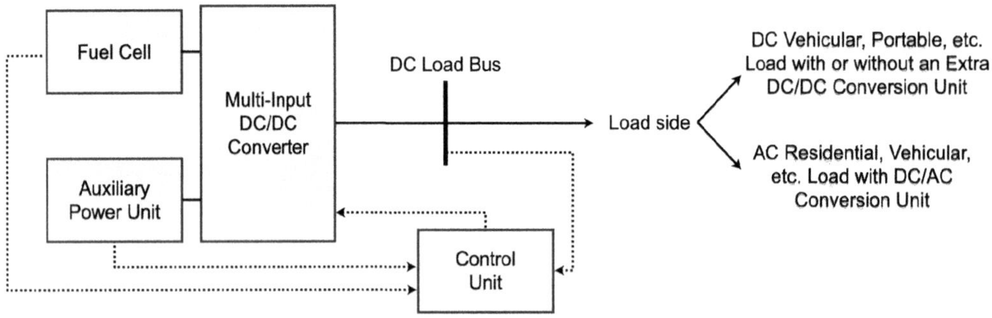

FIG. 13 Multi-input converter-based topology [15].

Combining the features of multiple converter-based topology, some researchers focused on utilizing multiple-input single-output design architecture as depicted in Fig. 13. This topology provides the advantage of reduced number of elements (multiple inputs of power sources can be connected to a DC bus via only one output capacitor) used in power electronics interface design. For example, Ferreira et al. [70], Perez et al. [71], and Napoli et al. [72] utilized the multi-input converter topology for an electrical vehicular system consisting of PEMFC, battery, and UC.

The design of the aforementioned single or multiinput converters can be realized as insulated or uninsulated, unidirectional or bidirectional according to the area of application. The basic uninsulated unidirectional and bidirectional DC/DC converter topologies are shown in Fig. 14. The proposed multiinput and bidirectional converter-based PCU offers the following advantages:

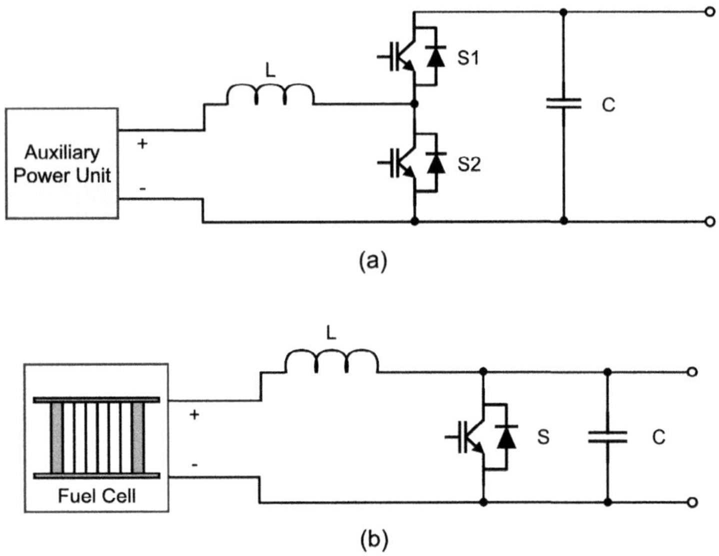

FIG. 14 Structures of the basic DC/DC converter topologies: (A) Bidirectional converter topology and (B) boost converter topology [15].

- Voltage levels of the input power sources can be different.
- The input sources can deliver the requested power individually and simultaneously.
- Electrically isolation prevents damage to input sources.
- The proposed PCU can work bidirectionally, i.e., the power can flow in either direction between individual input and output.

However, there are different advanced, more advantageous but complex topologies developed by power electronics researchers [15].

17.5.2 Vehicular power control strategy and energy management

Due to the different characteristics of multiple power sources, the efficiency and the fuel economy of hybrid systems mainly depend on proper EMS, which is very important for PEMFC-powered hybrid systems.

An FCEV energy control system must provide good acceleration and high recovery of braking energy. The power sharing between the FC system and energy storage system (battery or UC bank) is determined according to the load sharing and control algorithm. Main requirements of the control algorithm are satisfaction of the power demand and management of the power flow in accordance with efficient operation of the different power sources. Coupled with this, the control system should utilize the FC system and energy storage system to match the vehicle load profile by considering different features of power sources.

Since stable operation of the FC system is vital for efficiency, lifetime, and cost, the FC system should deliver the base load power without responding to peak power demand. Another purpose of the control system is to adopt the battery or UC bank in stabilizing the DC link voltage. This enables power transfer to/from the battery or UC bank and improves the driving performance.

Hybrid systems can provide improved efficiency when a well-designed energy splitting algorithm that considers the individual characteristics of each hybrid system is applied. There are many techniques for energy management that have been investigated in the literature, such as intelligence-based strategies, optimization-based strategies, and frequency decoupling strategies [73].

17.5.2.1 Intelligence-based energy management

Recently, many intelligence-based energy management strategies such as fuzzy logic and neural networks for use in PEMFC-powered systems have been reported in the literature [15]. Among them, fuzzy logic controller (FLC)-based methodology plays a dominant role due to its independence of a full mathematical plant model and training procedure. Nonlinearity and difficulties in proper identification of parameters of the plant mathematical models limit the use of model-based conventional control approaches.

Fuzzy logic provides a suitable structure compared to conventional control methods, especially for the systems comprising nonlinear behaviors where an overall mathematical model is difficult to obtain. Moreover, there is no need for historical data, which is an important advantage over other types of intelligent controllers such as neural networks.

In the literature, many researchers applied fuzzy logic approach for hybrid systems with PEMFC for different applications. Among them, Gao et al. [74] provided the power distribution in a hydrogen powered hybrid bus composed of PEMFC as the main power source and battery/UC combination as the auxiliary unit. In Erdinc and Uzunoglu [15], FLC utilized the SoC values of auxiliary units and the required power demand of the bus as inputs for determining the output power values of FC and UC systems. Erdinc et al. [53,66] regulated the energy management in PEMFC/UC and PEMFC/Battery/UC-powered hybrid vehicular systems using fuzzy logic. The aforementioned architectures [53,66] utilized the SoC values of battery and/or UC units and a transient-free form of the vehicular power demand, provided by a load sharing algorithm, based on standard drive cycles as inputs and gave the requested power value from the PEMFC system as output. Li et al. [75] proposed an FLC design for a PEMFC/Battery hybrid system with a similar structure to Erdinc et al. [66]; however, they optimized the parameters of their FLC for an optimal energy control. As a contribution to the studies of Erdinc et al. [53,66] and Li et al. [75], Eren et al. [76] inserted the control of internal dynamics of PEMFC system in the fuzzy logic-based energy management approach to provide a safe and efficient operation for PEMFC. Kisacikoglu et al. [77] directly controlled the duty cycles of the converters of PEMFC and UC as the outputs of the FLC in a PEMFC/UC hybrid vehicular system.

Another intelligent method used for the control of hybrid system is the neural network approach. Comparing neural network with one of the most popular intelligence-based technique, fuzzy logic, this quick response capability provides a significant advantage. Fuzzy logic approach does not need a training procedure; however, the rule-based fuzzy structure makes it longer to respond compared to neural network approach. Thus, neural network control may be a competitive approach with fuzzy logic for some applications. In addition, the aforementioned features would lead the neural networks to solve complex problems precisely and flexibly and provide a suitable basis for the control of complex systems. Consequently, neural network has demonstrated remarkable success in control area [15].

17.5.2.2 Optimization-based strategies for energy management

Power management and control system is very important for improving energy efficiency, vehicle performance, and for reducing the weight, size, and cost. Various flexible control methods and intelligent techniques are used for optimization, subject to system constraints. The power management system not only optimizes the available sources, but also manages the load dynamically, based on the need and priority of the demand. In these types of energy management approaches, the optimal reference power signals for onboard power sources are calculated by the minimization of a cost function. This cost function generally represents the fuel consumption or emissions, especially for vehicular applications. If this optimization is performed over a fixed driving cycle, a global optimum solution can be found. Many approaches such as optimal control theory, linear programming, dynamic or stochastic programming, genetic algorithm, simulated annealing, linear and nonlinear predictive model, and game theory are utilized for solving the aforementioned global optimization problem.

There is a significant number of studies dealing with the energy management of an FC-powered hybrid system utilizing an optimization-based methodology [78,79]. For example, Rodatz et al. [78] presented an experimentally applied equivalent fuel consumption minimization strategy aiming at minimizing the hydrogen consumption while maintaining drivability in a PEMFC/UC hybrid vehicular system. Xu et al. [79] performed a similar optimization approach for an FC/Battery hybrid bus. Similar fuel consumption minimization-based approaches were also utilized by Paladini et al. [80] for FC/Battery/UC hybridization for a vehicle in a simulated environment.

17.5.2.3 Frequency decoupling-based energy management strategies

During the daily operation of a system, load transitions may occur frequently. The load power profile is a nonstationary signal and consists of transients. The wavelet transform can be effectively used for analyzing power transients in FCEV power demand profile, because of its ability to deal with nonstationary and/or transient signals. A given signal can be decomposed into transients and desired characteristics can be extracted simultaneously in both the time and frequency domain by using the wavelet transform (WT). The WT decomposes into various components at different time intervals and frequency bands as a windowing technique. In wavelet analysis, long windows are used at low frequencies, and short windows are also used at high frequencies.

One of the most popular mother wavelets is Haar wavelet for detecting and localizing of transients. The Haar basis has the shortest filter length in the time domain in comparison to other wavelet bases. Because of the aforementioned features, the Haar wavelet transform is the most suitable functions for filtering HEV power demand load profile, which consists of transients corresponding to sharp peak power demand. The transients of the HEV power demand signal are clearly observed from the short-term frequency and component magnitude changes.

For some applications such as vehicular systems, these transient changes in load demand may occur significantly, which pose dynamic stress onto the FC membrane due to pressure oscillations and possible fuel starvation, which may reduce the lifetime of the FC system. Thus, to ensure FC lifetime prolongation by preventing the FC system from load changes with high frequencies, researchers applied different frequency decoupling techniques. The basic approach for this issue is inserting a transfer function for a delay with a suitable time constant (first-order low-pass filter). Liu et al. [81] proposed such a first-order filter for the PEMFC system. In addition, some conventional filtering techniques such as Butterworth high and low frequency filters, etc. have been utilized. These conventional filters decouple the main signal with a predefined cut-off frequency; thus, the FC system can be prevented from transient variations. However, the loss of important edge information is a significant phenomenon in conventional filtering techniques. Considering this issue, some researchers applied a wavelet transform-based frequency decoupling strategy for PEMFC-powered systems. Wavelet transform is a signal processing method that is well known for its capability to treat transient signals. The advantage of wavelet analysis, as opposed to conventional techniques, is that wavelet transform decomposes a signal into a series of short duration waves or local basis functions (wavelets) on the time axis, which allows the analysis of local phenomena in signals consisting of many transients. In addition, wavelet transform provides a quadratic mirror filtering application, and the loss of important edge information is minimized compared to conventional filtering techniques. Considering these features of wavelet transform, Uzunoglu and Alam [82] applied a wavelet-based EMS for a PEMFC/UC hybrid vehicular system. A similar wavelet-based approach was utilized by Zhang et al. [65] for a PEMFC/Battery/UC hybrid system.

With the aforementioned types of frequency decoupling strategies, the transients in the total power demand profile are captured, a safe operating condition for FC-based system can be provided and lifetime of the power sources can be extended [15].

17.5.2.4 Other methods for use in energy management

Existing methods in the literature in addition to the before mentioned techniques include basic linear proportional-integral (PI)-based approaches, complex adaptive control, robust control, and flatness-based control. A conventional linear PI controller is a basic approach that may be useful for use in real-time implementations of PEMFC systems. Many researchers utilized PI controller, especially with a main supervisory controller design. As an example, Thounthong et al. [83,84] utilized a multiobjective PI-based linear technique for SoC sustaining and DC link voltage stabilization. In addition, Payman et al. [85] applied PI controller for the power electronics converter duty cycle determination to regulate the desired values of the system parameters determined by a nonlinear flatness-based supervisory control approach.

Another common approach, adaptive control technique, involves modifying the control law used by a controller to cope with the fact that the parameters of the system being controlled are slowly time-varying or uncertain. For example, as an aircraft flies, its mass will slowly decrease as a result of fuel consumption; we need a control law that adapts itself to such changing conditions. Adaptive control does not need a priori information about the bounds on these uncertain or time-varying parameters. In addition, adaptive control is precisely concerned with control law changes. There are mainly two types of adaptive controllers: feedforward and feedback adaptive controllers. There are many types of feedback adaptive controllers such as model reference adaptive controllers (MRACs) and model identification adaptive controllers (MIACs) that are reported in the literature [15].

Robust control methods, which explicitly deal with uncertainty, are designed to function properly so long as uncertain parameters or disturbances are within a typical limit. Robust methods aim to achieve robust performance and/or stability in the presence of bounded modeling errors. In contrast with an adaptive control policy, a robust control policy is static; rather than adapting to measurements of variations, the controller is designed to work assuming that certain variables will be unknown [15].

17.5.3 Current status and future trends of FC-based vehicles

Leading automakers have started to make their light-duty hydrogen FCEVs for testing and commercial applications. Available FCEVs are on the road in selected regions like California, Canada, Europe, Iceland, Japan, and Singapore where access to hydrogen stations are available. FCEVs have come a long way; they now have quick refueling time of less than 10 min and driving range of around 300–600 miles [21,54,86–89].

As of June 2018, more than 6500 FCVs had been delivered to consumers all around the world. With approximately 6500 FCVs shipped to consumers, California was the top market for FCVs. Due to its large network of hydrogen refueling stations, California was the leading market for FCEVs, with nearly 3000 vehicles delivered [89]. The California Fuel Cell Partnership (CaFCP) plays a vital role in preparing to demonstrate and commercialize the vehicle technology of the 21st century. This organization is a unique collaborative of automakers, energy and FC technology companies, and government agencies. Currently, CaFCP is focused on the marketization and industrialization and while emphasizing the importance for market introduction of FCEVs [90]. Linde AG of Germany and Air Liquide of France have been collaborating throughout Europe to develop new products. Linde AG of Germany and Air Liquide of France have teamed up in Europe to raise Germany's station count to 400 by 2023 [89].

Although automakers have produced FC-based commercial vehicles, actually, they started to develop FCEV prototypes many years ago. For example, Toyota has developed various types of FCEVs, after it began its FC efforts in 1992. Ballard developed and demonstrated a hydrogen-fueled light-duty transit bus in 1993, followed by a 200-kW FC-powered heavy-duty transit bus in 1995. GM demonstrated an FC/battery hybrid FCEV in 1998. Nearly 200 B-Class FC-based Mercedes-Benz vehicles have been with customers in the USA and Europe since 2010 [88]. AC Transit has the most comprehensive program to obtain real-world driving impressions and experiences and realize the earlier phase of the hydrogen FCEVs in North America [17,90].

Some governments (e.g., United States, Japan) and automakers (e.g., Honda, Toyota, Daimler, Nissan, Ford) are working together to support hydrogen infrastructure, public awareness, and decrease operational costs of infrastructure firms. By cost reduction and mass production of competitive parts such as the FC stack, hydrogen, fueling stations, compact packaging, storage (battery and UCs), silicon carbide-based innovative power electronics, and electric motor technologies, these vehicles can be commercialized and introduced to the market for widespread use [18,21,37].

It is necessary to carry out further investigation on the performance, reliability, durability, and cost reduction of FC systems. FCEV design architecture plays a significant role while considering the performance, packaging, interior-space, efficiency, real-world performance, and durability issues. FCEV systems should be environmentally safe and commercially viable. Without doubt, technologies of a stable supply of highly pure hydrogen and inexpensive cell components are a prerequisite for market penetration of FC systems. In this case, since the number of renewable energy sources is increasing rapidly, new business models and marketing strategies can be developed. For example, the production of green hydrogen from renewable sources (biomass, biowaste, electrolyzers powered by PV and wind energy systems) and also other feasible and viable hydrogen production methods can be used to feed the FCEVs and to balance the electric and gas grids under the smart grid concept [5,20]. In addition, as a part of sustainable energy and transportation chain, FCEVs, and their modular infrastructures like power inverter and hydrogen stations can contribute to the efficiency of the smart energy systems (as V2G, V2H, flexible load, storage, CHP, grid balancing, energy exchange, feed-in tariffs and emerging application concepts). These kinds of EVs, like FCEVs, can also be used as an investment tool and as a major player of energy market in the future smart grids and smart cities.

The demonstration infrastructure should be adequate for the anticipated commercialization timeline [5]. Demonstration projects and proof-of-concept tests will serve as a springboard for tackling issues such as practical applications, data collection for standardization, FCEV durability, and increasing public awareness of hydrogen's utility in everyday life [5,17,21,91]. By the end of June 2021, there were more than 40,000 FCEVs on the road around the world. From 2017 to 2020, stocks expanded by an average of 70% each year, but in 2020, stock growth slowed to 40%, and new FC car registrations plummeted 15%, mirroring the broader car market downturn caused by the Covid-19 epidemic. In the first half of 2021, however, more than 8000 FCEVs are predicted to be sold, setting a new record. The IEA predicts that FC manufacturing may enable a stock of 6 million FCEVs by 2030, meeting about 40% of Net Zero Emissions by 2050 Scenario needs, based on current and stated capacity [91]. From the viewpoint of FCEVs market penetration, currently, there are limited commercial FCEVs, but in medium- and long-term periods, there will be huge numbers of FCEVs on the road depending on the aforementioned strategies of developed and developing countries. Thus, they can contribute significantly toward sustainable transportation and hydrogen chain.

17.5.4 Safety

Hydrogen FCs can revolutionize many sectors, since hydrogen is as safe as or even safer than other flammable fuels such as gasoline or natural gas. However, with the lack of professional design and management, hydrogen FCs pose several other risks which other flammable fuels do not.

Nowadays, FCEVs, hydrogen refueling stations, and other values of hydrogen chain started to operate in different regions, despite pending answers to potential questions (How safe is hydrogen? What are the potential hazards of hydrogen? Which improvements are required to avoid danger?) [5]. Hazards and associated risks for hydrogen FCs should be considered by all stakeholders such as designers, regulators, users, etc. in various situations when FCEVs are inoperable, in normal operation and in collision [5,16].

The main potential hazards can be listed as follows [5,16,21]:

- Hydrogen burns (rather than explosion)
- Higher flammability range
- Danger posed by invisible hydrogen flames (adding chemicals such as odorant for required luminosity can provide an easier detection)
- Hard detection of hydrogen leak
- Higher tendency to leak since hydrogen is the smallest molecule (professional design may prevent leaks)
- Potential of hydrogen to asphyxiate people
- Frostbite/cold burns of liquid hydrogen
- Refueling station accidents

These hazards may be prevented, or the impacts can be decreased with more R&D studies, administrative supports, efforts of all stakeholders, etc. For instance, Honda improved several designs to prevent FCEVs from collision impacts such as fastening bar as an impact resistant to prevent cells, straight frame structure to protect components, load path structure to disperse the collision load, and for efficient absorption of energy [21].

17.6 Conclusion

Almost all major companies have publicized ambitious plans to introduce hydrogen-powered vehicles in the market. Moreover, FCEVs are sold in selected regions with access to hydrogen stations. This vehicle type promises very clean operation and shows higher energy efficiency compared to conventional vehicles. When the FCEV technologies become mature, FCs can also be extended in many diverse areas from portable devices to silent heavy vehicles or aerospace applications. In addition, FCEVs and their infrastructure systems can also be used as a back-up or integrated system for renewable energy sources.

In the near future, as a part of smart grid, FCEVs and hydrogen energy chain should be considered for sustainable development.

References

[1] G. Sen, A.R. Boynuegri, M. Uzunoglu, O. Erdinc, J.P.S. Catalo, Design and application of a power unit to use plug-in electric vehicles as an uninterruptible power supply, Energies 9 (3) (2016) 171.

[2] K. Ahn, H.A. Rakha, Developing a hydrogen fuel cell vehicle (HFCV) energy consumption model for transportation applications, Energies 15 (2) (2022) 529.

[3] A. Ajanovic, R. Haas, Prospects and impediments for hydrogen and fuel cell vehicles in the transport sector, Int. J. Hydrog. Energy 46 (16) (2021) 10049–10058.

[4] B. Yagcitekin, M. Uzunoglu, A. Karakas, O. Erdinc, Assessment of electrically-driven vehicles in terms of emission impacts and energy requirements: a case study for Istanbul, Turkey, J. Clean. Prod. 96 (3910) (2015) 487–492.

[5] Y. Manoharan, S.E. Hosseini, B. Butler, H. Alzhahrani, B.T.F. Senior, T. Ashuri, J. Krohn, Hydrogen fuel cell vehicles; current status and future prospect, Appl. Sci. 9 (11) (2019) 2296.

[6] D. Margaritis, A. Anagnostopoulou, A. Tromaras, M. Boile, Electric commercial vehicles: practical perspectives and future research directions, Res. Transp. Bus. Manag. 18 (2016) 4–10.

[7] J. Brady, M. O'Mahony, The potential impacts of electric vehicles on climate change and urban air quality, Transp. Res. D 16 (2) (2011) 188–193.

[8] M. Browne, J. Allen, J. Leonardi, Evaluating the use of an urban consolidation centre and electric vehicles in central London, IATSS Res. 35 (1) (2011) 1–6.

[9] A. Perujo, B. Cuiffo, The introduction of electric vehicles in the private fleet: potential impact on the electric supply system and on the environment. A case study for the province of Milan, Italy, Energy Policy 38 (8) (2010) 4549–4561.

[10] P. Görbe, A. Magyar, K.M. Hangos, Reduction of power losses with smart grids fueled with renewable sources and applying EV batteries, J. Clean. Prod. 34 (2012) 125–137.

[11] M. Miotti, J. Hofer, C. Bauer, Integrated environmental and economic assessment of current and future fuel cell vehicles, Int. J. Life Cycle Assess. 22 (2017) 94–110.

[12] J. Köhler, W. Schade, G. Leduc, T. Wiesenthal, B. Schade, L.T. Espinoza, Leaving fossil fuels behind? An innovation system analysis of low carbon cars, J. Clean. Prod. 48 (2013) 176–186.

[13] A.A. Abd El Monem, A.M. Azmy, S.A. Mahmoud, Effect of process parameters on the dynamic behavior of polymer electrolyte membrane fuel cells for electric vehicle applications, Ain Shams Eng. J. 5 (1) (2014) 75–84.

[14] M. Uzunoglu, M.S. Alam, Dynamic modeling, design and simulation of a PEM fuel cell/ultra-capacitor hybrid system for vehicular applications, Energy Convers. Manag. 48 (5) (2007) 1544–1553.

[15] O. Erdinc, M. Uzunoglu, Recent trends in PEM fuel cell-powered hybrid systems: investigation of application areas, design architectures and energy management approaches, Renew. Sust. Energ. Rev. 14 (9) (2010) 2874–2884.

[16] F. Barbir, PEM Fuel Cells: Theory and Practice, Elsevier Academic Press, London, UK, 2005.

[17] W. Han, G. Zhang, J. Xiao, P. Bénard, R. Chahine, Demonstrations and marketing strategies of hydrogen fuel cell vehicles in China, Int. J. Hydrog. Energy 39 (25) (2014) 13859–13872.

[18] A. Veziroglu, R. Macario, Fuel cell vehicles: state of the art with economic and environmental concerns, Int. J. Hydrog. Energy 36 (1) (2011) 25–43.

[19] M.F. Felgenhauer, M.A. Pellow, S.M. Benson, T. Hamacher, Evaluating co-benefits of battery and fuel cell vehicles in a community in California, Energy 114 (1) (2016) 360–368.

[20] O.C. Onar, M. Uzunoglu, M.S. Alam, Modeling, control and simulation of an autonomous WT/PV/FC/UC hybrid power system, J. Power Sources 185 (2) (2008) 1273–1283.

[21] https://global.honda/innovation/FuelCell.html.

[22] P.P. Edwards, V.L. Kuznetsov, W.I.F. David, N.P. Brandon, Hydrogen and fuel cells: towards a sustainable energy future, Energy Policy 36 (12) (2008) 4356–4362.

[23] P. Corbo, F. Migliardini, O. Veneri, Hydrogen Fuel Cells for Road Vehicles, Springer-Verlag, London, 2011.

[24] M.H. Rashid, Modern Electric, Hybrid Electric, and Fuel Cell Vehicles Fundamentals, Theory, and Design, second ed., Taylor and Francis, 2010.

[25] R. Dell, P.T. Moseley, D.A.J. Rand, Towards Sustainable Road Transport, Elsevier, 2014.

[26] M. Uzunoglu, O.C. Onar, M.S. Alam, Modeling, control and simulation of a PV/FC/UC based hybrid power generation system for stand-alone applications, Renew. Energy 34 (3) (2009) 509–520.

[27] C.C. Elam, C.E.G. Padro, G. Sandrock, A. Luzzi, P. Lindblad, E.F. Hagen, Realizing the hydrogen future: the International Energy Agency's efforts to advance hydrogen energy technologies, Int. J. Hydrog. Energy 28 (6) (2003) 601–607.

[28] M. Momirlan, T.N. Veziroglu, Current status of hydrogen energy, Renew. Sust. Energ. Rev. 6 (1–2) (2002) 141–179.

[29] S. Sherif, F. Barbir, T. Veziroglu, in: Wind Energy and the Hydrogen Economy, Proceedings ASME Wind Energy Symposium, Reno, NV, 2003.

[30] I. Hadjipaschalis, A. Poullikkas, V. Efthimiou, Overview of current and future energy storage technologies for electric power applications, Renew. Sust. Energ. Rev. 13 (6–7) (2009) 1513–1522.
[31] A. Schell, H. Peng, D. Tran, E. Stamos, C.C. Lin, M.J. Kim, Modelling and control strategy development for fuel cell electric vehicles, Annu. Rev. Control. 29 (1) (2005) 159–168.
[32] T. Umegaki, J.M. Yan, X.B. Zhang, H. Shioyama, N. Kuriyama, Q. Xu, Boron- and nitrogen-based chemical hydrogen storage materials, Int. J. Hydrog. Energy 34 (2009) 2303–2311.
[33] D. Fraser, in: Solutions for Hydrogen Storage and Distribution, Proceedings of the PEI Wind-Hydrogen Symposium, PE, Canada, 2003, pp. 22–24.
[34] F.L. Darkrim, P. Malbrunot, G.P. Tartaglia, Review of hydrogen storage by adsorption in carbon nanotubes, Int. J. Hydrog. Energy 27 (2) (2002) 193–202.
[35] https://www.hydrogen.energy.gov.
[36] A. Kirubakaran, S. Jain, R.K. Nema, A review on fuel cell technologies and power electronic interface, Renew. Sust. Energ. Rev. 13 (9) (2009) 2430–2440.
[37] A. Emadi, M. Ehsani, J.M. Miller, Vehicular Electric Power Systems, Land, Sea, Air, and Space Vehicles, Marcel Dekker, 2004.
[38] A. Bocarsly, D. Michael, P. Mingos, Fuel Cells and Hydrogen Storage, Springer-Verlag, Berlin Heidelberg, 2011.
[39] C. Spiegel, PEM Fuel Cell Modeling and Simulation Using MATLAB®, Academic Press, 2008.
[40] Y. Wang, D.Y.C. Leung, J. Xuan, H. Wang, A review on unitized regenerative fuel cell technologies, part-A: unitized regenerative proton exchange membrane fuel cells, Renew. Sust. Energ. Rev. 65 (2016) 961–977.
[41] https://www.energy.gov/eere/fuel-cells/types-fuel-cells.
[42] J. Larminie, A. Dicks, Fuel Cell Systems Explained, John Wiley & Sons, 2003.
[43] J.M. Andujar, F. Segura, Fuel cells: history and updating. A walk along two centuries, Renew. Sust. Energ. Rev. 13 (9) (2009) 2309–2322.
[44] S. Mekhilef, R. Saidur, A. Safari, Comparative study of different fuel cell technologies, Renew. Sust. Energ. Rev. 16 (1) (2012) 981–989.
[45] Q. Cai, D.J.L. Brett, D. Browning, N.P. Brandon, A sizing-design methodology for hybrid fuel cell power systems and its application to an unmanned underwater vehicle, J. Power Sources 195 (19) (2010) 6559–6569.
[46] EG&G Technical Services, Inc., Fuel Cell Handbook, seventh ed., U.S. Department of Energy Office of Fossil Energy National Energy Technology Laboratory, 2004 (Under Contract No. DE-AM26-99FT40575).
[47] M. Uzunoglu, M.S. Alam, Dynamic modeling, design, and simulation of a combined PEM fuel cell and ultr capacitor system for stand-alone residential applications, IEEE Trans. Energy Convers. 21 (3) (2006) 767–775.
[48] J.H. Wee, Applications of proton exchange membrane fuel cell systems, Renew. Sust. Energ. Rev. 11 (8) (2007) 1720–1738.
[49] D.J.L. Brett, P. Aguiar, N.P. Brandon, R.N. Bull, R.C. Galloway, G.W. Hayes, K. Lillie, C. Mellors, C. Smith, A.R. Tilley, Concept and system design for a ZEBRA battery–intermediate temperature solid oxide fuel cell hybrid vehicle, J. Power Sources 157 (2) (2006) 782–798.
[50] L. Guo, K. Yedavalli, D. Zinger, Design and modeling of power system for a fuel cell hybrid switcher locomotive, Energy Convers. Manag. 52 (2) (2010) 1406–1413.
[51] I. Dincer, Technical, environmental and exergetic aspects of hydrogen energy systems, Int. J. Hydrog. Energy 27 (3) (2002) 265–285.
[52] C. Mi, M.A. Masrur, D.W. Gao, Hybrid Electric Vehicles: Principles and Applications with Practical Perspectives, first ed., John Wiley & Sons, 2011.
[53] O. Erdinc, B. Vural, M. Uzunoglu, A wavelet-fuzzy logic based energy management strategy for a fuel cell/battery/ultra-capacitor hybrid vehicular power system, J. Power Sources 194 (1) (2009) 369–380.
[54] https://www.hyundaiusa.com/tucsonfuelcell/index.aspx.
[55] B. Vural, O. Erdinc, M. Uzunoglu, Parallel combination of FC and UC for vehicular power systems using a multi-input converter-based power interface, Energy Convers. Manag. 51 (12) (2010) 2613–2622.
[56] N. Sulaiman, M.A. Hannan, A. Mohamed, E.H. Majlan, W.R. Wan Daud, A review on energy management system for fuel cell hybrid electric vehicle: issues and challenges, Renew. Sust. Energ. Rev. 52 (2015) 802–814.
[57] O.C. Onar, M. Uzunoglu, M.S. Alam, M.Y. El-Sharkh, A PEM fuel cell/battery hybrid power system under various load conditions, in: Proceedings International Hydrogen Energy Congress and Exhibition IHEC 2007 Istanbul, Turkey, 13–15 July, 2007.
[58] F. Villatico, F. Zuccari, Efficiency comparison between FC and ice in real urban driving cycles, Int. J. Hydrog. Energy 33 (12) (2008) 3235–3242.
[59] K. Ettihir, L. Boulon, K. Agbossou, S. Kelouwani, M. Hammoudi, Design of an energy management strategy for PEM fuel cell vehicles, in: 2012 IEEE International Symposium on Industrial Electronics, 28–31 May, IEEE, 2012.
[60] Z. Liu, L. Yang, Z. Maoa, W. Zhuge, Y. Zhang, L. Wang, Behavior of PEMFC in starvation, J. Power Sources 157 (1) (2006) 166–176.
[61] M.S. Bhaskar, et al., Survey of DC-DC non-isolated topologies for unidirectional power flow in fuel cell vehicles, IEEE Access 8 (2020) 178130–178166.
[62] Honda Fuel Cell Power FCX, December 2004. Available: http://world.honda.com/FuelCell/FCX/FCXPK.pdf. Press Information [Online].
[63] T.S. Key, H.E. Sitzlar, T.D. Geist, Fast Response, Load-Matching Hybrid Fuel Cell. Final Technical Progress Report, EPRI PEAC Corp, Knoxville, TN, 2003 (NREL/SR-560-32743).
[64] L. Gao, R.A. Dougal, S. Liu, Power enhancement of an actively controlled battery/ultracapacitor hybrid, IEEE Trans. Power Electron. 20 (1) (2005) 236–243.
[65] X. Zhang, C.C. Mi, A. Masrur, D. Daniszewski, Wavelet-transform-based power management of hybrid vehicles with multiple on board energy sources including fuel cell, battery and ultracapacitor, J. Power Sources 185 (2) (2008) 1533–1543.
[66] O. Erdinc, B. Vural, M. Uzunoglu, Y. Ates, Modeling and analysis of an FC/UC hybrid vehicular power system using a wavelet-fuzzy logic based load sharing and control algorithm, Int. J. Hydrog. Energy 34 (12) (2009) 5223–5233.
[67] P. Thounthong, V. Chunkag, P. Sethakul, B. Davat, M. Hinaje, Comparative study of fuel-cell vehicle hybridization with battery or supercapacitor storage device, IEEE Trans. Veh. Technol. 58 (8) (2009) 3892–3903.
[68] P. Thounthong, S. Rael, B. Davat, Energy management of fuel cell/battery/super-capacitor hybrid power source for vehicle applications, J. Power Sources 193 (1) (2009) 376–385.

[69] D. Feroldi, M. Serra, J. Riera, Energy management strategies based on efficiency map for fuel cell hybrid vehicles, J. Power Sources 190 (2) (2009) 387–401.

[70] A.A. Ferreira, J.A. Pomilio, G. Spiazzi, L.A. Silva, Energy management fuzzy logic supervisory for electric vehicle power supplies system, IEEE Trans. Power Electron. 23 (1) (2008) 107–115.

[71] A.M. Perez, W. Gao, J.J.F. Lozano, Fuzzy logic energy management strategy for fuel cell/ultracapacitor/battery hybrid vehicle with multiple-input DC/DC converter, in: IEEE Vehicle Power and Propulsion Conference, 7–10 Sept, IEEE, 2009.

[72] F. Akar, et al., A bidirectional nonisolated multi-input DC–DC converter for hybrid energy storage systems in electric vehicles, IEEE Trans. Veh. Technol. 65 (10) (2015) 7944–7955.

[73] A.K. Kaviani, G.H. Riahy, S.H.M. Kouhsari, Optimal design of a reliable hydrogen-based stand-alone wind/PV generating system, considering component outages, Renew. Energy 34 (11) (2009) 2380–2390.

[74] D. Gao, Z. Jin, Q. Lu, Energy management strategy based on fuzzy logic for a fuel cell hybrid bus, J. Power Sources 185 (1) (2008) 311–317.

[75] C.Y. Li, G.P. Liu, Optimal fuzzy power control and management of fuel cell/battery hybrid vehicles, J. Power Sources 192 (2) (2009) 525–533.

[76] Y. Eren, O. Erdinc, H. Gorgun, M. Uzunoglu, B. Vural, A fuzzy logic based supervisory controller for an FC/UC hybrid vehicular power system, Int. J. Hydrog. Energy 34 (20) (2009) 8681–8694.

[77] M.C. Kisacikoglu, M. Uzunoglu, M.S. Alam, Load sharing using fuzzy logic control in a fuel cell/ultra-capacitor hybrid vehicle, Int. J. Hydrog. Energy 34 (3) (2009) 1497–1507.

[78] P. Rodatz, G. Paganelli, A. Sciarretta, L. Guzzella, Optimal power management of an experimental fuel cell/supercapacitor-powered hybrid vehicle, Control. Eng. Pract. 13 (1) (2005) 41–53.

[79] L. Xu, J. Li, J. Hua, X. Li, M. Ouyang, Optimal vehicle control strategy of a fuel cell/battery hybrid city bus, Int. J. Hydrog. Energy 34 (17) (2009) 7323–7333.

[80] V. Paladini, T. Donateo, A. Risi, D. Laforgia, Super-capacitors fuel-cell hybrid electric vehicle optimization and control strategy development, Energy Convers. Manag. 48 (11) (2007) 3001–3008.

[81] G. Liu, J. Zhanga, Y. Sun, High frequency decoupling strategy for the PEM fuel cell hybrid system, Int. J. Hydrog. Energy 33 (21) (2008) 6253–6261.

[82] M. Uzunoglu, M.S. Alam, Modeling and analysis of an FC/UC hybrid vehicular power system using a novel wavelet based load sharing algorithm, IEEE Trans. Energy Convers. 23 (1) (2008) 263–272.

[83] P. Thounthong, S. Rael, B. Davat, Control algorithm of fuel cell and batteries for distributed generation system, IEEE Trans. Energy Convers. 23 (1) (2008) 148–155.

[84] P. Thounthong, S. Rael, B. Davat, Control strategy of fuel cell and supercapacitors association for a distributed generation system, IEEE Trans. Ind. Electron. 54 (6) (2007) 3225–3233.

[85] A. Payman, S. Pierfederici, F.M. Tabar, Energy control of supercapacitor/fuel cell hybrid power source, Energy Convers. Manag. 49 (6) (2008) 1637–1644.

[86] http://www.afdc.energy.gov/vehicles/fuel_cell.html.

[87] https://www.toyota.com/mirai/.

[88] https://www.mercedes-benz.com.sg/passengercars/mercedes-benz-cars/electromobility/f-cell/f-cell-intro.module.html.

[89] S. Hardman, G. Tal, Who are the early adopters of fuel cell vehicles? Int. J. Hydrog. Energy 43 (37) (2018) 17857–17866.

[90] http://cafcp.org.

[91] https://www.iea.org/reports/hydrogen.

CHAPTER

18

Monitoring battery status and facility conditions

Tambiara Tabassum and Mohammad Reza Khalghani

Electrical and Computer Engineering Department, Florida Polytechnic University, Lakeland, FL, United States

List of symbols

Symbols	Meaning
t_0, t_k	time
z_0, z_k	SoC at time instants t_0, t_k, respectively
η	Coulomb efficiency
Q/Q_{max}	maximum available capacity
I_L	load current
R_i	equivalent ohmic resistance
R_D	diffusion resistance
C_D	diffusion capacitance
U_D	voltage across C_D
U_t	battery terminal voltage
Q_n	nominal capacity
C_{loss}	capacity loss

18.1 Introduction: Importance of the topic, recent development (for revision chapter), potential applications, etc.

Batteries are one the main energy storages for modular and transportable applications in the modern society. Their wide applications can be ranged from electric fleets to electronic devices. Due to the environmental concerns, efficiency, and cost-effectiveness, the number of electric cars, including low-speed vehicles, hybrid electric vehicles, and fully electric vehicles, is significantly growing. The battery dependency in connected autonomous electric vehicles can be even more crucial since this mobility type needs more energy to be functional. We need to monitor the batteries' status to enable and enhance the battery energy management and controllability.

This chapter focuses on different aspects of battery monitoring systems and types.

18.2 Battery model and description

One of the main categories of estimation and monitoring methods is model-based techniques. These techniques require the mathematical model of the battery to estimate different operation parameters and variables, e.g., State of Charging (SoC), and State of Health (SoH). Therefore, this section focuses on a general explanation of the mathematical model of batteries.

Modeling the entire electrochemical mechanisms and behavior in different operating conditions is hard. In addition, since the common battery management systems have limited computation and calculation resources, a simplified and equivalent model of batteries is used for different estimation problems [1] (Fig. 1).

Handbook of Power Electronics in Autonomous and Electric Vehicles
https://doi.org/10.1016/B978-0-323-99545-0.00003-8

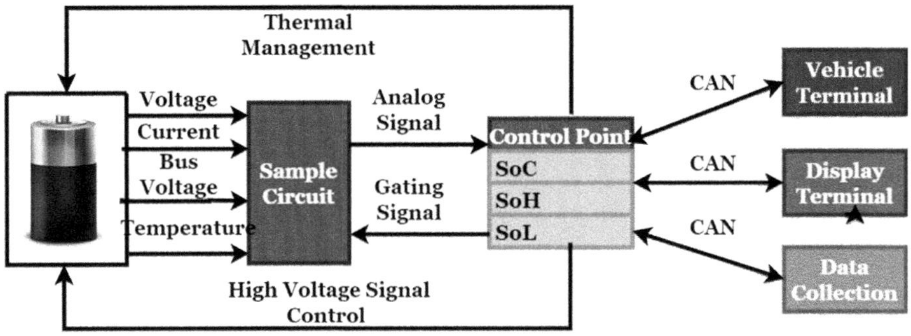

FIG. 1 Battery management system.

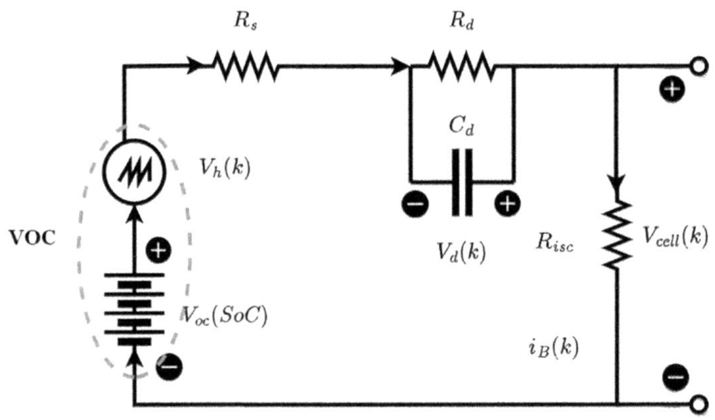

FIG. 2 Real-time battery model.

A real-time battery model is shown in Fig. 2. Here, two components make up the instantaneous open-circuit voltage (OCV), VOC; the first part $V_{oc}(SoC)$ stands for an equilibrium OCV that serves as a link between the SoC and the cell average OCV. The second component, V_h, is the hysteresis voltage, which captures the hysteresis effect of OCV. The RC circuit models the battery cell's transient response and IV properties; the series resistance R_s symbolizes the charging/discharging energy losses of the cell, the double-layer capacitance C_d and the charge transfer resistance R_c symbolize the short-term diffusion voltage V_d of the battery cell, and V_{cell} symbolizes the terminal voltage of the cell. The electrical circuit with the hysteresis effect can be modeled by discrete-time state-space equations [2].

$$x(k+1) = \begin{bmatrix} 1 & 0 & 0 \\ 0 & \alpha & 0 \\ 0 & 0 & H \end{bmatrix} x(k) + \begin{bmatrix} \frac{-\eta T_s}{C_{tot}} & 0 \\ \beta & 0 \\ 0 & H-1 \end{bmatrix} \begin{bmatrix} i_B(k) \\ sign(i_B(k)) \end{bmatrix} \tag{1}$$

$$y(k) = V_{oc}(SoC(k)) - V_d(k) - R_s i_B(k) + V_{hmax} V_h(k) \tag{2}$$

$$V_{oc}(SoC) = -a_0 \exp(-a_1 SoC) + a_2 + a_3 SoC - a_4 SoC^2 + a_5 SoC^3 \tag{3}$$

here, $y(k) = V_{cell}(k)$

$$x(k+1) = [SoC(k+1)\, V_d(k+1)\, V_h(k+1)]^T$$
$$H(i_B) = \exp(-\rho |i_B| T_s)$$

where k is the time index, $V_{cell}(k)$ is the measured terminal voltage, $x(k+1)$ is the state, η is the Coulomb efficiency ($\eta = 1$), C_{tot} is the total capacity of the battery cell [3], T_s is the sampling period, $i_B(k)$ is the instantaneous cell current ($i_B(k)$ is positive if the cell operates in discharging mode), V_h(max) is the maximum hysteresis voltage, $\alpha = \exp(-T_s/\tau)$ with $\tau = R_c C_d$, $\beta = R_c(1-\alpha)$, ρ is the hysteresis parameter characterizing convergence rate, and $a_j (j = 0, \ldots, 5)$ are the coefficients to parameterize $V_{oc}(SoC)$ function. For micro-shorted cell simulation, a short-circuit resistor R_{isc} can be added to the main circuit [4].

18.3 State-of charging (SoC) estimation

A BMS's main purpose is to ensure the efficient and safe use of the battery's energy, as well as to give the vehicle's energy management system accurate battery state information. By monitoring and managing the charging and discharging process of batteries, it should be able to provide the battery system with the necessary interventions if it is operating abnormally. SoC is referred to as the portion of the battery's maximum possible capacity that is still usable. In addition to aiding in providing information about the battery's current state of charge and available energy, precise estimation of battery SoC also ensures a dependable and secure mode of transportation. However, accurately estimating a battery's SoC is a difficult undertaking because batteries are intricate electrochemical systems with a specific nonlinear behavior dependent on many internal and external variables. Since a single cell embodies low voltage and energy, tens to thousands of cells must be connected in corresponding series and parallel to meet EV requirements. SoC calculations for resolving the inner inhabited states of battery packs remain extremely difficult due to the variable cell characteristics of the performance and operating conditions inside each battery pack. Moreover, aging, temperature variations, and charge-discharge cycles have a significant impact on battery performance, making it difficult to estimate an accurate SoC [5–7].

Numerous techniques have been suggested for precisely estimating the SoC of a battery, considering its crucial role in battery management. Furthermore, for more detailed classification, these methods have been divided into four groups.

1. Looking-up table-Based methods:

   The open-circuit voltage (OCV), impedance, and other static characteristic properties of batteries have a direct mapping relationship with SoC. Due to the correlations between SoC and one or more parameters, we can infer the SoC by measuring their values and then utilizing the looking-up table approach [8–11].

   - Open-circuit voltage

   Ref. [12] shows the relationship between battery SoC versus OCV for lithium-ion polymer battery (LiPB). The majority of battery management systems make use of this relationship to estimate SoC. Ref. [12] demonstrates that the OCV of a LiPB cell exhibits a monotonically rising pattern with its SoC. It is a useful tool for calibrating the flawed SoC. The measurement of battery OCV, however, necessitates turning off the power and letting the battery rest for a considerable amount of time, making it challenging to determine the correct OCV in real-time.

   - AC impedance

   This particular approach of SoC estimation is more suited for use in a laboratory. Thus, it cannot be implemented for running EVs since the measurement of battery impedance relies on the measurement device.

2. Ampere-hour integral method

   The ampere-hour integral approach can allow for the precise computation of the variation of the SoC when the maximum usable capacity of a battery is known, and its current can be measured accurately. Since there are no noticeable negative effects when using this approach normally, it is quite precise for batteries. The accurate SoC can be derived by ampere-hour integral method by using following equation:

$$z_k = z_0 - \int_{t_0}^{t_k} \frac{\eta I_L(t)dt}{Q} \tag{4}$$

where z represents battery SoC; z_0 and z_k denote SoC at time instant t_0 and t_k, respectively; η is Coulomb efficiency; $I_L(t)$ is load current of the battery; and Q is the nominal capacity. As operating conditions and aging status do not affect battery capacity, Q should not be taken as constant throughout the calculation and is used as maximum available capacity.

Here,

$$t_k = t_0 + k \times \Delta t$$

Δt is sampling interval.

However, there are three drawbacks of this method. First, the initial SoC must be known. Second, unpredictable disturbances like noise and temperature drift will inevitably cause measurement errors of battery current. Finally, due to changes in operation circumstances and battery aging, the Q needs to be recalibrated [12].

3. Model-based methods

   Fig. 3 shows a general flowchart of model-based estimation methods. It uses the state equation of batteries to combine the ampere-hour integral approach and the battery OCV parameters table-based looking-up method. Therefore, these methods can be considered as fusion of both ampere-hour integral and look-up table methods.

   - Equivalent circuit model (ECM)

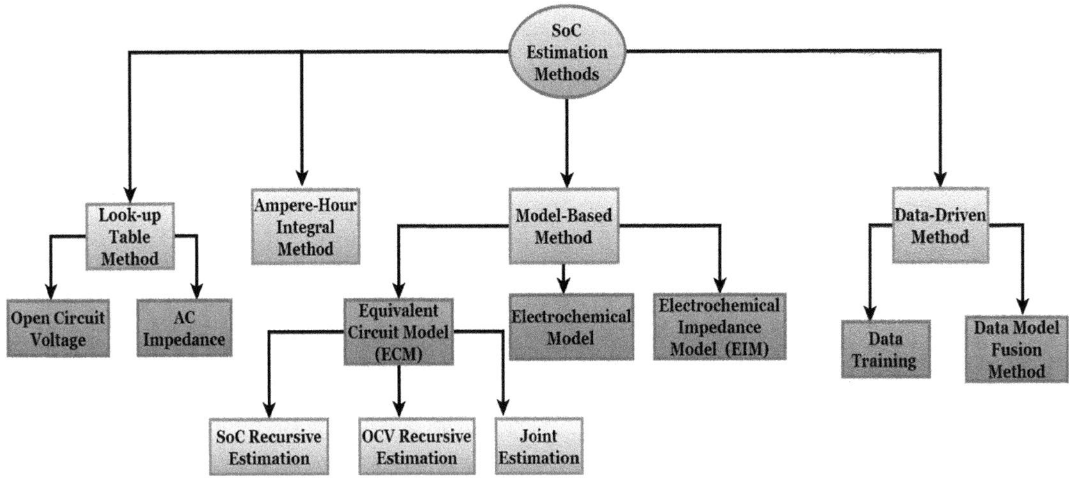

FIG. 3 SoC estimation techniques.

ECM is frequently used in BMS and vehicle energy management systems [13–40]. It constructs circuit networks using electrical circuit elements like resistors, capacitors, and voltage sources to characterize the terminal voltage of batteries. It is capable of precisely describing the battery's various dynamic behaviors. It can be utilized to effectively develop the model-based SoC estimate approach and has strong application and expansibility. Fig. 3 shows an ECM as NRC model with n-RC networks. It contains three components: voltage source that uses OCV, ohmic voltage across equivalent ohmic resistance R_i, and dynamic voltage behavior and the mass transport effects. R_D and C_D denote diffusion resistance and diffusion capacitance, respectively, and U_D is the voltage across C_D, U_t is battery terminal voltage, and I_L is the battery load current. Electrical behavior of NRC model can be described by the following equation:

$$\begin{cases} \dot{U}_{Di} = -\dfrac{1}{R_{Di}C_{Di}} U_{Di} + \dfrac{1}{C_{Di}} I_L \\ U_t = U_{oc} - I_L R_i - \displaystyle\sum_{i=1}^{n} U_{Di} \end{cases} \tag{5}$$

where $i = 1, 2, 3, \ldots\ldots\ldots\ldots, n$. R_{Di} denotes the ith equivalent diffusion resistance and C_{Di} denotes the ith equivalent diffusion capacitance [23]. It is worth mentioning that battery current is positive during discharge process and negative during charging process. ECM estimation can be divided into three subdivisions.

- SoC recursive estimation

The extended Kalman filter (EKF) continues to be the most popular state estimator in the field of battery modeling and state estimation for resolving both unconstrained and constrained state estimation issues. The EKF is the standard method in state estimation for achieving a recursive (approximate) maximum-likelihood estimation of the inhabited state. Based on data from the system model and speculations on uncertainties, it gives optimal state estimation with least mean square error [28]. The EKF algorithm updates the state estimation with the measurements to make corrections for the residual error of the model-based estimation process. It is worth mentioning that an important factor in EKF-based SoC estimation is the correlation between battery OCV and SoC. These SoC estimate approaches offer significant noise immunity because the EKF takes state and measurement noise characteristics into account. The model's prediction accuracy, however, determines how accurately the battery SoC is estimated using the recursive methods described previously. As a result, estimations based on a flawed or inappropriate battery model will be unrealistic. Ref. [37] presented a thorough examination of the SoC estimation findings taking into account the uncertain battery model characteristics.

- OCV recursive estimation

Recursive least squares (RLS) and the Kalman filter are the most frequently utilized methods for battery system identification. By minimizing the sum of squares of the terminal voltage prediction error based on the auto regressive exogenous (ARX) model, the RLS approach obtains the ideal matching results for battery parameters [29]. Based on its state equation, the Kalman filter minimizes the root mean square error between the desired output value and the actual output value to produce accurate real-time parameters.

An RLS-based online parameter identification approach using an incremental analysis-based ARX model has been presented and the trade-offs between model complexity and prediction accuracy have been carefully investigated and evaluated by [28–30] based on the NRC equivalent circuit model.

- Joint estimation

It is not ideal to estimate battery parameters and system states using the same time scale since battery parameters have a tendency to alter slowly over time while system states are more likely to fluctuate quickly over time. The same scale technique, on the other hand, hardly ever produces precise and reliable system estimates and significantly raises the computing cost of the control system. An essential requirement for battery SoC estimation is a precise and real-time computation of battery capacity. It is challenging to apply to the BMS the SoC estimation algorithm with known battery capacity. As a result, the multiscale EKF is created, using the macroscale to estimate battery parameters and the micro scale to estimate SoC, model parameters, and capacity where the macroscale and microscale mean the estimation under large and small calculated time intervals. According to verification results of the multiscale EKF, the maximum estimation errors of the battery's capacity and SoC against unpredictable operating conditions and degradation status are less than 2% [40]. According to the study previously, joint estimation approaches for battery parameters and states are more appropriate for use with EVs [12].

- Electrochemical model (EM)

In general, the EM specifies the mass, energy, and momentum transmission of each element for each phase and part of a battery cell. More specifically, the electrochemical model is able to describe the macroscopic quantities, such as cell current, voltage, and local distribution on a microscopic scale in terms of cell concentration, potential, current, and temperature [41,42]. With the advancement of battery technologies, the EM has been used in many applications. Its advantages include better forecasting of the battery's internal spatial and temporal states, including the concentration of the solid/electrolyte phase and the distribution of current and potential between the two electrodes. SoC estimation uses the one-dimensional (1D) model [41–43], the pseudo-two-dimensional (P2D) model [44], the quasi-three-dimensional full order physical model [45], and the first principal model [41]. The simplification of the P2D model, single-particle model [46–53], is the most widely used of these models. 1D electrochemical model has been used to build SoC estimator from time and again. However, to enhance the EM's ability to anticipate battery temperature, deterioration, and voltage behavior, P2D electrochemical model has been proposed afterward [54,55]. Moreover, due to its huge number of nonlinear equations, the P2D model frequently suffers from extended simulation times, making it computationally inefficient for simulating variables like cycling behavior and series/parallel configuration of stacked cells in battery packs. Later, the single-particle model was proposed as a solution to the low computing efficiency issue of P2D model, which makes it unsuitable for BMS application. The single-particle model only takes into account a single-solution resistance term and ignores the precise distribution of local concentration and potential in the solution phase. Utilizing the single-particle model, the EKF and unscented Kalman filter (UKF) are used to build the SoC estimator [50,56]. Estimation methods based on the EM can anticipate the battery's kinetic process and charge transfer mechanism. In addition to meeting the BMS's accuracy requirements, the estimations can also offer some guidelines for the battery's optimal design. Nevertheless, despite numerous simplifications, it can be challenging to pinpoint every characteristic. Furthermore, it necessitates a high level of professional experience, making it difficult to apply directly to the BMS.

4. Data-driven estimation methods

Data-driven control approaches create a controller by simply using the system's input-output data. In the following scenarios, the data-driven control strategy can demonstrate significant benefits [57].

- There is no known global mathematical model for the controlled system.
- The controlled system model has substantial uncertainty.
- A mathematical model for specifying a controlled system with an unknown operating process cannot be developed.
- It is impractical to study and build the controlled system's mechanism model because it is too complex, the number of orders is vast, or both.

It can be difficult to precisely estimate batteries using ECM and EIM because of the internal complicated chemical reaction process and unknown external operating conditions. The black-box model, which utilizes the nonlinear relationship of the input data to train the model, has several potential advantages, including parallel distributed processing, high computing speeds, fault tolerance, and the ability to adapt to this complex scenario. Black-box model is a common data-driven approach. The fuzzy controller [58,59], the neural network [57,60,61], the support vector machine [62,63], and a combination of these algorithms [64–66] are common algorithms that can be utilized for the black-box model. The black-box model approach can successfully address nonlinear modeling and state estimation issues while achieving high-prediction accuracy. Nonlinear statistical data modeling tools are more useful in terms of finding data patterns and modeling complex relationship between input and output. The neural network was utilized by [57] to create the SoC estimator, with the output layer being the voltage and the input layer, including the current, temperature, and SoC of the battery. Results demonstrate high computation accuracy. Nevertheless, these methods may not even converge due to their parametric sensitivity if the train data cannot completely reflect the current operating conditions.

18.4 State of health (SoH)

Over the course of the battery's lifespan, aging reduces the battery's capacity to store energy and deliver a certain power. Therefore, it is crucial to consider the state of health (SoH) as a significant indicator to evaluate the battery's aging status [67]. The battery should be taken out of the energy storage system once the health indicator hits the predetermined limit [68]. The most common SoH indicators are battery capacity and internal resistance, which represent energy and power capability, respectively [69,70]. According to research on internal aging mechanisms of various battery types, capacity loss is frequently caused by structural changes, the disintegration of active materials, and the loss of lithium stock [71–73]. The expansion of the solid electrolyte interface (SEI) layer is mostly responsible for the rise in resistance. According to battery test methods guidelines, a battery is considered unsuitable for device application when its capacity falls to 80% of its initial rated capacity during a certain test protocol. In this case, the battery must be replaced [74]. However, in some circumstances, the rise in internal resistance causing a decline in power will inevitably end in battery failure in advance. Therefore, for device application, taking these two factors jointly into account is crucial [67]. Charge curves, OCV (open-circuit voltage) curves, and other battery parameters that fluctuate considerably over time might also signal an increased risk of battery failure [75–77]. Much research has focused on model-based methodologies employing various types of estimating techniques to reveal the aging rule of these health-related parameters. Unfortunately, estimating health-related parameters and, by extension, the SoH, can be a very difficult undertaking. Because a variety of variables, including temperature and charge-discharge rates, can affect the battery aging process, long-term cycling tests and characteristic tests are required to examine the aging rule under these variables [67]. Various experimental and model-based SoH estimation techniques have been suggested in various research to follow these battery deterioration characteristics for online management of batteries in BMS [68,78,79]. According to Ref. [2], SoH estimation methods can be primarily divided into three types as shown in Fig. 4.

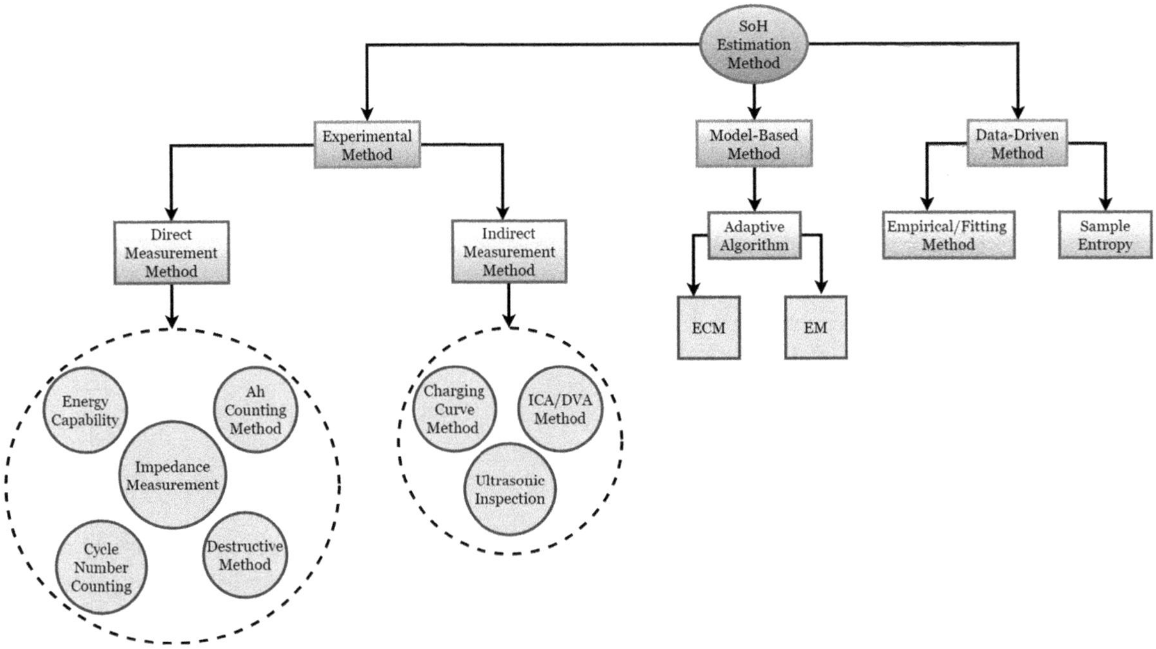

FIG. 4 SoH estimation techniques.

18.4.1 Experimental methods

To assess the battery aging behavior, experimental methods necessitate numerous laboratory tests. Some experimental approaches are challenging to accomplish onboard because they require specialized equipment and there is a big difference between real-world driving circumstances and lab environments. Direct measurement and indirect analysis are both used in experimental methods. The battery health state indicators are measured directly using direct measuring techniques, such as capacity testing, impedance measurements, and other tests. To determine the parameters associated to SoH, indirect analytical methods need to analyze and process data [67].

- Direct measurement techniques

Direct measurement techniques can be used in following forms:

- Energy capability
  The battery capacity defines the energy capabilities, which influences the mileage that EVs can travel. Battery capacity is a common SoH indication since it shows how much power can be stored in a fully charged battery. The SoH can be calculated directly if the current capacity of a battery can be measured precisely. There are sophisticated capacity testers on the market that can measure capacity precisely. Unfortunately, the method only works in a stable environment such as laboratory because it is challenging to halt a running electric vehicle and measure its entire charging capacity [67]
- AH counting method
  AH counting method plays a major role in SoH estimation. Researchers always use this method to verify the precision of capacity estimation results obtained from other methods [68]. Long-term monitoring and memorizing of the battery current is a necessary yet time- and energy-intensive technique for determining correct remaining capacity. The high-precision current sensor's ability to decrease accumulated error has a significant impact on the precise remaining capacity. When measuring capacity in a laboratory, the battery is typically fully charged and discharged with a tiny steady current at room temperature (20°C–25°C). Moreover, it is difficult to maintain such conditions in EVs. The remaining capacity can be calculated through high-precision measurement technique as,

$$SoH = \frac{Q_{max}}{Q_n} \times 100\%$$

where Q_n and Q_{max} are the nominal capacity and maximum available capacity of the battery [67]. Ah counting method is used in [80] for SoC and SoH estimation.

- Impedance measurement
  When current is applied, the internal resistance determines the voltage drop across the battery. The battery's current rate must be limited in order to run within a safe voltage range, which limits the amount of power that can be drawn from the battery. The battery's internal resistance and power capacity are tightly connected. Various methods of measuring resistance have been thoroughly researched as a widely used quantity to indicate the SoH of the battery. The polarization resistance and ohmic resistance make up the internal total resistance of a lithium battery. The voltage drop is mostly caused by ohmic resistance under normal operating condition and the voltage drop has a linear dependency on current magnitude. Ohmic resistance can be calculated by Ohm's law as,

$$R_o = \frac{\Delta U_t}{\Delta I_L}$$

where ΔU_t and ΔI_L denote pulse voltage and applied current pulse, respectively [81].

- Cycle number counting
  The number of battery cycle counts can be utilized as the foundation for the oH indication. Small electronic devices like phones and nanosatellites, as well as laptops, frequently use it to indicate the SoH of their batteries [82]. This method is straightforward and easy to use because it simply requires counting the number of charge and discharge cycles. By measuring the number of complete discharges, this technique basically calculates the SoH. The battery SoH can be estimated if the manufacturer provides the total number of life cycles for a battery and the current number of cycles the battery has undergone. The conversion coefficients are typically employed to superimpose and transform the charges and discharges of various levels into a full charge and discharge in the case of an operation that is not fully charging or discharging. The conversion coefficients can be determined through a different test.

- Destructive methods

  There are established, straightforward techniques for analyzing SoH and resolving the micromechanistic aspects of cell aging. The most popular techniques include Raman spectroscopy [83], X-ray diffraction [84], scanning electron microscopy [85], and so forth. These devices help to keep track of the microstructures during the aging process. A study involving the reasons behind battery aging at storage conditions is presented in Ref. [86]. It emphasizes that structural changes in the positive electrode, including the breakdown of active components and changes in stress, are the main contributors to aging at high SoC storage. However, the batteries must be disassembled in order to use these techniques to examine the aging mechanisms, which will permanently damage the batteries. Therefore, they are regarded as destructive methods. As a result, these techniques are only suitable for laboratory study.

- Indirect measurement techniques

  The indirect analysis approach is a common multi-step derivation technique that uses health indices relating to internal resistance or battery capacity decline. The battery SoH can be calculated once the relationship between the health indicators and capacity or resistance has been established.

  - Charging curve method

    Since the charging curve will change as the battery degrades, it can be used to describe the battery SoH. The CCCV charging mode, which stands for constant current and constant voltage with current limit, is frequently used for batteries. In [87], a SoH estimation method is developed using CV step as health indicator. The battery current is expressed as follows:

$$I(t, C_{loss}) = A(C_{loss})e^{-B(C_{loss})t} + C(C_{loss})$$

where C_{loss} is the capacity loss and A, B, C are capacity loss coefficients.

The SoH can also be calculated using constant current charging mode without CV period. By reconstructing charge voltage curves, quantitative analysis of the aging mechanism was carried out in [85,88]. The reconstructed charge voltage model includes five parameters defined by optimal algorithms. However, the estimation is only able to produce accurate results for temperature of 25°C because the effect of temperature on the charging curve was not included in these methods.

- ICA/DVA method

  Since voltage curves are not sufficient to obtain all internal information of the battery, incremental capacity analysis (ICA) and differential voltage analysis (DVA) are two other methods to process voltage data and get more detailed aging information. By integrating the battery's capacity at very low current rates during charging or discharging, IC curves are calculated for small voltage intervals (dQ/dV). On the other hand, the amount of electricity participating in two-phase transition in DV curves is calculated by estimating the distance between two peaks. In ICA, the two-phase transition voltage plateaus are transformed into distinguishable IC peaks. In those terms, DCA is more suitable for analyzing capacity quantitatively [88]. In Ref. [89], the local distance between the two peak points of the DV curve is used as SoH indicator. In [90–93], the position of the IC peaks is taken as a major indicator for battery aging. ICA was used in [94] where IC curves were obtained by support vector machine (SVM) with high accuracy.

  However, some drawbacks need to be overcome to be able to use ICA or DVA onboard. First, to assure correct understanding of the aging mechanisms and assess SoH, IC and DV curves are required to be excited by extremely low current. Moreover, the problem associated with that is, it is extremely hard to capture such low currents and large currents compromise accuracy. Second, typical techniques to derive IC and DV curves require complex computations. To overcome that problem, methods with low computation efforts have been developed, such as the point counting method [88] and the center least squared method [89]. Third, filtering and smoothing is very important for noise cancellation and to get smoother curves. The Gaussian method was used for filtering in Ref. [91].
- Ultrasonic inspection

  Ultrasonic inspection is very crucial for product maintenance and quality assurance. It is a practical method for finding all kinds of minor faults in materials. An ultrasonic inspection system typically consists of a monitor, a transducer, a receiver, and a pulser. Pulser generates waves and receiver receives it. The key battery interfaces, where degeneration processes occur, produce reflection, refraction, and wave pattern change. These features make it possible to diagnose medical conditions using the reflected wave from faulty interfaces. In some situations, including rail inspection [95] and the inspection of composite materials [96], the use of ultrasonic inspection offers a nondisassembling solution to the problem of internal fault detection. In Ref. [77], ultrasonic method to detect internal changes during battery aging by comparing the waveform change between a new battery and the battery under inspection.

18.4.2 Model-based methods

Although experimental approaches can yield comprehensive degradation information and precise SoH estimation results, online, real-time, and reliable measurement of battery health condition is more desirable for BMS. A model with adaptive filtering or data-driven algorithms can be used to estimate capacity, resistance, and other parameters, which can subsequently be used to quantify the degradation of lithium battery. The model-based estimation approach is an advancement of the indirect method that estimates or identifies the characteristic parameters through filtering or intelligent algorithms rather than only relying on the real-time measurable data. Closed-loop control and feedback are essential steps when utilizing adaptive algorithms that frequently use the EM and ECM. On the other hand, data-driven methods include black-box models, machine learning, and intelligent optimization algorithms.

Different adaptive filtering methods and data-driven methods have already been discussed in Section 18.3. Therefore, this section will just discuss some data-driven methods, which have not been introduced already.

- Data-driven methods
  - Empirical and fitting methods

    Empirical or fitting models typically provide faster processing due to their computational simplicity. Typical fitting models include polynomial, exponential, and power law. Without in-depth understanding of the electrochemical cell design and material characteristics, these methods forecast the future behavior of lithium batteries using the known degradation data. Using data from a large cycle-test matrix that covers three crucial factors, temperature, DOD, and discharge rate, Ref. [97] created a unified capacity loss model. Extensive experimental work served as the foundation for the construction of capacity loss models under various aging factors [97–101]. It is difficult to guarantee an absolute single variable since empirical and fitting models heavily rely on the findings of experiments.
  - Sample entropy

    In various domains, especially in the analysis of biomedical signals, the Approximate Entropy (ApEn) and Sample Entropy (SampEn) measures the complexity of time series and analyzes their characteristics [102,103]. SampEn is a more advanced version of ApEn. In comparison to ApEn, SampEn is less sensitive to record length and exhibits more consistency [104]. SampEn provides all the advantages of ApEn in addition to removing self-matches and accelerating computation. It could therefore be used as a signal for SoH estimate when applied to battery data that changes with battery aging [104]. SampEn characteristics were estimated in Ref. [105] using entire discharge voltage data at constant current, although EVs rarely operate under these conditions. The HPPC voltage sequence is frequently used to calculate SampEn as a SoH indicator since it is particularly sensitive to battery aging. In Ref. [106], a polynomial equation was used to fit the connection between capacity and SampEn from voltage sequence under the HPPC profile. SampEn and machine learning techniques can be combined to get better and more accurate outcomes. The learning algorithm uses SampEn as the data input and the SoH as the target vector. SampEn was integrated with SVM and relevance vector machine (RVM) and showed promising results [105].

18.5 Model estimation (impedance spectroscopy)

Impedance spectroscopy is one of the most promising techniques for collecting useful information about the battery. Electrochemical impedance spectroscopy or simply impedance spectroscopy is a measurement method that is frequently used in laboratories to assess the characteristics of fuel cells and battery cells [107,108]. The intricate electric transfer function of current and voltage recorded at the terminals is the impedance spectrum of a battery cell. The procedure uses external stimulation of the specimen to identify an unidentified system, making it nondestructive. Although a variety of excitation signals, such as statistically random noise and transients, can be applied to the device being tested, the most frequent type is a single-sine excitation [107]. Impedance data are used to study and characterize the underlying electrochemical processes and reactions. Several authors suggested implementing onboard impedance measurements in electric vehicles [109–113]. The methods are different in terms of the electronic platform, the excitation signal, the underlying battery system, and the necessary signal processing. Cost-effective architectures were developed to provide the excitation that use the battery charger [111], control the motor controller [112], or use the cell balancer [114]. Additionally, a few hardware and software architectures are created particularly to work with the aforementioned decentralized BMS framework [109,110]. Passive impedance spectroscopy proposes to use the current and voltage signals that occur purely from driving the vehicle to avoid additional energy consumption and disruption or interference with the regular mode of operation. This limits the task to gathering and analyzing current and voltage data.

Bohlen et al. [115] made one of the earliest attempts in the field of battery cells to forecast the pulse power capabilities of lead-acid starter batteries, although they only used the real part of the complex impedance calculation. Based on the Goertzel technique, Lohman et al. [116,117] assess the use of real driving data from automobiles for impedance spectroscopy; however, they noted that the higher frequency range has certain limitations. The battery testing device in [116] had a bandwidth restriction of 10 ms$\cong$50 Hz. In Ref. [117], the high-frequency white noise was added to the driver's excitation current. The system identification data set has been altered by the additional spectral components without the assurance of onboard scenario being accurately replicated. According to the research, passive impedance spectroscopy has a lot of potential for evaluating cell states inside electric car batteries. Nevertheless, a robust onboard approach requires both extensive experimental research and extra theoretical analysis of the estimating process. Moreover, very few studies have been reported in this regard. To develop the H_1 and particularly the H_2 technique based on power spectral density estimates with fast Fourier transforms (FFT), a framework for onboard passive impedance spectroscopy is presented in [114]. Experiments using parallel-connected battery cells validated theoretically derived ideal settings for passive impedance spectroscopy [114]. Kramers-Kronig relations were used to verify the efficacy of impedance spectra [114].

18.6 Degradation and performance monitoring (AC electricity impact)

The complex duty cycles imposed by electric machines and power electronics components are becoming a frequent aspect of battery service due to the rapidly expanding markets for electric vehicles and renewable energy systems. As a result, lithium-ion batteries, which are becoming more popular, must be able to handle alternating current (AC) that is superimposed at a variety of frequencies. The effects of alternating current (AC) on lithium batteries (LiBs) charge/discharge efficiency and cycle life have recently received a great deal of attention [118–120], with the methodologies used falling into three groups. First, several studies have examined into whether superimposed AC may improve the lifetime and charge-discharge performance of LIBs [121]. Second, batteries have been heated using AC-only current waveforms in extremely low ambient temperatures. The benefits include quick heating and great efficiency without material life reduction [119]. The third factor is that, even though battery requirements for service life and durability are only given in terms of DC currents, LIBs evaluated under most real-world circumstances are likely to have some AC component, such as ripple currents or AC harmonics superimposed on the principal DC [122]. The switching frequency of the power converters connected to the LIBs determines the frequency of ripple currents, which range from high-frequency (several kHz) to extremely low-frequency (caused by electric machines in hybrid electric vehicles [HEV] and electric vehicles [EV]) [123]. The main reason for exploring into this topic is that processes involved in how LIBs function respond on timelines that line up with these frequencies. It is well known that diffusion processes, charge transfer events, the creation of the solid electrolyte interphase (SEI), and double-layer charging all occur on timeframes ranging from microseconds to hours [124]. Given this, depending on the timeframe set by the frequency of the AC component, a cell exposed to AC may cause one or a combination of various electrochemical reactions [124,125]. Therefore, it is critical to know whether certain frequency ranges can accelerate the degradation of LIBs and whether certain frequencies may be associated with particular forms of degradation.

Several studies have been focused on the correlations between AC and the aging LIB. The life performance of 2 Ah-$LiCoO_2$ (LCO) that were subjected to AC without DC offsets with frequencies ranging from 1 Hz to 100 kHz were investigated in Ref. [126]. A significant degradation in cell capacity, which falls below 10 Hz AC frequency, were reported. However, the capacity loss was much lessened at higher frequencies, resembling the losses observed in control cells that were kept in open circuit. $LiNiCoAlO_2$ (NCA) cells that were subjected to 1200 charging/discharging cycles with coupled AC+DC current profiles with frequencies of 10 Hz, 55 Hz, 254 Hz, and 14.8 kHz were reported in Ref. [124]. It was shown that in comparison with the DC profile, capacity loss was more substantial at higher frequencies. It was claimed that the increased internal heat generation brought on by the superimposed AC was the primary cause of the rapid growth of surface film components and, as a result, the enhanced degradation rate. In another study of the same authors, EV batteries are made of LIB cells that are subjected to AC ripple with typical current harmonics spectrum of 59.94 Hz, 1.948 kHz, 2.008 kHz, 3.956 kHz, and 4.026 kHz. The difference between this study and the prior one was the use of a smaller AC amplitude in this one, which resulted in no noticeable decline of battery performance in this instance [124]. The effect of ripple current on twelve 28 Ah-$LiNiMnCoO_2$ (NMC) cells with frequencies of 1, 100, and 1000 Hz was investigated in Ref. [120] with no further deterioration on LIB was reported even with extremely low-frequency ripple current with 1 Hz. In Ref. [122], 2 Ah-NMC were exposed to 1500 cycles at a low-rate AC + DC profile, 0.225C ± 0.2C, at different frequencies. According to Uno's theory [126], AC with a frequency greater than the characteristic frequency (fc) degrades less quickly than AC with a lower frequency. Although no extra degradation for high-frequency ripple currents was reported in Ref. [122], the capacity loss for cells cycled with low-frequency AC was only increased by 1%–2%, which is not as significant as revealed in Uno's work [126]. In a study by Ref. [123] that focused solely on the impact of a DC/DC converter's switching frequency (5 kHz) on NMC cells, the findings showed that the ripple current caused by the converter did not significantly affect the capacity loss of these NMC cells. In a study of the effects of superimposed AC on a group of LCO cells in Ref. [127], it was found that the observed loss of capacity performance increased with the RMS value of a superimposed AC profile. This was attributed to a connection between induced Joule heating and RMS current. The deterioration rate was reported to be unaffected by other current parameters for the same RMS value, including frequency, waveform shape, and even DC value [127]. The effects of superimposed AC frequency on the aging of 24 LFP cells over a 200-day period were experimentally investigated in Ref. [128]. The findings showed that DC + AC (100 and 1000 Hz) not only did not negatively affect the LFP lifetime but also slightly improved the cell life performance compared to the DC cases for recommended C-rates of DC and AC ($\leq$0.5 C), where internal heat generation was not significantly affected by $|I_{RMS}|^2$.

18.7 State-of-life (remaining useful life (RUL)) estimation

Along with SoH, another equally important metric to monitor battery performance is remaining useful life (RUL). RUL estimation is necessary for Condition Based Maintenance (CBM) and battery health monitoring [129]. RUL can be defined as the period from the current observation time to a predetermined SoH threshold level since the battery's useful life is measured from its initial fresh state to a predetermined SoH threshold level. RUL is crucial for providing timely maintenance and battery system replacement [130,131]. The SoH threshold can be set at any level based on the user's needs, but it is typically set at SoH 0.8 in battery reliability studies like those in Refs. [132–134] due to the observation that after this point, the battery's capacity, and power both tend to decline much more quickly, making it unreliable and needing to be replaced [135]. However, even if the capacity falls below SoH 0.8, some battery types continue to degrade slowly [136]. System failure brought on by battery degradation can be avoided and the battery replacement schedule can be optimized if the degradation is precisely modeled, and RUL is properly estimated [137]. Additionally, the battery's useful cycle needs to be evaluated throughout the quality control process in battery manufacturing to determine whether a batch of batteries complies with the established standards [138]. Based on the aforementioned information, RUL estimation can be divided into experimental techniques and model-based techniques. However, experimental methods are not very feasible in modern research. They are very time-consuming and require a huge amount of charging/discharging cycles. Therefore, only model-based and model-free techniques will be discussed in this chapter.

Data-based approaches, which capture the degradation pattern from data about certain performance indicators, such as capacity [136,139], voltage [140], coulombic efficiency [134], impedance [140], and internal resistance [141], are a significant method in current research for degradation modeling and RUL prediction. This methodology, which can be further divided into machine learning method and statistical method in general, focuses on extracting features from degraded data, which is frequently real application data or experimental data. By simply applying the algorithm to degraded data without first making any assumptions or hypotheses, machine learning methods enable the model to learn from the data without being constrained by the user [142]. The data are transformed into a high-dimensional space using kernel approaches, which use specific kernel functions to find the best boundary for outputs. For example, flexible support vector regression (SVR)-based noniterative prediction models and multi-step prediction models to forecast the RUL were developed in Ref. [143], and SVR with a linear kernel was utilized in Ref. [144] to estimate SoH and predict the RUL. In addition to kernel techniques, neural network techniques—which decide the weight, bias parameters, and transfer functions between hidden layers [145]—have also been used in degradation modeling. Neural networks learn to create desirable outputs from given inputs by doing so [146]. For example, in Refs. [147,148], adaptive recurrent neural networks were used to forecast the RUL of batteries using the electrolyte and charge transfer resistance, while in Ref. [144], long short-term memory recurrent neural networks were used to predict the RUL using capacity data. A reliable measure of lithium-ion battery maintenance is the PF estimated value and probability distribution of the RUL [139,149,150]. Using the battery impedance data obtained by EIS, [150] concentrated on RUL prediction. This study offered a method for applying the PF in lithium-ion battery health prognostics.

The creation of prognostics tools is the foundation for many studies in estimating the health state of lithium-ion batteries [22,151–153]. Lithium-ion battery health condition analysis was conducted using regression methods, such as Support Vector Machine (SVM) [144] and Relevance Vector Machine (RVM) [105], which were able to determine the trend of battery performance degradation.

Researchers [144] are attempting to address this issue with some modification, in terms of dropouts such as randomly remove units and their connections from the neural network during training [154], or L1&L2 regularization. Moreover, there is still a general limitation that a relatively large amount of training data is required [155] to learn the data while avoiding over-fitting. Additionally, machine learning methods typically lack the ability to handle uncertainty [155], but some research types, like those in Refs. [144,148], maintain by assuming that the initial observation was random. In addition, unlike the machine learning approach, the statistical approach involves the formulation of hypotheses prior to the development of a model based on specific assumptions derived from our belief in the degradation data [142]. Due to the previous assumption or belief regarding degradation data, the statistical method is able to capture uncertainty [156] and requires far less training data than the machine learning method. Numerous studies have used statistical techniques, such as the quadratic model [157,158], the one-term exponential model [132], the two-term exponential model [139,159], the ensemble model with an exponential and a quadratic term [160], the logarithmic model [146], and filtering-based techniques, and they all adopted a deterministic model as the degradation model.

18.8 Communication networks and BMS systems

To maintain effective control over battery charging and discharging processes and ensure protection, a fast, precise, and proper BMS monitoring, and protection system is required. BMS is usually designed as a modular basis comprising two sections called Local and Central modules [161]. The purpose of the local module is to act as a data acquisition device for the implementation of control algorithms for battery cell protection as well as for measuring the battery's voltage, current, and temperature. The primary data logging controller is the Central module, and the collected data are used to estimate SoC, SoH, and so on. Central module usually has the communication interface with serial USB, TCP/IP to exchange monitored data with the local controller. Central module can be linked directly to an HDMI display, and as a user interface, remote desktop connection protocol can be used to access the display unit.

18.9 Emerging technologies in battery monitoring

The rapid integration of EVs increases the demand for raw materials, such as aluminum, nickel, and cobalt, which can result in an inflation of the price of these minerals and EV batteries [162]. According to researchers, a large number of batteries that will reach the end point of their useful life can either be disposed in landfills or reused/recycled. Unfortunately, due to the current stage of maturation, most of the batteries will end up getting disposed rather than being recycled. Most of the big industries are not geographically distributed; therefore, recycling these lithium-ion batteries involves high transport cost and legal restrictions. Recycling, however, is only economically feasible when the value of the recycled material exceeds or is equivalent to the price of extracting new chemical components [162]. Reusing EV batteries for secondary applications like charging infrastructure or integrating them with renewable energy generation systems is one alternative. Integrating second-life batteries with renewable energy generation systems can make them less volatile and enable the storage of extra energy for consumption during periods of low-energy production. The charging infrastructure for these vehicles can be improved by integrating these batteries into EV charging stations [163]. This demonstrates that the battery supply chain can also be circular since the customer can return the item for reuse or recycling. Effective supply chain management eliminates frauds, lowers indirect costs, aids in decision-making, boosts sales, and simplifies the production process. Tracking batteries along the supply chain is crucial in the context of the battery market to make sure that all market participants are aware of battery chemistry, authenticity, performance, life expectancy, and other information that may be necessary for specific regulatory standards. Technologies that aim to address industry-specific issues are emerging.

Blockchain is one of the technologies, which intends to disrupt various aspects of industrial and societal processes, by offering a distributed, immutable, and verifiable data repository that can follow the whole supply chain. The blockchain emerges as a viable mechanism for maintaining battery digital identities and tracking their lifecycle. Blockchain

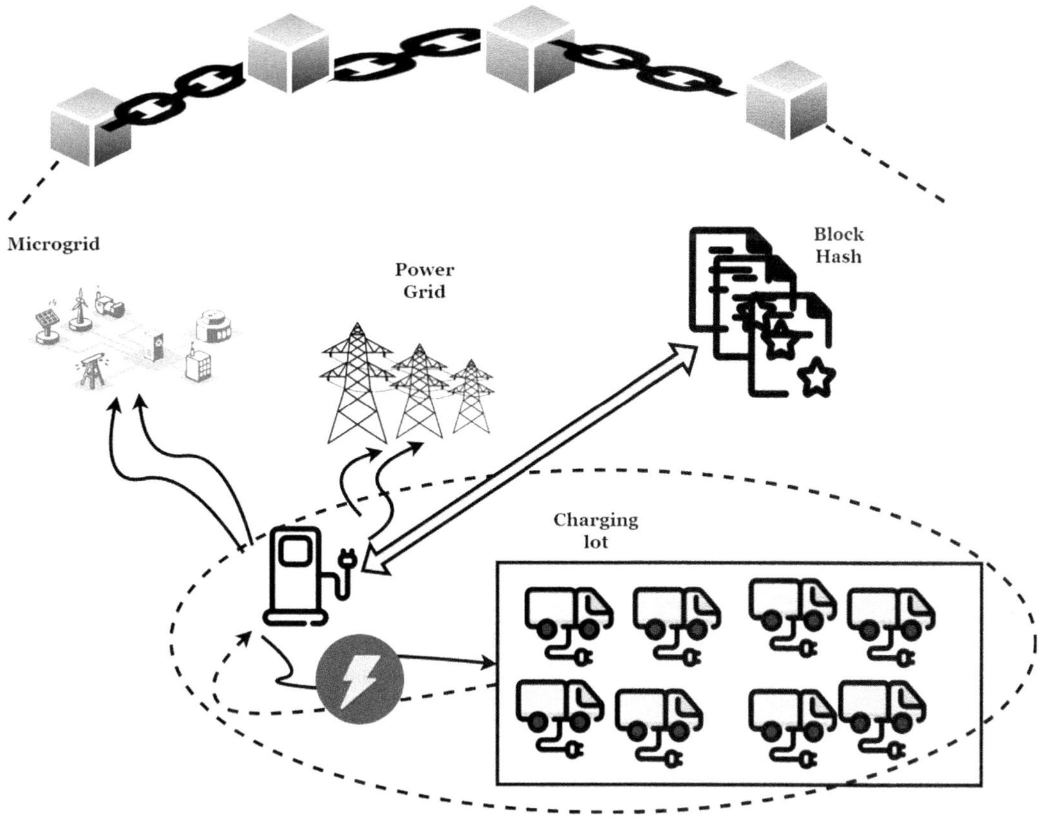

FIG. 5 Blockchain application in EVs.

will track and trace the evolution of these significant values, including their second lives, recycling, and disposal, from raw material extraction through production and battery use in electrical vehicles [164]. When there are conflicting interests among the various companies involved in an ecosystem, blockchain technology is ideally suited to address such issues. Blockchain technology makes it feasible to develop intelligent, unchangeable contracts that specify business rules and increase supply chain security and transparency. To prevent contaminants from contaminating their chemical components, the battery manufacturing process necessitates the use of a clean room. Batteries' performance, quality, and safety can all be affected by this contamination, which can affect a single battery or an entire batch of batteries. Using blockchain technology, a tracking system that enables the recall of one or more faulty batteries can be developed. A blockchain application scheme is shown in Fig. 5 [165].

18.10 Chapter summary

The state-of-the-art of the current battery state estimation, computation, and management approaches was comprehensively summarized in this study. Despite the fact that several state estimation techniques have been investigated, each technique has flaws and room for development. Additional research and tests are necessary. The authors of this publication intend for everyone with an interest in BMS to benefit from it. For engineers, best strategy can be chosen to evaluate the battery state based on the current demands. For researchers, some ideas might be acquired to further develop the state estimation approach.

References

[1] A. Farmann, W. Waag, D.U. Sauer, Adaptive approach for on-board impedance parameters and voltage estimation of lithium-ion batteries in electric vehicles, J. Power Sources 299 (2015) 176–188.
[2] B. Scrosati, J. Garche, Lithium batteries: status, prospects and future, J. Power Sources 195 (9) (2010) 2419–2430.
[3] G.L. Plett, Recursive approximate weighted total least squares estimation of battery cell total capacity, J. Power Sources 196 (4) (2011) 2319–2331.

[4] M. Ouyang, et al., Internal short circuit detection for battery pack using equivalent parameter and consistency method, J. Power Sources 294 (2015) 272–283.
[5] R. Yang, R. Xiong, H. He, H. Mu, C. Wang, A novel method on estimating the degradation and state of charge of lithium-ion batteries used for electrical vehicles, Appl. Energy 207 (2017) 331–340.
[6] S. Peng, C. Chen, H. Shi, Z. Yao, State of charge estimation of battery energy storage systems based on adaptive unscented Kalman filter with a noise statistics estimator, IEEE Access 5 (2017) 13202–13212.
[7] X. Hu, R. Xiong, B. Egardt, Model-based dynamic power assessment of lithium-ion batteries considering different operating conditions, IEEE Trans. Ind. Informat. 10 (3) (2014) 1948–1959.
[8] R. Xiong, J. Tian, H. Mu, C. Wang, A systematic model-based degradation behavior recognition and health monitoring method for lithium-ion batteries, Appl. Energy 207 (2017) 367–378.
[9] V. Pop, H.J. Bergveld, D. Danilov, P.P.L. Regtien, P.H.L. Notten, Battery Management Systems: Accurate State-of-Charge Indication for Battery-Powered Applications, Vol. 9, Springer, 2008.
[10] S. Rodrigues, N. Munichandraiah, A.K. Shukla, 'Review of state-of-charge indication of batteries by means of a.c. impedance measurements, J. Power Sources 87 (1–12) (2000) 12–20.
[11] X. Hu, F. Sun, Y. Zou, Comparison between two model-based algorithms for Li-ion battery SoC estimation in electric vehicles, Simul. Model. Pract. Theory 34 (2013) 1–11.
[12] R. Xiong, et al., Critical review on the battery state of charge estimation methods for electric vehicles, IEEE Access 6 (2017) 1832–1843.
[13] S. Lee, J. Kim, J. Lee, B.H. Cho, State-of-charge and capacity estimation of lithium-ion battery using a new open-circuit voltage versus state-of-charge, J. Power Sources 185 (2) (2008) 1367–1373.
[14] M. Thele, O. Bohlen, D.U. Sauer, E. Karden, Development of a voltage-behavior model for NiMH batteries using an impedance-based modeling concept, J. Power Sources 175 (1) (2008) 635–643.
[15] S. Wang, M. Verbrugge, J.S. Wang, P. Liu, Multi-parameter battery state estimator based on the adaptive and direct solution of the governing differential equations, J. Power Sources 196 (20) (2011) 8735–8741.
[16] Y. He, W. Liu, B.J. Koch, Battery algorithm verification and development using hardware-in-the-loop testing, J. Power Sources 195 (9) (2010) 2969–2974.
[17] M. Verbrugge, Adaptive, multi-parameter battery state estimator with optimized time-weighting factors, J. Appl. Electrochem. 37 (5) (2007) 605–616.
[18] M. Verbrugge, D. Frisch, B. Koch, Adaptive energy management of electric and hybrid electric vehicles, J. Electrochem. Soc. 152 (2) (2005) A333.
[19] H. He, X. Zhang, R. Xiong, Y. Xu, H. Guo, Online model-based estimation of state-of-charge and open-circuit voltage of lithium-ion batteries in electric vehicles, Energy 39 (1) (2012) 310–318.
[20] I. Snihir, W. Rey, E. Verbitskiy, A. Belfadhel-Ayeb, P.H.L. Notten, Battery open-circuit voltage estimation by a method of statistical analysis, J. Power Sources 159 (2) (2006) 1484–1487.
[21] C. Piao, W.-L. Fu, G.-H. Lei, C.-D. Cho, Online parameter estimation of the Ni-MH batteries based on statistical methods, Energies 3 (2) (2010) 206–215.
[22] Y.-H. Chiang, W.-Y. Sean, J.-C. Ke, Online estimation of internal resistance and open-circuit voltage of lithium-ion batteries in electric vehicles, J. Power Sources 196 (8) (2011) 3921–3932.
[23] F. Sun, R. Xiong, H. He, A systematic state-of-charge estimation framework for multi-cell battery pack in electric vehicles using bias correction technique, Appl. Energy 162 (2016) 1399–1409.
[24] R. Xiong, H. He, K. Zhao, 'Research on an online identification algorithm for a thevenin battery model by an experimental approach, Int. J. Green Energy 12 (3) (2015) 272–278.
[25] H. He, R. Xiong, J. Fan, Evaluation of lithium-ion battery equivalent circuit models for state of charge estimation by an experimental approach, Energies 4 (4) (2011) 582–598.
[26] R. Xiong, H. He, F. Sun, K. Zhao, Evaluation on state of charge estimation of batteries with adaptive extended Kalman filter by experiment approach, IEEE Trans. Veh. Technol. 62 (1) (2013) 108–117.
[27] R. Xiong, X. Gong, C.C. Mi, F. Sun, A robust state-of-charge estimator for multiple types of lithium-ion batteries using adaptive extended Kalman filter, J. Power Sources 243 (2013) 805–816.
[28] R.E. Kalman, A new approach to linear filtering and prediction problems, J. Basic Eng. 82 (1) (1960) 35.
[29] B.S. Bhangu, P. Bentley, D.A. Stone, C.M. Bingham, Nonlinear observers for predicting state-of-charge and state-of-health of lead-acid batteries for hybrid-electric vehicles, IEEE Trans. Veh. Technol. 54 (3) (2005) 783–794.
[30] J. Han, D. Kim, M. Sunwoo, State-of-charge estimation of lead acid batteries using an adaptive extended Kalman filter, J. Power Sources 188 (2) (2009) 606–612.
[31] F. Sun, X. Hu, Y. Zou, S. Li, Adaptive unscented Kalman filtering for state of charge estimation of a lithium-ion battery for electric vehicles, Energy 36 (5) (2011) 3531–3540.
[32] S. Sepasi, R. Ghorbani, B.Y. Liaw, Improved extended Kalman filter for state of charge estimation of battery pack, J. Power Sources 255 (2014) 368–376.
[33] Y. Shen, Hybrid unscented particle filter based state-of-charge determination for lead-acid batteries, Energy 74 (2014) 795–803.
[34] R. Xiong, F.-C. Sun, H.-W. He, Data-driven state-of-charge estimator for electric vehicles battery using robust extended Kalman filter, Int. J. Automot. Technol. 15 (1) (2014) 89–96.
[35] Z. Cheng, J. Lv, Y. Liu, Z. Yan, Estimation of state of charge for lithium-ion battery based on finite difference extended Kalman filter, J. Appl. Math. 2014 (2014) 348537.
[36] A. Baba, S. Adachi, SoC estimation of HEV/EV battery using series Kalman filter, Elect. Eng. Jpn. 187 (2) (2014) 53–62.
[37] F. Sun, R. Xiong, H. He, Estimation of state-of-charge and state-of-power capability of lithium-ion battery considering varying health conditions, J. Power Sources 259 (2014) 166–176.
[38] R. Xiong, F. Sun, H. He, T.D. Nguyen, A data-driven adaptive state of charge and power capability joint estimator of lithium-ion polymer battery used in electric vehicles, Energy 63 (2013) 295–308.

[39] D. Andre, C. Appel, T. Soczka-Guth, D.U. Sauer, Advanced mathematical methods of SoC and SOH estimation for lithium-ion batteries, J. Power Sources 224 (2013) 20–27.
[40] R. Xiong, F. Sun, Z. Chen, H. He, A data-driven multi-scale extended Kalman filtering based parameter and state estimation approach of lithium ion polymer battery in electric vehicles, Appl. Energy 113 (1) (2014) 463–476.
[41] A.P. Schmidt, M. Bitzer, Á.W. Imre, L. Guzzella, Experiment-driven electrochemical modeling and systematic parameterization for a lithium ion battery cell, J. Power Sources 195 (15) (2010) 5071–5080.
[42] S.K. Rahimian, S. Rayman, R.E. White, State of charge and loss of active material estimation of a lithium ion cell under low earth orbit condition using Kalman filtering approaches, J. Electrochem. Soc. 159 (6) (2012) A860.
[43] K.A. Smith, C.D. Rahn, C.-Y. Wang, Model-based electrochemical estimation of lithium-ion batteries, Control Syst. Technol. 18 (1) (2008) 714–719.
[44] X. Han, M. Ouyang, L. Lu, J. Li, Simplification of physics-based electrochemical model for lithium ion battery on electric vehicle. Part II: Pseudo-two-dimensional model simplification and state of charge estimation, J. Power Sources 278 (2015) 814–825.
[45] X. Li, S. Choe, State-of-charge (SoC) estimation based on a reduced order electrochemical thermal model and extended Kalman filter, in: Proc. Amer. Control Conf, 2013, pp. 1100–1105.
[46] H. Fang, Y. Wang, Z. Sahinoglu, T. Wada, S. Hara, Adaptive estimation of state of charge for lithium-ion batteries, in: Proc. Amer. Control Conf. (ACC), 2013, pp. 3485–3491.
[47] S.J. Moura, N.A. Chaturvedi, M. Krstic, PDE estimation techniques for advanced battery management systems—Part I: SoC estimation, in: Proc. Amer. Control Conf, 2012, pp. 566–571.
[48] S.J. Moura, N.A. Chaturvedi, M. Krstic, PDE estimation techniques for advanced battery management systems—Part II: SOH identification, in: Proc. Amer. Control Conf, 2012, pp. 566–571.
[49] S.J. Moura, N.A. Chaturvedi, M. Krstić, Adaptive partial differential equation observer for battery state-of-charge/state-of-health estimation via an electrochemical model, J. Dyn. Syst. Meas. Control. 136 (1) (2013) 11015.
[50] J. Scott Moura, M. Krstic, N.A. Chaturvedi, Adaptive PDE observer for battery SoC/SOH estimation, in: Proc. 5th Annu. Dyn. Syst. Control Conf. Joint JSME 11th Motion Vibrat. Conf, 2012, pp. 101–110.
[51] J. Lee, O. Nam, B.H. Cho, Li-ion battery SoC estimation method based on the reduced order extended Kalman filtering, J. Power Sources 174 (1) (2007) 9–15.
[52] S. Santhanagopalan, R.E. White, State of charge estimation using an unscented filter for high power lithium ion cells, Int. J. Energy Res. 34 (2) (2009) 152–163.
[53] D. Di Domenico, A. Stefanopoulou, G. Fiengo, Lithium-ion battery state of charge and critical surface charge estimation using an electrochemical model-based extended Kalman filter, J. Dyn. Syst. Meas. Control. 132 (6) (2010) 61302.
[54] M. Doyle, Comparison of modeling predictions with experimental data from plastic lithium ion cells, J. Electrochem. Soc. 143 (6) (1996) 1890.
[55] M. Doyle, T.F. Fuller, J. Newman, Modeling of galvanostatic charge and discharge of the lithium/polymer/insertion cell, J. Electrochem. Soc. 140 (6) (1993) 1526.
[56] S. Santhanagopalan, R.E. White, Online estimation of the state of charge of a lithium ion cell, J. Power Sources 161 (2) (2006) 1346–1355.
[57] Y. Shen, Adaptive online state-of-charge determination based on neurocontroller and neural network, Energy Convers. Manag. 51 (5) (2010) 1093–1098.
[58] A.J. Salkind, C. Fennie, P. Singh, T. Atwater, D.E. Reisner, Determination of state-of-charge and state-of-health of batteries by fuzzy logic methodology, J. Power Sources 80 (1) (1999) 293–300.
[59] P. Singh, R. Vinjamuri, X. Wang, D. Reisner, Design and implementation of a fuzzy logic-based state-of-charge meter for Li-ion batteries used in portable defibrillators, J. Power Sources 162 (2) (2006) 829–836.
[60] W.X. Shen, C.C. Chan, E.W.C. Lo, K.T. Chau, A new battery available capacity indicator for electric vehicles using neural network, Energy Convers. Manag. 43 (6) (2002) 817–826.
[61] C. Chan, E.W. Lo, S. Weixiang, The available capacity computation model based on artificial neural network for lead-acid batteries in electric vehicles, J. Power Sources 87 (1–2) (2000) 201–204.
[62] T. Hansen, C.-J. Wang, Support vector based battery state of charge estimator, J. Power Sources 141 (2) (2005) 351–358.
[63] W. Junping, C. Quanshi, C. Binggang, Support vector machine based battery model for electric vehicles, Energy Convers. Manag. 47 (7–8) (2006) 858–864.
[64] H. Cao, J. Yu, L. Kang, H. Yang, X. Ai, Modeling and prediction for discharge lifetime of battery systems using hybrid evolutionary algorithms, Comput. Chem. 25 (3) (2001) 251–259.
[65] M. Charkhgard, M. Farrokhi, State-of-charge estimation for lithium-ion batteries using neural networks and EKF, IEEE Trans. Ind. Electron. 57 (12) (2010) 4178–4187.
[66] T. Weigert, Q. Tian, K. Lian, State-of-charge prediction of batteries and battery–supercapacitor hybrids using artificial neural networks, J. Power Sources 196 (8) (2011) 4061–4066.
[67] R. Xiong, L. Li, J. Tian, Towards a smarter battery management system: a critical review on battery state of health monitoring methods, J. Power Sources 405 (2018) 18–29.
[68] M. Berecibar, I. Gandiaga, I. Villarreal, N. Omar, J. Van Mierlo, P. Van Den Bossche, Critical review of state of health estimation methods of Li-ion batteries for real applications, Renew. Sust. Energ. Rev. 56 (2016) 572–587.
[69] W. Waag, S. Käbitz, D.U. Sauer, Experimental investigation of the lithium-ion battery impedance characteristic at various conditions and aging states and its influence on the application, Appl. Energy 102 (2013) 885–897.
[70] G.L. Plett, Dual and joint EKF for simultaneous SOC and SOH estimation, in: Proc. 21st Electric Vehicle Symposium (EVS21), 2005, pp. 1–12.
[71] M. Wohlfahrt-Mehrens, C. Vogler, J. Garche, Aging mechanisms of lithium cathode materials, J. Power Sources 127 (2004) 58–64.
[72] R. Fu, S.Y. Choe, V. Agubra, J. Fergus, Modeling of degradation effects considering side reactions for a pouch type Li-ion polymer battery with carbon anode, J. Power Sources 261 (2014) 120–135.
[73] Y. Zhang, C.-Y. Wang, Cycle-life characterization of automotive lithium-ion batteries with $LiNiO_2$ cathode, J. Electrochem. Soc. 156 (2009) A527.

[74] T.Q. Duong, USABC and PNGV test procedures, J. Power Sources 89 (2000) 244–248.
[75] R. Mingant, J. Bernard, V. Sauvant Moynot, A. Delaille, S. Mailley, J.-L. Hognon, et al., EIS measurements for determining the SoC and SoH of Li-ion batteries, ECS Trans. 33 (2011) 41–53.
[76] J. Cannarella, C.B. Arnold, State of health and charge measurements in lithium-ion batteries using mechanical stress, J. Power Sources 269 (2014) 7–14.
[77] B. Sood, M. Osterman, M. Pecht, Health monitoring of lithium-ion batteries, in: 2013 10th Annual IEEE Symposium on Product Compliance Engineering (ISPCE), 2013, pp. 35–40.
[78] L. Tao, J. Ma, Y. Cheng, A. Noktehdan, J. Chong, C. Lu, A review of stochastic battery models and health management, Renew. Sust. Energ. Rev. 80 (2017) 716–732.
[79] A. Farmann, W. Waag, A. Marongiu, D. Uwe, Critical review of on-board capacity estimation techniques for lithium- ion batteries in electric and hybrid electric vehicles, J. Power Sources 281 (2015) 114–130.
[80] K.S. Ng, C.S. Moo, Y.P. Chen, Y.C. Hsieh, Enhanced coulomb counting method for estimating state-of-charge and state-of-health of lithium-ion batteries, Appl. Energy 86 (2009) 1506–1511.
[81] C. Zhang, J. Liu, S.M. Sharkh, C. Zhang, Identification of dynamic model parameters for lithium-ion batteries used in hybrid electric vehicles, Int. Symp. Electr. Veh. 1 (2009) 1–11.
[82] E.R. Wognsen, B.R. Haverkort, M. Jongerden, R.R. Hansen, K.G. Larsen, A score function for optimizing the cycle-life of battery-powered embedded systems, in: S. Sankaranarayanan, E. Vicario (Eds.), Formal Modeling and Analysis of Timed Systems. FORMATS 2015. Lecture Notes in Computer Science, vol. 9268, Springer, Cham, 2015, https://doi.org/10.1007/978-3-319-22975-1_20.
[83] L.J. Hardwick, H. Buqa, P. Novák, Graphite surface disorder detection using in situ Raman microscopy, Solid State Ionics 177 (2006) 2801–2806.
[84] K. Maher, R. Yazami, A study of lithium ion batteries cycle aging by thermodynamics techniques, J. Power Sources 247 (2014) 527–533.
[85] M. Ouyang, Z. Chu, L. Lu, J. Li, X. Han, X. Feng, et al., Low temperature aging mechanism identification and lithium deposition in a large format lithium iron phosphate battery for different charge profiles, J. Power Sources 286 (2015) 309–320.
[86] J. Li, J. Zhang, X. Zhang, C. Yang, N. Xu, B. Xia, Study of the storage performance of a Li-ion cell at elevated temperature, Electrochim. Acta 55 (2010) 927–934.
[87] A. Eddahech, O. Briat, J.M. Vinassa, Determination of lithium-ion battery state-of-health based on constant-voltage charge phase, J. Power Sources 258 (2014) 218–227.
[88] X. Han, M. Ouyang, L. Lu, J. Li, Y. Zheng, Z. Li, A comparative study of commercial lithium ion battery cycle life in electrical vehicle: aging mechanism identification, J. Power Sources 251 (2014) 38–54.
[89] L. Wang, C. Pan, L. Liu, Y. Cheng, X. Zhao, On-board state of health estimation of LiFePO4 battery pack through differential voltage analysis, Appl. Energy 168 (2016) 465–472.
[90] C. Weng, J. Sun, H. Peng, A unified open-circuit-voltage model of lithium-ion batteries for state-of-charge estimation and state-of-health monitoring, J. Power Sources 258 (2014) 228–237.
[91] Y. Li, M. Abdel-Monem, R. Gopalakrishnan, M. Berecibar, E. Nanini-Maury, N. Omar, et al., A quick on-line state of health estimation method for Li-ion battery with incremental capacity curves processed by Gaussian filter, J. Power Sources 373 (2018) 40–53.
[92] C. Weng, X. Feng, J. Sun, H. Peng, State-of-health monitoring of lithium-ion battery modules and packs via incremental capacity peak tracking, Appl. Energy 180 (2016) 360–368.
[93] J. Tian, R. Xiong, Q. Yu, Fractional order model based incremental capacity analysis for degradation state recognition of lithium-ion batteries, IEEE Trans. Ind. Electron. 46 (2018).
[94] C. Weng, Y. Cui, J. Sun, H. Peng, On-board state of health monitoring of lithium-ion batteries using incremental capacity analysis with support vector regression, J. Power Sources 235 (2013) 36–44.
[95] F.L. di Scalea, P. Rizzo, S. Coccia, I. Bartoli, M. Fateh, E. Viola, et al., Non-contact ultrasonic inspection of rails and signal processing for automatic defect detection and classification, Insight Non-Destr. Test Cond. Monit. 47 (2005) 346–353.
[96] T. D'Orazio, M. Leo, A. Distante, C. Guaragnella, V. Pianese, G. Cavaccini, Automatic ultrasonic inspection for internal defect detection in composite materials, NDT E Int. 41 (2008) 145–154.
[97] J. Wang, P. Liu, J. Hicks-Garner, E. Sherman, S. Soukiazian, M. Verbrugge, et al., Cycle-life model for graphite-LiFePO4 cells, J. Power Sources 196 (2011) 3942–3948.
[98] M. Petit, E. Prada, V. Sauvant-Moynot, Development of an empirical aging model for Li-ion batteries and application to assess the impact of vehicle-to-grid strategies on battery lifetime, Appl. Energy 172 (2016) 398–407.
[99] N. Omar, M.A. Monem, Y. Firouz, J. Salminen, J. Smekens, O. Hegazy, et al., Lithium iron phosphate based battery - assessment of the aging parameters and development of cycle life model, Appl. Energy 113 (2014) 1575–1585.
[100] M. Ecker, J.B. Gerschler, J. Vogel, S. Käbitz, F. Hust, P. Dechent, et al., Development of a lifetime prediction model for lithium-ion batteries based on extended accelerated aging test data, J. Power Sources 215 (2012) 248–257.
[101] Z. Li, L. Lu, M. Ouyang, Y. Xiao, Modeling the capacity degradation of LiFePO4/graphite batteries based on stress coupling analysis, J. Power Sources 196 (2011) 9757–9766.
[102] W. Chen, J. Zhuang, W. Yu, Z. Wang, Measuring complexity using FuzzyEn, ApEn, and SampEn, Med. Eng. Phys. 31 (2009) 61–68.
[103] Y. Kumar, M.L. Dewal, Complexity measures for normal and epileptic EEG signals using ApEn, Int. J. Comput. Commun. Technol. 2 (2011) 6–12.
[104] J.S. Richman, J. Randall Moorman, Physiological time-series analysis using approximate entropy and sample entropy, Am. J. Physiol. Heart Circ. Physiol. 278 (2000) H2039–H2049.
[105] A. Widodo, M.C. Shim, W. Caesarendra, B.S. Yang, Intelligent prognostics for battery health monitoring based on sample entropy, Expert Syst. Appl. 38 (2011) 11763–11769.
[106] H. Han, H. Xu, Z. Yuan, Y. Shen, A new SOH prediction model for lithium-ion battery for electric vehicles, in: 2014 17th Int Conf Electr Mach Syst, 2014, pp. 997–1002.
[107] E. Barsoukov, J.R. Macdonald, Impedance Spectroscopy: Theory, Experiment, and Applications, John Wiley & Sons, 2005, https://doi.org/10.1002/0471716243.

[108] A. Lasia, Electrochemical Impedance Spectroscopy and Its Applications, Springer, New York, 2014, https://doi.org/10.1007/978-1-4614-8933-7.
[109] A. Otto, S. Rzepka, T. Mager, B. Michel, C. Lanciotti, T. Günther, O. Kanoun, Battery management network for fully electrical vehicles featuring smart systems at cell and pack level, in: Advanced Microsystems for Automotive Applications 2012, Springer, 2012, p. 3, https://doi.org/10.1007/978-3-642-29673-4_1.
[110] L.H.J. Raijmakers, K.M. Shivakumar, M.C.F. Donkers, M.J.G. Lammers, H.J. Bergveld, Crosstalk interferences on impedance measurements in battery packs, IFAC-PapersOnLine 49 (11) (2016) 42–47, https://doi.org/10.1016/j.Ifacol.2016.08.007.
[111] R. Koch, R. Kuhn, I. Zilberman, A. Jossen, Electrochemical impedance spectroscopy for online battery monitoring-power electronics control, in: 2014 16th European Conference on Power Electronics and Applications, IEEE, 2014, pp. 1–10, https://doi.org/10.1109/EPE.2014.6910907.
[112] D.A. Howey, P.D. Mitcheson, V. Yufit, G.J. Offer, N.P. Brandon, Online measurement of battery impedance using motor controller excitation, IEEE Trans. Veh. Technol. 63 (6) (2013) 2557–2566, https://doi.org/10.1109/TVT.2013.2293597.
[113] E. Din, C. Schaef, K. Moffat, J.T. Stauth, A scalable active battery management system with embedded real-time electrochemical impedance spectroscopy, IEEE Trans. Power Electron. 32 (7) (2016) 5688–5698, https://doi.org/10.1109/TPEL.2016.2607519.
[114] B. Liebhart, L. Komsiyska, C. Endisch, Passive impedance spectroscopy for monitoring lithium-ion battery cells during vehicle operation, J. Power Sources 449 (2020) 227297.
[115] O. Bohlen, S. Buller, R.W. De Doncker, M. Gelbke, R. Naumann, Impedance based battery diagnosis for automotive applications, in: 2004 IEEE 35th Annual Power Electronics Specialists Conference (IEEE Cat. No. 04CH37551), Vol. 4, IEEE, 2004, pp. 2792–2797, https://doi.org/10.1109/PESC.2004.1355275.
[116] N. Lohmann, P. Weßkamp, P. Haußmann, J. Melbert, T. Musch, Electrochemical impedance spectroscopy for lithium-ion cells: test equipment and procedures for aging and fast characterization in time and frequency domain, J. Power Sources 273 (2015) 613–623, https://doi.org/10.1016/j.jpowsour.2014.09.132.
[117] N. Lohmann, P. Haussmann, P. Wesskamp, J. Melbert, T. Musch, Employing real automotive driving data for electrochemical impedance spectroscopy on lithium-ion cells, SAE Int. J. Altern. Powertrains 4 (2) (2015) 308–317, https://doi.org/10.4271/2015-01-1187.
[118] A. Tomaszewska, Z. Chu, X. Feng, S. O'Kane, X. Liu, J. Chen, et al., Lithium-ion battery fast charging: a review, eTransportation 1 (2019).
[119] X. Hu, Y. Zheng, D.A. Howey, H. Perez, A. Foley, M. Pecht, Battery warm-up methodologies at subzero temperatures for automotive applications: recent advances and perspectives, Prog. Energy Combust. Sci. 77 (2020) 100806.
[120] A. Bessman, R. Soares, O. Wallmark, P. Svens, G. Lindbergh, Aging effects of AC harmonics on lithium-ion cells, J. Storage Mater. 21 (2019) 741–749.
[121] L.-R. Chen, S.-L. Wu, D.-T. Shieh, T.-R. Chen, Sinusoidal-ripple-current charging strategy and optimal charging frequency study for Li-ion batteries, IEEE Trans. Ind. Electron. 60 (2013) 88–97.
[122] M.J. Brand, M.H. Hofmann, S.S. Schuster, P. Keil, A. Jossen, The influence of current ripples on the lifetime of lithium-ion batteries, IEEE Trans. Veh. Technol. 67 (2018) 10438–10445.
[123] P. Korth Pereira Ferraz, J. Kowal, A comparative study on the influence of DC/DCConverter induced high frequency current ripple on lithium-ion batteries, Sustainability 11 (2019).
[124] A. Jossen, Fundamentals of battery dynamics, J. Power Sources 154 (2006) 530–538.
[125] N. Kularatna, Dynamics and modeling of rechargeable batteries: what electrochemists? Work tells the electronic engineers, IEEE Power Electron. Mag. 1 (2014) 23–33.
[126] M. Uno, K. Tanaka, Influence of high-frequency charge–discharge cycling induced by cell voltage equalizers on the life performance of lithium-ion cells, IEEE Trans. Veh. Technol. 60 (2011) 1505–1515.
[127] K. Uddin, A.D. Moore, A. Barai, J. Marco, The effects of high frequency current ripple on electric vehicle battery performance, Appl. Energy 178 (2016) 142–154.
[128] L.W. Juang, P.J. Kollmeyer, A.E. Anders, T.M. Jahns, R.D. Lorenz, D. Gao, Investigation of the influence of superimposed AC current on lithium-ion battery aging using statistical design of experiments, J. Storage Mater. 11 (2017) 93–103.
[129] X. Si, W. Wang, C. Hu, D. Zhou, Eur. J. Oper. Res. 213 (2011) 1–14.
[130] D. Wang, Q. Miao, M. Pecht, Prognostics of lithium-ion batteries based on relevance vectors and a conditional three-parameter capacity degradation model, J. Power Sources 239 (2013) 253–264, https://doi.org/10.1016/j.jpowsour.2013.03.129.
[131] A. Eddahech, O. Briat, E. Woirgard, J.M. Vinassa, Remaining useful life prediction of lithium batteries in calendar ageing for automotive applications, Microelectron. Reliab. 52 (9–10) (2012) 2438–2442, https://doi.org/10.1016/j.microrel.2012.06.085.
[132] Y. Wu, W. Li, Y. Wang, K. Zhang, Remaining useful life prediction of lithium-ion batteries using neural network and bat-based particle filter, IEEE Access 7 (2019) 54843–54854.
[133] B. Long, W. Xian, L. Jiang, Z. Liu, An improved autoregressive model by particle swarm optimization for prognostics of lithium-ion batteries, Microelectron. Reliab. 53 (2013) 821–831.
[134] F. Yang, X. Song, G. Dong, K.-L. Tsui, A coulombic efficiency-based model for prognostics and health estimation of lithium-ion batteries, Energy 171 (2019) 1173–1182.
[135] Y. Zhang, R. Xiong, H. He, M.G. Pecht, Long short-term memory recurrent neural network for remaining useful life prediction of lithium-ion batteries, IEEE Trans. Veh. Technol. 67 (2018) 5695–5705.
[136] F. Yang, D. Wang, Y. Xing, K.-L. Tsui, Prognostics of Li (NiMnCo) O2-based lithium-ion batteries using a novel battery degradation model, Microelectron. Reliab. 70 (2017) 70–78.
[137] J. Zhang, J. Lee, A review on prognostics and health monitoring of Li-ion battery, J. Power Sources 196 (2011) 6007–6014.
[138] J.I. Park, S.H. Baek, M.K. Jeong, S.J. Bae, Dual features functional support vector machines for fault detection of rechargeable batteries, IEEE Trans. Syst. Man Cybern. Part C Appl. Rev. 39 (2009) 480–485.
[139] Q. Miao, L. Xie, H. Cui, W. Liang, M. Pecht, Remaining useful life prediction of lithium-ion battery with unscented particle filter technique, Microelectron. Reliab. 53 (6) (2013) 805–810, https://doi.org/10.1016/j.microrel.2012.12.004.
[140] J. Wu, C. Zhang, Z. Chen, An online method for lithium-ion battery remaining useful life estimation using importance sampling and neural networks, Appl. Energy 173 (2016) 134–140.

[141] S. Hoenig, H. Singh, T.G. Palanisamy, M. Eagan, Method and apparatus for predicting the available energy of a battery, Google Patents, 2003.
[142] P. Srikanth, Statistical Learning vs Machine Learning, 2014. https://medium.com/datascience-analytics/statistical-learning-vs-machine-learning-f9682fdc339f. (Accessed 1 February 2020).
[143] S. Wang, L. Zhao, X. Su, P. Ma, Prognostics of lithium-ion batteries based on battery performance analysis and flexible support vector regression, Energies 7 (2014) 6492–6508.
[144] A. Nuhic, T. Terzimehic, T. Soczka-Guth, M. Buchholz, K. Dietmayer, Health diagnosis and remaining useful life prognostics of lithium-ion batteries using data-driven methods, J. Power Sources 239 (2013) 680–688.
[145] D. An, N.H. Kim, J.-H. Choi, Practical options for selecting data-driven or physics-based prognostics algorithms with reviews, Reliab. Eng. Syst. Saf. 133 (2015) 223–236.
[146] H. Meng, Y.-F. Li, A review on prognostics and health management (PHM) methods of lithium-ion batteries, Renew. Sust. Energ. Rev. 116 (2019) 109405.
[147] J. Liu, A. Saxena, K. Goebel, B. Saha, W. Wang, An adaptive recurrent neural network for remaining useful life prediction of lithium-ion batteries, in: Annual Conference of the Prognostics and Health Management Society, 2010.
[148] M. Rezvani, M. AbuAli, S. Lee, J. Lee, J. Ni, A Comparative Analysis of Techniques for Electric Vehicle Battery Prognostics and Health Management (PHM), SAE International, 2011.
[149] W. He, N. Williard, M. Osterman, P. Michael, J. Power Sources 196 (2011) 10314–10321.
[150] B. Saha, K. Goebel, S. Poll, J. Christophersen, IEEE Trans. Instrum. Meas. 58 (2009) 291–296.
[151] G.K. Prasad, C.D. Rahn, Model based identification of aging parameters in lithium ion batteries, J. Power Sources 232 (2013) 79–85, https://doi.org/10.1016/j.jpowsour.2013.01.041.
[152] Z. Chen, C.C. Mi, Y. Fu, J. Xu, X. Gong, Online battery state of health estimation based on Genetic Algorithm for electric and hybrid vehicle applications, J. Power Sources 240 (2013) 184–192, https://doi.org/10.1016/j.jpowsour.2013.03.158.
[153] A.P. Schmidt, M. Bitzerb, A.W. Imre, L. Guzzella, Experiment-driven electrochemical modeling and systematic parameterization for a lithium-ion battery cell, J. Power Sources 195 (15) (2010) 7634–7638, https://doi.org/10.1016/j.jpowsour.2010.02.029.
[154] N. Srivastava, G. Hinton, A. Krizhevsky, I. Sutskever, R. Salakhutdinov, Dropout: a simple way to prevent neural networks from overfitting, J. Mach. Learn. Res. 15 (2014) 1929–1958.
[155] S.M. Rezvanizaniani, Z. Liu, Y. Chen, J. Lee, Review and recent advances in battery health monitoring and prognostics technologies for electric vehicle (EV) safety and mobility, J. Power Sources 256 (2014) 110–124.
[156] D. Liu, J. Zhou, Y. Peng, Data-driven prognostics and remaining useful life estimation for lithium-ion battery: a review, Instrumentation 1 (2014) 59–70.
[157] Y. Xing, E.W. Ma, K.-L. Tsui, M. Pecht, A case study on battery life prediction using particle filtering, in: Proceedings of the IEEE 2012 Prognostics and System Health Management Conference (PHM-2012 Beijing), 2012, pp. 1–6.
[158] M.V. Micea, L. Ungurean, G.N. Carstoiu, V. Groza, Online state-of-health assessment for battery management systems, IEEE Trans. Instrum. Meas. 60 (2011) 1997–2006.
[159] W. He, N. Williard, M. Osterman, M. Pecht, Prognostics of lithium-ion batteries based on Dempster-Shafer theory and the Bayesian Monte Carlo method, J. Power Sources 196 (2011) 10314–10321.
[160] Y. Xing, E.W. Ma, K.-L. Tsui, M. Pecht, An ensemble model for predicting the remaining useful performance of lithium-ion batteries, Microelectron. Reliab. 53 (2013) 811–820.
[161] I.N. Haq, et al., Development of battery management system for cell monitoring and protection, in: 2014 International Conference on Electrical Engineering and Computer Science (ICEECS), IEEE, 2014.
[162] A. Mayyas, D. Steward, M. Mann, The case for recycling: overview and challenges in the material supply chain for automotive li-ion batteries, Sustain. Mater. Technol. 19 (2019) e00087, https://doi.org/10.1016/j.susmat.2018.e00087.
[163] A.D. Gorbunova, I.A. Anisimov, Assessment of the use of renewable energy sources for the charging infrastructure of electric vehicles, Emerg. Sci. J. 4 (6) (2020) 539–550, https://doi.org/10.28991/esj-2020-01251.
[164] IEA, Global EV Outlook 2020 - Entering the Decade of Electric Drive?, 2020.
[165] Z. Su, et al., A secure charging scheme for electric vehicles with smart communities in energy blockchain, IEEE Internet Things J. 6 (3) (2018) 4601–4613.

CHAPTER

19

Battery characterization, life cycle, and modeling

Morteza Rezaei Larijani[a,b], *Shahin Hedayati Kia*[a], *MohammadReza Zolghadri*[b], *and Ahmed El Hajjaji*[a]

[a]University of Picardie Jules Verne, Amiens, France
[b]Sharif University of Technology, Tehran, Iran

19.1 Introduction

19.1.1 Review of battery technologies

Batteries possess high energy density, making them suitable for storing energy in electric vehicles (EVs). Several battery technologies are available, including lead-acid, nickel-metal hydride (Ni-MH), nickel-cadmium (Ni-Cd), lithium-ion (Li-ion), polymer Li-ion, lithium titanate oxide (LTO), and lithium-sulfur (Li-S), to name but a few [1,2]. The comparison between the four main battery technologies (lead-acid, Ni-Cd, Ni-MH, and Li-ion) in terms of volumetric and specific energy densities is shown in Fig. 1. The highly favorable option among existing battery technologies is the Li-ion battery, which is commonly used in EVs due to its high energy density. A brief review of the characteristics of these batteries is provided in this section.

19.1.1.1 Lead-acid battery

These batteries have been used since the 1900s, because of their durability, affordability, safety, and tolerance to temperatures. Nonetheless, they exhibit a drawback during overcharging because of hydrogen production resulting from water loss. To mitigate the problem, researchers have proposed adding distilled water to these batteries. Additionally, due to the use of lead content in their collector, they have low energy density and are susceptible to corrosion in contact with a sulfuric acid electrolyte. Because of all these limitations, along with substantial charging durations and heavyweight, lead-acid batteries are no longer employed in automotive applications [4–8].

19.1.1.2 Ni-Cd battery

Ni-Cd batteries use nickel oxide hydroxide and metallic cadmium as electrodes. These batteries have a higher energy density compared to lead-acid batteries. However, these batteries have a high self-discharge rate and need a complex charging method to prevent overcharging. Additionally, Ni-Cd batteries are known for the "memory effect," wherein the battery may lose capacity if it is not fully discharged before being recharged. While this effect has been mitigated in newer Ni-Cd batteries, it remains a challenging issue. Furthermore, the presence of cadmium raises a serious environmental concern since cadmium is a toxic and heavy metal. Therefore, proper disposal methods are crucial to prevent environmental contamination. Nowadays, Ni-Cd batteries have been largely replaced by other battery technologies such as Ni-MH and Li-ion batteries, which offer higher energy density and are more environmentally friendly [7–11].

19.1.1.3 Ni-MH battery

Ni-MH batteries use nickel oxyhydroxide as a cathode, and its negative electrode is made of a hydrogen-absorbing alloy and an alkaline electrolyte. These batteries are a good substitute for lead-acid batteries due to their high energy density, and a good replacement for Ni-Cd because they do not contain toxic metals like cadmium, making them more environmentally friendly and recyclable. Ni-MH batteries can operate at high voltage within a wide range of

Handbook of Power Electronics in Autonomous and Electric Vehicles
https://doi.org/10.1016/B978-0-323-99545-0.00001-4

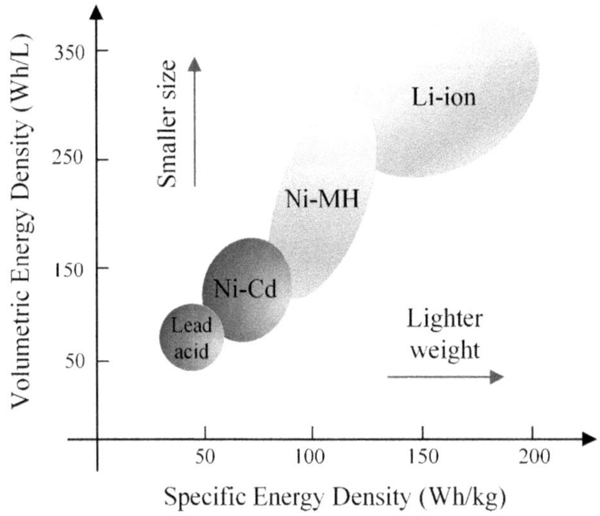

FIG. 1 Battery technologies in terms of volumetric and specific energy density [3].

temperatures and can withstand continued overcharging and discharging compared to lead-acid batteries [7]. This battery technology has a self-maintenance feature, utilizing excess energy to split and recombine water during overcharging, eliminating the need for regular maintenance. Nevertheless, charging these batteries with a high C-rate results in high heat and hydrogen buildup, causing cell rupture. Besides, overdischarging will reverse the polarization of the cell, reducing the capacity of the battery cell. Moreover, the battery exhibits poor performance and loses its capacity at low temperatures [12]. Ni-MH batteries have a higher self-discharge rate, meaning that they can lose their state-of-charge (SoC)[a] over time, even when not in use. Finally, despite their energy density, these batteries take a longer time to charge, and unfortunately, they are less recyclable [5,7,8,13].

19.1.1.4 *Li-ion battery*

Li-ion batteries use lithium ions as the charge carriers, and they are increasingly recognized as the ideal solution in the automobile industry due to their higher volumetric and specific energy densities (250–350 Wh/L and 100–250 Wh/kg), higher voltage per cell, greater power density, exceptional performance in high temperatures, and minimal self-discharge compared to alternative batteries [14–17]. These batteries offer high energy-to-weight ratios, lack memory effects, and a minimum rate of self-discharge. Li-ion batteries can function over a wide range of SoC while maintaining a long cycle life [18]. Despite these advantages, they are in desperate need of a protection circuit, known as a battery management system [19]. Another type, lithium-sulfur (Li-S) battery, boasts a higher energy density (500 Wh/kg) and greater safety as compared to Li-ion batteries. Nevertheless, Li-S has not been commercialized and is not used for EV applications due to restrictions in self-discharge and a loss of capacity with more cycling occurring [8,17,20].

19.2 Lithium-ion battery

19.2.1 Characteristics

Li-ion batteries are widely used in portable electronics, EVs, and renewable energy storage systems because of their numerous advantages. Li-ion batteries can be categorized into various types, each different from the others in terms of their cathode material. Table 1 categorizes Li-ion batteries, among which lithium cobalt oxide ($LiCoO_2$), lithium iron phosphate ($LiFePO_4$), lithium manganese oxide ($LiMn_2O_4$), lithium nickel cobalt aluminum oxide ($LiNi_{1-x-y}Co_xAl_yO_2$, i.e., $LiNi_{0.8}Co_{0.15}Al_{0.05}O_2$), and lithium nickel cobalt manganese oxide ($LiNi_{1/3}Mn_{1/3}Co_{1/3}O_2$) are widely used in industrial automotive [17,21,22].

Four typical Li-ion batteries are LCO, LFP, NCA, and NMC whose characteristics are given in Table 2 in terms of capacity, voltage, discharge current, mass, cost, energy density, and operating temperature.

[a] Battery state-of-charge (SoC) expresses how much the battery is charged.

TABLE 1 Li-ion battery characteristics in terms of their cathode/anode materials [21,23].

Lithium-ion battery	
Cathode	$LiCoO_2$ (LCO)
	$LiNiO_2$
	$LiMn_2O_4$ (LMO)
	$LiMn_2O_4$ (LMO)
	$LiFePO_4$ (LFP)
	LiV_2O_5
	$LiTiS_2$
	$LiWO_3$
	$NaCoO_2$
	$LiNi_{0.8}Co_{0.15}Al_{0.05}O_2$ (NCA)
	$LiNi_{0.8}Co_{0.2}O_2$
	$LiNi_{1/3}Mn_{1/3}Co_{1/3}O_2$ (NMC)
	$LiNi_{0.7}Co_{0.3}O_2$
	$Li_{1.156}Mn_{1.844}O_4$
Anode	LiC_6a (graphite)
	$Li_4Ti_5O_{12}$ (LTO)
	TiO_2

TABLE 2 Li-ion battery technologies specifications [18,24,25].

Battery	LCO	LFP	NCA	NMC
Anode	Graphitic carbon	Nanocarbon	Graphitic carbon	Graphitic carbon
Cathode	$LiCoO_2$	$LiFePO_4$	$LiNi_xCo_yAl_{1-x-y}O_2$	$LiNi_{0.8}Mn_{0.15}Co_{0.05}O_2$
Nominal capacity(Ah)	2.5	1.1	2.9	3
Nominal voltage (V)	2.5	1.1	3.6	3.6
Maximum discharge current (A)	20	30	6	20
Measured mass (g)	44.9	39.4	44.8	44.9
Cost per cell ($)	7.5	9.95	7.85	6
The volumetric energy density (Wh/L)	533.4	212.1	676	612.1
Specific energy density (Wh/kg)	195.8	88.9	243	224.8
Operating temperature range (°C)	0 to 50	−30 to 60	0 to 45	0 to 50

Within the Li-ion battery, when discharge occurs, ions of lithium migrate from the anode (composed of carbon) through the separator and electrolyte to the cathode, which determines the battery type as discussed earlier. If these ions move quickly, the battery is under a high C-rate current, which generates considerable heat. One of the popular Li-ion technologies, which has a low self-discharge current, and high energy density is NCA Li-ion batteries that are used in the automotive industry, like 4680-type, as shown in Fig. 2, used by Tesla, Inc. [18,26–30].

Also, one of the commercial NCA Li-ion batteries is the NCR18650B[b] Li-ion battery, as shown in Fig. 3, and its characteristics are provided in Table 3. The modeling of the battery cell is supposed to be based on this battery.

[b] Note that NCA is a Li-ion technology in which the cathode is lithium-nickel-cobalt-aluminum-oxide, and NCR is the abbreviation for nickel-cobalt-rechargeable, which is a commercial name for this type of NCA Li-ion battery.

FIG. 2 NCA 4680 battery of Tesla, Inc. [26].

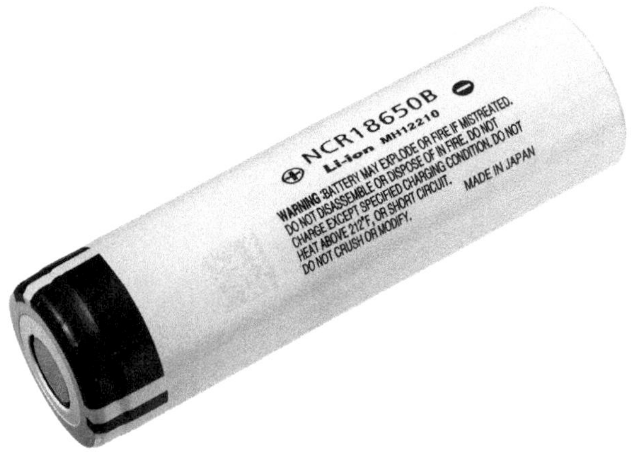

FIG. 3 NCR18650B Li-ion battery cell, 3250 mAh, 3.6 V, 243 Wh/kg [31].

TABLE 3 Characteristics of NCR18650B Li-ion battery (NCA) [32].

Maximum voltage	$V_{b_{max}}$	4.2 V
Nominal voltage	$V_{b_{nom}}$	3.6 V
Minimum voltage (cut-off)	$V_{b_{min}}$	2.5 V
Typical capacity	Q_b	3.25 Ah
Nominal current	I_{b_n}	3.25 A
Maximum pulse discharge current	$I_{b_{max}}$	13 A
Energy density	E_b	243 Wh/kg

19.2.2 Life cycle

The life cycle of a Li-ion battery refers to the number of charge and discharge cycles it can undergo before its capacity significantly degrades. Each cycle consists of a full charge and discharge, and the battery performance gradually diminishes over time, this results in reducing the battery capacity which is known as capacity fade. Battery degradation is influenced by several factors, among which, usage patterns, discharging rates, temperature, and the battery depth-of-discharge[c] are of great importance [7]. For instance, extreme temperatures negatively impact the longevity of Li-ion batteries, resulting in reduced battery life. The recommended operational temperature for batteries is approximately 25°C [33–35]. The charge and discharge rates, often expressed as the C-rate, can affect battery lifespan. Charging or discharging a battery at higher C-rates generates more heat and can accelerate degradation. Slower charging and

[c] How much of the battery capacity is utilized during each cycle.

discharging rates are generally better for battery longevity. For example, for the NCR18650B Li-ion battery cell, whose specification is provided in Table 3, the C-rate is 3.25 A. Charging and discharging the battery more than 3.25 A significantly decreases the lifetime of this type of battery. Due to the frequent start/stop operation of EVs, the discharging and charging profile of the energy storage varies significantly. Li-ion and Ni-MH batteries show that if they are repeatedly discharged at high load currents, their lifespan is reduced to about 200–300 cycles [36,37]. The best operation performance is obtained when they are discharged between 20% and 50% of the nominal capacity. In [37], it is shown that the aging effect of Li-ion batteries can be further accelerated and influenced by high temperatures under high load and low SoC, especially when the load current changes dramatically.

Battery degradation models fall into different categories based on the factors they consider and the methods they employ to simulate the degradation process, among which the empirical model is a commonly used method [7,33,34,38–40]. An empirical model for battery degradation is a mathematical representation that describes the relationship between various factors and the degradation of a physical battery over time. Empirical models are based on observed data rather than underlying physical or chemical principles. These models use statistical techniques to capture the patterns and trends seen in experimental data. They may use statistical methods to correlate degradation patterns with operational parameters [41–44]. The choice of the model category depends on the specific application, available data, and the desired level of detail in understanding and predicting battery degradation. Researchers and engineers often use a combination of these models to develop a comprehensive understanding of the complex processes involved in battery aging. The most commonly used method is the dynamic battery degradation capacity model based on a semiempirical approach, which relies on the Arrhenius degradation model, as provided by Eq. (1).

$$Q_{loss} = Be^{-\left(\frac{E_a + A \cdot C - rate}{RT_{bat}}\right)}(Ah)^z \tag{1}$$

where Q_{loss} is the battery capacity loss, A is the correction coefficient, B is the exponential coefficient, E_a is the activation energy ($J.mol^{-1}$), R is the gas constant ($8.3145\,J.mol^{-1}.K^{-1}$), T_{bat} is the absolute temperature of the battery in Kelvin (K), Ah is the Ah-throughput, z is the law factor, and *C-rate* is the discharge rate. Given the capacity fade model, the discrete equation of the battery degradation model is given in Eq. (2):

$$Q_{loss,t_{k+1}} - Q_{loss,t_k} = \Delta Ah\, A^{\frac{1}{z}} e^{\left(-\frac{E_a + A \cdot C - rate}{RT_{bat}}\right)} \left(Q_{loss,t_k}\right)^{\frac{z-1}{z}} \tag{2}$$

where $Q_{loss,t_{k+1}}$ and Q_{loss,t_k} are the accumulated battery capacities at time $k+1$ and k, and ΔAh is the Ah-throughput which is given in Eq. (3):

$$\Delta Ah = \frac{1}{3600} \int_{t_k}^{t_{k+1}} |I_{bat}| dt \tag{3}$$

Using capacity loss to evaluate battery degradation can facilitate comparisons of different energy management strategies in EV applications [45]. In the following section, the battery cell modeling in the EV applications will be presented.

19.3 Modeling battery cell

The model of the battery cell should be accurate enough and operate in real-time as far as the EV applications are concerned. Battery as the principal energy storage component plays a key role in the evaluation and analysis of energy management strategies in EVs. Li-ion batteries have nonlinear behavior, making the modeling a challenging issue [1,7,17,46–48]. Generally, there are three methods in modeling batteries: mathematical model, electrochemical model, and equivalent circuit model (ECM) [17]. Among these methods, the ECM is precise enough to represent dynamic behaviors of the battery cell, and it is not too complex to be implemented in real-time digital simulators (RTDS) [49,50]. The ECM which consists of double RC branches, as shown in Fig. 4, is suitable for representing the dynamic behaviors of the battery cell [2].

The open-circuit voltage is v_{oc}, R_0 represents the series resistance for the modeling voltage drop, R_1C_1, and R_2C_2 are used for modeling the effect of the electrode surface and diffusion process. All these parameters depend on the battery SoC, which relies on the battery cell current, i_b. SoC is computed by the Coulomb-counting method, as given in Eq. (4).

$$\frac{d}{dt}(SoC(t)) = -\frac{i_b(t)}{3600 \times Q_b} \tag{4}$$

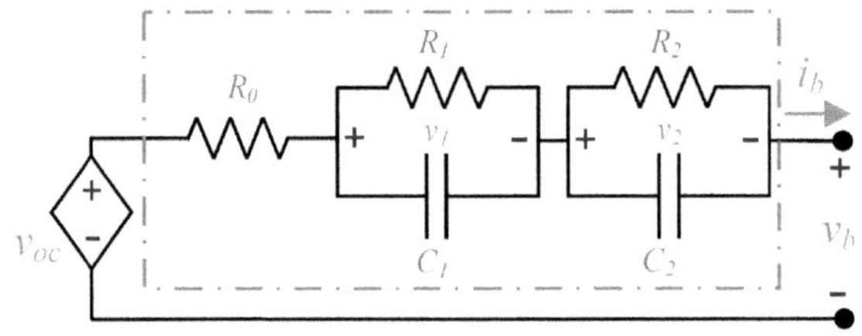

FIG. 4 Equivalent circuit model (ECM) of the battery cell [47].

Given the circuit schematic, the battery terminal voltage, v_b, and RC branch voltages can be computed based on the first-order equations which are given in Eqs. (5)–(7).

$$v_b = V_{oc} - (R_0 i_b + v_1 + v_2) \tag{5}$$

$$\frac{v_1}{R_1} + C_1 \frac{dv_1}{dt} = i_b \tag{6}$$

$$\frac{v_2}{R_2} + C_2 \frac{dv_2}{dt} = i_b \tag{7}$$

19.3.1 ECM parameter estimation

The ECM parameters can be estimated by discharging the battery cell using a pulse current and measuring its terminal voltage in relaxation time [16]. In this chapter, instead of using a physical battery, the electrochemical model of the battery cell in MapleSim is used as a high-fidelity model for parameter extraction. MapleSim obviates the necessity for laboratory experiments, ensuring safety and removing the environmental impact during testing [23]. The electrochemical model of the battery cell is shown in Fig. 5, of which the most important parameters are its capacity and cathode type (positive electrode), which is 3.25 Ah and NCA, respectively, based on the NCR18650B. The amplitude of the discharge pulse current, as shown in Fig. 6, is C-rate (3.25 A), and the pulse duration, t_p, and period, T, is 72 s and 1000 s, respectively. During each C-rate pulse current lasting 72 s, there is a linear reduction of 2% in SoC, as shown in Fig. 7, according to Eq. (4), while the voltage decreases exponentially, given the RC circuit. Then, when the pulse current is zero, the terminal voltage increases exponentially, and the capacitors are discharged in the steady state. Hence, the open-circuit voltage can be computed by measuring the voltage curve at the end of each pulse period.

The open-circuit voltage, v_{oc}, versus SoC is plotted in Fig. 8.

The data of open-circuit voltage can be fitted to a polynomial function or stored in a lookup table to be used in simulation. Similarly, R_0 is obtained by measuring the voltage drop of terminal voltage when the pulse current is applied, and the RC branch parameters are obtained, given in Eqs. (5)–(7), by measuring the terminal voltage curve

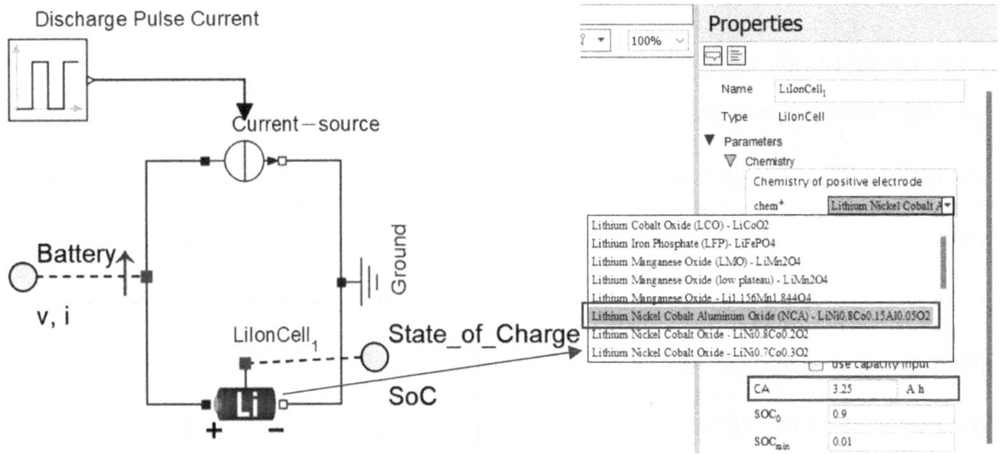

FIG. 5 Electrochemical model of the NCA battery cell in MapleSim under discharge pulse current.

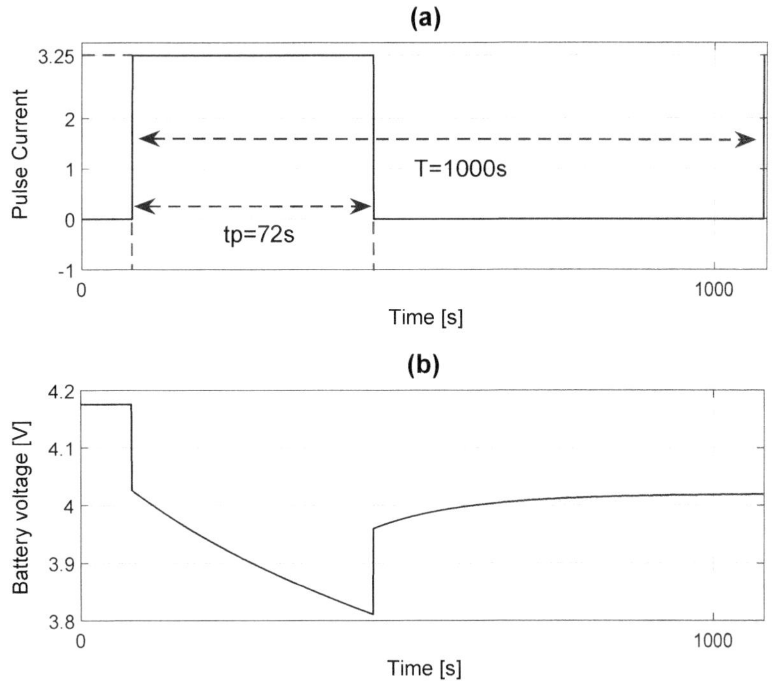

FIG. 6 Pulse test. (A) Discharge current. (b) Battery cell voltage for each pulse.

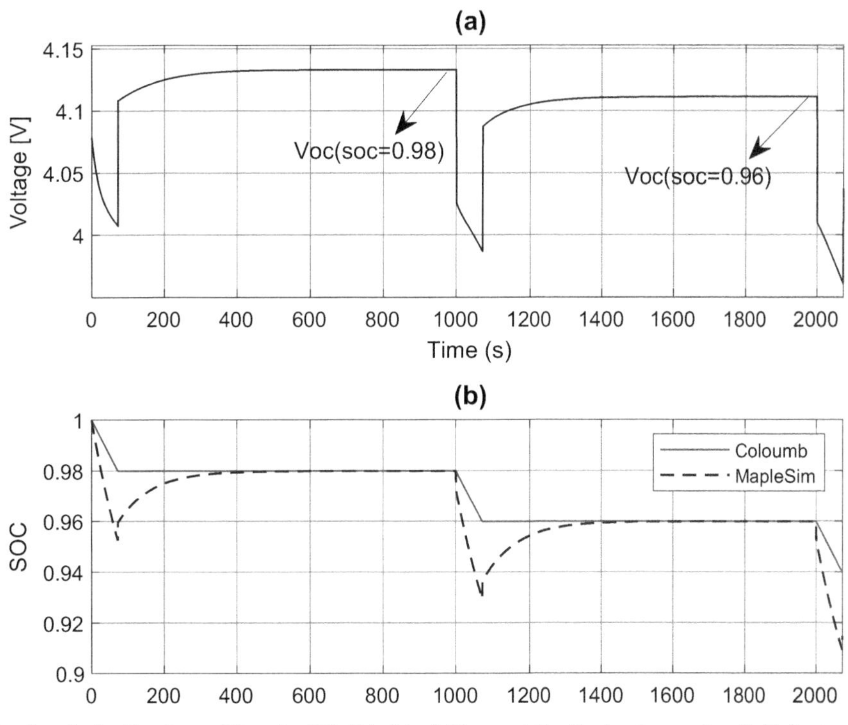

FIG. 7 (A) Battery electrochemical cell voltage, (B) and cell SoC in MapleSim, and the Coulomb-counting SoC, for two discharge pulse currents.

in the transient state when the pulse current has been zero. Since the ECM has two-time constants, a curve fitting tool of MATLAB is used on the data of resistance and capacitance to obtain their function versus SoC. Resistance and capacitances are fitted to exponential and fourth-order Gaussian functions, respectively, which are shown in Figs. 9 and 10. A detailed procedure of the parameter estimation and their values are provided in [16]. The equation of resistances and capacitances is presented in Eqs. (8) and (9). The coefficient of the capacitances is given in Table 4.

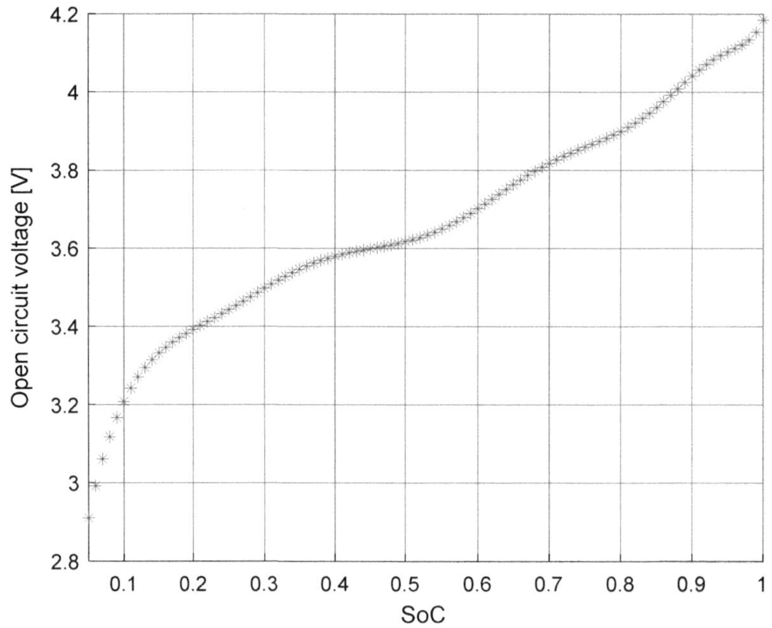

FIG. 8 Open-circuit voltage, v_{oc}, of the NCA Li-ion battery cell versus SoC.

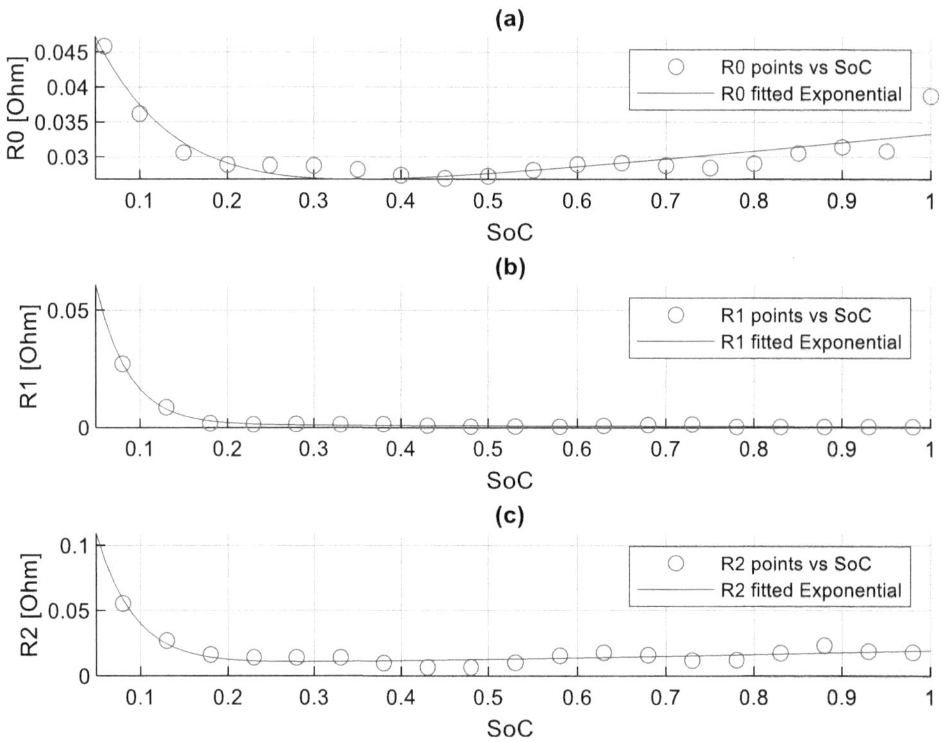

FIG. 9 Resistances of the ECM as a function of SoC. (A) R_0. (B) R_1. (C) R_2 [16].

$$
\begin{aligned}
R_0(SoC) &= 0.022e^{0.39SoC} + 0.042e^{-10.39SoC} \\
R_1(SoC) &= 0.234e^{-27.50SoC} + 0.001e^{-1.26SoC} \\
R_2(SoC) &= 0.326e^{-23.62SoC} + 0.008e^{-0.84SoC}
\end{aligned}
\tag{8}
$$

$$
C_j(SoC) = \alpha_1 e^{-\left(\frac{soc-\beta_1}{\delta_1}\right)^2} + \alpha_2 e^{-\left(\frac{soc-\beta_2}{\delta_2}\right)^2} + \alpha_3 e^{-\left(\frac{soc-\beta_3}{\delta_3}\right)^2} + \alpha_4 e^{-\left(\frac{soc-\beta_4}{\delta_4}\right)^2}
$$
$$
j \in \{1,2\} \tag{9}
$$

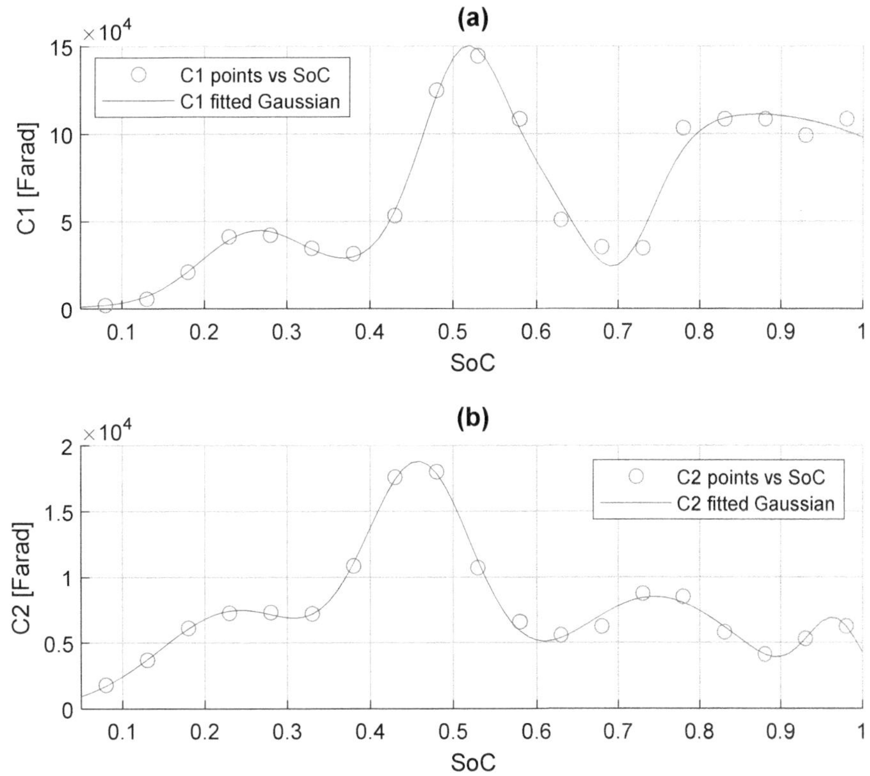

FIG. 10 Capacitances of the ECM as a function of SoC. (A) C_1. (B) C_2 [16].

TABLE 4 Coefficients of Gaussian function for representing capacitances [16].

Coefficients	C_1	C_2
α_1	2.23×10^7	7412
β_1	0.524	0.238
δ_1	0.139	0.131
α_2	1.16×10^5	1.813×10^4
β_2	0.837	0.460
δ_2	0.388	0.092
α_3	8.66×10^4	5888
β_3	0.354	0.967
δ_3	0144	0.050
α_4	-2.22×10^7	8487
β_4	0.524	0.747
δ_4	0.140	0.151

The schematic of the battery cell model is shown in Fig. 11. After simulation in MATLAB/Simulink, the ECM model is implemented in dSPACE for real-time simulation. The terminal voltage of the ECM and MapleSim electrochemical model is shown in Fig. 12.

In Fig. 12, the voltage response of the ECM is indicated by the red line, and the blue dash line belongs to the electrochemical high-fidelity model of the battery cell in MapleSim. The ECM model is implemented in the dSPACE SCALEXIO 9HE with a time-step simulation of 0.5 s. The error between the ECM and the electrochemical model in RTDS is

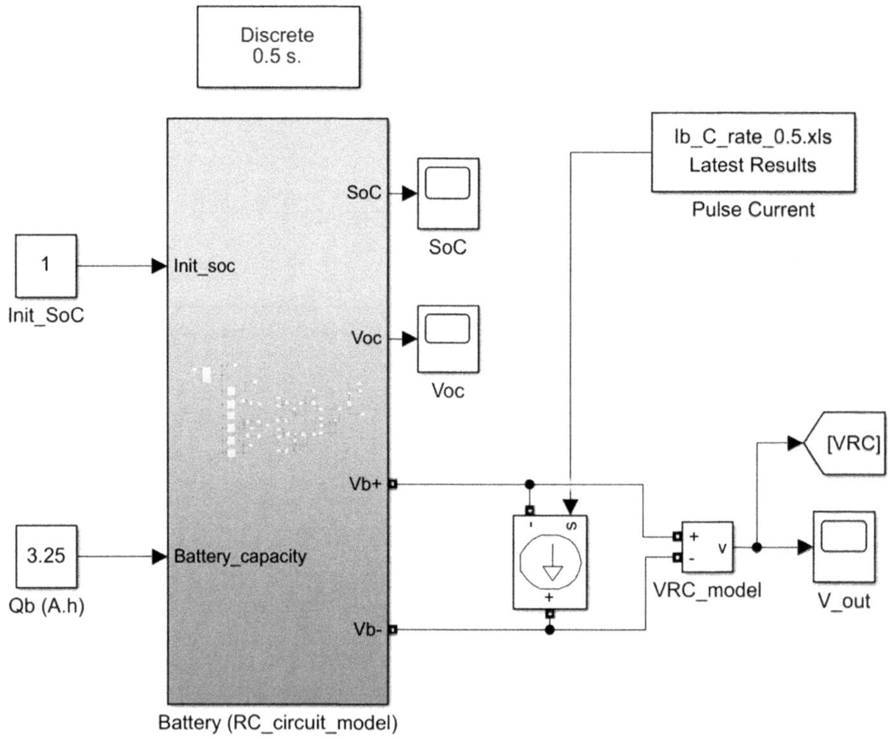

FIG. 11 Schematic of the battery cell model in MATLAB/Simulink.

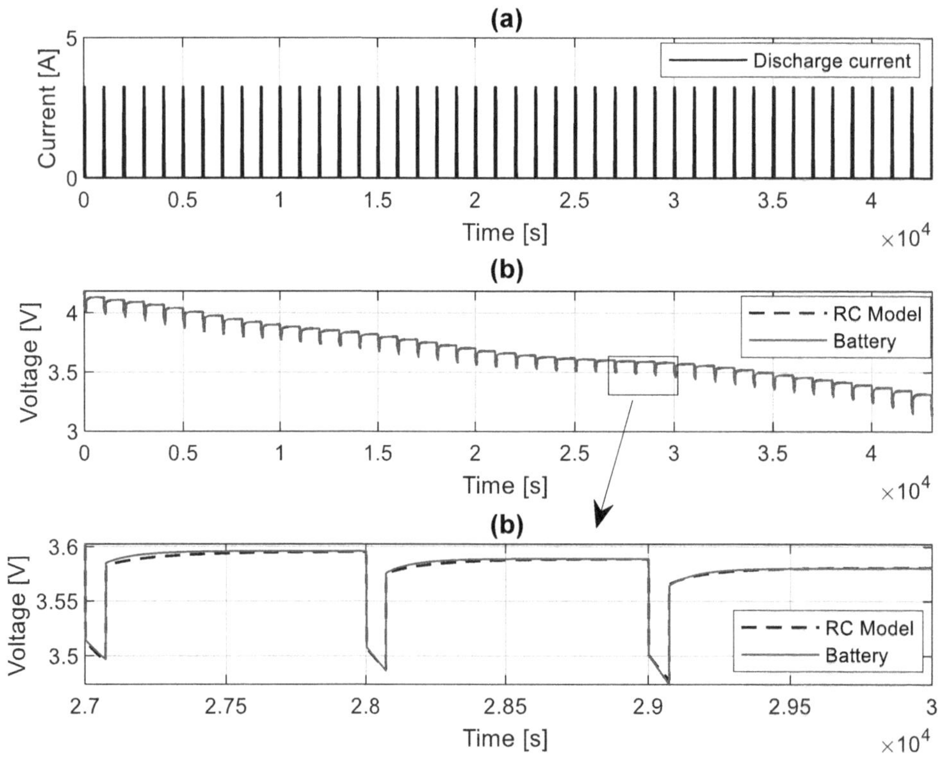

FIG. 12 Battery cell pulse current and voltage under C-rate discharge pulse current.

measured by two factors of root-mean-square-percentage-error (RMSPE), given in Eq. (10), and mean-absolute-percentage-error (MAPE), given in Eq. (11). These values are computed at 0.208% and 0.061%, respectively [16]. Small errors prove that the ECM method is accurate enough for modeling the dynamic behaviors of the Li-ion batteries.

$$\mathrm{RMSPE}(\%) = 100 \times \frac{\sqrt{\frac{1}{N}\sum_{i=1}^{N}\left(v_b(i) - \widehat{v}_b(i)\right)^2}}{\frac{1}{N}\sum_{i=1}^{N} v_b(i)} \tag{10}$$

$$\mathrm{MAPE}(\%) = \frac{100}{N}\sum_{i=1}^{N}\left|\frac{v_b(i) - \widehat{v}_b(i)}{v_b(i)}\right| \tag{11}$$

where v_b is the terminal voltage of the electrochemical model of the battery cell and $\widehat{v}_b$ is the terminal voltage of the battery model obtained by the ECM method.

19.4 Conclusion

The field of battery technology for EVs encompasses various technologies, including lead-acid, Ni-MH, Li-ion, and Li-S batteries. Lead-acid batteries, once prevalent for their durability and affordability, have become obsolete in automotive applications due to drawbacks such as low energy density and environmental concerns. Ni-Cd batteries have a higher energy density as compared to lead-acid, but due to the usage of cadmium, they are harmful to the ecosystem and human beings. Ni-MH batteries emerged as a suitable alternative with higher energy density and recyclability but faced challenges related to heat generation, overdischarging, and slower charging times. However, Li-ion batteries have risen as the most promising and widely adopted option in EVs. They offer high energy density, compact size, lightweight nature, and excellent performance across a range of temperatures. The battery life cycles, which are influenced by factors such as usage patterns, temperature, and charging rates, play a crucial role. A common method for predicting degradation is the dynamic battery degradation capacity model based on a semi-empirical Arrhenius degradation approach. For accurate real-time modeling of Li-ion batteries in EVs, the equivalent circuit model has proven effective. The model double RC branches within the ECM method accurately represents dynamic battery behaviors, whose parameters are estimated by discharge pulse current on the electrochemical high-fidelity model of the NCA battery cell in MapleSim. The battery cell model is implemented in dSPACE RTDS with a time step of 0.5 s. The small error in the terminal voltage between the ECM and the electrochemical model demonstrates the excellent performance of the ECM method for modeling the battery cell.

References

[1] A. Seaman, T.-S. Dao, J. McPhee, A survey of mathematics-based equivalent-circuit and electrochemical battery models for hybrid and electric vehicle simulation, J. Power Sources 256 (2014) 410–423, https://doi.org/10.1016/j.jpowsour.2014.01.057.

[2] M. Chen, G.A. Rincon-Mora, Accurate electrical battery model capable of predicting runtime and I-V performance, IEEE Trans. Energy Convers. 21 (2) (2006) 504–511, https://doi.org/10.1109/TEC.2006.874229.

[3] Battery Comparison of Energy Density - Cylindrical and Prismatic Cells. Accessed 20 November 2023. [Online]. Available: https://www.epectec.com/batteries/cell-comparison.html..

[4] P. Lailler, J.-F. Sarrau, C. Sarrazin, Comparative study for '36 V' vehicle applications: advantages of lead-acid batteries, J. Power Sources 95 (1) (2001) 58–67, https://doi.org/10.1016/S0378-7753(00)00638-8.

[5] X. Fan, et al., Battery technologies for grid-level large-scale electrical energy storage, Trans. Tianjin Univ. 26 (2) (2020) 92–103, https://doi.org/10.1007/s12209-019-00231-w.

[6] C.B. Zhu, M. Coleman, W.G. Hurley, State of charge determination in a lead-acid battery: combined EMF estimation and Ah-balance approach, in: 2004 IEEE 35th Annual Power Electronics Specialists Conference (IEEE Cat. No.04CH37551), Vol. 3, 2004, pp. 1908–1914, https://doi.org/10.1109/PESC.2004.1355409.

[7] Battery Degradation in Electric and Hybrid Electric Vehicles: A Survey Study | IEEE Journals & Magazine | IEEE Xplore. Accessed 8 November 2023. [Online]. Available: https://ieeexplore.ieee.org/document/10109713..

[8] I. Buchmann, Batteries in a Portable World: A Handbook on Rechargeable Batteries for Non-Engineers, fourth ed., Cadex Electronics Inc., Richmond, BC, 2017.

[9] Z. Johnson, R.A. Flake, Sealed nickel-cadmium batteries, in: IEEE 35th International Power Sources Symposium, 1992, pp. 106–109, https://doi.org/10.1109/IPSS.1992.282018.

[10] Z. Johnson, T.A. Edgar, Maintenance-free nickel-cadmium batteries, in: Proceedings of the Sixth Annual Battery Conference on Applications and Advances, 1991, pp. 97–107, https://doi.org/10.1109/BCAA.1991.761503.

[11] U. Koehler, Chapter 2 - General overview of non-Lithium battery systems and their safety issues, in: J. Garche, K. Brandt (Eds.), Electrochemical Power Sources: Fundamentals, Systems, and Applications, Elsevier, 2019, pp. 21–46, https://doi.org/10.1016/B978-0-444-63777-2.00002-5.

[12] B. Pierozynski, On the low temperature performance of nickel-metal hydride (NiMH) batteries, Int. J. Electrochem. Sci. 6 (4) (2011) 860–866, https://doi.org/10.1016/S1452-3981(23)15040-1.

[13] M.A. Hannan, F.A. Azidin, A. Mohamed, Hybrid electric vehicles and their challenges: a review, Renew. Sust. Energ. Rev. 29 (2014) 135–150, https://doi.org/10.1016/j.rser.2013.08.097.

[14] Clean Energy Institute. n.d. Lithium-Ion Battery, [Online]. Available: https://www.cei.washington.edu/science-of-solar/battery-technology/.

[15] U.-H. Kim, S.-B. Lee, N.-Y. Park, S.J. Kim, C.S. Yoon, Y.-K. Sun, High-energy-density Li-ion battery reaching full charge in 12 min, ACS Energy Lett. 7 (11) (2022) 3880–3888, https://doi.org/10.1021/acsenergylett.2c02032.

[16] M. Rezaei Larijani, M. Zolghadri, S. Hedayati Kia, A. El Hajjaji, Battery cell dynamic modeling using the RC equivalent circuit for RTDS frameworks, in: IECON 2021 – 47th Annual Conference of the IEEE Industrial Electronics Society, IEEE, Toronto, ON, 2021, pp. 1–6, https://doi.org/10.1109/IECON48115.2021.9589552.

[17] A. Fotouhi, D.J. Auger, K. Propp, S. Longo, M. Wild, A review on electric vehicle battery modelling: from Lithium-ion toward Lithium–Sulphur, Renew. Sust. Energ. Rev. 56 (2016) 1008–1021, https://doi.org/10.1016/j.rser.2015.12.009.

[18] H.M. Barkholtz, A. Fresquez, B.R. Chalamala, S.R. Ferreira, A database for comparative electrochemical performance of commercial 18650-format lithium-ion cells, J. Electrochem. Soc. 164 (12) (2017) A2697–A2706, https://doi.org/10.1149/2.1701712jes.

[19] S.M. Lukic, J. Cao, R.C. Bansal, F. Rodriguez, A. Emadi, Energy storage systems for automotive applications, IEEE Trans. Ind. Electron. 55 (6) (2008) 2258–2267, https://doi.org/10.1109/TIE.2008.918390.

[20] M. Shaibani, et al., Expansion-tolerant architectures for stable cycling of ultrahigh-loading sulfur cathodes in lithium-sulfur batteries, Sci. Adv. 6 (1) (2020) eaay2757, https://doi.org/10.1126/sciadv.aay2757.

[21] Targray. n.d. Cathode Active Materials: NCA, NMC, LFP, LMO, LCO. [Online]. Available: https://www.targray.com/li-ion-battery/cathode-materials/cathode-active-materials.

[22] L. Lu, X. Han, J. Li, J. Hua, M. Ouyang, A review on the key issues for lithium-ion battery management in electric vehicles, J. Power Sources 226 (2013) 272–288, https://doi.org/10.1016/j.jpowsour.2012.10.060.

[23] MapleSim Battery Library - Battery Modeling - Maplesoft. [Online]. Available: https://www.maplesoft.com/products/toolboxes/battery/..

[24] Types of Lithium-ion Batteries – Battery University. Accessed 20 August 2020. [Online]. Available: https://batteryuniversity.com/learn/article/types_of_lithium_ion..

[25] Wikipedia, Lithium Nickel Cobalt Aluminium Oxides, February 24, 2021, [Online]. Available: https://en.wikipedia.org/w/index.php?title=Lithium_nickel_cobalt_aluminium_oxides&oldid=1008749833.

[26] N. Williams. n.d. 4680 Battery: Everything You Need to Know About These New Cells, History-Computer. Accessed 19 November 2023. [Online]. Available: https://history-computer.com/4680-battery-cells/.

[27] Y. Ding, Z.P. Cano, A. Yu, J. Lu, Z. Chen, Automotive Li-ion batteries: current status and future perspectives, Electrochem. Energy Rev. 2 (1) (2019) 1–28, https://doi.org/10.1007/s41918-018-0022-z.

[28] Tesla. n.d. Electric Cars, Solar & Clean Energy, Accessed 19 March 2021. [Online]. Available: https://www.tesla.com/.

[29] MotorTrend. n.d. 2012 Tesla Model S First Drive - Motor Trend, Accessed 19 March 2021. [Online]. Available: https://www.motortrend.com/news/2012-tesla-model-s-first-drive/.

[30] T. Waldmann, R.-G. Scurtu, K. Richter, M. Wohlfahrt-Mehrens, 18650 vs. 21700 Li-ion cells – a direct comparison of electrochemical, thermal, and geometrical properties, J. Power Sources 472 (2020) 228614, https://doi.org/10.1016/j.jpowsour.2020.228614.

[31] Batterie-Boutique.fr. n.d. Batterie Li-ion Panasonic NCR18650B sans tête, flattop 65.3x18.5mm 3.6-3.7V 3400mAh pour les applications basse tension, Accessed 19 March 2021. [Online]. Available: https://batterie-boutique.fr/fr/batterie-li-ion-panasonic-ncr18650b-sans-tete-flattop-653x185mm-36-37v-3400mah-pour-les-applications-basse-tension/.

[32] Batterie-Boutique.fr. n.d. Batterie Li-ion Panasonic NCR18650B sans tête, flattop 65.3x18.5mm 3.6-3.7V 3400mAh pour les applications basse tension, Accessed 19 March 2021. [Online]. Available: https://batterie-boutique.fr/fr/batterie-li-ion-panasonic-ncr18650b-sans-tete-flattop-653x185mm-36-37v-3400mah-pour-les-applications-basse-tension/.

[33] M. Dubarry, C. Truchot, B.Y. Liaw, Synthesize battery degradation modes via a diagnostic and prognostic model, J. Power Sources 219 (2012) 204–216, https://doi.org/10.1016/j.jpowsour.2012.07.016.

[34] S.M. Rezvanizaniani, Z. Liu, Y. Chen, J. Lee, Review and recent advances in battery health monitoring and prognostics technologies for electric vehicle (EV) safety and mobility, J. Power Sources 256 (2014) 110–124, https://doi.org/10.1016/j.jpowsour.2014.01.085.

[35] Dragonfly Energy. n.d. What Is a Battery Life Cycle?, Dragonfly Energy. Accessed 12 November 2023. [Online]. Available: https://dragonflyenergy.com/battery-life-cycle/.

[36] H. Ahmed Sher, K.E. Addoweesh, Power storage options for hybrid electric vehicles—a survey, J. Renew. Sustain. Energy 4 (5) (2012) 052701, https://doi.org/10.1063/1.4759457.

[37] N. Omar, et al., Lithium iron phosphate based battery – assessment of the aging parameters and development of cycle life model, Appl. Energy 113 (2014) 1575–1585, https://doi.org/10.1016/j.apenergy.2013.09.003.

[38] A. Carnovale, X. Li, A modeling and experimental study of capacity fade for lithium-ion batteries, Energy AI 2 (2020) 100032, https://doi.org/10.1016/j.egyai.2020.100032.

[39] G. Ning, B. Haran, B.N. Popov, Capacity fade study of lithium-ion batteries cycled at high discharge rates, J. Power Sources 117 (1) (2003) 160–169, https://doi.org/10.1016/S0378-7753(03)00029-6.

[40] L. Wu, X. Fu, Y. Guan, Review of the remaining useful life prognostics of vehicle lithium-ion batteries using data-driven methodologies, Appl. Sci. 6 (6) (2016) 6, https://doi.org/10.3390/app6060166.

[41] V. Sangwan, R. Kumar, A.K. Rathore, An empirical capacity degradation modeling and prognostics of remaining useful life of Li-ion battery using unscented kalman filter, in: 2018 8th IEEE India International Conference on Power Electronics (IICPE), 2018, pp. 1–6, https://doi.org/10.1109/IICPE.2018.8709470.

[42] M. El-Dalahmeh, I. Bashir, M. Al-Greer, M. El-Dalahmeh, Lithium-ion batteries capacity degradation trajectory prediction based on decomposition techniques and NARX algorithm, in: 2022 57th International Universities Power Engineering Conference (UPEC), 2022, pp. 1–6, https://doi.org/10.1109/UPEC55022.2022.9917597.

[43] X. Cong, C. Zhang, J. Jiang, W. Zhang, Y. Jiang, A hybrid method for the prediction of the remaining useful life of lithium-ion batteries with accelerated capacity degradation, IEEE Trans. Veh. Technol. 69 (11) (2020) 12775–12785, https://doi.org/10.1109/TVT.2020.3024019.

[44] A. Guha, A. Patra, State of health estimation of lithium-ion batteries using capacity fade and internal resistance growth models, IEEE Trans. Transp. Electrification 4 (1) (2018) 135–146, https://doi.org/10.1109/TTE.2017.2776558.

[45] M. Rezaei Larijani, S. Hedayati Kia, M. Zolghadri, A. El Hajjaji, A. Taghavipour, Linear parameter-varying model predictive control for intelligent energy management in battery/supercapacitor electric vehicles, IEEE Access 12 (2024) 51026–51040, https://doi.org/10.1109/ACCESS.2024.3385861.

[46] X. Hu, W. Liu, X. Lin, Y. Xie, A.M. Foley, L. Hu, A control-oriented electrothermal model for pouch-type electric vehicle batteries, IEEE Trans. Power Electron. 36 (5) (2021) 5530–5544, https://doi.org/10.1109/TPEL.2020.3027561.

[47] A. Hentunen, T. Lehmuspelto, J. Suomela, Time-domain parameter extraction method for Thévenin-equivalent circuit battery models, IEEE Trans. Energy Convers. 29 (3) (2014) 558–566, https://doi.org/10.1109/TEC.2014.2318205.

[48] B. Schweighofer, K.M. Raab, G. Brasseur, Modeling of high power automotive batteries by the use of an automated test system, IEEE Trans. Instrum. Meas. 52 (4) (2003) 4, https://doi.org/10.1109/TIM.2003.814827.

[49] M.R. Larijani, M. Zolghadri, S.H. Kia, A. El Hajjaji, Performance analysis of the lithium-ion battery RC equivalent circuit model using EPA drive cycles, in: 2022 13th Power Electronics, Drive Systems, and Technologies Conference (PEDSTC), Tehran, Iran, IEEE, 2022, pp. 393–397, https://doi.org/10.1109/PEDSTC53976.2022.9767430.

[50] M. Chen, G.A. Rincon-Mora, Accurate electrical battery model capable of predicting runtime and I-V performance, IEEE Trans. Energy Convers. 21 (2) (2006) 2, https://doi.org/10.1109/TEC.2006.874229.

CHAPTER

20

Sustainability and environmental impacts of electric vehicles

Anushree Ramanath
Schneider Electric, San Jose, CA, United States

20.1 Introduction

The advent of electric vehicles (EVs) marks the initiation of a critical exploration into the transformative potential of these vehicles within the broader context of environmental responsibility and technological evolution. As societies grapple with the repercussions of climate change and seek alternatives to conventional fossil fuel-dependent transportation, EVs have emerged as a promising solution. This paradigm shift represents a departure from traditional internal combustion engine (ICE) vehicles, offering a pathway to mitigate the environmental impact of transportation and reduce dependence on finite fossil fuel resources.

The impacts of EVs across their entire life cycle are multifaceted. The environmental implications span from the extraction of raw materials for battery production to end-of-life considerations, requiring a comprehensive life cycle assessment. EVs hold the potential to curtail greenhouse gas (GHG) emissions significantly, enhance air quality in urban environments, and contribute to the broader sustainability agenda. However, achieving these benefits necessitates a nuanced examination of manufacturing processes, resource consumption, and the integration of EVs with renewable energy sources.

The societal and economic dimensions of this transition are equally consequential. The adoption of EVs not only redefines the automotive landscape but also sparks discussions on job creation, economic restructuring, and the role of government policies in shaping sustainable transportation systems. As EVs become more accessible to consumers, understanding the factors influencing their adoption, such as cost considerations and charging infrastructure, becomes pivotal. It is crucial to consider a holistic exploration to understand the intricate dynamics, possible challenges and opportunities, and future trajectories of EVs in the quest for a more sustainable and environmentally conscious mode of transportation.

20.2 Environmental impacts

The environmental impacts of EVs have become a focal point in the quest for sustainable transportation solutions [1]. With the growing awareness of climate change and the need to reduce GHG emissions, EVs have emerged as a promising alternative to traditional ICE vehicles. Unlike their fossil fuel counterparts, EVs produce zero tailpipe emissions during operation, contributing to improved air quality and mitigating the environmental footprint of transportation [2]. However, the comprehensive evaluation of the environmental benefits of EVs extends beyond their usage phase, necessitating an examination of the life cycle, including manufacturing processes, resource extraction for batteries, and end-of-life considerations.

20.2.1 Reduction of GHG emissions and air quality improvement

The reduction of GHG emissions and the concurrent improvement of air quality represent paramount advantages of EVs in the pursuit of sustainable transportation solutions. One of the primary environmental benefits of EVs lies in

Handbook of Power Electronics in Autonomous and Electric Vehicles
https://doi.org/10.1016/B978-0-323-99545-0.00020-8

their capacity to dramatically decrease GHG emissions associated with traditional ICE vehicles [3]. By shifting the propulsion source from fossil fuels to electricity, EVs can operate with lower or even zero emissions, particularly when charged using renewable energy sources like solar or wind power. This transition holds immense significance in the global effort to mitigate climate change, aligning with international goals to curb carbon emissions and limit the detrimental effects of anthropogenic activities on the environment.

Beyond the reduction in GHGs, the inherent design of EVs contributes substantially to the improvement of air quality, particularly in densely populated urban areas. Unlike conventional vehicles emitting pollutants directly into the atmosphere, EVs produce zero tailpipe emissions during operation. This characteristic makes electric mobility a key player in combatting air pollution, a critical concern for public health [4]. The adoption of EVs can alleviate the concentration of harmful pollutants such as nitrogen oxides and particulate matter, thereby fostering cleaner and healthier urban environments. As governments and communities increasingly recognize the health implications of poor air quality, the transition to EVs emerges as a strategic and imperative move toward sustainable transportation.

The shift toward electric mobility represents a crucial step in fostering cleaner and more sustainable transportation systems, addressing both local air quality concerns and the broader challenge of reducing the transport sector's carbon footprint on a global scale. As technological advancements and infrastructure development continue to support the transition to EVs, the dual benefits of diminished GHG emissions and enhanced air quality position EVs as key contributors to a more environmentally conscious and healthier future.

While the reduction of GHG emissions and improved air quality are notable advantages, the full realization of these benefits hinges on factors such as the energy sources used for electricity generation and the life cycle analysis of vehicle production [5]. Policymakers, industry stakeholders, and consumers must collaboratively address these considerations to maximize the positive environmental impact of EVs and ensure a sustainable, holistic approach to the future of transportation.

20.2.2 Comparative environmental impact with ICE vehicles

Comparing the environmental impact of EVs with ICE vehicles is a multifaceted analysis that considers various stages of the vehicle life cycle. From a GHG emission perspective, EVs generally exhibit a lower carbon footprint, especially when charged using electricity from renewable sources. While ICE vehicles emit carbon dioxide during operation, EVs, even when considering the energy used in battery production, tend to offer a more environmentally friendly alternative over their lifetime. However, it's crucial to acknowledge that the sustainability of EVs depends significantly on the energy mix of the grid from which they draw power, as grids powered by fossil fuels may diminish the overall emission benefits [6].

Comparing the environmental impact of EVs with ICE vehicles is a multifaceted analysis that considers various stages of the vehicle life cycle. From a GHG emission perspective, EVs generally exhibit a lower carbon footprint, especially when charged using electricity from renewable sources. While ICE vehicles emit carbon dioxide during operation, EVs, even when considering the energy used in battery production, tend to offer a more environmentally friendly alternative over their lifetime. However, it's crucial to acknowledge that the sustainability of EVs depends significantly on the energy mix of the grid from which they draw power, as grids powered by fossil fuels may diminish the overall emission benefits [7].

In terms of local air quality, the comparative advantage of EVs becomes more pronounced. ICEs release pollutants such as nitrogen oxides, particulate matter, and volatile organic compounds directly into the atmosphere, contributing to smog formation and respiratory issues. In contrast, EVs produce zero tailpipe emissions during operation, significantly reducing the immediate environmental impact in urban areas [8]. This characteristic positions EVs as a cleaner and healthier choice, particularly in regions struggling with air quality concerns and aiming to mitigate the adverse health effects associated with vehicular emissions. However, it is critical to assess the entire life cycle of EVs from production to maintenance to ensure sustainability at every stage.

20.3 Resource use and manufacturing considerations

The key aspect of resource use and manufacturing considerations for an EV is the production of high-capacity batteries. The extraction and processing of key battery components, such as lithium, cobalt, and nickel, raise concerns about resource depletion, environmental impact, and ethical mining practices. Moreover, the energy-intensive manufacturing processes involved in creating EVs contribute to the overall ecological footprint. As the automotive industry undergoes a transformative shift toward electrification, understanding and mitigating the environmental implications of resource extraction and manufacturing are imperative for realizing the full potential of EVs as a sustainable transportation solution.

20.3.1 Battery production processes

The battery production processes for EVs play a pivotal role in shaping both the environmental and economic dimensions of the electric mobility landscape. Central to the heart of EV technology is the lithium-ion battery, a sophisticated energy storage system that empowers these vehicles. The production of lithium-ion batteries involves several intricate steps, starting with the extraction and processing of raw materials. Lithium, cobalt, nickel, and manganese are essential components, and their extraction raises concerns about environmental impact, resource depletion, and, in certain cases, unethical mining practices. Striking a balance between the increasing demand for these materials and ensuring responsible sourcing remains a key challenge in promoting the sustainability of EV battery production.

Once the raw materials are procured, the manufacturing process entails complex chemical engineering processes, involving the precise assembly of electrodes, separators, and electrolytes to form battery cells. The assembly demands a controlled environment to maintain the integrity and efficiency of the battery. The energy intensity of this manufacturing phase is a noteworthy consideration, as it contributes to the overall carbon footprint of EVs. Strategies aimed at reducing the energy consumption in manufacturing, optimizing recycling processes, and exploring alternative battery chemistries are actively pursued by researchers and manufacturers to enhance the sustainability of battery production.

The scale and standardization of battery production are critical aspects influencing the economic feasibility of EVs. Economies of scale can lead to cost reductions, making EVs more accessible to consumers. However, achieving this requires substantial investments in production facilities and infrastructure, with considerations for the environmental impact of constructing and operating such facilities. As the EV market continues to expand, advancements in battery production technologies, coupled with strategic sustainability measures, are crucial for addressing the challenges associated with resource extraction, energy consumption, and economic viability, ultimately steering the electric mobility sector toward a more sustainable future.

20.3.2 Battery recycling and disposal

Lithium-ion batteries, commonly employed in EVs, contain valuable and sometimes scarce materials, including lithium, cobalt, nickel, and manganese. Efficient recycling processes can help recover these materials, reducing the demand for new mining and minimizing environmental impact. However, the challenge lies in developing robust and cost-effective recycling methods that can handle the complexity of lithium-ion batteries. The diverse chemistries, varying form factors, and safety considerations make recycling a technically intricate process.

Battery disposal at the end of its life cycle poses environmental and safety concerns. If not managed properly, batteries can release hazardous materials into the environment, leading to soil and water contamination. Responsible disposal methods are crucial to prevent environmental degradation and protect human health. As the number of EVs on the road continues to rise, establishing streamlined and standardized battery recycling and disposal practices becomes imperative. Governments and industry stakeholders need to collaborate to create regulatory frameworks that incentivize and ensure proper disposal procedures, while innovations in recycling technologies are crucial for optimizing the recovery of valuable materials and minimizing the environmental footprint of the electric mobility sector.

The economic viability of battery recycling is closely tied to the market demand for recycled materials [9]. As recycling technologies advance and the circular economy gains prominence, recycled materials from EV batteries can find applications in various industries, including new battery production. This closed-loop approach contributes not only to resource conservation but also to the economic sustainability of EVs. A holistic strategy encompassing effective recycling processes, stringent disposal regulations, and market-driven incentives is essential to maximize the environmental and economic benefits of battery recycling for the long-term success of the EV industry.

20.3.3 Sustainable manufacturing

At the core of sustainable manufacturing for EVs is the need to address the energy intensity associated with the production of its components, particularly high-capacity batteries. To minimize the carbon footprint, manufacturers are increasingly incorporating renewable energy sources into their production processes. Adopting solar and wind power for manufacturing facilities not only reduces GHG emissions but also contributes to the overall sustainability of EVs [10].

Beyond energy considerations, sustainable manufacturing involves responsible resource management and ethical sourcing practices. As the demand for EVs escalates, ensuring a secure and ethical supply chain for critical materials like lithium, cobalt, and nickel becomes paramount. Industry stakeholders are exploring alternative battery chemistries and materials to reduce dependency on scarce or environmentally impactful resources. Additionally, circular economy principles are gaining traction, encouraging the reuse and recycling of materials in the manufacturing process. This shift toward sustainable resource management not only reduces the environmental footprint but also enhances the long-term resilience of the EV industry.

Collaboration among industry players, governments, and research institutions is fundamental to driving innovations in sustainable manufacturing. Continuous research and development efforts focus on optimizing production processes, improving energy efficiency, and minimizing waste generation. The integration of environmentally friendly materials, such as recycled plastics and lightweight composites, further contributes to the sustainability of EV manufacturing. By embracing these multifaceted approaches, the automotive sector can align with broader environmental goals, ensuring that the transition to electric mobility is not only technologically advanced but also environmentally responsible and economically viable.

20.4 Charging infrastructure and its economic impacts

The development of robust charging infrastructure is a linchpin in the widespread adoption and economic success of EVs. As the EV market expands, the availability and accessibility of charging stations become critical factors influencing consumer confidence and acceptance. Establishing a comprehensive charging network not only addresses range anxiety but also stimulates economic growth. Investment in charging infrastructure leads to job creation, stimulates local economies, and fosters innovation. Moreover, it facilitates a transition toward cleaner energy sources, reinforcing the economic benefits of sustainable transportation. Thus, it is crucial to closely examine the economic impacts of charging infrastructure on the EV ecosystem, exploring its role in shaping consumer behavior, market dynamics, and broader economic transformations.

20.4.1 Impact on electricity grids

The growing adoption of EVs poses challenges to electricity grids, particularly in densely populated urban areas. This calls for strategic planning to ensure the seamless integration of EVs into the existing power systems. The increased demand for electricity during peak charging times could strain the grid if not managed effectively. To address this, smart charging solutions and advanced grid management technologies are being implemented. These systems enable dynamic load balancing, optimizing charging times based on grid capacity and electricity demand fluctuations, thus mitigating potential disruptions [11].

The geographical distribution of charging stations also plays a crucial role in grid impact. Concentrated clusters of charging stations in specific regions may strain local grids, requiring substantial upgrades. Conversely, a well-distributed and strategically located charging infrastructure ensures a more even load on the grid. Governments and utility providers are actively investing in grid enhancements to accommodate the surge in EV charging demand. Integrating renewable energy sources, such as solar and wind power, into the grid can also enhance the sustainability of EV charging, reducing the overall carbon footprint associated with electric mobility.

As EV adoption continues to grow, the long-term success of EV charging infrastructure depends on robust planning and collaboration between stakeholders. This includes not only grid operators but also automakers, charging station providers, and government entities. A harmonized approach that considers grid capacity, renewable energy integration, and innovative technologies is essential to create an EV charging ecosystem that is not only convenient for users but also sustainable and supportive of the broader goals of transitioning to cleaner transportation.

20.4.2 Grid upgrades and smart grid technologies

Grid upgrades and the implementation of smart grid technologies are indispensable components of ensuring the efficient and resilient integration of EV charging infrastructure with existing power systems. Traditional electricity grids were not originally designed to handle the increased demand posed by widespread EV charging. Consequently, grid upgrades become essential to accommodate the additional load, prevent overloads during peak charging times, and enhance overall reliability. Investments in grid infrastructure include the reinforcement of transmission and distribution networks, the installation of high-capacity transformers, and the development of smart grid solutions that enable real-time monitoring and control of energy flows [12].

Smart grid technologies play a pivotal role in optimizing the performance of EV charging infrastructure. These technologies allow for demand response mechanisms, enabling the grid to communicate with charging stations and adjust their operation based on grid conditions. Time-of-use pricing models, facilitated by smart grids, incentivize EV owners to charge their vehicles during periods of lower electricity demand, thus balancing the load on the grid. Additionally, advanced metering systems enable precise tracking of electricity consumption, enhancing billing accuracy and facilitating the implementation of flexible pricing structures. The integration of smart grid technologies not only enhances the reliability of the grid but also supports the sustainability goals associated with EV adoption by promoting efficient energy use and reducing the environmental impact of charging operations [13].

To ensure the successful deployment of EV charging infrastructure and its harmonious integration with smart grids, collaboration among governments, utilities, and technology providers is paramount. Regulatory frameworks that incentivize grid upgrades and the implementation of smart grid technologies, coupled with ongoing innovation in energy management systems, contribute to building a resilient and adaptive infrastructure capable of meeting the evolving demands of electric mobility.

20.4.3 Economic opportunities and challenges

The development of charging infrastructure for EVs presents a spectrum of economic opportunities and challenges that have far-reaching implications for various industries and stakeholders. On the positive side, the construction and maintenance of EV charging stations create job opportunities in sectors ranging from construction to technology. This job creation extends beyond the immediate installation phase, encompassing ongoing maintenance, customer support, and the development of new technologies to enhance charging efficiency and accessibility. Moreover, as the EV market expands, there are opportunities for entrepreneurs and businesses to invest in and operate charging stations, contributing to economic growth and innovation.

However, economic challenges accompany these opportunities, necessitating careful navigation for sustained success. One challenge is the initial capital investment required for building a comprehensive charging network. The establishment of charging infrastructure demands significant financial commitments, and the return on investment may not be immediate. To address this, partnerships between governments, private businesses, and investors are often crucial to share the financial burden and incentivize the expansion of charging networks. Additionally, uncertainties related to the future demand for EVs and the rapid evolution of technology pose challenges for investors and businesses, requiring a dynamic and adaptable approach to charging infrastructure development.

The economic landscape of charging infrastructure is influenced by regulatory policies and market dynamics. Governments play a pivotal role in shaping the economic viability of EV charging through incentives, subsidies, and regulatory frameworks. Balancing the need for profitability with the broader societal and environmental goals of sustainable transportation requires careful policy considerations [14]. Striking this balance is crucial for fostering a competitive and economically sustainable market for EV charging infrastructure, ensuring that it remains a vibrant and integral component of the evolving automotive ecosystem.

20.5 Policy and incentives

The landscape of EV adoption is intricately shaped by policy frameworks and incentives designed to encourage the transition to sustainable transportation. Governments around the world are implementing a variety of measures to accelerate the adoption of EVs, recognizing their role in mitigating climate change and reducing dependence on fossil fuels [14]. Policy initiatives include financial incentives such as tax credits, rebates, and subsidies to make EVs more economically attractive for consumers. Additionally, regulatory measures, such as emission standards and zero-emission vehicle mandates, play a crucial role in driving automakers toward the production of EVs [15]. This necessitates a closer examination of the multifaceted policy and incentive landscape, exploring its impact on consumer behavior, industry dynamics, and the overall trajectory of electric mobility.

20.5.1 Perspectives on EV policy

The perspectives on EV policy are diverse and often reflect a complex interplay of environmental, economic, and social considerations. From an environmental standpoint, supportive policies are seen as imperative in mitigating climate change and improving air quality. Governments worldwide are adopting measures such as emissions standards, fuel efficiency regulations, and direct incentives to encourage the adoption of EVs, with the aim of reducing GHG emissions from the transportation sector. These policies not only align with global sustainability goals but also contribute to addressing local air pollution concerns, especially in urban areas.

On an economic front, perspectives on EV policy vary based on industry involvement and regional considerations. Automotive manufacturers view supportive policies as crucial for stimulating market demand and fostering innovation in EV technologies. In contrast, concerns about economic feasibility and the potential impact on traditional automotive industries may influence certain stakeholders to approach EV policies with caution. The economic benefits, however, extend beyond the automotive sector, with job creation opportunities arising in areas such as EV manufacturing, charging infrastructure development, and research and development of advanced technologies.

From a social perspective, the success of EV policies relies heavily on public acceptance and accessibility. Governments and policymakers often face the challenge of striking a balance between incentivizing EV adoption and addressing concerns related to affordability, charging infrastructure availability, and range anxiety. Perspectives on EV policy also consider the inclusivity of the transition, ensuring that benefits are accessible to a broad demographic. Addressing disparities in access to EVs and charging infrastructure is crucial for fostering widespread adoption and garnering public support for the transition to electric mobility. As perspectives continue to evolve, finding common ground among stakeholders is essential for shaping effective and inclusive EV policies that drive sustainable transportation initiatives globally.

20.5.2 Incentives for EV adoption

Incentives for EV adoption are key drivers in encouraging consumers to make the transition to more sustainable modes of transportation. Financial incentives, such as tax credits and rebates, are among the most prevalent tools used by governments to stimulate EV adoption. These incentives directly reduce the upfront costs of purchasing EVs, making them more economically attractive for consumers. Subsidies for EV charging infrastructure installation also contribute to the growth of EV adoption, addressing concerns related to charging accessibility and reinforcing the overall economic feasibility of electric mobility.

Beyond direct financial incentives, nonmonetary benefits further incentivize consumers to choose EVs. Incentives may include preferential treatment in high-occupancy vehicle (HOV) lanes, reduced toll fees, or exemptions from certain vehicle-related taxes and fees. These perks not only provide tangible advantages to EV owners but also contribute to a positive public image of EVs, fostering a broader cultural shift toward sustainable transportation choices.

Incentives are not limited to individual consumers; fleet incentives also play a crucial role in promoting electric mobility. Governments and businesses often collaborate to offer incentives for companies to integrate EVs into their fleets. These incentives may include grants for purchasing EVs, installing charging infrastructure at corporate facilities, or receiving tax benefits for operating an eco-friendly fleet. By extending incentives to commercial entities, governments aim to accelerate the overall adoption of EVs and contribute to a more sustainable and environmentally conscious transportation ecosystem.

20.6 Consumer adoption

Consumer adoption of EVs marks a pivotal paradigm shift in the automotive landscape, reflecting a growing awareness of environmental sustainability and a desire for innovative, cost-effective transportation solutions [16]. As advancements in EV technology continue to address concerns related to range anxiety and charging infrastructure, consumers are increasingly considering EVs as viable alternatives to traditional ICE cars. Key aspects for consideration include factors influencing consumer attitudes and behaviors regarding EV adoption, encompassing considerations such as cost, charging infrastructure, and overall driving experience. Understanding these dynamics is essential for policymakers, manufacturers, and industry stakeholders seeking to accelerate the widespread acceptance and integration of EVs into mainstream automotive culture.

20.6.1 Cost considerations for consumers

Cost considerations play an important role in shaping consumer attitudes toward EVs, thus influencing the adoption and mainstream integration of this transformative technology. One of the primary factors influencing consumer decisions is the upfront cost of EVs compared to traditional ICE vehicles. While the overall price of EVs has been decreasing over time, the initial purchase price remains a significant consideration for many consumers. Government incentives, such as tax credits and rebates, are instrumental in alleviating this financial barrier, making EVs more accessible and financially competitive.

Beyond the upfront cost, the total cost of ownership is a crucial consideration for consumers evaluating the economic feasibility of EVs. Lower operating and maintenance costs characterize EV ownership due to fewer moving parts, reduced reliance on expensive ICE components, and the generally lower cost of electricity compared to traditional fuels [17]. The long-term economic benefits, including potential savings on fuel and maintenance, contribute to the overall attractiveness of EVs, offsetting the initial purchase price over time.

Charging infrastructure and the cost of installing home-charging stations also factor into consumer decisions. The availability and convenience of charging stations influence the perceived practicality of EV ownership. Governments and utility companies are increasingly investing in expanding charging infrastructure, addressing consumer concerns about range anxiety, and ensuring that EV owners have convenient access to charging facilities. As charging infrastructure becomes more widespread and affordable, it contributes to enhancing the overall appeal of EVs, further incentivizing consumers to embrace this sustainable mode of transportation. In this dynamic landscape, the careful consideration of these cost-related factors is pivotal for stakeholders aiming to foster widespread consumer adoption of EVs.

20.6.2 Consumer perception and awareness

In recent years, there has been a notable shift in consumer attitudes toward EVs, driven by increasing environmental consciousness and a desire for more sustainable transportation options. However, challenges persist in dispelling lingering misconceptions and enhancing overall awareness [18]. Some consumers still perceive EVs as expensive or believe in the myth of limited range, commonly known as range anxiety. Addressing these concerns requires comprehensive education campaigns to inform consumers about the technological advancements that have mitigated these issues, such as improved battery efficiency and the expansion of charging infrastructure [19].

Positive consumer experiences and word-of-mouth recommendations are powerful tools in shaping perceptions of EVs. As more individuals experience the benefits of driving EVs, including lower operating costs, reduced environmental impact, and a smooth and quiet driving experience, positive narratives emerge, contributing to a favorable perception of electric mobility. Additionally, initiatives aimed at showcasing real-world EV success stories and dispelling common myths can play a crucial role in altering perceptions, fostering a more accurate understanding of the capabilities and advantages of EVs.

Government policies, incentives, and public awareness campaigns further play a role in shaping consumer perceptions. Subsidies, tax credits, and other financial incentives are instrumental in incentivizing consumers to explore EV options. Moreover, targeted awareness campaigns can disseminate accurate information about the benefits of EVs, addressing concerns related to charging infrastructure, battery life, and overall vehicle performance. In this dynamic landscape, efforts to enhance consumer perception and awareness are pivotal for achieving widespread adoption, fostering a positive and informed consumer base that actively contributes to the ongoing transition to electric mobility.

20.6.3 Overcoming barriers to adoption

Overcoming barriers to the widespread adoption of EVs is crucial for realizing the full potential of sustainable transportation. One significant barrier is the perceived lack of charging infrastructure, leading to range anxiety among potential EV buyers. Governments and private entities are actively investing in expanding charging networks, strategically locating charging stations along major highways and in urban centers. As charging infrastructure becomes more robust and accessible, concerns about range limitations are alleviated, paving the way for increased consumer confidence in adopting EVs.

Cost considerations are another prominent barrier, with the initial purchase price of EVs often perceived as higher than traditional ICE vehicles. Governments worldwide are addressing this barrier by offering financial incentives, tax credits, and rebates to make EVs more economically attractive to consumers. Additionally, as battery technology advances and economies of scale come into play, the overall cost of EVs is decreasing, making them more competitive in the automotive market. Transparent communication about the long-term cost savings associated with EV ownership, including reduced fuel and maintenance expenses, further contributes to overcoming cost-related barriers.

Consumer education and awareness are essential for dispelling myths and addressing misconceptions about EVs. Many potential buyers still harbor concerns about battery life, maintenance costs, and the overall performance of EVs.

Educational campaigns that highlight technological advancements, real-world success stories, and the environmental benefits of electric mobility play a crucial role in altering perceptions. By actively engaging with consumers to provide accurate and comprehensive information, stakeholders in the EV ecosystem contribute to breaking down barriers and fostering a more informed and receptive audience for sustainable transportation solutions.

20.7 Integration with renewable energy

The integration of EVs with renewable energy sources marks a visionary convergence in the pursuit of sustainable transportation solutions. As the world shifts toward cleaner energy alternatives, the coupling of EVs with renewable sources, such as solar and wind power, holds the promise of a harmonious and environmentally conscious future. This vision is encapsulated in Fig. 1. The integration of EVs with renewable energy not only contributes to mitigating climate change but also represents a strategic leap toward a more resilient and sustainable mobility paradigm.

20.7.1 Renewable energy and EV synergies

The synergies between EVs and renewable energy sources form a powerful alliance at the forefront of sustainable transportation. At the heart of this integration is the potential to mitigate the environmental impact of both the transportation and energy sectors. Renewable energy, derived from sources such as solar, wind, and hydropower, provides a clean and abundant alternative to traditional fossil fuels. When harnessed to power EVs, it ensures that the entire life cycle of transportation is characterized by reduced GHG emissions, aligning with global efforts to combat climate change.

The technological advancements in both EVs and renewable energy systems contribute to a mutually reinforcing relationship. Smart charging solutions enable EVs to charge during periods of high renewable energy generation, optimizing the use of clean power. Vehicle-to-grid (V2G) technology allows EVs to not only draw power from the grid but also return excess energy back, creating a dynamic and flexible energy ecosystem [20]. Moreover, advancements

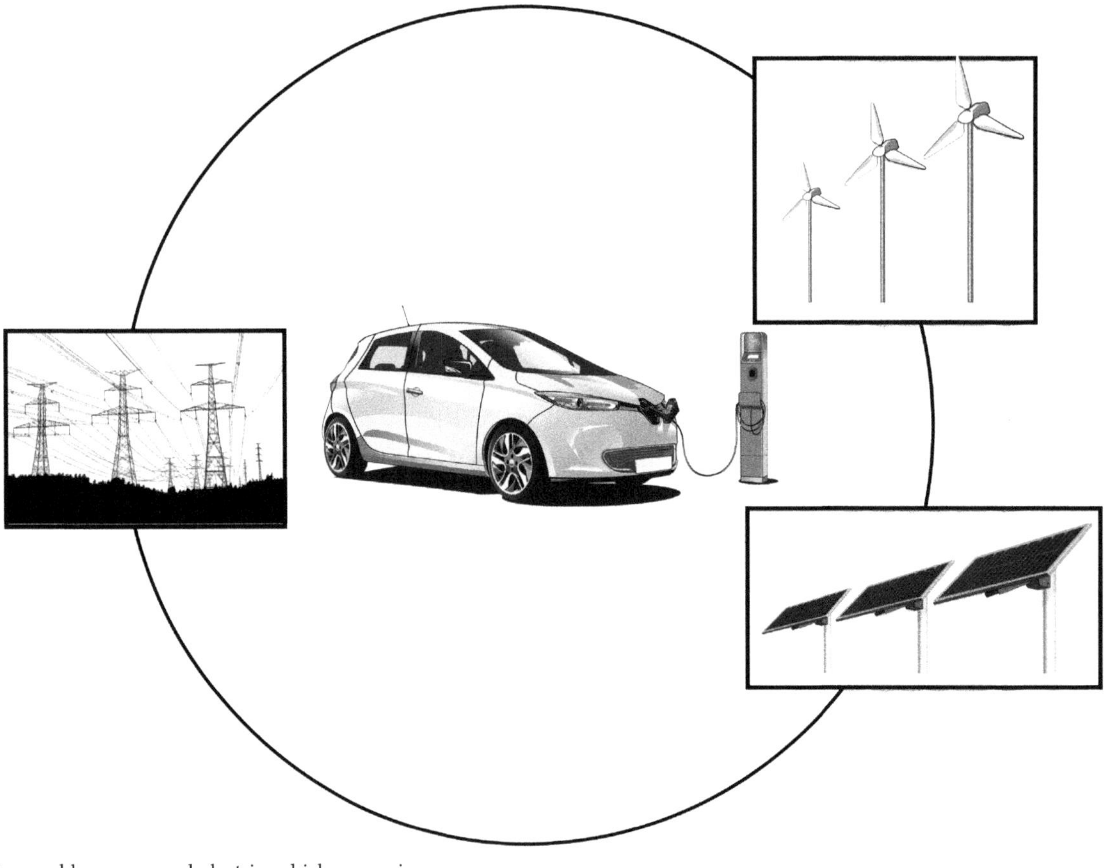

FIG. 1 Renewable energy and electric vehicle synergies.

in energy storage technologies, such as home battery systems, enable EV owners to store excess renewable energy for later use, fostering greater energy independence and resilience.

Beyond environmental benefits, the integration of EVs with renewable energy offers economic advantages and contributes to energy democratization. By decentralizing power generation and encouraging the adoption of solar panels and home charging solutions, individuals and communities become active participants in the energy transition. Governments and businesses are recognizing these synergies, investing in renewable energy infrastructure and promoting policies that incentivize the coupling of EVs with clean energy sources. This convergence of technologies marks a pivotal step toward a sustainable energy future, redefining the way we power our vehicles and reshaping the broader landscape of the energy ecosystem.

20.7.2 Sustainability of charging

The sustainability of charging EVs encompasses a multifaceted approach that considers environmental, economic, and societal dimensions. One key aspect is the source of electricity used for charging. As the electricity grid shifts toward renewable energy sources, the charging process becomes inherently more sustainable. Charging EVs with electricity generated from solar, wind, or other clean sources significantly reduces the carbon footprint associated with the vehicle's operation, contributing to broader efforts to combat climate change. Policies and initiatives promoting the integration of renewable energy into the grid play a pivotal role in ensuring the sustainability of EV charging [21].

Charging infrastructure development is another critical factor in enhancing the sustainability of EVs. Strategic placement of charging stations, incorporating energy-efficient technologies, and utilizing smart charging solutions contribute to the overall environmental impact. Moreover, investments in expanding charging infrastructure not only address range anxiety but also stimulate economic growth by creating jobs and supporting the electric mobility ecosystem. Sustainable charging also involves considerations of energy efficiency during the transmission and distribution phases, minimizing energy losses and optimizing the overall effectiveness of the charging process.

Public awareness and education about sustainable charging practices are paramount in fostering responsible EV ownership. Encouraging off-peak charging, utilizing home-charging solutions with renewable energy sources, and embracing innovative technologies like V2G contribute to a more sustainable charging ecosystem [22]. As the EV market continues to expand, the sustainability of charging practices becomes integral to the overall success of the transition to electric mobility, ensuring that the benefits of EVs extend beyond individual consumers to positively impact the environment and society at large.

20.8 Technological advances

The technological advances in EVs have propelled this transformative mode of transportation into the forefront of automotive innovation [23]. Fig. 2 highlights some of the key technological drivers in the advancement of EVs. Rapid strides in battery technology, charging infrastructure, and overall vehicle design have revolutionized the landscape of sustainable mobility. From extended battery range and faster charging capabilities to smart connectivity features, these advancements not only address the traditional challenges associated with EVs but also redefine the driving experience, shaping a future where EVs are at the forefront of efficiency, performance, and environmental consciousness.

20.8.1 Battery technology innovations

The rapid evolution of battery technology stands as a cornerstone in the transformative journey of EVs, thus revolutionizing the efficiency, range, and overall appeal of these sustainable modes of transportation. One of the most significant innovations is the advancement in lithium-ion battery chemistry, which forms the backbone of EV energy storage. Continuous research and development have led to improvements in energy density, enabling batteries to store more energy in a smaller and lighter form factor. This not only extends the driving range of EVs but also contributes to the reduction of their overall weight, enhancing performance and energy efficiency [24].

Solid-state batteries represent a groundbreaking innovation poised to redefine the landscape of EVs. Compared to traditional liquid electrolytes, solid-state batteries offer higher energy density, faster charging times, and improved safety. With no liquid components, these batteries are less prone to overheating and have the potential to address some of the safety concerns associated with lithium-ion batteries [25]. As research and development in solid-state technology

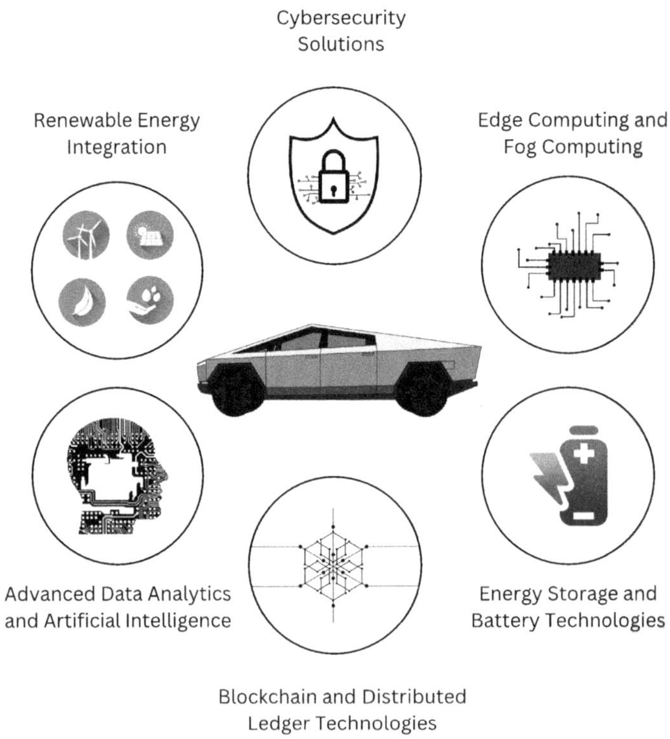

FIG. 2 Technological advances in electric vehicles.

progress, the commercialization of these batteries could mark a significant leap forward in the practicality and widespread adoption of EVs.

The advancements in battery management systems (BMSs) contribute to the longevity and reliability of EV batteries [26]. Intelligent BMS technologies optimize charging and discharging cycles, monitor cell health, and prevent issues like overcharging or overheating. Machine learning algorithms integrated into BMSs further enhance predictive maintenance, providing real-time insights into battery health and performance. These innovations collectively contribute to addressing key concerns related to the lifespan, safety, and efficiency of EV batteries, propelling EVs into a future where battery technology plays a pivotal role in reshaping the dynamics of sustainable transportation.

20.8.2 Advances in charging infrastructure

The rapid advances in charging infrastructure for EVs are instrumental in overcoming one of the key barriers to widespread adoption—the accessibility and speed of charging. High-power charging stations, often utilizing direct current (DC) fast-charging technology, represent a significant leap forward. These stations can provide a substantial amount of energy in a short period, dramatically reducing the time required to charge an EV. With power levels continually increasing, some fast-charging stations can replenish a significant portion of an EV's battery capacity in just a matter of minutes, enhancing the convenience and practicality of long-distance travel for EV owners [14].

The development of smart and connected charging infrastructure is another noteworthy innovation. Intelligent charging stations can communicate with EVs to optimize charging times based on factors such as grid demand, electricity prices, and the vehicle's energy needs. This not only promotes grid efficiency but also allows EV owners to take advantage of cost-effective charging opportunities. Furthermore, smart infrastructure facilitates remote monitoring, diagnostics, and maintenance, ensuring the reliability and functionality of charging stations. Integration with mobile apps and digital platforms also enables EV users to locate, reserve, and pay for charging services seamlessly, enhancing the overall user experience.

Wireless charging technologies are emerging as a futuristic solution, eliminating the need for physical cables and connectors [27]. Inductive charging systems allow EVs to charge simply by parking over specially designed pads embedded in the ground. As this technology advances, it holds the potential to revolutionize the charging experience, making it even more convenient and user-friendly. The collective progress in charging infrastructure not only addresses the practical concerns of EV users but also aligns with the broader vision of an integrated, efficient, and widespread electric mobility ecosystem.

20.8.3 Integration of artificial intelligence and smart technologies in EVs

The integration of artificial intelligence (AI) and smart technologies into EVs represents a paradigm shift in the automotive industry, ushering in a new era of efficiency, safety, and user experience. This is in addition to the model-based development [28] and software built for electrification [29]. Modeling of EVs is accomplished by battery chemistry, vehicle dynamics, and transmission performance [30], whereas AI plays a pivotal role in enhancing the overall intelligence of EVs, optimizing various aspects of their operation. Advanced driver-assistance systems (ADASs) powered by AI enable features such as lane-keeping assistance, adaptive cruise control, and automated emergency braking, contributing to improved safety and accident prevention. Moreover, AI algorithms continuously analyze real-time data from sensors [31], cameras, and other sources to make split-second decisions, enhancing the vehicle's ability to navigate complex road conditions. This paves way for building fully self-driving EVs in the near future.

Smart technologies are reshaping the user experience within EVs, fostering connectivity and convenience. Voice-activated controls, touchscreens, and gesture recognition systems are becoming integral components of EV interiors, offering intuitive interfaces for users to control various functions, from entertainment to climate control. AI-driven personalization features learn from user preferences and behavior, adjusting settings such as seat position, climate control, and infotainment options to create a tailored and comfortable driving experience. Additionally, integration with smart home devices and IoT ecosystems enables EV owners to remotely monitor and control their vehicles, check charging status, and even initiate preconditioning before heading to a charging station.

AI is also optimizing battery management and energy efficiency in EVs. Predictive analytics and machine learning algorithms analyze driving patterns, environmental conditions, and battery health to optimize energy consumption. This ensures that EVs are not only more energy-efficient but also have enhanced range predictions, addressing concerns related to range anxiety. Furthermore, AI facilitates predictive maintenance, identifying potential issues with the vehicle or its components before they become critical, leading to increased reliability and reduced maintenance costs. As the synergy between AI and smart technologies continues to evolve, the integration of these innovations into EVs is poised to redefine the driving experience, making electric mobility more intelligent, connected, and user-centric.

20.9 Future prospects and challenges

The future prospects of EVs are poised at the intersection of innovation, sustainability, and global mobility transformation. Looking ahead, the prospects for EVs include advancements in battery technology, expanding charging infrastructure, and a broader societal shift toward eco-friendly transportation. However, along with these promising developments come challenges such as addressing infrastructure gaps, ensuring energy sustainability, and managing the transition for traditional automotive industries. This demands a comprehensive exploration of the future trajectory of EVs, analyzing the potential breakthroughs and obstacles that will shape the landscape of sustainable transportation in the years to come.

20.9.1 Emerging technologies and upcoming trends

Emerging technologies and upcoming trends in EVs are steering the automotive industry toward a transformative and sustainable future. One prominent trend is the continued evolution of battery technology. Researchers are exploring alternative battery chemistries, such as solid-state batteries, which promise higher energy density, faster charging times, and improved safety compared to traditional lithium-ion batteries. These advancements hold the potential to address current limitations, further extending the driving range of EVs and enhancing overall performance. Moreover, innovations in energy storage and management systems are paving the way for more efficient and reliable EVs.

Connected and autonomous features are becoming integral to the evolution of EVs. With advancements in AI and connectivity, EVs are transitioning into smart, interconnected vehicles. Enhanced driver-assistance systems, vehicle-to-everything (V2X) communication, and predictive analytics are shaping the landscape of autonomous driving. This not only improves safety but also contributes to optimized traffic flow and energy efficiency. As 5G networks become more prevalent, the seamless integration of connected technologies is poised to redefine the driving experience, making EVs not just eco-friendly but also technologically sophisticated.

Charging technology is another arena witnessing rapid developments. High-power charging stations, capable of delivering substantial energy in a short time, are becoming more prevalent, reducing charging times and enhancing the practicality of long-distance EV travel. Wireless charging technologies, utilizing inductive charging pads embedded in roads or parking spaces, aim to streamline the charging process further. These trends collectively indicate a future where EVs are not only environmentally sustainable but also technologically advanced, promising a dynamic and efficient driving experience for users worldwide.

20.9.2 Anticipated challenges and solutions

The future development of EVs is accompanied by anticipated challenges, and addressing these obstacles is critical for the continued growth and widespread adoption of sustainable transportation. One significant challenge revolves around charging infrastructure. While advancements have been made, the development of a comprehensive and universally accessible charging network remains a priority. Governments, businesses, and industry stakeholders must collaborate to invest in infrastructure expansion, especially in urban areas and along major transportation routes. Innovative solutions such as high-speed charging stations, standardized charging protocols, and incentives for private and public investments can significantly contribute to overcoming this challenge.

Another challenge lies in the production and sourcing of materials for EV batteries. The growing demand for EVs places increased pressure on the availability and ethical sourcing of materials like lithium, cobalt, and nickel. Future developments need to focus on sustainable mining practices, recycling initiatives, and the exploration of alternative battery chemistries that minimize reliance on scarce resources. Government regulations and industry standards can play a crucial role in incentivizing ethical material sourcing, ensuring the long-term environmental and social sustainability of the EV supply chain [32].

The transition to electric mobility also poses challenges related to workforce skills and education. The automotive industry's shift toward EVs necessitates a skilled workforce capable of designing, manufacturing, and maintaining these advanced vehicles. Governments and educational institutions need to collaborate with the industry to develop training programs, ensuring that workers are equipped with the necessary skills for the evolving automotive landscape. By addressing these challenges collectively, stakeholders can contribute to the sustained and successful development of EVs, fostering an inclusive, environmentally conscious, and technologically advanced automotive future.

20.9.3 Vision for the future of EVs

The vision for the future of EVs encompasses a transformative and sustainable paradigm in the global automotive landscape. At the forefront of this vision is the widespread adoption of EVs as a mainstream mode of transportation, leading to a substantial reduction in GHG emissions and a shift away from traditional fossil fuel reliance [33]. Governments, industries, and consumers alike are envisioned to actively participate in creating a greener and more environmentally conscious transportation ecosystem.

An integral part of the future vision for EVs involves continued advancements in battery technology. Breakthroughs in energy density, charging speed, and overall battery performance are anticipated, addressing current limitations and enabling EVs to surpass the range and efficiency of ICE vehicles. This technological evolution will contribute not only to increased consumer confidence but also to the viability of EVs in various sectors, including commercial transportation and public transit. Moreover, the development of a circular economy for EV batteries, encompassing recycling and repurposing initiatives, is envisioned to minimize environmental impact and ensure responsible end-of-life management.

The vision for the future of EVs also includes the integration of smart technologies, AI, and renewable energy sources. EVs are expected to become integral components of smart cities, contributing to a connected and efficient urban mobility network. V2G technology is envisioned to enable bidirectional energy flow, allowing EVs to not only draw power from the grid but also contribute excess energy back, thereby enhancing grid stability. Additionally, the synergy between EVs and renewable energy sources, such as solar and wind power, is anticipated to further reduce the carbon footprint of transportation, fostering a future where electric mobility is synonymous with sustainability, innovation, and a cleaner, healthier planet.

20.10 Conclusion

The impact and sustainability of EVs represent a transformative force in shaping the future of transportation. The environmental benefits, ranging from reduced GHG emissions to improved air quality, underscore the pivotal role EVs play in mitigating climate change. As technology advances, the life-cycle considerations of EVs, including resource use, manufacturing, and end-of-life management, are evolving toward more sustainable practices. The integration of EVs with renewable energy sources further propels the vision of a cleaner, greener mobility ecosystem. However, realizing the full potential of EVs necessitates addressing challenges such as charging infrastructure, material sourcing, and workforce education. With continued innovation, collaborative efforts, and a commitment to sustainable practices, the journey toward widespread EV adoption promises not only a revolution in transportation but a fundamental shift toward a more environmentally conscious and resilient future.

20.10.1 Summary of key findings

The examination of the impact and sustainability of EVs reveals a compelling narrative of positive environmental strides and ongoing technological advancements. Key findings underscore the pivotal role of EVs in reducing GHG emissions and improving air quality, contributing significantly to the global fight against climate change. With zero tailpipe emissions, EVs stand out as a sustainable alternative to traditional ICE vehicles, aligning with the imperative to transition toward cleaner transportation options. Additionally, the exploration of life cycle considerations emphasizes the need for responsible resource use, innovative manufacturing processes, and effective end-of-life management to enhance the overall sustainability of EVs.

In the realm of technology, the study reveals a dynamic landscape of continuous innovation, particularly in battery technology and charging infrastructure. Advances in battery chemistry, energy density, and management systems contribute to the increasing efficiency and range of EVs, addressing concerns related to performance and range anxiety. The growth of smart technologies and the integration of AI not only enhance the driving experience but also optimize energy consumption and support the seamless integration of EVs into broader smart ecosystems. Moreover, the symbiotic relationship between EVs and renewable energy sources emerges as a key finding, emphasizing the potential for a holistic approach to sustainable mobility that extends beyond the vehicle itself.

However, challenges persist, necessitating a comprehensive approach to fully unlock the potential of EVs. The need for robust charging infrastructure, ethical material sourcing for batteries, and workforce education are identified as critical areas for attention. As these findings unfold, they underscore the imperative for collaborative efforts among governments, industries, and consumers to foster a sustainable and resilient future where EVs play a central role in redefining the landscape of transportation and environmental stewardship.

20.10.2 Implications for sustainability

The implications for the sustainability of EVs are profound, heralding a transformative shift in the automotive industry and beyond. At the forefront is the positive impact on environmental sustainability. The adoption of EVs significantly reduces the carbon footprint associated with transportation, mitigating climate change and improving air quality. As the world grapples with the imperative to transition away from fossil fuels, the widespread adoption of EVs emerges as a key strategy for achieving sustainable mobility, aligning with global efforts to meet climate goals and transition toward a more eco-friendly transportation ecosystem.

The implications extend to broader societal and economic aspects as well. The growth of the EV market fosters job creation and economic opportunities, particularly in sectors related to EV manufacturing, charging infrastructure development, and advancements in battery technology. Governments and industries that invest in the EV ecosystem are not only contributing to environmental sustainability but also driving economic growth and innovation. The transition to sustainable transportation aligns with evolving consumer preferences, as individuals increasingly prioritize environmentally conscious choices. This shift in consumer behavior, coupled with supportive government policies and incentives, reinforces the societal implications of sustainability, fostering a culture of responsible and eco-friendly transportation choices.

However, challenges must be navigated to realize the full implications of EVs for sustainability. The development of robust charging infrastructure, ethical sourcing of materials for batteries, and addressing disparities in access to EVs are crucial considerations. These challenges underscore the need for a comprehensive and integrated approach to sustainability, involving collaboration among governments, industries, and consumers. As EVs continue to evolve, their implications for sustainability resonate not only in the context of cleaner transportation but also in reshaping societal values, economic landscapes, and the overall trajectory toward a more sustainable and resilient future.

20.10.3 Recommendations for future research

Future research in EVs should focus on key areas to drive innovation, address challenges, and further enhance the sustainability and viability of electric mobility. Firstly, advancements in battery technology remain paramount. Researchers should explore novel battery chemistries, materials, and designs to enhance energy density, reduce production costs, and extend the lifespan of batteries. Investigating alternative energy storage solutions, such as solid-state batteries or emerging postlithium technologies, could pave the way for the next generation of high-performance, sustainable, and safe EV batteries. Additionally, understanding and optimizing battery recycling processes are critical for minimizing environmental impact and ensuring responsible end-of-life management.

Secondly, charging infrastructure development requires continued attention. Future research should delve into efficient and scalable charging solutions, exploring faster-charging technologies, higher-capacity stations, and innovative grid integration strategies. Investigating the impacts of widespread EV adoption on electricity grids and proposing smart grid technologies to manage the increased demand intelligently will be essential. Furthermore, research should address standardization and interoperability issues, ensuring a seamless and user-friendly charging experience. Exploring decentralized and renewable energy-based charging solutions, such as home solar charging or community microgrids, could contribute to a more resilient and sustainable charging infrastructure.

Lastly, socioeconomic aspects of EV adoption warrant in-depth investigation. Research should examine the socioeconomic implications of EVs, including their impact on job markets, supply chains, and local economies [34]. Evaluating the effectiveness of government incentives and policies in promoting EV adoption and addressing equity concerns is crucial for ensuring an inclusive transition to electric mobility. Moreover, understanding consumer behaviors, preferences, and potential barriers to EV adoption can guide the development of targeted educational campaigns and policies. By focusing on these research areas, scholars and industry experts can contribute significantly to the continued evolution and successful integration of EVs into mainstream transportation systems.

References

[1] S. Habib, et al., A comprehensive study of implemented international standards, technical challenges, impacts and prospects for electric vehicles, IEEE Access 6 (2018) 13866–13890, https://doi.org/10.1109/ACCESS.2018.2812303.

[2] A. Ramanath, Electrical machines in electrified transportation, in: EEPower, 2021.

[3] D. Çelık, M.E. Meral, M. Waseem, A new area towards to digitalization of energy systems: enables, challenges and solutions, in: 2022 14th International Conference on Electronics, Computers and Artificial Intelligence (ECAI), 2022, pp. 1–6, https://doi.org/10.1109/ECAI54874.2022.9847313.

[4] B. Caroleo, et al., Assessing the impacts of electric vehicles uptake: a system dynamics approach, in: 2017 IEEE 41st Annual Computer Software and Applications Conference (COMPSAC), vol. 2, 2017, pp. 790–795, https://doi.org/10.1109/COMPSAC.2017.67.

[5] A. Tintelecan, A.C. Dobra, C. Mar⊠iş, LCA indicators in electric vehicles environmental impact assessment, in: 2019 Electric vehicles international conference (EV), 2019, pp. 1–5, https://doi.org/10.1109/EV.2019.8892893.

[6] T. Aljohani, G. Alzahrani, Life cycle assessment to study the impact of the regional grid mix and temperature differences on the GHG emissions of battery electric and conventional vehicles, in: 2019 SoutheastCon, 2019, pp. 1–9, https://doi.org/10.1109/SoutheastCon42311.2019.9020666.

[7] R. Kaushik, et al., Sustainability of electric vehicle in India, in: 2022 International Conference on Inventive Computation Technologies (ICICT), 2022, pp. 664–667, https://doi.org/10.1109/ICICT54344.2022.9850638.

[8] S. Dang, et al., Sustainable energy management: an analysis report of the impacts of electric vehicles, in: 2014 14th International Conference on Environment and Electrical Engineering, 2014, pp. 318–322, https://doi.org/10.1109/EEEIC.2014.6835886.

[9] S. Jain, B.N. Singh, Environmental impacts of reclaiming, recycling, and reusing (R3) parts of electric vehicles' powertrain, in: 2022 IEEE International Conference on Power Electronics, Drives and Energy Systems (PEDES), 2022, pp. 1–6, https://doi.org/10.1109/PEDES56012.2022.10080569.

[10] A. Ramanath, Introduction to transportation electrification, in: EEPower, 2021.

[11] W. Shuai, P. Maillé, A. Pelov, Charging electric vehicles in the smart city: A survey of economy-driven approaches, in: IEEE Transactions on Intelligent Transportation Systems 17.8, 2016, pp. 2089–2106, https://doi.org/10.1109/TITS.2016.2519499.

[12] G.M. Silva, et al., Recharging electric vehicles: impact on the load curve of a low-voltage distribution network, in: 2019 IEEE PES Innovative Smart Grid Technologies Conference—Latin America (ISGT Latin America), 2019, pp. 1–6, https://doi.org/10.1109/ISGT-LA.2019.8895307.

[13] Z. Zhang, et al., Application research of an electric vehicle DC fast charger in smart grids, in: 2012 IEEE 6th International Conference on Information and Automation for Sustainability, 2012, pp. 258–261, https://doi.org/10.1109/ICIAFS.2012.6419913.

[14] International Energy Agency, Global EV outlook 2020, in: IEA, 2020.

[15] E.M. Tabares, et al., Analysis of electric vehicle policies in the Colombian market: the city of Medellín as a case study, in: 2020 Congreso Internacional de Innovaci'on y Tendencias en Ingenier'ıa (CONIITI), 2020, p. 1, https://doi.org/10.1109/CONIITI51147.2020.9240379.

[16] A.A. Kebede, et al., A comprehensive review of stationary energy storage devices for large scale renewable energy sources grid integration, Renew. Sust. Energ. Rev. 159 (2022), https://doi.org/10.1016/j.rser.2022.112213.

[17] E. Apostolaki-Iosifidou, et al., Electric vehicle charge management for lowering costs and environmental impact, in: 2020 IEEE Conference on Technologies for Sustainability (SusTech), 2020, pp. 1–7, https://doi.org/10.1109/SusTech47890.2020.9150524.

[18] M. Scott, D. Hopkins, J. Stephenson, Understanding sustainable mobility: the potential of electric vehicles, in: 2014 IEEE 15th International Conference on Mobile Data Management, vol. 2, 2014, pp. 27–30, https://doi.org/10.1109/MDM.2014.63.

[19] A.G. Boulanger, et al., Vehicle electrification: status and issues, Proc. IEEE 99 (6) (2011) 1116–1138, https://doi.org/10.1109/JPROC.2011.2112750.

[20] I.T. Vadium, et al., Electric vehicle carbon footprint reduction via intelligent charging strategies, in: 2019 8th International Conference on Modern Power Systems (MPS), 2019, pp. 1–6, https://doi.org/10.1109/MPS.2019.8759783.

[21] Y. Zheng, et al., Electric vehicle battery charging/swap stations in distribution systems: comparison study and optimal planning, IEEE Trans. Power Syst. 29 (1) (2014) 221–229, https://doi.org/10.1109/TPWRS.2013.2278852.

[22] K. Joshi, A. Lakum, Assessing the impact of plug-in hybrid electric vehicles on distribution network operations using time-series distribution power flow analysis, in: 2014 IEEE International Conference on Power Electronics, Drives and Energy Systems (PEDES), 2014, pp. 1–6, https://doi.org/10.1109/PEDES.2014.7042120.

[23] S. Patra, P. Pathak, P. Kulkarni, Cascading impacts of electric vehicles (EVs) and automated vehicles (AVs) on different sectors, in: 2022 international interdisciplinary humanitarian conference for sustainability (IIHC), 2022, pp. 136–142, https://doi.org/10.1109/IIHC55949.2022.10059771.

[24] A. Ramanath, Utilizing electrified powertrains in electrified transportation, in: EEPower, 2021.

[25] M.A. Hannan, et al., State-of-the-art and energy management system of lithium-ion batteries in electric vehicle applications: issues and recommendations, IEEE Access 6 (2018) 19362–19378, https://doi.org/10.1109/ACCESS.2018.2817655.

[26] A. Ramanath, Energy storage systems in electrified transportation, in: EEPower, 2021.

[27] B.J. Limb, et al., Economic and environmental feasibility, architecture optimization, and grid impact of dynamic charging of electric vehicles using wireless power transfer, in: 6th Hybrid and Electric Vehicles Conference (HEVC 2016), 2016, pp. 1–6, https://doi.org/10.1049/cp.2016.0988.

[28] D. Doerffel, S.A. Sharkh, System modeling and simulation as a tool for developing a vision for future hybrid electric vehicle drivetrain configurations, in: 2006 IEEE Vehicle Power and Propulsion Conference, 2006, pp. 1–6, https://doi.org/10.1109/VPPC.2006.364315.

[29] A. Ramanath, Introduction to transportation electrification part 2, in: EEPower, 2022.

[30] A. Ramanath, Electric vehicle modeling considerations, in: EEPower, 2021.

[31] A. Ramanath, Understanding the importance of power electronics in electrified transportation, in: EEPower, 2021.

[32] N.C. Onat, N.N.M. Aboushaqrah, M. Kucukvar, Supply chain linked sustainability assessment of electric vehicles: the case for Qatar, in: 2019 IEEE 6th International Conference on Industrial Engineering and Applications (ICIEA), 2019, pp. 780–785, https://doi.org/10.1109/IEA.2019.8715205.

[33] S. Gajanayake, et al., Study of the impact of electric vehicles on fuel consumption and carbon dioxide emission scenarios in Sri Lanka, in: 2020 Moratuwa Engineering Research Conference (MERCon), 2020, pp. 494–498, https://doi.org/10.1109/MERCon50084.2020.9185248.

[34] M. Neaimeh, et al., Integrating smart meter and electric vehicle charging data to predict distribution network impacts, in: IEEE PES ISGT Europe 2013, 2013, pp. 1–5, https://doi.org/10.1109/ISGTEurope.2013.6695238.

Index

Note: Page numbers followed by *f* indicate figures and *t* indicate tables.

Q

R

S

9780323995450